A COURSE IN DATA STRUCTURES USING C

Deepak Gupta
R.K. Tiwari

A COURSE IN
DATA STRUCTURES USING C

(For Engineering Students)

By

DEEPAK GUPTA (B.TECH, ME)
Bhagwan Parshuram Institute of Technology, Delhi

R.K.TIWARI (MCA, ME)
Bhagwan Parshuram Institute of Technology, Delhi

S.K. KATARIA & SONS®

Publisher of Engineering Books
4885/109, Prakash Mahal, Dr. Subhash Bhargav Lane,
Daryaganj, New Delhi-110002 (India)
Phone: +91-11-43551243; Mobile: +91-9871775858, 7678425209
e-mail: katariabook@gmail.com, katariabooks@yahoo.com
Website : www.skkatariaandsons.com
(e-books, subscriptions and bulk buying also available on our website)

Published by :

S.K. Kataria & Sons®

4885/109, Prakash Mahal, Dr. Subhash Bhargav Lane,
Opposite Delhi Medical Association, Daryaganj, New Delhi-110002
Phone: +91-11-43551243; Mobile: +91-9871775858, 7678425209
e-mail: katariabook@gmail.com, katariabooks@yahoo.com

First Edition : 2010, Reprint : 2025

Price: ₹ 695/-

Laser Typesetting by : Arpit Printographers

PREFACE

Data Structure is way to organize the data. It is that base of computer science through which one can design the environment with strong background. For almost every field it is required to have knowledge of data applications and data structure can play a very important role in that area.

This book can provide an easy going method to fulfill the requirements of all students who really want to make themselves comfortable in the field of data management.

This book is divided into 14 chapters. The first chapter includes analysis of algorithms, the second chapter how the recurrence relation works, the third chapter consists of working of Arrays ini the field of data storage, forth will focus on Structures, Fifth can introduce to pointers. The other chapters have a sequence as that expected by teachers & students.

The programs illustrated in the book have been accumulated, experienced & tested by author during their teaching of the subject. In addition, numerous questions are adapted from university question papers.

This book consists of many Solved Theoretical as well as Practical University Level Questions according to the prescribed format of the univerties.

Deepak Gupta

R.K. Tiwari

ACKNOWLEDGEMENTS

This is the grace of Lord Shiva, Guruji, our parents and respected teachers that inspired us to do a challenging work in the field of education. We have tried our best level to create a different path for those students who do not understand the exact applications in the field of Engineering and Technology.

Our main aim behind this book is to motivate all the students, so that they will be having high confidence level in their respective fields. Data Structure is not very simple but if the training in on the right way, one can create a miracle. We have covered almost every thing from the beginning to the end with all the important aspects in simple to understand manner.

This book is dedicated to our previous, current and future students.

Our sincere thanks goes to our respected Director Prof. A.K.Tandon, Dr.(Mrs.) Daya Gupta, Head of the Department, Department of Computer Engineering, Delhi College of Engineering, Mrs. Rajni Jindal, Assistant Professor, Department of Computer Engineering, Delhi College of Engineering for their valuable guidance and worthy suggestions which pushed us forward to complete the text for which we are greatly indebted to them.

A special word of thanks goes to Mr. Sanjay Kansal, B.Tech Electronics and Communication, from IIT Roorkee, Director, Raj's Maths Academy, who helped us to bring this book to the present form.

My big thanks go to M/s S.K. Kataria and Sons, Delhi for providing me a platform to bring this book to life.

For any suggestions & feedback, please feel free to send your comments to

Deepak Gupta **R.K. Tiwari**

CONTENTS

LISTS OF ALGORITHMS

LISTS OF PROGRAMS

CHAPTER 1

INTRODUCTION TO DATA STRUCTURE

1.1. DATA STRUCTURES

INSIDE THIS CHAPTER

Data structures is a way for structuring data *i.e.* way of storing data in a computer so that it can be used efficiently. Data structures usually consists of two things: First one is the type of structure that we are going to use to store data and second how efficiently we perform operations onto that data stored in structure so that it results in less execution time and amount of memory space they require. Some of the operations may be insertion, deletion, searching, etc.

If the data structure is carefully chosen then the underlying algorithm that makes use of data structures will be an efficient one. The underlying idea behind data structure is the choice of Abstract Data Type (ADT) (discussed later). Data structures are implemented by a programming language as data types and the references and operations they provide.

1.1.1. Few Definitions

Before we give you actual definition of the data structure in terms of ADT, we provide you some definitions that will be helpful in setting the foundation for the data structures.

- **DATA :** Data is the basic raw facts. For example name of a person, age of an employee, chair color, numbers, strings, etc. Information is the processed data. In computers data can be numeric, nonnumeric, binary, character data etc.
- **ALGORITHM :** An Algorithm is a finite sequence of instructions, each of which have a clear meaning and can be executed with a finite amount of effort in finite time. Whatever the input values, an algorithm will definitely terminate after executing finite number of instructions.
- **DATA TYPE :** Data type of a variable is the set of values that the variable may assume, For example, the basic data types in C are int, char, float, double etc.
- **ABSTRACT DATA TYPE (ADT) :** An ADT is a set of elements with a collection of well defined operations. The operations can take as operands not only instances of the ADT but other types of operands or instances of other ADTs. Similarly results need not be instances of the ADT. Object-oriented languages such as C++ and Java provide explicit support for expressing ADTs by means of classes. Examples of ADTs include List, stack, queue, set, tree, graph, etc.

"**DATA STRUCTURES** is an implementation of an ADT is a translation into statements of a programming language, the declarations that define a variable to be of that ADT type and the operations defined on the ADT (using procedures of the programming language)".

Example : A "Queue" is an ADT which can be defined as a sequence of elements with operations such as null(Q), empty(Q), enqueue(x, Q), and dequeue(Q). This can be implemented using data structures such as: array, singly linked list, doubly linked list, circular array etc.

DATA STRUCTURE is the implementation of ADT in terms of the data types or other data structures. A data structure is modeled in Java by a class. Classes specify how operations are performed. In Java, to implement an ADT as a data structure, an interface is implemented by a class.

Abstraction and representation help us understand the principles behind large software systems. Information-hiding can be used along with abstraction to partition a large system into smaller subsystems with simple interfaces that makes them easier to understand and use.

1.1.2. Types of Data Structures

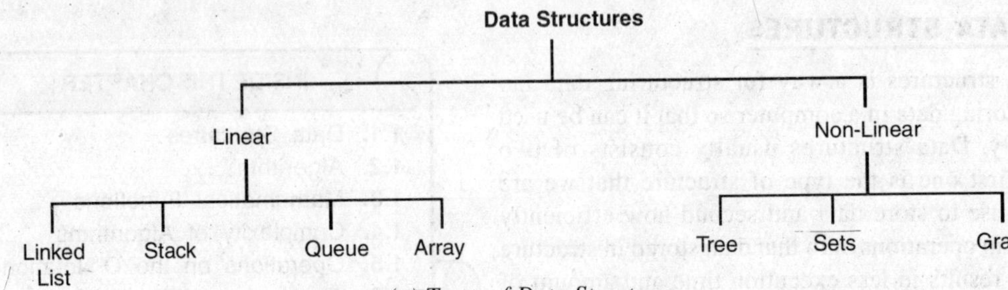

(a) Types of Data Structures

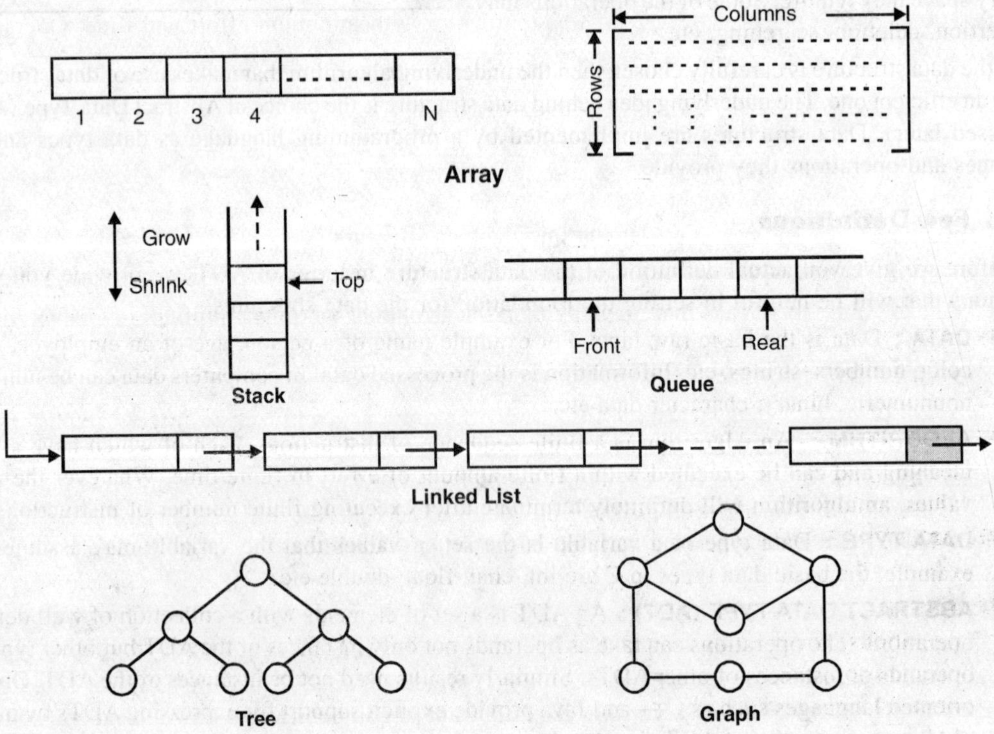

(b) Types of Data Structures

Fig. 1.1

Primarily two categories of data structures exist. One is linear and second is non-linear. When the elements are accessed or placed in contiguous memory locations then data structure is known as linear data structure. Arrays, stacks, queues and linked list are example of linear data structure. In linear data structures we have to first allocate memory for the number of elements like in arrays. If stacks and queues are implemented through arrays then we have to first assume the size of stack and queue in advance. This sometimes leads to wastage of memory.

In non-linear data structures elements are not accessed in sequential manner. The individual elements are linked through pointers and may be placed any where in the memory. Examples of non-linear data structures are tree, graphs, sets etc.

When talking about accessing of elements in any type of data structures we can say the data structure is linear and while talking of non-contiguous memory allocation of elements we can say data structure is non-linear. Thus in terms of accessing the elements of linked list it is linear data structure and while talking of allocation of individual nodes of linked list it is non-linear. Same analogy applies to stack and queue when implemented using linked list. Array in both senses is a linear data structure.

The simple reason behind so many data structures is that no single data structure works well for all purposes, and so it is important to know the strengths and limitations of several of them. Different kinds of data structures are suited to different kinds of applications, and some are highly specialized to certain tasks. For example, B– trees and B+ trees are particularly well-suited for implementation of databases.

1.1.3. Operation on Data Structures

The various operations that can be performed over data structures are as given below. Note that not all the operations may be carried out efficiently over any of the data structures. For some operations a specific data structure may be suitable and for others operation some other data structure may be suited. In fact while choosing a specific data structure for our solution of given problem, we have to keep in mind what operations can be performed over this data structure with minimum effort and time. Operations given below are the most common and frequently employed operations:

1. **Traversing :** Accessing each element of the data structure once and doing some processing over the data.
2. **Searching :** Searching for a specific element.
3. **Sorting:** Sorting the data in ascending or descending order.
4. **Insertion :** Insertion of new element into the data structure.
5. **Deletion :** Deletion of a specific element.

The other possible operations may be copying, concatenation, merging, splitting, reversing etc.

1.1.4. Application of Data Structures

Application of various data structures like arrays, linked list, stacks, queues, trees, graphs etc are covered in their respective chapters, and once you have gone each of the chapters and got the idea of various data structures, it will be very easy for you to enumerate numbers of applications of those data structures in our daily life and as well as to various fields of computer science, IT, electronics, electrical and various other fields of engineering and non engineering disciplines.

1.2. ALGORITHM

In mathematics and computing, an algorithm is a procedure (a finite set of well-defined instructions) for accomplishing some task which will terminate in a defined end-state. The computational complexity and efficient implementation of the algorithm are important in computing, and this depends on suitable data structures.

The word algorithm comes from the name of the 9th century Persian mathematician **Abu Abdullah Muhammad ibn Musa al-Khwarizmi.** The word algorithm originally referred only to the rules of performing arithmetic using Hindu-Arabic numerals but evolved via European Latin translation of al-Khwarizmi's name into algorithm by the 18th century. The word evolved to include all definite procedures for solving problems or performing tasks.

DO YOU KNOW

ABU JA'FAR MOHAMMED IBN MUSA AL-KHOWARIZMI (C.780–C.850) is an astronomer and mathematician, was a member of the House of wisdom, an academy of scientists in Baghdad. The name al-khowarizmi means "from the town of kowarzizm," which is now called Khiva and is part of Uzbekistan. Al-khowarizmi wrote books on mathematics, astronomy, and geography. Western Europeans first learned about algebra from his works. The word *algebra* comes from al-jabr, part of the title of his book *Kitab al-jabr w'al muquabala*. This book was translated into Latin and was a widely used textbook. His book on the use of Hindu numerical describes procedures for arithmetic operations using these numerals. European authors used a Latin corruption of his name, which later evolved to the word *algorithm*, to describe the subject of arithmetic with Hindu numerals.

Knuth has given a list of five properties that are widely accepted as requirements for an algorithm:

1. **Finiteness :** "An algorithm must always terminate after a finite number of steps"
2. **Definiteness :** "Each step of an algorithm must be precisely defined; the actions to be carried out must be rigorously and unambiguously specified for each case"
3. **Input :** ". . . quantities which are given to it initially before the algorithm begins. These inputs are taken from specified sets of objects"
4. **Output :** ". . . quantities which have a specified relation to the inputs"
5. **Effectiveness :** ". . . all of the operations to be performed in the algorithm must be sufficiently basic such that they can in principle be done exactly and in a finite length of time by a man using paper and pencil"

1.2.1. Expressing Algorithms

Algorithms can be expressed in many kinds of notation, including natural languages, pseudocode, flowcharts, and programming languages. Natural language expressions of algorithms tend to be verbose and ambiguous, and are rarely used for complex or technical algorithms. Pseudocode and flowcharts are structured ways to express algorithms that avoid many of the ambiguities common in natural language statements, while remaining independent of a particular implementation language. Programming languages are primarily intended for expressing algorithms in a form that can be executed by a computer, but are often used as a way to define or document algorithms.

Examples

One of the simplest algorithms is to find the largest number in an (unsorted) list of numbers, The solution necessarily requires looking at every number in the list, but only once at each. From this follows a simple algorithm, which can be stated in English as:

1. Assume the first item is largest.
2. Look at each of the remaining items in the list and if it is larger than the largest item so far, make a note of it.
3. The last noted item is the largest in the list when the process is complete.

Here is a more formal coding of the algorithm in pseudocode:

Algorithm 1.1 ....

Largest Number

Input: A non.empty list of numbers L.

Output: The largest number in the list L.

1. largest ← L_0
2. for each item in the list $L_{i=1}$, do
   if the item > largest, then
3. largest ← the item
4. return largest

☞ **NOTE :**

1. "← is loose shorthand for "changes to". For instance, with "largest ← the item", it means that the largest number found so far changes to this item.
2. "return" terminates the algorithm and outputs the value listed behind it.

1.2.2. Types of Algorithms

1. Recursive
2. Iterative
3. Divide and Conquer
4. Dynamic Programming
5. Greedy Approach
6. Backtracking
7. Branch and Bound
8. Brute Force Approach
9. Approximation Algorithms

1.3. MATHEMATICAL FUNCTIONS

Mathematical functions are useful in the creation and analysis of algorithms. In this section, some of the most basic and most commonly used functions with their properties are listed.

(*i*) **Floor of *x* :** The greatest integer less than or equal to *x*, where *x* is any real number.

Notation : x

For example, 3.14 = 3 1/2 = 0 –1/2 = –1

(*ii*) **Ceiling of *x* :** This is the smallest integer greater than or equal to *x*, where *x* is any real number.

Notation : x

For example, 3.14 = 4 1/2 = 1 –1/2 = 0

(*iii*) **Modulo :** Given any two real numbers *x* and *y*, *x* mod *y* is defined as

$$x \bmod y = x \quad \text{if } y = 0$$
$$= x - y * x / y \quad \text{if } y <> 0$$

For example, 10 mod 3 = 1 24 mod 8 = 0 –5 mod 7 = 2

1.3.1. Identities

The following are the identities related to the mathematical functions defined above:

- $x = x$ if and only if x is an integer
- $x = x$ if and only if x is not an integer
- $-x = -x$
- $x + y <= x + y$
- $x = x + x \bmod 1$
- $z (x \bmod y) = zx \bmod zy$

1.4. COMPLEXITY OF ALGORITHMS

Several algorithms could be created to solve a single problem. These algorithms may vary in the way they get, process and output data. Hence, they could have significant difference in terms of *performance* and *space utilization*. It is important to know how to analyze the algorithms, and knowing how to measure the efficiency of algorithms helps a lot in the analysis process.

It is very convenient to classify algorithms based on the relative amount of time or relative amount of space they require and specify the growth of time/space requirements as a function of the input size.

Thus, we have the notions of :

Time Complexity : Running time of the program as a function of the size of input.

Space Complexity : Amount of computer memory required during the program execution, as a function of the input size.

1.4.1. Algorithm Efficiency

Algorithm efficiency is measured in two criteria: space utilization and time efficiency. *Space utilization* is the amount of memory required to store the data while *time efficiency* is the amount of time required to process the data.

Before we can measure the time efficiency of an algorithm we have to get the execution time. **Execution time** is the amount of time spent in executing instructions of a given algorithm. It is dependent on the particular computer (hardware) being used. To express the execution time we use the notation :

***T*(*n*)**, where T is the function and n is the size of the input.

There are several factors that affect the execution time. These are :

- input size
- instruction type
- machine speed
- quality of source code of the algorithm implementation
- quality of the machine code generated from the source code by the compiler

To make the difference clearer, let's compare based on the execution time where n=100000 and time unit = 1 msec:

F(n)	*Running Time*
$\log_2 n$	19.93 microseconds
n	1.00 seconds
$n \log_2 n$	19.93 seconds
n^2	11.57 days
n^3	317.10 centuries
2^n	Eternity

1.4.2. Asymptotic Notation

To analyze the time complexity of an algorithm, we have to rely on notations. When we say we are analyzing algorithms using notations, we say we are analyzing the rsource usage of an algorithm within a constant multiple. We say "within constant multiple" because other constant multiples creep in when translating an algorithm to executable code. These are:

(*i*) Programmer ability and effectiveness

(*ii*) Programming language

(*iii*) Compiler

(*iv*) Computer Hardware

The order of growth or rate of growth of the running time of an algorithm is a notion that gives a simple characterization of the algorithm's efficiency. It is simply a function $f(n)$ where n is the input size. Using the order of growth we want to examine how fast the running time of an algorithm increases when n increases. To understand this, we take few values for $f(n)$ and n and see how they increases as n increases. This is shown in Fig. 1.2.

n \ f_n	$\log_2 n$	n	$n \log_2 n$	n^2	n^3	2^n
5	3	5	15	25	125	32
10	4	10	40	100	10^3	10^3
100	7	100	700	10^4	10^6	10^{30}
1000	10	10^3	10^4	10^6	10^9	10^{300}

Fig. 1.2. Order of growth of some functions.

For the comparison of the relative performance of several algorithms, order of growth is an important concept. This we will see more while dealing with the types of notations.

When we talk about asymptotic efficiency of algorithms it means we are concerned with how the running time of an algorithm increases with the size of the input in the limit, as the size of the input increases without bound. All the asymptotic notations that we are going to discuss now describe the asymptotic running time of an algorithm in terms of functions whose domains are the set of natural numbers

$$N = \{0, 1, 2, ...\}.$$

1. Big-oh (*O*) Notation

Let n be the size of the input and $f(n)$, $g(n)$ be positive functions of n. For a given function $g(n)$, we denote by $O(g(n))$ (pronounced "big-oh or oh of g of n") the set of functions

Formal definition : $O(g(n)) = \{ f(n)$: there exist positive constants c and n_0 such that $O <= f(n) <== cg(n)$ for all $n \varepsilon n_0.\}$

Informal definition : $f(n)$ is $O(g(n))$ if f grows at most as fast as g.

The Big-Oh notation clearly tells us that $f(n)$ is bounded by some multiple of $g(n)$ for almost all n. The "Oh" notation specifies asymptotic upper bounds. $O(1)$ refers to constant time. $O(n)$ indicates linear time; $O(n^k)$ (k frxed) refers to polynomial time; $O(\log n)$ is called logarithmic time; $O(2^n)$ refers to exponential time, etc.

DO YOU KNOW

PAUL GUSTAV HEINRICH BACHMANN (1837–1920), the son of a Lutheran pastor, shared his father's pious lifestyle and love of music. His mathematical talent was discovered by one of his teachers, even though he had difficulties with some of his early mathematical studies. After recuperating from tuberculosis in Switzerland, Bachmann studied mathematics, first at the University of Berlin and later at Gottingen, where he attended lectures presented by the famous number theorist Dirichlet. He received his doctorate under the German number theorist Kummer in 1862, his thesis was on group theory. Bachmann was a professor at Breslau and later at Munster. After he retired from his professorship, he continued his mathematical writing, played with piano, and served as a music critic for newspapers. Bachmann's mathematical writings include a five-volume survey of results and methods in number theory, a two-volume work on elementary number theory, a book on irrational numbers, and a book on the famous conjecture known as Fermat's Last Theorem. He introduced big-O notation in his 1892 book *Analytische Zahlentheorie.*

Example 1.1.

(*i*) $5n^2 — 6n + 2 = O(n^2)$

(*ii*) $6n^3 — 2n^2 = O(n^3)$

(*iii*) $5. 2^n + 6 = O(2^n)$

(*iv*) $5n + 7 = O(n)$

Example 1.2.

(*i*) Let $f(n) = n^2 + n + 5$ then $f(n)$ is $O(n^2)$, $f(n)$ is $O(n^3)$ but $f(n)$ is not $O(n)$

(*ii*) Let $f(n) = 3^n$ then $f(n)$ is $O(4^n)$ but $f(n)$ is not $O(2^n)$

(*iii*) If $f_1(n)$ is $O(g_1(n))$ and $f_2(n)$ is $O(g_2(n))$, then $f_1(n) + f_2(n)$ is $O(\max(g_1(n), g_2(n)))$

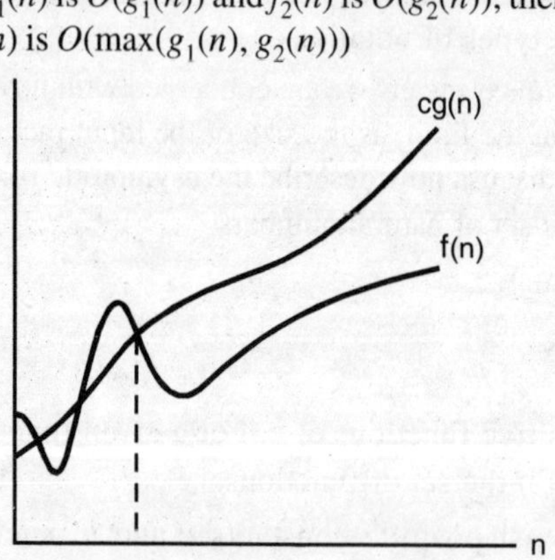

Fig. 1.3.

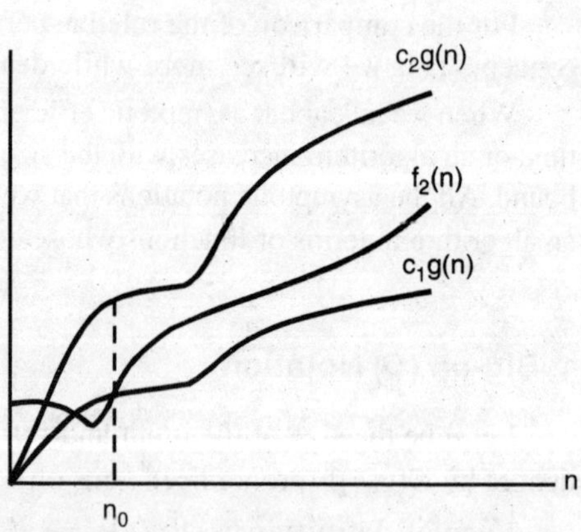

Fig. 1.4.

From Fig. 1.3. it is clear that for all values of $n_0 > n$, the value of $f(n)$ always lies on or below $cg(n)$.

2. Theta (θ) Notation

For a given function $g(n)$, we denote by $\Theta(g(n))$ the set of functions.

Formal Definition : $\theta(g(n)) = \{f(n)$: there exist positive constants c_1, c_2, and n_0 such that $0 <= c_1 g(n) <= f(n) <= c_2\, g(n)$ for all $n \,\varepsilon\, n_0\}$.

Informal definition : $f(n)$ is $\theta(g(n))$ if f is essentially the same as g, to within a constant multiple. The formal definition says that a function $f(n)$ belongs to the set $\theta(g(n))$ if there exist positive constants c_1 and c_2 such that it can be put in between $c_1 g(n)$ and $c_2 g(n)$, for sufficiently large n. From this it is clear that θ notation is used when the function $f(n)$ is bounded both from above and below by the function $g(n)$. We say that $g(n)$ is an asymptotically tight bound for $f(n)$ or in other words that θ notation specifies asymptotically tight bounds. Figure 1.4 shows the θ notation.

Example 1.3.

Consider a function $P(n) = \frac{1}{2} n^2 — 3n$, show that $P(n) = \theta(n^2)$.

Solution. Following the definition of θ notation we can write

$$c_1 n^2 < \frac{1}{2} n^2 — 3n <= c_2 n^2$$

for all $n >= n_0$

Dividing by n^2, equation becomes: $c_1 <= \frac{1}{2} - \frac{3}{n} <= c_2$,

by choosing $c_1 = \frac{1}{14}$, $c_2 = \frac{1}{2}$, and $n_0 = 7$, we can show that $\frac{1}{2} n^2 - 3n = \theta(n^2)$.

3. Big-Ohm (Ω) Notation

For a given function $g(n)$, we denote by $\Omega(g(n))$ (pronounced "big ohm of g of n") the set of functions.

$\Omega(g(n)) = \{f(n)$: there exist positive constants c and n_0 such that $0 <= cg(n) <= f(n)$ for all $n >= n_0$. Ω-notation provides an **asymptotic lower bound on** a function. Figure 1.5 given below shows Ω-notation:

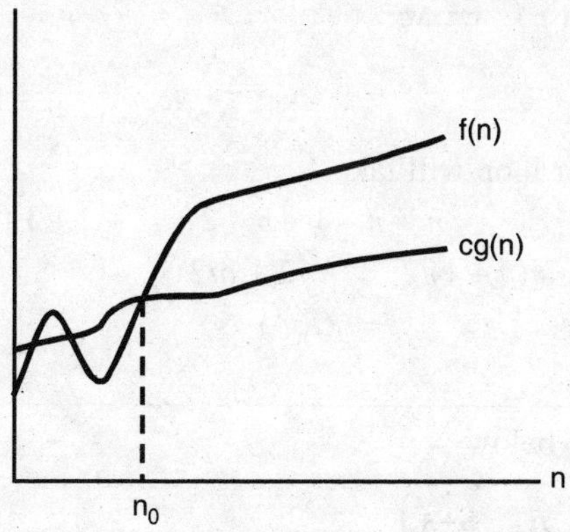

Fig. 1.5.

From figure note that for $n_0 > n$ the value of $f(n)$ always lies on or above cg(n). Thus Ù-notation gives a lower bound for a function to within a constant factor.

1.5. OPERATIONS ON THE O-NOTATION

Rule for Sums

Suppose that $T_1(n) = O(f(n))$ and $T_2(n) = O(g(n))$. Then, $t(n) = T_1(n) + T_2(n) = O(\max(f(n), g(n)))$.

Proof. By definition of the O-notation,

$T_1(n)$	$c_1 f(n)$	for n	n_1 and
$T_2(n)$	$c_2\, g(n)$	for n	n_2

Let $n_0 = \max(n_2, n_2)$. Then

$T_1(n) + T_2(n)$	$c_1\ f(n) + c_2, g(n)$	n	n_0.
	$(c_1 + c_2) \max(f(n), g(n))$	n	n_0.
	$c \max(f(n), g(n))$	n	n_0.

Thus, $$T(n) = T_1(n) + T_2(n) = O(\max(f(n), g(n))).$$

For example,

1. $T(n) = 3n^3 + 5n^2 = O(n^3)$
2. $T(n) = 2n + n^4 + n \log_2 n = O(2n)$

Rule for Products

Suppose that $T_1(n) = O(f(n))$ and $T_2(n) = O(g(n))$. Then, $T(n) = T_1(n) * T_2(n) = O(f(n) * g(n))$.

Example 1.4.

Consider the algorithm below:

```
for(int i=1; i<n-1; i++){
for(int i=1; i<=n; i++){
steps taking O(1) time
}
}
```

Since the steps in the inner loop will take

$$n + n - 1 + n - 2 + ... + 2 + 1 \text{ times,}$$

then
$$n(n+1)/2 = n^2/2 + n/2$$
$$= O(n^2)$$

Example 1.5.

Consider the code snippet below :

```
for (i=1; i <= n, i++)
for (j=1; j <= n, j++)
// steps which take O(1) time
```

Since the steps in the inner loop will take $n + n - 1 + n - 2 + ... + 2 + 1$ times, then the running time is

$$n(n+1)/2 = n^2/2 + n/2$$
$$= O(n^2)$$

Example 1.6.

Linear Search Algorithm

```
1 found = false;
2 loc  = 1;
3 while ((loc <= n) && (!found)){
4 if (item == a[loc]found = true;
5 else loc = loc + 1;
6 }
```

STATEMENT	*# of times executed*
1	1
2	1
3	$n + 1$
4	n
5	n

$T(n) = 3n + 3$ so that $T(n) = O(n)$ Since $g(n) <= c\, f(n)$ for $n >= n\,0$, then

$$3n + 3 \quad <= c\, n$$

$$(3n + 3)/n \quad <= c \quad = 3 + 3 / n <= c$$

Thus $c = 4$ and $n_0 = 3$.

1.5.1. General Rules

The following are the general rules on determining the running time of an algorithm :

FOR loops

At most the running time of the statement inside the for loop times the number of iterations.

NESTED FOR loops

Analysis is done from the inner loop going outward. The total running time of a statement inside a group of for loops is the running time of the statement multiplied by the product of the sizes of all the for loops.

Consecutive Statements

The statement with the maximum running time.

IF/ELSE

Never more than the running time of the test plus the larger of the running times of the conditional block of statements.

1.6. TYPES OF ANALYSIS

1. Worst case Running Time

The behavior of the algorithm with respect to the worst possible case of the input instance. The worst-case running time of an algorithm is an upper bound on the running time for any input. Knowing it gives us a guarantee that the algorithm will never take any longer. There is no need to make an educated guess about the running time. For expressing worst case running time of an algorithm Big-Oh notation is used.

2. Average case Running Time

The expected behavior when the input is randomly drawn from a given distribution. The average-case running time of an algorithm is an estimate of the running time for an "average" input. Computation of average-case running time entails knowing all possible input sequences, the probability distribution of occurrence of these sequences, and the running times for the individual sequences. Often it is assumed that all inputs of a given size are equally likely. For expressing average case running time of an algorithm θ notation is used.

3. Best case Running Time

The behavior of the algorithm when input is in already in order. For exmple in sorting, if elements are already sorted for a specific algorithm. The best case running time rarely occurs in practice comparatively with the first and second case. For expressing best case running time of an algorithm Big- Ohm notation is used.

4. Amortized Running Time

Here the time required to perform a sequence of (related) operations is averaged over all the operations performed. Amortized analysis can be used to show that the average cost of an operation is small, if one averages over a sequence of operations, even though a simple operation might be expensive. Amortized analysis guarantees the average performance of each operation in the worst case.

Example 1.7.

Table shown worst case and average case of some of the operations/algorithms

Operation/Algo	Worst Case	Average Case
Searching for mim or max element in a list	$O(n)$	$O(n)$
Quick sort	$O(n^2)$	$O(n \log_2 n)$
Merge Sort	$O(n \log_2 n)$	$O(n \log_2 n)$
Heap Sort	$O(n \log_2 n)$	$O(n \log_2 n)$
Bubble Sort	$0(n^2)$	$O(n^2)$
Searching an element in Binary search tree	$0(n)$	$O(n \log_2 n)$

SOLVED PROBLEMS

1. What is the rule of summation for Big-Oh Notation ?

Ans. Consider time complexity $T_1(n)$ for an algorithm A is $O(f(n))$ and for another algorithm B time complexity $T_2(n)$ is $O(g(n))$. Then $T_1(n) + T_2(n)$ is known as rule of summation of Big-Oh notations. It can be easily proved as :

Following the definition of O notation we can write :

$T_1(n) <= c_1 f(n)$ and $T_2(n) <= c_2 f(n)$ for some positive constants c_1, c_2 and n_1, n_2 such that $n >= n_1$ (for algorithm A) and $n >= n_2$ (for algorithm B) then

$T_1(n) + T_2(n) <= c_1 f(n) + c_2 g(n)$ for $n > n_0$

$T_1(n) + T_2(n) <= (c_1 + c_2) \max(f(n), g(n))$ for $n > n_0$

Thus $T_1(n) + T_2(n) = O(\max(f(n), g(n)))$

2. Consider time complexity $T_1(n)$ for an algorithm *A* is $O(n^2)$ and for another algorithm *B* time complexity $T_2(n)$ is $O(\log n)$. What is the time complexity of $T_1(n) + T_2(n)$?

Ans. Using the result from question 1,

$$T_1(n) + T_2(n) \text{ is } O(n^2) \text{ as } O(n^2) > O(\log n).$$

3. What is the rule of multiplication for Big-Oh Notation ?

Ans. Consider time complexity $T_1(n)$ for an algorithm *A* is $O(f(n))$ and for another algorithm *B* time complexity $T_2(n)$ is $O(g(n))$. Then $T_1(n).\ T_2(n)$ is known as rule of multiplication of Big-Oh notations. It can be easily proved as:

Following the definition of *O* notation we can write

$T_1(n) <= c_1.f(n)$ and $T_2(n) <= c_2.f(n)$ for some positive constants c_1, c_2 and n_1, n_2 such that $n >= n_1$ (for algorithm *A*) and $n >= n_2$ (for algorithm *B*) then $T_1(n)\ T_2(n) <= c f(n)\ g(n)$ for $n > n_0$ (where $c = c_1.c_2$)

Thus $\quad \boldsymbol{T_1(n).T_2(n) = O(f(n).g(n))}$

4. Consider time complexity $T_1(n)$ for an algorithm *A* is $O(n^2)$ and for another algorithm B time complexity $T_2(n)$ is $O(n)$. What is the time complexity of $T_1(n).T_2(n)$?

Ans. Using the result from question 2, $T_1(n).\ T_2(n)$ is $O(n^3)$.

5. Write an algorithm to find GCD of two numbers.

Ans. Algorithm Find *GCD* (*X,Y*)

The algorithms finds GCD of two positive integer numbers passed as arguments.

(1) Perform division *X/Y*

(2) Set *Q* = *X/Y* *i.e.* quotient to *Q* and *R* = *X* mod *Y* (remainder)

(3) if *R* equal to 0 goto step 7

(4) else Repeat step 5 and 6

(5) Set *Y*=*X*, *R*=*Y*

(6) goto step 2

(7) Print Y as GCD of *X* and *Y*

6. When can we say that time complexity of an algorithm is $O(n \log_2 n)$?

Ans. If an algorithm solves a problem by dividing it into number of small sub problems; each sub problem is solved independently and solution of the main problem is found out by combining the solution of sub problems then we can say that time complexity of algorithm will be $\boldsymbol{O(n \log_2 n)}$

7. Arrange the following in the increasing order of their time complexity ?

$$O(1), O(2^n), O(n^2), O(n \log_2 n), O(n^3), O(n), O(\log_2 n)$$

Ans. The required order is

$$O(1) < O(\log_2 n) < O(n) < O(n \log_2 n) < O(n^2) < O(n^3) < O(2^n)$$

8. List some of the rules of Big-OH (*O*) notation.

Ans. (*i*) $n^m = O(n^p)$ where $p >= m$

(*ii*) $O(f(n)) + O(g(n)) = O(|f(n)| + |g(n)|)$

(*iii*) $C.O(f(n)) = O(f(n))$

(*iv*) $O(O(f(n)) = O(f(n))$

(*v*) $O(f(n)\, O(g(n))) = O(f(n).g(n))$

(*vi*) $O(f(n)\, g(n)) = f(n)\, O(g(n))$

9. What will be the time complexity of expression :

$$\frac{n^3}{3} + \frac{n^2}{2} + \frac{n}{4}.$$

Ans. Consider the first rule from above we have:

$$= O(n^3) + O(n^3) + O(n^3)$$
$$= O(n^3)$$

10. Does every algorithm have a C++ program ?

Ans. We must assume that by "algorithm" we mean something composed of steps are of a nature that they can be performed by a computer. If so, than any algorithm can be expressed in **C**$^{++}$. In particular, if an algorithm can be expressed in any other computer programming language, then it can be expressed in **C**$^{++}$, since all (sufficiently general) computer programming languages compute the same set of functions.

11. Consider the design for a spelling checker program meant to run on a home computer. The spelling checker should be able to handle quickly a document of less than twenty pages. Assume that the spelling checker comes with a dictionary of about 20,000 words. What primitive operations must be implemented on the dictionary, and what is a reasonable time constraint for each operation?

Ans. The primitive operations are (1) adding new words to the dictionary and (2) searching the dictionary for a given word. Typically, dictionary access involves some sort of pre-processing of the word to arrive at the "root" of the word.

A twenty page document (single spaced) is likely to contain about 20,000 words. A user may be willing to wait a few seconds between individual "hits" of mis-spelled words, or perhaps up to a minute for the whole document to be processed. This means that a check for an individual word can take about 10-20 ms. Users will typically insert individual words into the dictionary interactively, so this process can take a couple of seconds. Thus, search must be much more efficient than insertion.

12. Imagine that you are given an array of records that is sorted with respect to some key field contained in each record. Give two different algorithms for searching the array to find the record with a specified key value. Which one do you consider "better" and why?

Ans. Students at this level are likely already familiar with binary search. Thus, they should typically respond with sequential search and binary search. Binary search should be described as better since it typically needs to make fewer comparisons (and thus is likely to be much faster).

13. How many total orderings can be defined on a set with n elements? Explain your answer.

Ans. A total ordering can be viewed as a permuation of the elements. Since there are $n!$ permuations of n elements, there must be $n!$ total orderings.

14. Difference between Linear and nonlinear Data Structures.

Ans.

Linear Data Structure	Nonlinear Data Structures
A data structure is linear if every item is related (or attached) to its previous and next item (*e.g.* array, linked list, stack & queues)	A data structure is non-linear if every item is attached to many other items in specific ways to reflect relationships (e.g, n-ary tree or graphs).
In linear data structure data items are arranged in a linear sequence	In non-linear data structure data items are not in a sequence
In programming, linear means that they are described by one (single) series of data *i e.* each data item has at most one predecessor and at most one successor	All other are Non-linear
Direct access is available for a linear data structure by a single numeric index.	Random access is much faster for a large (very large) non-linear data structure. Non Linear (trees) are faster to search.

15. Difference between Top down & Bottom up Approach?

Ans.

Top Down	Bottom Up
You can imagine a tree like structure, in which when you are following top-down approach you move from root node to leaf node	When you follow bottom-up approach you follow leaf node to root node.
In this an overview of the system is first formulated, specifying but not detailing any first-level subsystems. Each subsystem is then refined in yet greater detail, Sometimes in many additional subsystem levels, until the entire specification is reduced to base elements.	In this approach the individual base elements of the system are first specified in great detail. These elements are then linked together to form larger subsystems, which then in turn are linked, sometimes in many levels, until a complete top-level system is formed.
Top down design proceeds from the abstract entity to get to the concrete design.	Bottom up design proceeds from the concrete design to get to the abstract entity.
Top down design is most often used in designing brand new systems,	Bottom up design is sometimes used when one is reverse engineering a design; i.e. when one is trying to figure out what somebody else designed in an existing system.
top to down approach :-c You write first the main function, that calls stubs, then you subdivise these stubs in smaller stubs until a real work has to be done, that you code in the final files.	bottom to up approach :-c++ You write (or gather) small components that do basic actions that are required to do the full program work. Then you assemble them by custom code, then assemble these parts to write the main method.

16. What does it mean when someone says that the running time of an algorithm is O(N log N) ?

Ans. That is an example of a notation known as big-Oh notation. We write $f(N)$ is $O(g(N))$ if there exists a constant c such that $f(N) <= cg(N)$ for all N. We also say that the running time of an algorithm is $O(g(N))$ if the running time is $O(g(N))$ for all possible inputs. This notation is widely used by theoretical computer scientists to prove theorems about algorithms, so you are sure to see it if you take a course in algorithms and data structures. It provides a worst-case performance guarantee.

17. What is data structure ?

Ans. A data structure is a way of organizing data that considers not only the items stored, but also their relationship to each other. Advance knowledge about the relationship between data items allows designing of efficient algorithms for the manipulation of data.

18. Find the time complexity of the following program fragment in terms of big "Oh" notation :

```
XYZ (a, b, c, m, n)
{
    for i ← 1 to m do
    for j ← 1 to n do
        c(i, j) ← a[i, j] + b[i, j];
}
```

Ans. The big "Oh" notation denotes or represents the worst time or maximum time to solve or execute an algorithm or program.

The above program fragment named XYZ have two for loop, first loop for number of rows and second for loop for number of columns. Hence,

```
XYZ (a, b, c, m, n)
{
    for i ← 1 to m do                              ...(1)
    {
        for j ← 1 to n do                          ...(2)
        {
            c[i, j] ← a[i, j] + b[i, j]            ...(3)
        }
    }
}
```

The program fragment consists of three statements i.e. we have to sum up all the iterations of every statement to get the total time i.e.

```
For i ← 1 to m runs (m + 1) times (OUTER LOOP)
    for j ← 1 to n runs m(n + 1) times (INNER LOOP)
    c[i, j] ← a[i, j] + b[i, j] runs m : n times
```

Therefore, the total time required is

$$T(n) = (m+1) + m(n+1) + mn$$
$$= 2mn + 2m + 1$$

The total time in terms of big 'oh' is

(1) if $m = n$ then $T(n) = O(n^2)$

(2) if $m \neq n$ then $T(n) = O(m\,n)$

19. List the various data structure operations ?

Ans. Large number of operations can be performed on data structures. Some of the important operations are spelled out below :

(*a*) CREATION

(*b*) INSERTION

(*c*) MODIFICATION

(*d*) TRAVERSING

(*e*) SEARCHING

(*f*) DELETION

(*g*) SORTING

(*h*) MERGING

(*i*) COPYING

(*j*) CONCATENATING

(*k*) SPLITTING

20. What is Abstract data type (ADT) ?

Ans. ADT is a useful tool for specifying the logical properties of a data type. A data type is a collection of values and a set of operations on those values.

The specification of an ADT includes :

(1) The domain of value,

(2) The operation on the domain

Objects such as lists, graphs, etc. along with their operations, can be viewed as abstract data type.

21. What are the two main factors for measuring the performance of algorithm ?

Ans. (1) No. of inputs, (2) No. of comparisons.

SUMMARY

1. Data structures is a way for structuring data *i.e.* way of storing data in a computer so that it can be used efficiently.
2. An Algorithm is a finite sequence of instructions, each of which have a clear meaning and can be executed with a finite amount of effort in finite time. Whatever the input values, an algorithm will definitely terminate after executing finite number of instructions.
3. Abstract Data Type (ADT): An ADT is a set of elements with a collection of well defined operations. The operations can take as operands not only instances of the ADT but other types of operands or instances of other ADTs. Similarly results need not be instances of the ADT.
4. When the elements are accessed or placed in contiguous memory locations then data structure is known as linear data structure.

5. In non-linear data structures elements are not accessed in sequential manner. The individual elements are linked through pointers and may be placed any where in the memory.
6. **Time Complexity :** Running time of the program as a function of the size of input.
7. **Space Complexity :** Amount of computer memory required during the program execution, as a function of the input size.
8. **Big-oh (*O*) notation :** $O(g(n)) = \{ f(n)$: there exist positive constants c and n_0 such that $0 <= f(n) <== cg(n)$ for all n å n_0.}
9. **Theta (θ) Notation:** $\theta(g(n)) = \{f(n)$: there exist positive constants c1, c2, and n_0 such that $0 <= c_1 g(n) <= f(n) <= c_2 g(n)$ for all n å $n_0\}$.
10. **Big-Ohm (Ω) Notation :** $\Omega(g(n)) = \{f(n)$: there exist positive constants c and n_0 such that $0 <= cg(n) <= f(n)$ for all $n >= n_0$.

EXERCISES

1. Define data structures?
2. What do you understand by ADT?
3. How will you decide whether a given data structure is linear or nonlinear, homogenous or hetegeneous?
4. What is an algorithm? What are the good properties of an algorithm?
5. What do you understand by space and time complexity of an algorithm?
6. What do you understand by analyzing an algorithm?
7. What is the importance of notations in the analysis of goithms?
8. Describe all 3 main notations with neat figure?
9. Which notation can be used for placing a (*i*) Lower bound (*ii*) Upper bound (*iii*) Tight bound over an algorithm?
10. Discuss in brief all different types of analysis (*i*) Worst case (*ii*) Average Case (*iii*) Best case
11. Consider the following C function

```
int demo(int p)
{
if (!(p%2))
        return p/2;
else
        demo(demo(3*p+1));
}
```

Show that the function terminates for all integers.

❑❑❑

CHAPTER **2**

RECURSION

2.1. INTRODUCTION

Recursion may have the following definitions:

(*i*) The nested repetition of identical algorithm is recursion.

(*ii*) It is a technique of defining an object/process by itself.

(*iii*) Recursion is a process by which a function calls itself repeatedly until some specified condition has been satisfied.

INSIDE THIS CHAPTER

2.2. WHEN TO USE RECURSION

Recursion can be used for repetitive computations in which each action is stated in terms of previous result. There are two conditions that must be satisfied by any recursive procedure.

1. Each time a function calls itself it should get nearer to the solution.
2. There must be a decision criterion for stopping the process.

In making the decision about whether to write an algorithm in recursive or non-recursive form, it is always advisable to consider a tree structure for the problem. If the structure is simple then use non-recursive form. If the tree appears quite bushy, with little duplication of tasks, then recursion is suitable.

2.3. AN EXAMPLE

The recursion algorithm for finding the factorial of a number is given below :

Algorithm 2.1. Factorial (int n) ....

Input : *n*, the number whose factorial is to be found.

Output : *f*, the factorial of *n*

```
if(n=0)
```

```
  f=1 & return
else
f=factorial(n-1)  * n
if end
algorithm ends.
```

The general procedure for any recursive algorithm is as follows :

1. Save the parameters, local variables and return addresses.
2. If the termination criterion is reached perform final computation and goto step 3 otherwise perform final computations and goto step 1.
3. Restore the most recently saved parameters, local variable and return address and goto the latest return address.

2.4. ITERATION V/S RECURSION

2.4.1. Demerits of Recursive Algorithms

1. Many programming languages do not support recursion, hence recursive mathematical function is implemented using iterative methods.
2. Even though mathematical functions can be easily implemented using recursion it is always at the cost of execution time and memory space. For example, the recursion tree for generating 6 numbers in a fibonacci series generation is given in Fig. 2.1. A fibonacci series is of the form 0, 1, 1, 2, 3, 5, 8, 13, ... etc, where the third number is the sum of preceding two numbers and so on. It can be noticed from the Fig. 2.5 that, $f(n-2)$ is computed twice, $f(n-3)$ is computed thrice, $f(n-4)$ is computed 5 times.
3. A recursive procedure can be called from within or outside itself and to ensure its proper functioning it has to save in some order the return addresses so that, a return to the proper location will result when the return to a calling statement is made.
4. The recursive programs needs considerably more storage and will take more time.

2.4.2. Demerits of Iterative Methods

1. Mathematical functions such as factorial and fibonacci series generation can be easily implemented using recursion than iteration.
2. In iterative techniques looping of statement is very much necessary.

Recursion is a top down approach to problem solving. It divides the problem into pieces or selects out one key step, postponing the rest.

Iteration is more of a bottom up approach. It begins with what is known and from this constructs the solution step by step. The iterative function obviously uses time that is $O(n)$ where as recursive function has an exponential time complexity.

It is always true that recursion can be replaced by iteration and stacks. It is also true that stack can be replaced by a recursive program with no stack.

S.No.	*Iteration*	*Recursion*
1	Iteration involves four clear cut steps initialization, condition, execution, and updation.	There must be an exclusive if statement inside the recursive functions.
2	Any recursive problem can be solved iteratively	Not all problems have recursive solution
3	It is more efficient in terms of memory utilization and execution speed.	Recursion is generally a worse option to go for simple problems.

2.5. MEASURING RESOURCE REQUIREMENTS

If we have devised an algorithm, it possible to implement it by writing a computer program1 and then to assess its efficiency by performing experiments. For example, for different inputs to the algorithm, we might measure the amount of memory space required, or the processor time required, to compute the outputs. Note that the latter measure is not usually the same as the actual time required, since the processor is likely to be doing other things as well as running the algorithm. You will have a little experience of experimentation in Practical One, where the time function will be used to measure the processor time required to evaluate various functions on various arguments. In some cases, this will yield unexpected observations worthy of further explanation.

The basic problems with an experimental approach to checking efficiency (and correctness, for that matter) are threefold. First, it is necessary to implement the algorithm: this may be time-consuming and error-prone. Second, details of the implementation may affect efficiency, and obscure the essential behaviour of the algorithm. Third, it is not normally possible to test the algorithm on all possible inputs, which may mean missing good or bad special cases.

2.5.1. Analysis of Factorial Algorithm

The factorial algorithm can be expressed using simple recursion:

```
Factorial( int n )
{
    if ( n = = 1)
    {
        return 1;
    }
    else
    {
        return (n * factorial(n-1) );
    }
}
```

Looking at the computation that has to be done, we might identify three things that will consume time: integer multiplication, integer subtraction and recursive function calls. There are other possibilities but, if we try to take every little detail into account, we are unlikely to get any kind of simple analysis. Let us ignore the cost of making recursive calls, and suppose that the cost of a multiplication is M and that that of a subtraction is S. We can then define a function $T(n)$, meaning "the time required by the algorithm to compute n!", in a very similar form to that of the actual algorithm :

$$T(0) = 0$$

$$T(n) = M + S + T(n-1) \text{ for } n > 0$$

From this,

$$\begin{aligned} T(n) &= (M + S) + T(n-1) \\ &= 2(M + S) + T(n-2) \\ &= \ldots \\ &= n(M + S) + T(n-n) \\ &= n(M + S) \end{aligned}$$

If we regard M and S as being constants, this expression indicates that the time to compute $n!$ is proportional to n so, for example, computing $(2n)!$ will take twice as long as computing $n!$ does. The acid test of such an analysis is to check its predictions against observed reality. Times reported from the SML function call time fac [512,1024,2048,4096] were 2, 11, 53 and 228 — more than quadrupling as the value of n is doubled. To explain this discrepancy, we must both check our idealised analysis for flaws and also consider whether the SML implementation might be introducing unexpected behaviour.

In this case, one obvious problem is an over-simplistic view of the cost of performing arithmetic. Regarding the subtraction time S as a constant is reasonable here, since n is a 'normal size' integer, *i.e.*, fits into 16 or 32 bits, and so $n - 1$ can be computed in one processor step. However, this is not the situation for multiplication. Employing some knowledge of the problem to be solved, we discover that $n!$ grows very rapidly as n increases: Stirling's approximation for the factorial function reveals that $n!$ grows at roughly the same rate as nn, so the number of bits needed to represent $n!$ (that is $\lg n!$ where lg is the base-two logarithm function) grows at roughly the same rate as $\lg nn = n \lg n$. Thus, it is not reasonable to charge a constant time for every multiplication: the number of bits used for the right-hand operand in n * fact $(n - 1)$ roughly proportional to $(n - 1) \lg (n - 1)$ will rapidly become larger than 16, 32 or whatever limit our processor has on normal integers.

Had we implemented the algorithm in C, we could not have used multiplication in such a cavalier fashion, since integers are not allowed to be bigger than the size that the processor can handle easily. To further understand our SML experiment, we need to establish how much time such a long multiplication takes, in terms of individual processor steps. Our SML system employs a straightforward algorithm (similar to long multiplication as learnt at school, in fact). This means that the time to multiply a normal size integer by a b-bit long integer is proportional to b. Thus, a more accurate expression for $T(n)$ would be :

$$T(n) = (n-1)\lg(n-1)\,M + S + T(n-1)$$

(where M is still some constant). Using appropriate solution techniques, we can establish that $T(n) \rightarrow n^2 \lg n$ For example, $(2n)!$ will take just over four times as long to compute as $n!$ does. This corresponds rather better to the experimental results. Normally, we should hope to be able to analyse algorithms without having to probe details of possible implementations, because we assume that they will use 'reasonable' computational operations. Here, the SML system was being a little 'unreasonable' perhaps, since it was suggesting to us that an expensive operation was available at a modest charge.

2.5.2. Analysis of Fibonacci Algorithms

In the obvious Fibonacci algorithm, there are two conspicuous things that consume time: integer addition and recursive function calls. Letting A be a constant representing the time required for a simple addition, we can write down a function $T(n)$ meaning "the time required by the algorithm to compute the n-th Fibonacci number" :

$$\begin{aligned} T(0) &= 0 \\ T(1) &= 0 \\ T(n) &= A + T(n-2) + T(n-1) \quad \text{for } n > 1 \end{aligned}$$

Using appropriate solution techniques, we discover (to our slight horror) that $T(n)$ is roughly proportional to $2n$.

This is a very fast rate of growth, and helps to explain why our SML implementations run very slowly, even for modest values of n. The problem is the pair of recursive calls, which duplicate much work. A little thought leads to the alternative fastfib algorithm that eliminates one of the recursions, and so has :

$$T(n) = A + T(n-1)$$

which is of similar style to the earlier factorial time analysis. Thus, the time requirement is proportional to n — a dramatic improvement. Again, as with the factorial algorithm, we might worry about whether the addition operations can be done simply. However, this is a lesser concern when just comparing the two different Fibonacci algorithms, since both perform similar kinds of addition operations.

The time complexity of an algorithm gives an idea of the number of operations needed to solve it. While we shall discuss the time complexity issues in more detail later, in this section we work out the number of operations for the some of the recursive functions that we have developed so far. To do this we make use of recurrence relations. A recurrence relation, also known as a difference equation, is an equation which defines a sequence recursively: each term of the sequence is defined as a function of the preceding terms.

2.5.3. Complexity of the Recursive Binary Search Function Rbinary

```
B_Search(arr, lb, ub, x)
{
    if ( lb > ub)
    {
        return;
    }
    else
    {
        mid = (lb + ub)/2;
        if (arr[mid] == x)
        {
            print "element found";
            exit(0);
        }
        elseif (arr[mid] > x)
        {
            B_Search(arr, mid+1, ub, x)
        }
        else
        {
            B_Search(arr, lb, mid-1, x)
        }
    }
}
```

The time complexity of the algorithm can be obtained by noting that every time the function is called, the number of elements to be handled reduces to half the previous value. Assuming it takes one operation for each of the 3 IF statements, and 2 operations for computing mid, the total number of operations can be expressed through the recurrence relation with

$$T(n) = T(n/2) + 5$$
$$T(1) = 1$$

Note this relation is very similar to the one that we had for the efficient version of the exponentiation algorithm, and it can be shown that the complexity of this algorithm is also logarithmic.

2.6. SOLUTION OF A RECURRENCE RELATION

Example 2.1.

$$T(n) = T(n/2) + 3 \quad ...(1)$$

$$T(1) = 1 \quad ...(1)$$

Solution.

To solve this recurrence relation, we note that $T(n/2)$ can be expressed as

$$T(n/2) = T(n/4) + 3 \quad ...(2)$$

Substituting for $T(n/2)$ from (2) in relation (1) ,we get

$$T(n) = T(n/4) + 3(2)$$

By repeated substitution process explained above, we can solve for $T(n)$ as follows :

$$\begin{aligned} T(n) &= T(n/2) + 3 \\ &= [T(n/4) + 3\,] + 3 \\ &= T(n/4) + 3(2) \\ &= [T(n/8) + 3] + 3(2) \\ &= T(n/8) + 3(3) \end{aligned}$$

Now we know that there is a relationship between 3 and 8, and we can rewrite the recurrence relation as

$$T(n) = T(n/\,2^3) + 3(3)$$

We can continue the process one step further, and rewrite the relation as

$$T(n) = \mathrm{T}(\mathrm{n}/\,2^4) + 3(4)$$

Now we can see a pattern running through the various relations and we can write the generalized relation as

$$T(n) = \mathrm{T}(\mathrm{n}/\,2^k) + 3(k)$$

Since we know that $T(1) = 1$, we find a substitution such that the first term on the right hand side reduces to 1. The following choice will make this possible

$$2^k = n$$

We can get the value of k by taking log of both sides to base 2, which yields

$$k = \log n$$

Now substituting this value in the above relation, we get

$$T(n) = 1 + 3 \log n$$

Thus we can say that this algorithm runs in LOGARITHMIC TIME, and obviously is much more efficient than the previous algorithm.

2.7. DIVIDE AND CONQUER

The *divide-and-conquer* strategy solves a problem by:

1. Breaking it into *subproblems* that are themselves smaller instances of the same type of problem.
2. Recursively solving these subproblems
3. Appropriately combining their answers

The real work is done piecemeal, in three different places: in the partitioning of problems into subproblems; at the very tail end of the recursion, when the subproblems are so small that they are solved outright; and in the gluing together of partial answers. These are held together and coordinated by the algorithm's core recursive structure.

Divide-and-conquer algorithms often follow a generic pattern: they tackle a problem of size n by recursively solving, say, a subproblems of size n/b and then combining these answers in $O(n^d)$ time, for some a; b; $d > 0$. Their running time can therefore be captured by the equation $T(n) = aT(n/b) + O(n^d)$. We next derive a closed-form solution to this general recurrence so that we no longer have to solve it explicitly in each new instance.

Example 2.2.

Binary search

The ultimate divide-and-conquer algorithm is, of course, *binary search:* to find a key k in a large file containing keys $z[0; 1; : : : ; n - 1]$ in sorted order, we first compare k with $z[n/2]$, and depending on the result we recurse either on the first half of the file, $z[0, ..., n/2 - 1]$, or on the second half, $z[n/2, ..., n-1]$. The recurrence now is $T(n) = T(n/2) + O(1)$, which is the case $a = 1$; $b = 2$; $d = 0$. Plugging into our master theorem we get the familiar solution: a running time of just $O(\log n)$.

The algorithm of binary search is shown below :

Algorithm 2.2. Binary search

```
B_Search(arr, lb, ub, x)
{
    if ( lb > ub)
    {
        return;
    }
    else
    {
        mid = (lb + ub)/2;
        if (arr[mid] == x)
        {
            print "element found";
            exit(0);
        }
        elseif (arr[mid] > x)
        {
            B_Search(arr, mid+1, ub, x)
        }
        else
        {
            B_Search(arr, lb, mid-1, x)
        }
    }
}
```

SOLVED PROBLEMS

1. Find space and time complexity of the following algorithm :

```
ADD (a, b)
{
c <- 0.0;
c <- a * b;
return c;
}
```

Ans. For finding the total space required by the algorithm, we have to search how many variables are declared in our algorithm.

In the above algorithm, we have three variables *i.e.* a, b, *c* but the size or the data type of the variables are not defined in the algorithm, therefore we can assume the lowest size *i.e.* of char *i.e.* one byte.

Thus the total space complexity is

$$S_{ADD(a, b)} >= 3$$

For finding the total computation time required by an algorithm, we have to check how much times each statement of the algorithm is executed or the number of comparisons done by the algorithm.

In the above algorithm, each line of the algorithm executes only once therefore the total time required by the algorithm is

$$T_{ADD(a, b)} = 3$$

2. Find space and time complexity of the following iterative algorithms:

(*i*)
```
ADD (a, b)
{
c <- 0.0;
for i <- 1 to b do
        c <- c + a;
return c;
}
```

(*ii*)
```
Matadd (a, b, c, m, n)
{
  for i <- 1 to m do
       for j <- 1 to n do
             c[i, j] = a[i, j] + b[i, j];
  return c;
}
```

Ans. (*i*) For finding the total space required by the algorithm, we have to search how many variables are declared in our algorithm.

In the above algorithm, we have four variables *i.e.* a, b, *c*, i but the size or the data type of the variables are not defined in the algorithm, therefore we can assume the lowest size *i.e.* of char *i.e.* one byte.

Thus the total space complexity is

$$S_{ADD(a,b)} >= 4$$

For finding the total computation time required by an algorithm, we have to check how much times each statement of the algorithm is executed or the number of comparisons done by the algorithm.

In the above algorithm, the first line is executed only once, the second line containing for loop which runs from i=1 to b *i.e.* b times, where it is true and once when it is false *i.e.* for loop or the second line will run b+1 times, third line will run when the second line is true *i.e.* it runs b times and the last line will be executed once

Therefore	$c <- 0.0$	1time
	for i <- 1 to b	b+1 times
	$c <- c + a$	b times
	return *c*	1 time
		$T_{ADD(a, b)} = 2b + 3$

(*ii*) For finding the total space required by the algorithm, we have to search how many variables are declared in our algorithm.

In the above algorithm, we have seven variables *i.e.* *a*, *b*, *c*, *i*, *j*, *m*, *n* where *a*, *b*, and *c* are matrices of size *m***n* *i.e.* having mn elements in each matrix but the size or the data type of the variables are not defined in the algorithm, therefore we can assume the lowest size *i.e.* of char *i.e.* one byte.

Thus the total space complexity is

$$S_{Matadd(a,\, b,\, c,\, m,\, n)} >= 3mn + 4$$

For finding the total computation time required by an algorithm, we have to check how much times each statement of the algorithm is executed or the number of comparisons done by the algorithm.

In the above algorithm, the first line is a for loop and runs *m*+1 times due to this the second and third line will executes m times. Now, the second line is again a for loop therefore it will run *n*+1 times due to this the third line will run *n* times and the last line will executes once

Therefore	for $i <- 1$ to m	*m*+1times
	for $j <- 1$ to n	*m* (*n*+1) times
	$c[i,j] <- a[i,j] + b[i,j]$	*m n* times
	return *c*	1 time
		$T_{Matadd(a,\, b,\, c,\, m,\, n)} = 2mn + 2m + 2$

3. Find space and time complexity of the following recursive algorithms:

(*i*)
```
ADD (a, n)
  {
  if (n = = 0)
return 0;
  else return (ADD (a. n-1) + a[n])
  }
```

(*ii*) **FACT (*n*)**

```
    {
      if (n = = 1)
return 1;
      else return ( n * FACT (n-1) )
    }
```

Ans. (*i*) To find the space complexity of a recursive algorithm, we have to see the recursive return statement *i.e.*

```
return (ADD (a, n-1) + a[n] );
```

The above statement contains three things i.c rcturn address, parameters of the function and lastly an operation *i.e.* addition operation.

return address	1
value of *n*	1
addition operation	1

$S_{ADD(a, n)}$ >= 3 (*n*+1) as we start the recursive loop from the value *n* downto 0 *i.e.* we have the rucursive call *n*+1 times.

For time complexity of the recursive function we have

$$T_{ADD(a,\ n)} = \begin{cases} 2 & n=0 \\ T_{ADD(a,\ n-1)} + 2 & n>0 \end{cases}$$

i.e. when *n*=0 the first and second line only be executed and when *n*>0 the first and third line will be executed and in third statement we have recursion ADD(*a*,*n*–1) and an addition operation.

$$\begin{aligned} T_{ADD(a,n)} &= T_{ADD(a,\,n\text{-}1)} + 2 \\ &= T_{ADD(a,\,n\text{-}2)} + 2 + 2 \\ &= T_{ADD(a,\,n\text{-}3)} + 2 + 2 + 2 \\ &= T_{ADD(a,\,n\text{-}n)} + 2n \\ &= 2n + 2 \end{aligned}$$

(*ii*) To find the space complexity of a recursive algorithm, we have to see the recursive return statement *i.e.*

```
return (FACT (n-1) * n );
```

The above statement contains three things i.e return address, parameters of the function and lastly an operation *i.e.* multiply operation.

return address	1
value of *n*	1
multiply operation	1

$S_{FACT(n)}$ >= 3 (*n*) as we start the recursive loop from the value *n* downto 1 *i.e.* we have the rucursive call *n* times.

For time complexity of the recursive function we have

$$T_{FACT(n)} = \begin{cases} 2 & n=1 \\ T_{FACT(n\text{-}1)} + 2 & n>1 \end{cases}$$

i.e. when n=1 the first and second line only be executed and when n>1 the first and third line will be executed and in third statement we have recursion FACT(n-1) and a multiply operation.

$$\begin{aligned} T_{FACT(n)} &= T_{FACT(n-1)} + 2 \\ &= T_{FACT(n-2)} + 2 + 2 \\ &= T_{FACT(n-3)} + 2 + 2 + 2 \\ &= T_{FACT(n-(n-1))} + 2\,(n-1) \\ &= 2\,(n-1) + 2 = 2n \end{aligned}$$

4. Write a function to print all of the permutations for the elements of an array containing *n* distinct integer values.

Ans.

```
void rpermute(int *array, int n) {
     swap(array, n-1, Random(n));
     rpermute(array, n-1);
  }
```

5. Prove (using induction) that the recurrence *T*(*n*) = *T*(*n* – 1) + *n*; *T*(1) = 1 has as its closed-form solution *T*(*n*) = *n*(*n* + 1)/2.

Ans. Base case : $T(n-1) = 1 = 1(1+1)/2.$

Induction hypothesis : $T(n-1) = (n-1)(n)/2.$

Induction step :

$$\begin{aligned} T(n) &= T(n-1) + n \\ &= (n-1)(n)/2 + n \\ &= n(n+1)/2. \end{aligned}$$

Thus, the theorem is proved by mathematical induction.

6. Arrange the following expressions by growth rate from slowest to fastest.

$4n^2 \quad \log_3 n \quad n! \quad 3^n \quad 20n \quad 2 \quad \log_2 n \quad n^{2/3}$

Ans. $2 \quad \log_3 n \quad \log_2 n \quad n^{2/3} \quad 20n \quad 4n^2 \quad 3n \quad n!$

7. (*a*) Find a growth rate that squares the run time when we double the input size. That is, if *T*(*n*) = *X*, then *T*(2*n*) = x^2.

(*b*) Find a growth rate that cubes the run time when we double the input size. That is, if *T*(*n*) = *X*, then *T*(2*n*) = x^3.

Ans. (*a*) These questions are quite hard. If $f(n) = 2^n = x$, then $f(2n) = 2^{2n} = (2^n)^2 = x^2$.

(*b*) The answer is $2^{(n\log_2 3)}$. Extending from part (*a*), we need some way to make the growth rate even higher. In particular, we seek some way to make the exponent go up by a factor of 3. Note that, if $f(n) = n^{\log_2 3} = y$, then $f(2n) = 2^{\log_2 3}\, n^{\log_2 3} = 3x$. So, we combine this observation with part (*a*) to get the desired answer.

8. Using the definition of big-Oh, show that 1 is in $O(1)$ and that 1 is in $O(n)$.

Ans. First, we need to find constants c and no such that $1 = c \times 1$ for $n > n_0$. This is true for any positive value $c < 1$ and any positive value of n_0 (since n plays no role in the equation).

Next, we need to find constants c and n_0 such that $1 = c \times n$ for $n > n_0$. This is true for, say, $c = 1$ and $n_0 = 1$.

9. Determine for the following code fragments in the average case. Assume that all variables are of type int.

(*a*) **a = b + c;**
```
d = a + e;
```
(*b*) **sum = 0;**
```
for (i=0; i<3; i++)
for (j=0; j<n; j++)
sum++;
```
(*c*) **sum=0;**
```
for (i=0; i<n*n; i++)
sum++;
```
(*d*) **for (i=0; i < n-1; i++)**
```
for (j=i+1; j < n; j++) {
tmp = A[i][j];
A[i][j] = A[j][i];
A[j][i] = tmp;
}
```
(*e*) **sum = 0;**
```
if (EVEN(n))
for (i=0; i<n; i++)
sum++;
else
sum = sum + n;
```

Ans. (*a*) This fragment is $\Theta(1)$.

(*b*) This fragment is $\Theta(n)$ since the outer loop is executed a constant number of times.

(*c*) This fragment is $\Theta(n^2)$ since the loop is executed n^2 times.

(*d*) This fragment is $\Theta(n^2 \log n)$ since the outer for loop costs $n \log n$ for each execution, and is executed n times. The inner loop is dominated by the call to sort.

(*e*) One branch of the if statement requires $\Theta(n)$ time, while the other requires constant time. By the rule for if statements, the bound is the greater cost, yielding $\Theta(n)$ time.

10. So can I use the fact that the running time of an algorithm is *O*(*N* log *N*) or *O*(*N*^2) to predict performance?

Ans. No, because the actual running time might be much less. Perhaps there is some input for which the running time is proportional to the given function, but perhaps the that input is not found among those expected in practice. Mathematically, big-Oh notation is less precise than the tilde notation we use: if $f(N) \sim g(N)$, then $f(N)$ is $O(g(N))$, but not necessarily vice versa. Consequently, big-Oh notation cannot be used to predict performance. For example, knowledge that the running time of one algorithm is *O*(*N* log *N*) and the running time of another algorithm is *O*(*N*^2) does not tell you which will be faster when you run implementations of them. Generally, hypotheses that use big-Oh notation are not useful in the context of the scientific method because they are not falsifiable.

11. Is the loop for (int i = N-1; i >= 0; i––) more efficient than for (int i = 0; i < N; i++)?

Ans. Some programmers think so (because it simplifies the loop continuation expression), but in many cases it is actually less efficient. Don't do it unless you have a good reason for doing so.

12. What is the value of x after running the following code fragment?

```
int x = 0;
for (int i = 0; i < N; i++)
    for (int j = i + 1; j < N; j++)
        for (int k = j + 1; k < N; k++)
            x++;
```

Ans. N choose 3 = $N(N-1)(N-2)/6$.

13. What is the data structures used to perform recursion?

Ans. Stack. Because of its LIFO (Last In First Out) property it remembers its 'caller' so knows whom to return when the function has to return. Recursion makes use of system stack for storing the return addresses of the function calls.

Every recursive function has its equivalent iterative (non-recursive) function. Even when such equivalent iterative procedures are written, explicit stack is to be used.

14. A recursive definition of multiplication is as follows :

$$a * b = \begin{cases} a, & \text{if } b = 1 \\ a * (b-1) + a, & \text{if } b > 1 \end{cases}$$

How many times this recursion will be called for multiplying 9 * 4 ?

Ans. To multiply 9 * 4, we call this definition four times :

$$\begin{aligned} & 9 * 4 && \text{(first call)} \\ &= 9 * 3 + 9 && \text{(second call)} \\ &= 9 * 2 + 9 + 9 && \text{(third call)} \\ &= 9 * 1 + 9 + 9 + 9 && \text{(fourth call)} \\ &= 9 + 9 + 9 + 9 \\ &= 36 \end{aligned}$$

15. Write a program to calculate the binomial co-efficient to nCr of two numbers ?

Ans.

```
#include <stdio.h>
#include <conio.h>
#include <process.h>
int recbioff (int, int);

void main ( )
{
    int n, r, ncr;
    clrscr ( );
    printf ("Enter values of n and r");
    scanf ("%d %d", &n, &r);
    ncr = recioff (n, r);
    printf ("Binomial co-efficient of
            c(%d, %d) is %d, n, r, ncr);
    getch();
{
```

```
int recbioff (int x, int y)
{
    if (y > x)
    {
    printf ("The value of n can't be less than r");
        getch( );
        exit (1);
    }
if (y = = 0 || y = = x)
    {
        return (1);
    }
    else
    {
    return (recbioff (x - 1, y - 1) + recbioff (x - 1, y));
    }
}
```

Output of the Program :

```
Enter values of n and r 6
2
Binomial co-efficient of C(6, 2) is 15
```

16. Write a recursive C routine that find a / B i.e. quotient when a is divided by B.

Ans.
```
int Quotient (int a, int b)
{
if(a<b)
return 0;
else
return Quotient (a-b,b) +1;
}
```

The code is simple to understand.

SUMMARY

1. Recursion is a process by which a function calls itself repeatedly until some specified condition has been satisfied.
2. There are two conditions that must be satisfied by any recursive procedure.
   (*a*) Each time a function calls itself it should get nearer to the solution.
   (*b*) There must be a decision criterion for stopping the process.
3. Iteration is more efficient in terms of memory utilization and execution speed while recursion is generally a worse option to go for simple problems.
4. The *divide-and-conquer* strategy solves a problem by :
   (*a*) Breaking it into *subproblems* that are themselves smaller instances of the same type of problem
   (*b*) Recursively solving these subproblems
   (*c*) Appropriately combining their answers.

❑❑❑

CHAPTER **3**

ARRAYS

INSIDE THIS CHAPTER

3.1. INTRODUCTION

An array is a data structure which can store multiple elements of same type under one name. In another way we can say that array is a collection of variables of same type all of which are referred by a common name. Specific element of an array is accessed by an index. A reference to array elements in a program often includes one or more subscript depending upon what is the dimension of the array. The element of array may be of primitive type or user defined types, but all must be of same types. The two most common from of array is: One dimensional and two dimensional. Another name for one dimensional array is Vector. Two dimensional arrays are known as matrix.

3.2. ONE DIMENSIONAL ARRAY

In C 1-D array can be declared as

```
int arr[5];
```

By the above statement we are declaring an array named as arr and which can store 5 intger elements. Note regardless of which programming Languages we use for the declaration of array, while declaring array we need to give three piece of information: data type, array name and its size. The size is specified in [ ] square brackets (subscript operator). Size must be given when an array is declared. Array name is an identifier which follows the rules of writing an identifier. Because such each integer element takes 2 bytes so this array declaration reserves 10 bytes of memory for the array when the array elements are initialized. All the elements of an array are stored in contiguous locations with the first element accessed as **arrayyname [0], second as arrayname [1]** and so on. In our case It will be **arr[0], arr[2] up to arr[4]. The first element of an array starts at index 0 and last element at index arraysize-1.** The starting index for the array *i.e.* 0 here is known as Lower Bound (LB) and higher index- 1 is considered as Upper Bound (UB). Different programming languages support different values for LB and UB. For example on FORTRAN lower bound starts at 1. Similarly the value of UB may be the size of the array *i.e.* for an array of size 5, LB may be 1 be 1 and UB may be 5. Apart from 1,0 for lower bound and size, other values of LB and UB may be there. This we will shall see later.

Number of elements in array is given by the following formula:

N=UB-LB+1

For example in C for the following declaration: **int arr [10],** number of elements in the array will be:

N=9-0+1=10

Specific element of an array array are referenced by means of a two level syntactic mechanism, where the first part is the name of the array and second part consisting of one or more items known as indexes or subscripts. For example in C **arr[4]** refers the 5th element of the array named arr.

The syntax of array name is common to almost all programming languages. The array name is following by the list of indexes, which is surrounded by either square brackets ([ ]) in C, C++, Java or parenthesis (()) in Visual Basic and in FORTRAN.

To more understand arrays in C, we see one short program:

Program 3.1. ....

```
# include < stdio. h>
#include < conio. h>
# define S 5
main ( )
{
int a [S];
int i;
clrscr ( );
 for (i=0; i<S; i++)
{
printf("\n Enter a [%d] element : ", i);
scanf("%d", &a[i]);
}
printf ("Array elements are \n");
for (i = 0; i<S; i++)
printf ("a [%d] = %d\n", i, a [ i ]) ;
getch ( ) ;
}
```

This program declares an array of size 5 of name a and loads the array with 5 values entered by the user. To store the five value we have used a for loop which runs from 0 to 4. On each iteration it prompts you to enter the next element of the array and it stores at the location which is denote by its address and a [i] where i will run from 0 to 4. Then all the elements of array are displayed one by one through for loop. In the program we have used a macro temple S which denotes size of array. It is preferable to use macro when giving size of an array because later when you want to change size of the array you'll have to go to every place in the program where you used the array and change the size explicitly. Macro saves this effort because you can simply change the value of S which will be automatically expanded to the places where you have used it. Note if arrays are not initialized they contain garbage values.

3.2.1. Initializing 1-D Array

The array elements are initialized as:

```
int a[ ] = {34, 56, 78, 89, 67};
```

We initialize the array within opening and closing braces { } and separate the element by comma. The size of array in subscript operator [ ] is optional. The size of the array is not given in the above code but we can calculate the size as:

```
const int S = sizeof (a) / sizeof (int);
```

Size of array a will be 10 as total 5 elements are there in the array and array is of int type and size of int is 2 in turbo C so S = 10/2= 5.

One most important concept to note while dealing with arrays in C that array boundary in not checked while accessing an element of array using subscript notation *i.e.* if our array is declared as int arr[5] and we try to access any element beyond 0 and 4 then it will not be checked either at compile time or run time. This may sometimes result in crashing the program.

3.2.2. Array Traversal

Traversing an array is extremely simple. The simplicity comes from from the fact that all array element are indexed *i.e.* can be accessed using an index that lies within LB and UB (inclusive both). For traversing we simply have to take a loop that runs from lower bound to upper bound and say control variable is k, then we can process each element of the array by using expression arr[k], arr[k + 1] and so on. This we have in the program give earlier where we displayed array using a for loop that run from 0 to 4.

DO YOU KNOW

George Boole (1815-1864) Son of a cobbler was born in lincoln, England, in November 1815. Because of his family's difficult financial situation, Boole had to struggle to educate himself while supporting his family. Nevertheless, he became one of the most important mathematicians of the 1800s. Although he considered a career as a clergyman, he decided instead to go in teaching and soon afterwards opened a school of his own. In his preparation of teaching mathematics, Boole – unsatisfied with textbooks of his day – decided to read the works of the great mathematicians. While reading papers of the great French mathematician Lagrange, Boole made discoveries in the calculus of variations, the branch of analysis dealing with finding curves and surfaces optimizing certain parameters.

In 1848 Boole published *The Mathematical Analysis of Logic*, the first of his contributions to symbolic logic. In 1849 he was appointed professor of mathematics at Queen's College in Cork, Ireland. In 1854 he published *The Laws of Thought*, his most famous work. Boole introduced what is now called *Boolean agebra* in his honor. Boole married in 1855; his wife was the niece of the professor of Greek at Queen's College.

Algorithm for Traversing Linear Arrays

Let A be a collection of data elements stored in the memory of the computer. Supose we want to print the contents of each element of A or suppose we want to count the number of elements of A with a given property. This can be accomplished by *traversing* A, that is, by accessing and processing (frequently called *visiting*) each element of A exactly once.

The following algorithm traverses a linear array LA. The simplicity of the algorithm comes from the fact that LA is a linear structure. Other linear structures, such as linked lists, can also be easily traversed. On the other hand, the traversal of nonlinear structures, such as trees and graphs, is considerably more complicated.

Algorithm 3.1. ....

(Traversing a Linear Array) Here LA is a linear array with lower bound LB and upper bound UB. This algorithm traverses LA applying an operation PROCESS to each element of LA.

1. [Initialize counter] Set K := LB.
2. Repeat steps 3 and 4 while K ≤ UB.
3. [Visit element] Apply PROCESS to LA[K]
4. [Increase counter.] Set K:= K + 1

   [End of Step 2 loop]
5. Exit

We also state an alternative form of the algorithm which uses a repeat-for loop instead of the repeat-while loop.

Algorithm 3.2. ....

(Traversing a Linear Array) This algorithm traverses a linear array LA with lower bound LB and upper bound UB.

1. Repeat for K = LB to UB:

   Apply PROCESS to LA[K].

   [End of loop]
2. Exit

Caution : The operation PROCESS in the traversal algorithm may use certain variables which must be initialized before PROCESS is applied to any of the elements in the array. Accordingly, the algorithm may need to be preceded by such an initialization step.

3.2.3. Insertion and Deletion

Insertion and deletion operations on array are not the easy one but before that we tell you what exactly insertion and deletion means. Inserting an element into array means adding an element at some specified index in the array. Similarly deleting an element from array means removing an element from the array at given inex.

Consider an array of size 10. Assume out of 10 locations in the array, 8 locations are filled and two are empty *i.e.* location number 9 and 10 are empty. Now if we want to insert an element at the end of the array, then this task is very simple but inserting the element at any index between 1 to 8 means shifting all elements downwards (to higher index) so that, the index at which the new element is to be inserted can be freed. Again shifting can also be done when array is not full *i.e.* all locations are not occupied. If array is full then there is no chance of adding an element to the array. Unless otherwise stated we assume during insertion that our array is not full. In case array is full and we still have to add new element to the array to new array. This causes overhead on insertion.

As element have to be shifted during insertion and other problems as discussed above, insertion in array is a costly affair. The complete C Program for the insertion is given below:

Program 3.2. /* Program insertion into array */ ....

```
# include <stdio. h>
# include <conio. h>
# include < stdlib. h>
#define S 6
main ( )
{

int arr [10];
int i, pos, num, j;
clrscr ( );
for (i = 0; i<S; i++)
{
printf ("\nEnter arr [%d] element :" i ) ;
scanf (" %d", &arr [i] );
}
clrscr ( ) ;
printf ("\n\nArray is \n\n") ;
for (i = 0; i <S; i + +)
printf (" % 2d" , arr [i]) ;
printf ("\nEnter the element which u want to insert \n");
scanf("%d", &num);
printf ("\nEnter the position (first is 0) \n");
scanf("%d", &pos);
if (pos> = 0 & & pos < = 9)
{
for (j = S; j > pos ; j–);
   arr [j] = arr [j-1];
arr [j] = num;
}
else
{
printf (" wrong position selected \n");
getch ( ) ;
exit (1);
 }
printf (" Array after insertion \n") ;
for (i = 0; i<S+1; i++)
   printf ("%d", arr [i]);
getch ( ) ;
}
```

The program inserts an element in the array. Initial size of the array a is 10 but we store only 6 elements in it. You may store maximum 9 elements as to run the program correctly space for one element should be there in the array. In order to insert the element we are taking the position pos where the user wants to insert the element along with the element to be inserted. For insertion at the supplied position we have to shift all elements downward by pos number of times starting from number of elements actually stored in the array. When the loop quits we store element at j position.

We have seen how insertion takes place in array, now we consider how deletion takes place in array. If the element to be deleted is at the end of array, it is simple task. But again similar to insertion, if we have to delete an element have to be shifted upward. Note one thing here is that element is not deleted physically that is we cannot deallocate the memory for the deleted element. Given below is the code for deleting an element from the array.

Program 3.3. /* Program Deletion From an array */ ....

```
#include<stdio. h>
#include<conio. h>
#include<stdlib. h>
#define S 6
main ( )
{
int arr [S];
int i, pos, num, j;
clrscr( );
for (i = 0; i<S; i++)
{
printf ("\nEnter a [%d] element : ", i);
scanf ("%d", &arr[i] );
}
printf ("\n\mArray is \n\n");
for (i = 0; i<S;i ++)
printf ("%2d ", arr [ i ] ;

printf ("\nEnter the position of the element");
printf (" which u want to delete \n");
scanf ("%d", & pos);
if (pos > = 0 && pos < S)
{
for (j = pos ; j<S; j ++)
    arr [ j ] = arr [ j + 1 ] ;
}
else
{
printf (" Wrong position selected\n" );
getch ( ) ;
exit ( 1 ) ;
}
```

```
printf ("Array after deletion\n" ) ;
for (i = 0; i <S-1;i ++ )
   printf ("%d", arr [ i ]) ;
getch ( ) ;
}
```

The logic for deletion is similar to logic for insertion in the array. Starting from the pos given by the user of element to be deleted. We shift all elements upward as to cover up the place of the deleted element. Note after deletion, for displaying elements we run loop for S-1 times.

We can one more logic for deletion in an array provided ordering of elements in array does not matter. The logic is simple: We copy the last element of the array at the index of the element to be deleted and last position is filled with some unwanted value. Assuming all empty locations in array are contagious, this logic requires that we must have an access to the last element of the array. We do not give code for this, I think can do this at your own.

Algorithm for Inserting and Deleting

Let A be a collection of data elements in the memory of the computer. "Inserting" refers to the operation of adding another element to the collection A, and "deleting" refers to the operation of removing one of the elements from A. This section discusses inserting and deleting when A is a linear array.

Inserting an element at the "end" of a linear array can be easily done provided the memory space allocated for the array is large enough to accommodate the additional element.

Algorithm 3.3. (Inserting into a Linear array) ....

INSERT (LA, N, K, ITEM) Here LA is a linear array with N elements and K is a positive integer such that K ≤ N. This algorithm inserts an element ITEM into the Kth position in LA.

1. [Initialize counter.] Set J:= N
2. Repeat Steps 3 and 4 while J ≥ K
3. [Move Jth element downward.] Set LA[J + 1]:= LA[J].
4. [Decrease counter.] Set J:= J – 1.

   [End of Step 2 loop.]
5. [Insert element.] Set LA[K] := ITEM.
6. [Reset N.] Set N:= N + 1.
7. Exit

The following algorithm deletes the Kth element from a linear array LA and assigns it to a variable ITEM.

Algorithm 3.4. (Deleting from a Linear Array) ....

DELETE (LA, N, K, ITEM) Here LA is a linear array with N elements and K is a positive integer such that K ≤ N. This algorithm deletes the Kth element from LA.

1. Set ITEM:= LA[K]
2. Repeat for J = K to N – 1:

   [Move J + 1st element upward.] Set LA[J] := LA[J + 1]

   [End of loop.]
3. [Reset the number N of elements in LA.] Set N:= N – 1.
4. Exit.

Remark. We emphasize that if many deletions and insertions are to be made in a collection of data elements, then a linear array may not be the most efficient way of storing the data.

3.2.4. Searching an Element in Array

For searching an element in array we have search and binary search technique. Both the techniques have been discussed in the chapter "Searching".

3.2.5. Address Calculation

A single dimension array is a list of memory cells that are contagious. Suppose the array ARR is defined to have a lower bound 1. The access formula or function for ARR is often of the form:

```
address (ARR [K]) = address (ARR [ 1 ] + (k-1) * element_size
```

For example in C suppose array is declared as int ARR [10]; The address in memory are as shown

1020	1022	1024	1026	1028	1030	1032	1034	1036	1038

To find address of say fifth element (array elements in C starts at index 0)

```
address (ARR [5 ] = address (ARR [1 ] + (5 - 1 ) * sizeof (int );
               = 1020 + 4 * 2
               = 1020 + 8
               = 1028
```

3.2.6. Few Programs Using 1-d Array

In this section we present few good programs using 1-D array. Explanation for the program is given following the program.

Program 3.4. /* program counting frequencies of elements if array */ ....

```
# include <conio. h>
#include <stdio. h>
#defin S 6
main ( )
{
int a [S], freq [S] ;
int i, j, k, n = S;
clrscr( );
for (i = 0, i< S; i++)
{
printf ("\n Enter a [ %d ] element: ", i);
scanf("%d" , &a [i]) ;
freq [ i] = 1;
}
clrscr ( );
printf ("Original Array \n");
for (i = 0; i< S; i++)
```

```
printf ("%d\t", a [i]) ;
/* main logic starts here */
for (i = 0;i <n;i ++)
for (j = i + 1; j<n; j ++)
{
if (a [i] = = a [j])
{
for (k = j; k< n; k++)
a [ k ] = a [ k + 1 ];
freq [ i ] ++;
n-;
}
}
/* main logic ends here */
printf ("\nArray with freq\n");
printf ("\nElement\ tFreq\n");
for (i = 0; i<n; i++)
printf ("%d\t%d\n", a [ i ], freq [ i ] ;
getch ( );
}
```

OUTPUT

```
Enter a[0] element:3
Enter a[1] element:2
Enter a[2] element:3
Enter a[3] element:11
Enter a[4] element:2
Enter a[5] element:5
Original Array
3  2  3  11  2  5
Array Freq
3    2
2    2
11   1
5    1
```

The logic is similar to one which we do in general sorting with few additions. We start with the first element of the array and compare this element with all the other elements of the array. If any other element of the array is same (say at position j) we shift all elements upwards starting with this j position and decrement size of array by one. Initially we assume that frequency of each element is 1 and for that an array freq is used of same size as of thc original array. If a[i] matches with any of the a[j] element then freq[i] is incremented by one. In the end we display the array along with their frequencies.

For example we take array : 3 2 3 11 2 5

And initially freq array : 1 1 1 1 1 1

n = 6; for i = 0 and j = 2 a[il and a[jl are equal so freq[0] is incremented by one and array elements are shifted upward so

array now becomes : 3 2 11 2 5 5

freq array becomes: 2 1 1 1 1 1

n = 5;

next no element matches with a[0] so we now we compare a[1] with all remaining elements. a[1] matches with a[3]

array now becomes : 3 2 11 2 5 5

freq array becomes: 2 2 1 1 1 1

n=4;

next for a[2] and a[3] there will be no matching and their frequency will remain one.

Program 3.5. /* program checking and removal of duplicate elements of array*/ ....

```
#include<conio. h>
#include<stdio. h>
#include<stdlib. h>
#define S 6
#define true 1
#define false 0
main ()
{
int a[S];
int i, j, k, n S, dup found = false;
clrscr 0;
for(i =0;i<S;i++)
{
printf("\nEnter a[%d] element:", i);
scanf("%d", &a[i]);
}
clrscr();
printf ("Original Array\n");
for(i = 0;i<S;i++)
printf("%d" , a[i]);
for(i = 0;i<n;i++)
for(j = i+1;j<n;j++)
{
if(a[i] = = a[j])
{
for(k = j;k<n;k++)
```

```
a[k] = a[k+1];
n--;
dup_found = true;
}
}
if(dup_found = = true)
printf("\nArray contains duplicate element\n");
else
{
printf("\nArray does not contains duplicate element\n");
getch();
exit (0);
}
printf("\n Array after removing duplicate elements\n");
for(i = 0;i<n;i++)
printf("%d"", a[i]);
getch();
}
```

OUTPUT ....

```
Enter a[0] element:3
Enter a[1] elemerit:2
Enter a[2] element:3
Enter a[3] element:11
Enter a[4] element:2
Enter a[5] element:5
Original Array
3  2  3  11  2  5
Array contains duplicate elements
Array after removing duplicate elements
3  2  11  5
```

The program is quite similar to previous program. We have removed the frequency array and to check whether array contains duplicate element we have taken one variable dup_found whose initial value is false. If control reaches to if block its value is changed to true.

3.3. TWO DIMENSIONAL ARRAY

A two dimension array can be thought of a matrix with rows and columns. For example if we declare a 2D array in C as:

```
int arr[3] [4];
```

Then this declaration means a 2D array arr of 3 rows and 4 columns which can be represented as:

	Col 1	Col 2	Col 3	Col 4
Row 1	arr[0][0]	arr[0][1]	arr[0][2]	arr[0][3]
Row 2	arr[1][0]	arr[1][1]	arr[1][2]	arr[1][3]
Row 3	arr[2][0]	arr[2][1]	arr[2][2]	arr[2][3]

The first row and column starts at 0. All rows are listed horizontally and columns are listed vertically. Inter section of row and column becomes the position of element of the 2D array **for e. g. first row(0) and second column(1) element will be arr[0][1].** For accessing an element in 2-D array we require two subscript: first is for row and second is for column. Using his any element of the matrix can be accessed. For example for any ith row and jth column in a matrix named arr, element can be accessed by writing arr[i, j] or arr(i, j) (in some programming languages like FORTRAN, Visual Basic) Just because of two subscripts the array is known as two dimensional arrays. More then two dimensional will be discussed in the next section. In the array declared above as: arr[3] [4], the first dimension is for rows, and second for the column. Here the array arr wifi be of size 3 by 4. In general we say matrix is m by n where m is number of rows and n is number of columns. Sometimes we also say that length of first dimension is m and length of second dimension is n. Total number of elements in the matrix is 3*4 = 12 or in general mn. Many programming languages (like Visual Basic) allow integer ranges for rows and columns instead of fixed integers as seen just now. For example array declaration can be of type:

```
ARR(10:15, -2:2)
```

Here 10 is the LB for the first dimension and 15 is the UB. For second dimension -2 is the LB and 2 is the UB. LB and UB are separated by:. In this examples we will have rows : 10, 11, 12, 13, 14, 15 and for each of the preceding rows we will columns as: –2, –1, 0, 1, 2. Total rows and columns are calculated as:

NR=UB–LB+ 1=15-10+1=6

NC=UB–LB+ 1=2-(-2)+ 1=5

Thus matrix arr is of size 6 by 5.

Number of elements = NR * NC = 6 * 5 = 30.

3.3.1. Initialization of 2 Array

Initialization of 2-D array is simple. This is done as in C:

```
int arr[3][3] = {{1, 2, 3}, {4, 3, 6), {2, 3, 4));
```

A two dimensional array is a collection of 1D array so I have enclosed each 1D array in braces though this is not necessary but provides better understandability i. e. you may write in this manner also

```
int arr[][3] = {1, 2, 3, 4, 3, 6, 2, 3, 4};
```

You can omit row size while declaring a 2D array but you cannot leave column size field why? Simple by counting number of elements for each row ie column size the number of row can be calculated but by without knowing how many elements for each row a 2D array cannot be allocated. This is because in C, a 2-D array is stored in row major order (discussed later). So the following will result in an error.

```
int arr[3][ ] = {{1, 2, 3}, {4, 3, 6}, {2, 3, 4}};
```

3.3.2. Few Programs On 2-D Array

To understand 2-D array in C, we take few programs

Program 3.6. /* Program demo of 2-D array */ ....

```
#include<stdio. h>
#include<conio. h>
main ()
{
int arr[3] [4];
int i, j;
clrscr ();
for(i = 0;i<3;i++)
{
for(j=0;j<4;j++)
{
printf("\n Enter the arr[%d] [%d] element::", i, j);
scanf("%d", &arr[i] [j] )
}
}
printf("Array elements are\n");
for(i = 0;i<3;i++)
{
for(j = 0;j<4;j++)
printf("%d ", arr[i][j]);
printf("\n");
}
getch ();
}
```

As we are dealing with two dimensional array we have to first store all the elements (depend upon number of columns) for first row, then for second up to total number of rows. For this we require two loops one which controls rows and one column. Through this we have first taken the values for 2D array and later displayed them in the form of matrix.

Program 3.7. /* Program transpose of a mattix */

```
#include<stdio. h>
#include<conio. h>
main ()
{
int arr[3] [2];
int i, j;
clrscr ();
printf ("Enter the 6 elements of matrix\n");
for(i = 0;j<3;i++)
for(j = 0;j<2;j++)
scanf("%d", &arr[i] [j]);
printf ("The Matrix is\n"),
for(i = 0;i<3;i++)
{
for(j = 0;j<2;j++)
printf("%d", arr[i] [j]);
printf ("\n");
}
printf ("The Transpose Matrix is\n");
for(i = 0;i<2;i++)
{
for(j =0;j<3;j++)
printf("%d", arr[j] [i]);
printf ("\n");
}
getch ();
}
```

The transpose of a matrix is matrix with its row and column interchanged. For example if the matrix is written in C

```
int a [ 3] [2] = { {1, 2}, {4, 5}, {6, 8} };
```

Diagrammatically as

1	2
4	5
6	8

the transpose matrix is int a[2) [3) = { {1, 4, 6), {2, 5, 8 } };

1	4	6
2	5	8

To print transpose of a matrix simply change the row and column and while printing interchange i and j.

Program 3.8. /* Program.multiplication of two matrices */ ....

```
#include<stdio.h>
#include<conio. h>
main ()
{
int a[3] [3], b[3] [3], c[3] [3]
int i, j;
clrscr ();
printf ("Enter elements of first matrix\n");
for(i = 0;i<3;i++)
{
for(j = 0;j<3;j++)
{
printf("\nEnter a[%d] [%d] element: ", i, j);
scanf("%d", &a[i] [j]);
}
}
clrscr();
printf ("Enter elements of second matrix\n");
for(i = 0;j<3;i++)
{
for(j = 0;j<3;j++)
{
printf("\nEnter b[%d] [%d] element:", i, j);
scanf("%d", &b[i] [j]);
}
}
clrscr () ;
printf("\n\nFirst Matrix is\n");
for(i =0;j<3;j++)
{
for(j = 0;j<3;j++)
{
printf("%2d" , a[i] [j]);
printf ("\n");
}
```

```
}
clrscrO;
printf("\n\nSecond Matrix is\n");
for(i = 0;i<3;i++)
{
for(j = 0;j<3;j++)
{
printf("%2d", b[il [j]);
printf("\n");
}
}
/* main logic starts here*/
for(i = 0;j<3;i++)
{
for(j = 0;j<3;j++)
{
c[i][j) = 0;
for(k = 0;k<3;k++)
c[i] [j] = c[il [j] + a[i] [k]*b[kl [j];
}
}

/* main logic ends here*/
printf('\n\n Multiplication of two Matrices is\n");
for(i= 0;i<3;i++)
{
for(j = 0;j<3;j++)
printf("%2d" , c[i] [j]);
printf("\n");
}
getch ();
}
```

For matrix multiplication of two matrices the number of col of first row must equal number of row of second matrix *i.e.* if dimension of first matrix is *m* by *n* then dimension of second matrix must be *n* by *p* and the resultant matrix will be of order *m* by *p*. The logic is as follows: We take two matrices of order 3 by 3. Suppose order of first matrix is *m* by *n* and that of second is *n* by *p*.

1	2	3		2	4	5		29	11	23
4	3	2		6	2	3	=	36	24	37
2	5	6		5	1	4		64	24	49

In the main logic the first loop runs from 0 to *m*. Second inner loop runs from 0 to *p* and third inner loop runs from 0 to n(common column of first matrix and row of second matrix). As the direction of arrow shows initially first row is chosen for value of $i = 0$ then for value of $j = 0$, k runs from 0 to 3 so we get

```
c[0] [0] = 0;
i = 0, j = 0, k = 0;
c[0][0] = c[0] [0]+a[0] [0]*b[0] [0];
        = 0+ 1*2;
        = 2;
k= 1
c[0] [0] = c[0] [0]+a[0] [1]*b[1] [0];
         = 2+ 2*6;
         = 2+12;
         = 14;
k= 2
C[0] [0] = c[0] [0]+a[0] [2]*b[2] [0];
         = 14+ 3*5;
         = 14+15;
         = 29;
```

This is the value of first row and first column of resultant matrix.

After this third inner loop finishes. Now we have to choose second column of second matrix so j becomes 1 and for value of $i = 0, j = 1$, k again runs from 0 to 3.

```
C[0] [1] = 0;
i = 0, j = 1, k = 0;
c[0] [1] = c[0] [1]+a[0] [0]*b[0] [1];
         = 0+ 1*4;
         = 4;
k=1
c[0] [1] = c[0] [1]+a[0] [1]*b[1] [1];
         = 4+ 2*2;
         = 4+4;
         = 8;
k=2
c[0] [1] = c[0] [1]+a[0] [2]*b[2] [1];
         = 8+ 3*1;
         = 8+3;
         = 11;
```

This is the value of first row and second column of resultant matrix.

After this third inner loop finishes. Now we have to choose third column of second matrix so j becomes 2 and for value of $i = 0, j = 2$, k again runs from 0 to 3.

```
c[0] [2] = 0;
i =0, j = 2, k = 0;
c[0] [2] = c[0] [2]+a[0] [0]*b[0] [2];
         = 0+ 1*5;
         = 5;
```

```
k= 1
c[0] [2]= c[0] [2]+a[0] [1]*b[1] [2];
        = 5+ 2*3;
        = 5+6;
        = 11;
k= 2
c[0] [2] = c[0] [2]+a[0] [2]*b[2] [2];
         = 11+ 3*4;
         = 11+12;
         = 23;
```

This is the value of first row and third column of resultant matrix.

After this value of j exhausts and value of i now become 1 and process repeats.

3.3.3. Memory Representation of 2-D Array

2-D arrays are implemented in a linear manner in memory. There are mainly two methods used for the mapping of 2-D arrays in single dimensional arrays in memory.

1. Row major order : In this method the elements of the array that have first subscript, the lower bound value of that subscript are stored first, followed by the elements of the second value of the first subscript and so on. In short in row order rows are listed on the basis of number of columns. For example the 2-D array **int arr[3] [4]** in C is represented in memory in row order as

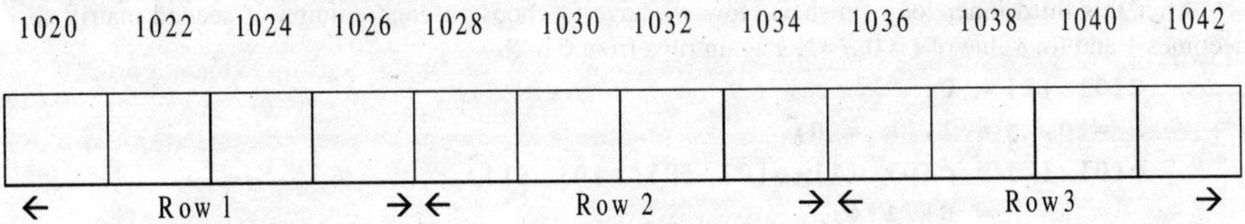

To calculate the address of any 2-D array element is given as

```
Address(arr[ i ] [j ]) = base address of arr + sizeof(data
         type of arr) (no. of cols x  i +j)
```

Assuming matrix is of size M by N and size of data type of array is S and base address of array is array name itself then we have

```
Address (arr[i) [j]) = arr + S(N * i + j)
```

As an example consider an int type array arr[3][4] and size of int is 2. Assume array index starts at 0. To find the address of element at 3rd row and 4th column (3rd row means index 2, 4th column means index 3)

```
Address(arr[2][3]) = 1020 + 2 x (4 x 2 + 3)
                   = 1020 + 2 x (11)
                   = 1042
```

2. Column Major Order : In column major order column are listed on the of number of rows. For example the 2-D array **arr[3] [4]** is represented in memory in column major order as

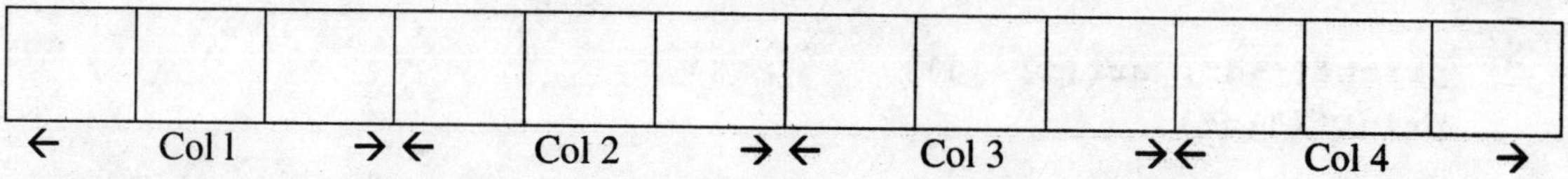

In column major order address of any element can be computed as:

```
Address (arr[ i ] [ j ]) = base address of arr + sizeof (data
      type of arr) (no. of rows x  j +i)
```

or

```
Address(arr[i] [j]) = arr + S(M * j + i)
```

(here arr[i] [j] means ith column and jth row)

```
arr[3] [2] = 1020 + 2 x (3 x 3 + 2)
            = 1020 + 2 x (11)
            = 1042
```

Note we have shown you the above formula considering the fact array indexes start at 0. In case array index start at 1, we have the modified formula:

(*i*) For row major order :

```
Address(arr[i][j]) = arr + S(N * (i-i) ÷ (j-1))
```

(*ii*) For column major order :

```
Address(arr[i)[j]) = arr + S(M * (j-1) + (i-i))
```

A program is given below that displays array elements in row major and column major order.

Program 3.9. /*Program Printing array elements in row major and column major order*/

```
#include<stdio.h>
#include<conio. h>
void main()
{
int arr[3] [4] = {1, 2, 3, 5, 4, 3, 6, 7, 2, 3, 4, 0};
int i, j;
printf("Array elements in row major order\n");
for(i = 0;i<3;i++)
{
for(j= 0;j<4;j++)
printf("%d", arr[i] [j]);
printf ("\n");
}
printf("Array elements in column major order\n");
for(j =0;j<4;j±+)
{
for(i = 0;i<3;i++)
```

```
printf('%d", arr[i] [j]);
printf("\n");
}
getch ();
}
```

OUTPUT ....

```
Array elements in row major order
1 2 3 5
4 3 6 7
2 3 4 0
Array elements in column major order
1 4 2
2 3 3
3 6 4
5 7 0
```

3.4. MULTIDIMENSIONAL ARRAY

We have seen that a 2-D array is a collection of 1-D array, similarly a 3-D array is collection of 2-D array, a 4-D array is a collection of 3-D array and so on. We will restrict ourselves for the discussion of 3-D array only. A 3-D array in C can be declared as:

```
int arr[2] [3] [4];
```

The above declaration creates a 3-D array of size 2 by 3 by 4 in C. It can simply be assumed as collection of 2-D array as stated earlier, The first 2-D array is arr[0][j][k] and second 2-D array is arr[1][j][k]. For both the array i varies from 0 to 2 and j varies from 0 to 3. The array can also be initialized as:

```
int a[2] [3] [2] = {{{1, 2}, {2, 3}, {3, 4}}, {{3, 2}, {2, 5}, {7, 2}}};
```

Here {l, 2} represents l-D array with 2 elements. {{l, 2), {2, 3}, {3, 4} } represents 1st 2-D array. Any element of this array will be represented by a[0] [j] [k] (j from 0 to 2 and k from 0 to 1). Similarly any element of 2nd 2-D array will be represented as a[1] [j] [k] (j from 0 to 2 and k from 0 to 1).

The above array can easily be traversed as :

```
for(i =0;i<2;i++)
{
printf("2D Array %d\n", i+1);
for(j = 0;j<3;j++)
{
for(k = 0;k<2;k++)

printf("%d", a[i] [j] [k]);
printf ("\n");
}
```

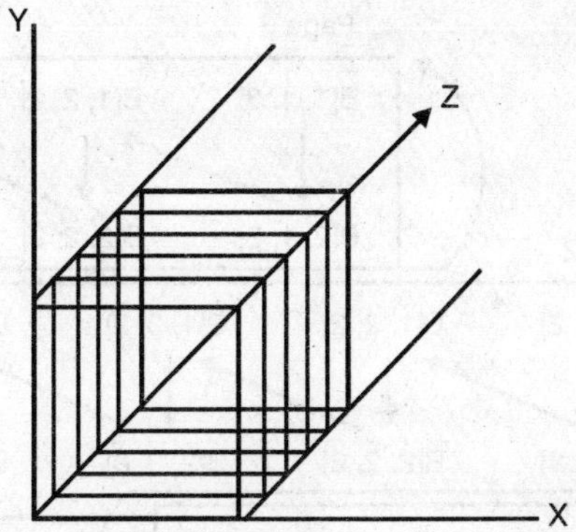

Fig. 3.1. 3D Array

In Figure 3.1, 3d array is a collection of 2D arrays which is represented by stack of squares.

3.4.1. General Multidimensional Arrays

General multidimensional arrays are defined analogously. More specifically, an n-dimensional $m_1 \times m_2 \times ... \times m_n$ array B is a collection of $m_1 . m_2 ... m_n$ data elements in which each element is specified by a list of n integers—such as $K_1, K_2, ..., K_n$ called *subscripts,* with the property that

$$1 \le K_1 \le m_1, \quad 1 \le K_2 \le m_2, \quad ..., \quad 1 \le K_n \le m_n$$

The element of B with subscripts $K_1, K_2, ..., K_n$ will be denoted by

$$B_{K_1, K_2, ..., K_n} \text{ or } B[K_1, K_2, ..., K_N]$$

The array will be stored in memory in a sequence of memory locations. Specifically, the programming language will store the array B either in row-major order or in column-major order. By *row-major order,* we mean that the elements are listed so that the subscripts vary like an automobile odometer, *i.e.*, so that the last subscript varies first (most rapidly), the next-to-last subscript varies second (less rapidly), and so on. By *column-major order,* we mean that the elements are listed so that the first subscript varies first (most rapidly), the second subscript second (less rapidly), and so on.

Example 3.1.

Suppose B is a three-dimensional $2 \times 4 \times 3$ array. Then B contains $2 . 4 . 3 = 24$ elements. These 24 elements of B are usually shown in Fig. 3.2, *i.e.*, they appear in three layers, called *pages,* where each page consists of the 2×4 rectangular array of elements with the same third subscript. (Thus the three subscripts of an element in a three-dimensional array are called, respectively, the *row, column* and *page* of the element). The two ways of storing B in memory appear in Fig. 3.3. Observe that the arrows in Fig. 3.2 indicate the column-major order of the elements.

The definition of general multidimensional arrays also permits lower bounds other than 1. Let C be such an n-dimensional array. As before, the index set for each dimension of C consists of consecutive integers from the lower bound to the upper bound of the dimension. The length L_i of dimension i of C is the number of elements in the index set, and L_i can be calculated, as before, from

$$L_i = \text{Upper bound} - \text{Lower bound} + 1 \qquad ...(i)$$

For a given subscript K_i, the effective index E_i of L_i is the number of indices preceding K_i in the index set, and E_i can be calculated from

$$E_i = K_i - \text{Lower bound}$$

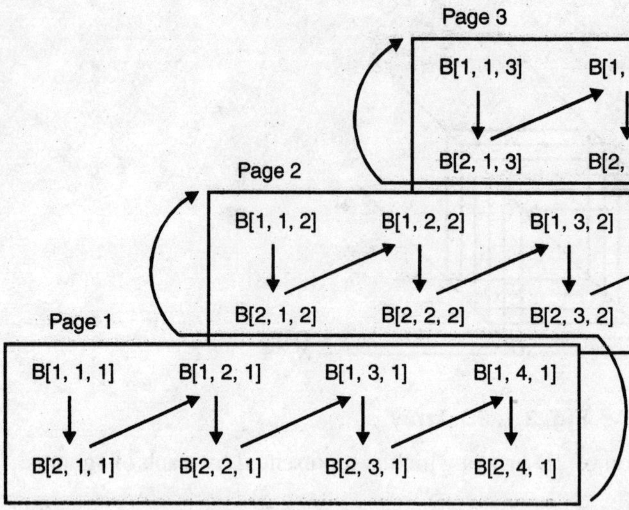

Fig. 3.2.

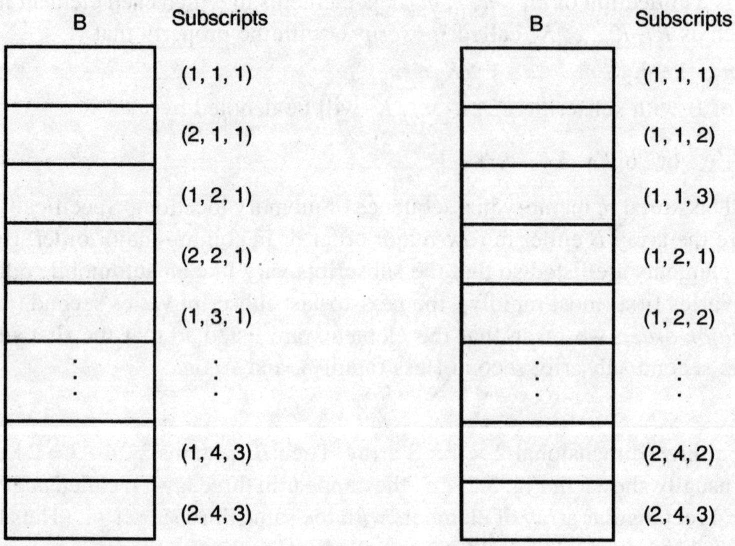

(a) Column-major order *(b) Row-major order*

Fig. 3.3.

Then the address LOC($C[K_1, K_2, ..., K_N]$ of an arbitrary element of C can be obtained from the formula

$$Base(C) + w[(((...(E_N L_{N-1} + E_{N-1}) L_{N-2}) + ... + E_3) L_2 + E_2) L_1 + E_1]$$

or from the formula

$$Base(C) + w[(((...(E_1 L_2 + E_2) L_3 + E_3) L_4 + ... + E_{N-1}) L_N + E_N]$$

According to whether C is stored in column-major or row-major order. Once again, $Base(C)$ denotes the address of the first element of C, and w denotes the number of words per memory location.

Example 3.2.

Suppose a three-dimensional array MAZE is declared using

MAXE (2:8, –4:1, 6:10)

Then the lengths of the three dimensions of MAZE are, respectively,

$$L_1 = 8 - 2 + 1 = 7, \qquad L_2 = 1 - (-4) + 6, \qquad L_3 = 10 - 6 + 1 = 5$$

Solution. Accordingly, MAZE contains $L_1 . L_2 . L_3 = 7 . 6 . 5 = 210$ elements.

Suppose the programming language stores MAZE in memory in row-major order, and suppose Base(MAZE) = 200 and there are w = 4 words per memory cell. The address of an element of MAZE—for example, MAZE[5, –1, 8]—is obtained as follows. The effective indices of the subscripts are, respectively,

$$E_1 = 5 - 2 = 3, \quad E_2 = -1 - (-4) = 3, \quad E_3 = 8 - 6 = 2$$

Using Eq. (4.9) for row-major order, we have :

$$E_1L_2 = 3 . 6 = 18$$
$$E_1L_2 + E_2 = 18 + 13 = 21$$
$$(E_1L_2 + E_2)L_3 = 21 . 5 = 105$$
$$(E_1L_2 + E_3)L_3 + E_3 = 105 + 2 = 107$$

Therefore, LOC(MAZE[5, –1, 8] = 200 + 4(107) = 200 + 428 = 628

Example 3.3.

Suppose multidimensional arrays *A* and *B* are declared using

$A(-2{:}2, 2{:}22)$ and $B(1{:}8, -5{:}5, -10{:}5)$

(*a*) Find the length of each dimension and the number of elements in *A* and *B*.

(*b*) Consider the element *B*[3, 3, 3] in B. Find the effective indices E_1, E_2, E_3 and the address of the element, assuming Base(*B*) = 400 and there are w = 4 words per memory location.

Solution. (*a*) The length of a dimension is obtained by :

$$\text{Length} = \text{Upper bound} - \text{Lower bound} + 1$$

Hence the lengths L_i of the dimensions of *A* are :

$$L_1 = 2 - (-2) + 1 = 5 \text{ and } L_2 = 22 - 2 + 1 = 21$$

Accordingly, *A* has 5 . 21 = 105 elements. The lengths L_i of the dimensions of *B* are :

$$L_1 = 8 - 1 + 1 = 8, \quad L_2 = 5 - (-5) + 1 = 11, \quad L_3 = 5 - (-10) + 1 = 16$$

Therefore, B has 8 . 11 . 16 = 1408 elements.

(*b*) The effective index E_i is obtained from $E_i = k_i - \text{LB}$, where k_i is the given index and LB is the lower bound. Hence

$$E_1 = 3 - 1 = 2, \quad E_2 = 3 - (-5) = 8 \quad E_3 = 3 - (-10) = 13$$

The address depends on whether the programming language stores *B* in row-major order or column-major order. Assuming *B* is stored in column-major order, we use Eq. (4.8) :

$$E_3L_2 = 13 . 11 = 143$$
$$E_3L_2 + E_2 = 143 + 8 = 151$$
$$(E_3L_2 + E_2)L_1 = 151 . 8 = 1208$$
$$(E_3L_2 + E_2)L_1 + E_1 = 1208 + 2 = 1210$$

Therefore, LOC(*B*[3, 3, 3]) = 400 + 4(1210) = 400 + 4840 = 5240

SOLVED PROBLEMS

1. Suppose the two arrays A_1 and A_2 are declared as :

$A_1(-3{:}3, 3{:}33)$ and $A_2(1{:}9, -4{:}4, -10{:}5)$

Find the length of each dimension and number of elements in A_1 and A_2.

Ans. (*i*) First dimension length L_1 in $A_1 = 3 - (-3) + 1 = 7$
Second dimension length L_2 in $A_1 = 33 - 3 + 1 = 31$
Number of elements in $A_1 = L_1 * L_2 = 7*31 = 217$

(*ii*) First dimension length L_1 in $A_2 = 9 - 1 + 1 = 9$
Second dimension length L_2 in $A_2 = 4 - (-4) + 1 = 9$
Third dimension length L_3 in $A_2 = 5 - (-10) + 1 = 16$
Number of elements in $A_2 = 9 * 9 * 16 = 1296$

2. Write a C routine for reversing an array by swapping element A[0] with A[N-l], A[1] with A[N-2] and so on.

Ans.

```
void reverse(int arr[],int N)
{

int i,t;
for(i=0;i<N/2;i++)

{

t=arr[i];
arr[i]=arr [n-i-1];
arr [n–i–1] =t;

}

}
```

3. Consider the linear arrays AAA(5:50), BBB(–5:10) and CCC(18).

(*a*) Find the number of elements in each array.

(*b*) Suppose Base(AAA) = 300 and w = 4 words per memory cell for AAA. Find the address of AAA[15], AAA[35] and AAA[55].

Ans. (*a*) The number of elements is equal to the length; hence use the formula

$$\text{Length} = \text{UB} - \text{LB} + 1$$

Accordingly, Length (AAA) = 50 – 5 + 1 = 46
Length (BBB) = 10 – (–5) + 1
Length (CCC) = 18 – 1 + 1 = 18
Note that Length (CCC) = UB, since LB = 1.

(*b*) Use the formula

$$\text{LOC}(\text{AAA[K]}) = Base(\text{AAA}) + w(\text{K} - \text{LB})$$

Hence : LOC(AAA[15]) = 300 + 4(15 – 5) = 340
LOC(AAA[35]) = 300 + 4(35 – 5) = 420

AAA[55] is not an element of AAA, since 55 exceeds UB = 50.

4. Suppose *A* is a sorted array with 200 elements, and suppose a given element x appears with the same probability in any place in *A*. Find the worst-case running time *f* (*n*) and the average-case running time *g*(*n*) to find *x* in *A* using the binary search algorithm.

Ans. For any value of k, let n_k, denote the number of those elements in A that will require k comparisons to be located in A. Then :

k:	1	2	3	4	5	6	7	8
n_k:	1	2	4	8	16	32	64	73

The 73 comes from the fact that $1 + 2 + 4 + ... + 64 = 127$ so there are only $200 - 127 = 73$ elements left. The worst-case running time $f(n) = 8$. The average-case running time $g(n)$ is obtained as follows :

$$g(n) = \frac{1}{n}\sum_{k=1}^{\infty} k.n_k$$

$$= \frac{1\cdot1 + 2\cdot2 + 3\cdot4 + 4\cdot8 + 5\cdot16 + 6\cdot32 + 7\cdot64 + 8\cdot73}{200}$$

$$= \frac{1353}{200} = 6.765$$

Observe that, for the binary search, the average-case and worst-case running times are approximately equal.

5. "the space required by the array-based list implementation is (*n*), but can be greater." Explain why this is so.

Ans. The space required by the array-based list implementation is fixed. It must be at least n spaces to hold n elements, for a lower bound of $\Omega(n)$. However, the actual number of elements in the array (n) can be arbitrarily small compared to the size of the list array.

6. Consider a 2-D array $A(V_1, V_2)$ with V_1 rows and each row have V_2 elements. Suppose this array is stored in memory in row-major order and base address is a. What will be the address of elements $A[i,j]$?

Ans. As the above given array consists of V_1 number of rows and V_2 number of columns as shown below :

$$\begin{array}{c} \\ 1 \\ 2 \\ 3 \\ \vdots \\ V_1 \end{array}
\begin{array}{ccccc} 1 & 2 & 3 & \cdots & V_2 \\ a_{11} & a_{12} & a_{13} & \cdots & a_{1V_2} \\ a_{21} & \cdots & \cdots & \cdots & 2 \\ \cdots & \cdots & \cdots & \cdots & 3 \\ \vdots & \vdots & \vdots & \cdots & \vdots \\ a_{V_1} & \cdots & \cdots & \cdots & a_{V_1V_2} \end{array} \; V_1 \times V_2$$

As the array is stored in row-major order, therefore the elements of the above array is stored row-by-row starting from 1 to V_1 and each row consists of V_2 number of element, hence to retrieve the location of *ith row* and *jth column* follow the procedure below :

(1) As we have to retrieve ith row hence all the rows above ith row have been stored with all the V_2 elements in each row *i.e.*

$(i - 1)\ V_1$ = Total number of element above ith row, which have been already stored.

(2) As we have to retrieve the jth location in ith row, hence there are total of $(j - 1)$ elements before jth column in ith row.

Therefore the total effective address is :

$$V_1(i - 1) + (j - 1)$$

To find the actual address,

$$A[i,j] = \text{Base address} + w(\text{Effective Add})$$

i.e. $$A[i,j] = a + w[V_1(i-1) + (j-1)]$$

where w is the total width of each element in the array.

7. Given an n-square lower-diagonal matrix A that has elements on the main diagonal and below the diagonal as non-zero. Find the formula that will give L in terms of J & K such that B[L] = A[J, K], so that one can access the value of A[J, K] from array B, where B is a linear array. Assume row-major ordering ?

Ans. As given in the problem above the n-square matrix is lower triangular sparse matrix as shown below :

$$\begin{pmatrix} a_{11} & & & & \\ a_{21} & a_{22} & & & \\ a_{31} & a_{32} & a_{33} & & \\ \vdots & \vdots & \vdots & & \\ a_{n1} & a_{n2} & \dots & \dots & a_{nn} \end{pmatrix}_{n \times n}$$

As the matrix consists of very large number of zero's, hence this matrix is first converted into one-dimensional array to reduce the memory wastage.

Therefore, we will create a 1-D array B with n number of elements and only the non-zero elements of lower triangular sparse matrix are stored in array B according row-major ordering.

If we want to accessing the location of any element stored using row-major ordering, say A[I, J], then we have to add up all the non-zero elements of each row above *I*th row and in addition to all the elements in Ith row upto *J*th column *i.e.*

$$\frac{I(I-1)}{2} + (J-1) = \text{Effective address}$$

As every row have number of elements equivalent to row number *i.e.*

$$1 + 2 + 3 \dots + (I-1) = \frac{I(I-1)}{2}$$

Hence the address retrieval is

$$A[I,J] = \text{Base address} + w\left[\frac{I(I-1)}{2} + (J-1)\right]$$

8. A multi-dimensional array is declared as *z*[3:30; –5:15; 10:20]. Base (*z*) = 1000, and there are 4 words per memory location ?

(*i*) Find the length of each dimension and number of elements in *z*.

(*ii*) Find the address of *z*(5, 10, 15) assume *z* is stored in row-major.

Ans. (*i*) Length of each dimension of array *z* is

$$L_1 = UB_1 - LB_1 + 1 = 30 - 3 + 1 = 28$$

$$L_2 = UB_2 - LB_2 + 1 = 15 - (-5) + 1 = 21$$

$$L_3 = UB_3 - LB_3 + 1 = 20 - 10 + 1 = 11$$

Total number of elements in $z = L_1 \,.\, L_2 \,.\, L_3$

$$= 28 \,.\, 21 \,.\, 11$$

$$= 6468$$

(*ii*) $L_1 = 28,\quad L_2 = 21,\quad L_3 = 11$

$E_1 = 5 - 3 = 2\quad E_2 = 10 - (-5) = 15\quad E_3 = 15 - 10 = 5$

Effective address

$$((E_1 . L_2 + E_2) L_3 + E_3)$$

$$\begin{aligned} A[5, 10, 15] &= B.A. + w((E_1 . L_2 + E_2) L_3 + E_3) \\ &= 1000 + 4 ((2 . 21 + 15) 11 + 5) \\ &= 1000 + 4 (632) \\ &= 3528 \end{aligned}$$

9. Consider the following multi-dimensional array Y[3:10; 1:15; 10:20]

(*i*) Find the number of elements in Y.

(*ii*) Suppose base address for Y = 400 and size of each element in the array is 2. Find the effective address of Y [5, 10, 15] assuming Y is stored in row-major order ?

Ans. It is a 3-dimensional array *i.e.*

$$Y[3:10;\ 1:15;\ 10:20]$$

(*i*) The size of each array is

$$\begin{aligned} L_1 &= UB_1 - LB_1 + 1 = 10 - 3 + 1 = 8 \\ L_2 &= UB_2 - LB_2 + 1 = 15 - 1 + 1 = 15 \\ L_3 &= UB_3 - LB_3 + 1 = 20 - 10 + 1 = 11 \end{aligned}$$

Total number of elements in

$$Y = L_1 * L_2 * L_3 = 8 * 15 * 11 = 1320$$

(*ii*) Base Address (B.A.) = 400

Size (width, *w*) = 2

$L1 = 8,\quad L_2 = 15,\quad L_3 = 11$

$E_1 = K_1 - LB_1 = 5 - 3 = 2\quad E_2 = K_2 - LB_2 = 10 - 1 = 9\quad E_3 = K_3 - LB_3 = 15 - 10 = 5$

where E_i is the effective index *i.e.*

$$E_i = K_1 - LB_1$$

As array *Y* is stored in row-major order.

$$\begin{aligned} Y[5, 10, 15] &= B.A. + W((E_1.L_2 + E_2)L_3 + E_3) \\ &= 400 + 2((2 . 15 + 9) 11 + 5) \\ &= 400 + 2(434) = 1268 \end{aligned}$$

10. In an array SPACE (–1 : 10; 5 : 14). Find

(i) The length in each direction.

(ii) The number of elements that can be stored.

(iii) If the starting address of the array is START and size of each element is SIZE. What is the address of the element SPACE (5, 8) ?

Ans. (*i*)

$$\begin{aligned} L_1 &= UB_1 - LB_1 + 1 = 10 - (-1) + 1 = 12 \\ L_2 &= UB_2 - LB_2 + 1 = 14 - 5 + 1 = 10 \end{aligned}$$

(*ii*) Total number of elements are

$$L_1 * L_2 = 12 * 10 = 120$$

(*iii*) (*a*) If the array is stored in row-major order i.e.

$$\text{SPACE}(5, 8) = \text{START} + \text{SIZE}(L2[5-1]+[8-1])$$
$$= \text{START} + \text{SIZE}(47)$$

(*b*) If the array is stored in column-major order *i.e.*

$$\text{SPACE}(5, 8) = \text{START} + \text{SIZE}(L1[8-1]+[5-1])$$
$$= \text{START} + \text{SIZE}(88)$$

11. A 4-D array a [3] [4] [5] [6] is stored using column-major order. What will be the address of a[2] [3] [4] [5], if base address of array is a and β is the size of each element ? Assume that array indices starts from zero ?

Ans. As the above 4D array is stored using column major, and hence the array stored in memory column-by-columnb.

As array indices starts from zero, hence,

$L_1 = 3$ $L_2 = 4$ $L_3 = 5$ $L_4 = 6$
$E_1 = 2$ $E_2 = 3$ $E_3 = 4$ $E_4 = 5$

As arrya is stored column wise, hence the effective address is

$$(((E_4 . L_3 + E_3) L_2 + E_2) L_1 + E_1)$$

$$(((5.5+4)4+3)3+2) = 359$$

Hence the address is

$$a[2]\,[3]\,[4]\,[5] = \text{BA} + w(\text{E}\,.\,\text{A})$$
$$= a + \beta(359)$$
$$= a + 359\beta$$

SUMMARY

1. An array is a data structure which can store multiple elements of same type under one name.
2. A single dimension array is a list of memory cells that are contagious. Suppose the array ARR is defined to have a lower bound 1.
3. **Row Major Order :** In this method the elements of the array that have first subscript, the lower bound value of that subscript are stored first, followed by the elements of the second value of the first subscript and so on. In short in row order rows are listed on the basis of number of columns.
4. **Column Major Order :** In column major order column are listed on the of number of rows.

EXERCISES

1. Array name contains base address of the array? Can we change the base address of array?
2. Can we store value and address the same array? Explain.
3. What are array? Why do we need them?
4. How do we initialize array elements?
5. What do you understand by 2D arrays? Also discuss multidimensional array.
6. How to calculate address for 2-D array when array is represented as row major order and column majororder?
7. Write a program to read 10 integers in an array. Sort them out on the basis of number of digits in each element.
8. Write a program to shift all zero elements in an array toward end of the array.

9. Consider a matrix of 0's and 1's A Sequence of successive 1's along a row, or a column or a diagonal is defined to be on a line. The number of 1's in a line is called its length. Write a C program to find out the line with the maximum length in such a matrix.

10. Write a C routine to find maximum and second maximum element in an array. Also find out the time complexity of your routine.

11. Write a C routine that finds determinant of a square matrix. Discuss the time complexity of your code also.

12. How can you add and multiply each of 10 digits long? Write the C routines for the same.

13. Suppose you have to find out if there is an element A[i] [j] in an M by N matrix of integers such that A[i] [j] is the smallest value in the ith row and largest value in the jth column. Write a C routine for this task. Also discuss time complexity of your function.

14. Write a C function that finds mode of an array. The mode of an array is defined as an element with the maximum number of frequencies. I f there are more then one such element then mode does not exist.

15. Write a C function that finds median for an array.The median for an array of odd number of elements is defined as: The median of a List of integers is one element in the list such that number of element less then and greater then median is same. For example we have a list of numbers as:

{11, 6, 2, 3, 7, 91, 55, 18, 20}

The median of list is 11 as there are 4 elements (6, 2, 3,7) that are less then 11 and also there are 4 elements that are greater then 11 (91, 55, 18, 20).

❑❑❑

CHAPTER 4

STRUCTURES

INSIDE THIS CHAPTER

4.1. INTRODUCTION

A structure is a derived data type which consists of collection of different data types such as int, char, float and even structure also. We see example of structure in our daily life. For keeping information about student we have to store his/her name, registration number, age, address, class etc. Similarly for keeping information about employee we may have his/her employee code, date of joining, salary etc. All these attributes related to student or employee cannot be stored in array alone. Even if we wish to store these attribute using array we will have to use separate array for each attribute say one for salary, one for employee code and so on. Again managing this array separately won't be an easy task. In such situations we can use structures. In some of the programming languages like Pascal and Visual Basic structures are also known as records.

The general syntax of creating a structure is

```
struct tagname
{
member variable1;
member variable2;
member variable3;
......
member variablen;
};
```

The struct is a keyword that is used to create a structure and tag name is the name of the structure that conforms to the rule of writing identifier. For understanding number of points about the structure we write a small program and tell you various features of structure.

Program 4.1.

```
struct person
{
char name[10];
int age;
};

main ()
{
struct person p;
clrscr();
printf ("Enter name for person\n");
scanf("%s", p. name);
printf ("Enter age for person\n");
scanf("%d", &p. age);
printf("Name = %s\n", p. name);
printf ("Age %d\n", p. age);
getch();
}
```

OUTPUT

```
Enter name for person
Mridul
Enter age for person
23
Name = Mridul
Age=23
```

1. Structure is a heterogeneous data structure. Remember array is a homogenous data structure *i.e.* all the elements of an array must be of the same type. You cannot mix int with float or float with char. This is not the case with structure ie structure can have there elements of any data type even they have other user defined data types as there elements.
2. Structure is defined by struct keyword. Following struct any valid identifier which conforms to identifier writing rules is allowed. In this program we have declared a structure named person. We could have chosen any other name also say student or employee.
3. The structure is opened using { and closed with } followed by a semicolon (;)
4. The structure declared above contains two member elements: first one is character array name of length 10 and second one is age of type int. As know int takes two bytes in memory (for DOS) so total size for this structure is 12 bytes.
5. The declaration alone does not reserve memory space. It simply tells the compiler about the prototype of the structure. The actual definition *i.e.* creation of a variable of structure data type allocate memory.
6. The structure is an Abstract Data Type *i.e.* following is only like a template.

```
struct person
{
    char name [10];
    int age;
};
```

In order to use structure we have to create a variable of type structure just like we create variable of any other data type.

7. From statement struct person p; we are actually creating a variable *p* of type struct person type. We have to use struct keyword each time we create a variable of type person *i.e.* you just cannot write person *p*; before doing some changes which we wifi see later in the coming programs.
8. At the time of compilation 12 bytes will be allocated for this variable which consists of two parts: 10 bytes for name and 2 bytes for age. We can access these member elements by using membership operator or dot operator or period operator (.)

   (Exact allocation of 12 bytes is compiler dependent)
9. Structure can be declared before main *i.e.* global or it can be local to main or any other function or block.
10. The structure cannot be read as a whole. We have to store values into the individual members of the structure.
11. Diagrammatically we can show the structure as

Name	Age	person

12. The structure can be type defined using the typedef keyword as

```
typedef struct person Per;
```

DO YOU KNOW

RONALD RIVEST (BORN 1948) received a B.A. from Yale in 1969 and his Ph.D. in computer science from Stanford in 1974. Rivest is a computer science professor at M.I.T. and was a cofounder of *RSA Data Security*, which held the patent on the RSA cryptosystem that he invented together with Adi Shamir and Laonard Adleman. Areas that Rivest has worked in besides cryptography include machine learning, VLSI design, and computer algorithms. He is a coauthor of a populat text on *Introduction to Algorithms*

The above statement gives new name Per to struct person. Now we can use Per any where we want to use struct person statement. It could be written in the following manner also:

```
typedef struct

{

char name[10];
```

```
int age;
}person ;

meaning the whole
struct
{
char name[10];
int age

};
```

has been given a new name person. So now you can use simply person to declare variable as person type structure.

You can also declare structure variables directly just after the structure declaration as shown below:

```
struct person
{
char name[10];
int age;
}p1, p2;
```

Note we can declare variable just before putting semicolon to terminate structure declaration. Here we have created two variables p1 and p2.

4.2. DECLARATION AND INITIALIZATION OF STRUCTURE

Structure variables can be declared and initialized at the same time. This is shown below in the code. The string members have to be enclosed within double quotes " ",

```
struct person
{
char name[10];
int age;
};
struct person p1 = {"Avanindra", 21};
struct person p2 = {"Anshul", 22};
```

Note the following will result into error.

```
struct person p1;
p1 = {"Avanjndra", 21};
```

Program 4.2. /* Program Declaration and initialization of structures separately*/.....

```
main ()
{
struct item
{
char iname[10];
int icode;
int iprice;
};
clrscr( );
struct item i1;
strcpy(i1. name, "Mouse");
i1. icode = 101;
i1. iprice = 250;
printf(" Item Details\n");
printf("Name %s\tCode = %d", i1. name, i1. icode);
printf("\tPrice = %d\n", , i1. iprice);
getch ();
}
```

OUTPUT

```
Item Details
Name = Mouse Code = 101 Price = 250
```

In the code given earlier we declared and defined the members of structure at the same place. In this program we have initialized each member separately.

4.3. COPYING STRUCTURES VARIABLES

One structure can be copied to another simply using assignment operator. When this is done all the members from one structure variables are copied to other structure variables and both exist independently. Note there is one copy of pointer members is shared between the two or more structure variables when they are issigned to each other.

Program 4.3. /* Program copy one structure to another */ ....

```
main ( )

{
typedef struct
{
char cname[20];
char model[10];
int price;
```

```
}mobile;
clrscr ();
mobile m1 = { "Nokia", "6600", 12000};
mobile m2, m3;
m2 = m1;

strcpy(m3. cname, "Sony_Err");
strcpy(m3. model, "K700i");
m3. price = m1. price;
printf(" Mobile Details\n");
printf ("Company Name = %s\tModel = %s", m1. cname, m1.model);

printf("\tprice = %d\n", m1. price);

printf(" company Name = %s\tmodel = %s", m2. cname, m2.model);
printf ("\tprice = %d\n", m2. price);

printf (" Company Name =%s\tmodel = %s", m3. cname, m3.model);

printf("\tprice = %d\n", m3. price);

getch( );

}
```

OUTPUT

```
Mobie Details
Company Name = Nokia Model = 6600 Price 12000
Company Name = Nokia Model = 6600 Price = 12000
Company Name = Sony_Err Model = K700i Price = 12000
```

In the mobile structure we have taken three members cname for company, model and price. Variable ml is initialized directly. The statement m2 = ml copies structure ml to m2. All the member values of ml are assigned to m2 in just a single statement. For m3 we have given new values to cname and model but price. We have copied price from price member of ml.

4.4. STRUCTURE ARITHMETIC

Arithmetic on structures can be done by doing it on its data members. Any type of operation using the fundamental operators can be performed. Given below few programs that performs some arithmetic involving structures.

Program 4.4. /* Program sum of two durations (year, month, days) maintained as structures as stored in another structure*/

```
typedef struct
{
int day;
int month;
int year;
}duration;

main()

{
duration d1, d2, d3;
int d,m, y;
clrscr();
printf ("Enter years, months and days for first duration\n");
scanf("%d%d%d", &d1. year, &d1. month, &d1. day);
printf("Enter years, months and days for sec duration\n");
scanf("%d%d%d", &d2. year, &d2. month, &d2. day);
d = (d1. day+d2. day)%30;
m = d1. month+d2. month+(d1. day+d2. day)/30;
y = d1. year+d2. year+m/12;
d3. day = d;
d3. month m%12;
d3. year = y;
printf("\nFirst duration\n");
printf("%d years %d months ", d1. year, d1. month);
printf ("and %d days\n", d1. day);
printf("\nSecond duration\n");
printf("%d years %d months ", d2. year, d2. month);
printf ("and %d days\n", d2. day);
printf("\nSum of two durations\n");
printf("%d years %d months ", d3. year, d3. month);
printf ("and %d days\n", d3. day);
getch ( );
}
```

OUTPUT

```
Enter years, months and days for first duration
2 7 20
Enter years, months and days for second duration
4 5 15
```

```
First duration
2 years 7 months 20 days
Second duration
4 years 5 months 15 days
Sum of two durations
7 years 1 month 5 days
```

We have declared structure duration with three members' year, month and day of type integer. We input year, month and day for two structure variables and find out sum of two durations. We assume each month of 30 days.

Our calculation is based on following

1. We sum days of two structures and the mod of sum by 30 gives us total number of days.
2. Integer division of sum by 30 from step 1 gives us extra months.
3. We add months of two structures with extra months from step 2 and mod by 12 this sum gives us total number of months.
4. Integer division of sum by 12 in step 3 gives us extra years.
5. We add years of two structures with extra years from step 4 and sum of these gives us total number of years.

On the similar basis try to write a program where duration consist of hours, minutes and seconds.

Program 4.5. /* Program Complex number arithmetic (sum & mul) */ ....

```
typedef struct
{
float real ;
float imag;
} complex;

main( )
{
complex c1, c2, c3, c4;
float a, b;

clrscr( );
printf (" Enter the real part for first complex number\n");
scanf ("%f", & a);
printf ("Enter ther imag part for first comlex number\n");
scan ("%f", & b),
c1. real = a;
c1. imag =b;
printf ("Enter the real part for sec complex number\n");
scanf ("%f", & a);
printf ("Enter the imag part for sec complex number\n");
```

```
scanf ("%f", &b);
c2. real = a;
c2. imag =b;
c3. real = c1. real + c2. real;
c3. imag = c1. imag + c2. imag;
c4. real = c1. real*c2. real – c1. imag*c2. imag;
c4. imag = c1. real*c2. imag+c1. imag*c2. real;
printf("\tComplex Number 1\n")

printf("\t%5. 2f + j%5. 2f\n",c1.real, c1. imag);

printf("\tComplex Number 2\n");

printf("\t%5. 2f + j%5. 2f\n", c2.real, c2. imag);

printf('\tSum of Ci & C2\n");

printf("\t%5.2f + j%5. 2f\n", c3.real, c3. imag);
printf("tMul of C1 & C2\n");
printf("\t%5.2f + j%5. 2f\n", c4.real, c4. imag);
getch();
}
```

OUTPUT

```
Enter real part for first complex number
    1.2
Enter imag part for first complex number
    1. 0
Enter real part for first complex number
    2.1
Enter imag part for sec complex number
    1. 4
Complex Number 1
    1. 20 + j1. 40
Complex Number 2
    2.10 + j1.40
Sum of C1 & C2
    3. 30 + j2. 40
Mul of C1 & C2
    1. 12 + j3. 78
```

In the program we have declared a complex structure. A complex number has two parts one is real and second is imaginary. Imaginary part is recognized by placing j or i before number and is called iota. When two complex numbers are added / subtracted their real parts are added/subtracted separately as well as imaginary part. When two complex numbers are multiplied the resultant complex is obtained as follows:

Assume two complex numbers are $(x_1 + jy_1)$ and (x_2+jy_2). Here j is called iota and $\sqrt{j}$ is –1 so $j * j$ will be –1. Now multiplication will be :

$$(x_1 + jy_1) * (x_2 + jy_2) = x_1x_2 + j\,x_1y_2 + jy_1x_2 - y_1\,y_2$$
$$= (x_1\,x_2 - y_1\,y_2) + j(x_1\,y_2 + x_2\,y_1)$$

(real part) (imag. part)

On the basis of this discussion I think it will be easy to understand the program presented above.

4.5. ARRAY OF STRUCTURE

Similar to array of any data type an array of structure can also be created and will be very useful when processing of structure like book, person, employee, vehicle, item etc.

To declare an array of structure first you have to create a structure first. For example

```
struct vehicle
{
char vname[15];
int now;
char fuel_type;
};
struct vehicle varr[5] ;
```

This creates an array varr of structure type vehicle. Total 13 *5 bytes memory will be allocated during compilation. The first structure variable can accessed as varr[0], second as varr[1] and so on. The individual data member of structures can be accessed as varr[0]. vname, varr[0]. now etc.

Program 4.6. /* Program demo of Array of structures */...................................

```
main ( )
{
typedef struct student
{
char sname [10];

int rollno;
char cname[10];
}student;
clrscr( );
student sarr[5];
int i;
for(i = 0;i<5;i++)
{
```

```
printf("\n Enter name and class name for stu. no %d:",i+1);
scanf("%s%s", sarr[i] . sname, sarr[i] . cname);

printf ("Enter the roll no. \n");

scanf("%d", &sarr[i] . rollno);
}
printf ("\nStudent' S Details\n");
printf ("\SName\tCName\tRollN\n");
for(i = 0;i<5;i++)

{

printf("%s\t%s\t", sarr[i]. sname, sarr[i]. cname);

printf("%5d\n",  sarr[i]. rollno);
}
getch( );
}
```

OUTPUT

```
Enter name and class name for stu. no 1
   Namit Nineth
Enter the roll no.
   902
Enter name and class name for stu. no 2
   Khushi Seventh
Enter the roll no.
   714
Enter name and class name for stu. no 3
   Love Fifth
Enter the roll no.
   511
Enter name and class name for stu. no 4
   Runjhun Sixth
Enter the roll no.
   625
Enter name and class name for stu. no 5
   Mohit Eighth
Enter the roll no.
   823
Student's Details
```

```
SName       CName       RoliNo
Narnit      Nineth      902
Khushi      Seventh     714
Love        Fifth       511
Runjhun     Sixth       625
Mohit       Eighth      823
```

The declaration student sarr[5] creates an array of 5 element of type structure student. The members of first element of array are accessed as sarr[0]. sname, sarr[0]. rollno and sarr[0]. cname. At the time of compilation 22 *5 = 110 bytes are allocated (22 is the size of structure). We show this in tabular format.

In the parenthesis we have written number of bytes allocated.

As clear from the diagram 22 bytes are allocated for each array element.

	sname (10)	rollno(2)	cname(10)
sarr [0]	Na mit	902	Nineth
sarr[1]	Khushi	714	Seventh
sarr[2]	Love	511	Fifth
sarr[3]	Runjhun	625	Sixth
sarr[4]	Mohit	823	Eighth

Program 4.7. /* Program generating Merit list of students */

```
#define S 3
main ( )
{
typedef struct student
{
char sname[10];
int rollno;
int ml, m2, m3;
float per;

} student;
clrscr ( );
student sa[S];
int  rno = 101, i, s;
for(i = 0;i<S;i++)
{
printf("\n Enter name and marks in three subjects\n");
scanf("%s%d%d%d", sa[i] .sname, sa[i] . ml, sa[i]. m2, sa[i]. m3);
sa[i] . rolno = rno;
```

```
rno++;
s = sa[i]. m1 +sa[i]. m2+sa[i]. m3;
sa[i] . per = s/3. 0;
}

for(i = 0;i<S;i++)
{
for(j =i+1;j<S;j++)
if(sa[i]. per<sa[j]. per)
{
temp sa[i];
sa[i] = sa[j];
sa[j] = temp;
   }

}
printf("Student's Merit List\n");
print  ("\Name\tRollNo\tper\n");
for(i = 0;i<S;i++)
{
printf("%s\t%d\t", sa[i] .sname, sa[i] . rollno);
printf("%5. 2f\n", sa[i]. per);
}
getch ()
}
```

OUTPUT

```
Enter name and marks in three subjects
   Ajay 89 97 90
Enter name and marks in three subjects
   Ajit 81 91 86
Enter name and marks in three subjects
   Kashish 88 97 92
Student's Merit List
   Name        RollNo     Per
   Kashish     103        92.34
   Ajay        101        92.00
   Ajit        102        86.00
```

In the program we have modified the student structure and added three members for marks and one for percentage per. We assume all the marks are from 100. We input name and marks for each student and find out percentage and assign to per field of each array element. We sort the array elements on the basis of per field. We apply the same logic as we applied to sort a simple 1D array. We have also taken a temporary structure variable temp to save array element while student structures are being swapped.

4.6. NESTING OF STRUCTURES

A structure can be nested within a structure; If this is done then nested structure can act as a member of containing structure. To understand this, consider the declaration of two structures as given below:

```
struct address
{
char h_no[6];
int s_no;
char city[10];
char state[10];
};
struct person
{
char name[10];
int age;
struct address addr;
};
```

A variable addr of type struct address is the member of structure person. Thus address is nested within person or to say in other words person has address. Now if we declare a variable p of person type as:

```
struct person p;
```

then data member of address structure can be accessed as

```
p. addr. h_no, p. addr. s_no, p. addr. City, p. addr. state
```

and data members of structure person p can be

accessed in usual manner.

```
p. name, p. age
```

In the address structure we have stored the house number, street number, city and state. A variable of this structure is a member of person structure. Thus to access the members of address structure we have to use dot membership twice say for example p. addr. city. As p is a structure variable of person type and because of addr as a member of structure person we can write p. addr. This p. addr acts as a variable of structure address type and we can access any of its member using dot operator. Thus the following statements would be valid.

```
struct address ar = p. addr;
```

After the above assignment, data members of address structure can be accessed as

```
ar. h_no, ar. s_no, ar. city and ar. state
```

When you initialize the data members of structure, data members of nested structure has to be explicitly initialized. This is shown below:

```
struct address
{
char h_no[6];
int s_no;
char city[10];
char state[10];
};
struct person
{
char name[10];
int age;
struct address addr;
};
typedef struct person;
person p = {"Manav", 25, {"623/4", 2, "Aiwar", "Rajasthan"}};
```

Program 4.8. /* Program Nesting of structure date within employee structure */

```
struct date
{
int day;
int month;
int year;
};
struct employee
{
char ename[15];
char dept[15];
struct date doj;
char qual[10];
};

main ()
{
struct employee e;
clrscr () ;
printf ("Enter the emp name, qual. and dept. \n");
```

```
scanf("%s%s%s", e. ename, e. qual, e. dept);
printf ("Enter the date of joining (day month year)\n");

scanf("%d%d%d", &e. doj. day, &e. doj. month, &e. doj. year);
printf ("Employee Details\n\n");
printf("Narne:\t%s\tDeptt:\t%S\n", e. ename, e. dept);

printf("Qual:\t%s", e. qual);
printf("\tDOJ:\t%2d/%2d/%2d\f1", e. doj. day, e. doj. month,e.
     doj. year);
getch ();
}
```

OUTPUT

```
Enter the emp name, qual. and dept.
   Tanmay MTech CE
Enter the date of joining (day month year)
   07 07 03
Emploee Details
   Name: Tanmay Deptt:CE
Qual: MTech
DOJ: 07/07/03
```

The program is quite simple. For date of joining of employee we have created a separate structure date which is act as a member of structure employee. No type checking has been put for date so even if entered date is invalid program will work

4.7. STRUCTURE AND POINTERS

Just as we create pointers for int, float, and other in built data types we can create pointers for the structures. For example for the structure given below as

```
struct person
{
char name[10];
int age;
};
```

We write

```
struct person per, *ptr;
ptr = &per;
```

To access any of the member we can write ptr->age or (*ptr). age See few program presented below for detailed explanation.

Program 4.9. /* Program demo of structure with pointer */

```
typedef struct
{
int icode;
char iname[10];
int quan;
int uprice;
} item;

main()
{
item i1;
item *p;
clrscr ();
p = &i1;
printf("Enter item code, item name, quan. and unit price\n");
scant ("%d%s%d%d", &p->i_no, p->iname, &p->quan, &p->uprice);
printf ("\nItem Code %d\n", (*p) . i_no);
printf ("Item Name = %s\n", p->iname);
printf ("Item Quan %d\n", (&i1)->quan);
printt ("Unit Price %d\n", i1. uprice);
getch ( );
```

OUTPUT

```
Enter item code, item name, quan. And unit price
101 Keyboard 10 125
Item Code =101
Item Name=Keyboard
Item Quan = 10
Unit Price = 125
```

In the item structure we have created four members item code, item name, number of items *i.e.* quantity and unit price. The statement item *p creates a pointer of item type and in it we assign the address of variable i1 of type item. To access the members of structure using this pointer p we use -> operator called pointer to member operator. To access say icode we can follow any of the two approaches:

(*a*) p ->icode

(*b*) (*p). icode

We can also write **(&il)->icode** or our usual method **i1. icode.**

If we want we can create the pointer to structure in this manner also.

```
struct item
{
int icode;
```

```
char iname[10];
int quan;
int uprice;
} i1, *p;
```

Program 4.10. /*Program demo of structure with pointer to nested structure */.......

```
struct date
{
int day;
int month;
int year;
};
struct employee
{
char ename[15];
char dept[15];
struct date doj;
char qual[10];
};
main ()
{
struct employee e = {"Navin", "Comp", {17, 07, 2003},
"MTech"};
struct date *d = &e. doj;

clrscr ( );
printf ("Employee Details\n\n");
printf("Name:\t%s\tDeptt:\t%s\n", e. ename,e. dept);
printf("Qualification:\t%s" e. qual);
printf("\tDOJ:\t %2d/%2d/%2d \n", d->day,d->month, d->year);
getch ( );
}
```

OUTPUT

```
Employee Details
Name: Navin
Deptt: Comp
Qual: Mtech
DOJ: 17/07/03
```

We have the structure employee and date created earlier in programs. We have defined a pointer to date structure variable doj which is embedded in employee structure and printed the date of joining using this pointer d.

4.8. STRUCTURE AND FUNCTION

We can pass structure to function as parameters and have structure as return type for functions. We can also pass addresses of structures to the functions.

```
struct employee
{
int ecode;
char ename[10];
int esal;
};
void input(struct employee*);
void show(struct employee);
struct employee modify(struct employee);
```

The prototype of input function tells to compiler that this function takes a pointer to structure employee. If ernp a variable of type struct employee then function can be called as

```
input(&emp);
```

The show function accepts a variable of type struct employee. The third functions modify takes an argument of struct employee type and return a variable of type struct employee.

Program 4.11. /* Program demo of structure with function and pointer */...............

```
struct emp
{
int ecode;
char ename[10];
int esal;
};
main ( )
{
struct emp e = {1013, "Mukul", 20000};
void modify(struct emp*);
modify(&e);
printf ("Details of Employee\n");
printf("Name:\t%s\n", p. ename);
printf("ECode:\t%d\n", p. ecode);
printf("Sal :\t%d\n", p. esal);
getch ( );
}
void modify(emp *p)
{
p->ecode =1022;
p->ename = "Tanvi";
```

```
    (*p). esal =15000;
    }
```

OUTPUT

```
    Name: Tanvi
    ECode.1022
    Sal :15000
```

The statement void modify (struct emp*) tells the compiler that modify is a function which accepts address of variable of struct emp type. As we call the function by reference whatever changes performed inside the function they persist even if control comes out from the function so the next few printf statement displays the modified fields of the structure emp.

SUMMARY

1. A structure is a derived data type which consists of collection of different data types such as int, char, float and even structure also.
2. Structure is a heterogeneous data structure.
3. Structure is defined by struct keyword.
4. The declaration alone does not reserve memory space. It simply tells the compiler about the prototype of the structure. The actual definition *i.e.* creation of a variable of structure data type allocate memory.
5. One structure can be copied to another simply using assignment operator.
6. Arithmetic on structures can be done by doing it on its data members.
7. Just as we create pointers for int, float, and other in built data types we can create pointers for the structures.

EXERCISES

Answer the followings :

1. What is a structure in C and how it is declared?
2. How structure elements are stored in memory?
3. How structure elements are accessed using pointer? Which operator is used?.
4. How arrays of structure are defined? How are they helpful to programmer.
5. Distinguish between an array and a structure?
6. What is the need for a tag in a structure declaration? Demonstrate with segment of program.
7. Write a program to define structure with tag state with field's state name, no of districts and total population. Read and display the data.
8. Create user defined data type equivalent to int. Declare three variables of its type. Perform arithmetic operations using these variables.

❑❑❑

CHAPTER 5 POINTERS

5.1. INTRODUCTION

Pointers are a powerful concept in C and add to its strength. Mastery over pointers empowers a C programmer to deal with system level junks. The pointers enable us to.

1. Write efficient and concise programs
2. Establish inter-program data communication
3. Dynamically allocate and de-allocate memory
4. Optimize memory space usage
5. Deal with hardware components
6. Pass variable number of arguments to functions

Let us discuss the organization of memory before stepping into the concept of pointers. Memory is organized as an array of bytes. A byte is a basic storage and accessible unit in memory. Each byte is identifiable by a unique number called address. Suppose we have 1 KB of memory. Since 1 KB = 1024 bytes, the memory can be viewed as an array of locations of size 1024 with the subscript range [0-1023]. 0 represents the address of the first location; I represents the address of the second location and henceforth the location is identified by the address 1023.

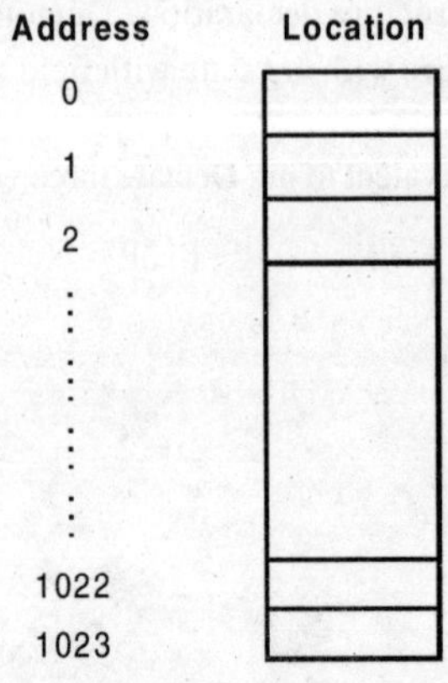

We know that variables are to be declared before they are used in a program. Declaration of a variable conveys two things to the compiler. As a result, the compiler.

1. Allocates a location in memory. The number of bytes in the location depends on the data type of the variable.
2. Establishes a mapping between the address of the location and the name of the variable.

Example :

The declaration int i; tells the compiler to reserve a location in memory. We know that the size of a variable of int type is two bytes. So the location would be two bytes wide.

If the location is :

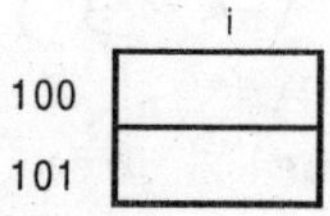

Variable nameAddress of the location

i 100

A mapping between the variable name and the address of the location is established. Note that the address is of the first byte of the location the variable. The address of a variable is also a number, numeric data item. It can also be retrieved and stored in another variable. A variable which can store the address of another variable is called a pointer.

5.2. POINTER OPERATORS - & AND *

C provides two special operators known as pointer operators. They are & and *. & stands for 'Address of' and it is used to retrieve the address of a variable. * stands for 'value at address' and it is used to access the value at a location by means of its address.

Since a pointer is also a variable, it should be declared before it is used. The syntax of declaring a pointer variable is written as :

SYNTAX

```
data-type *variable_name;
```

data-type is any valid data type supported by C or any user-defined type and variable-name is the name of the pointer variable. The presence of * before variable_name indicates that it is a pointer variable.

DO YOU KNOW

A pointer is a dynamic data structure i.e. memory allocation to a pointer variable is done at run-time.

Example :

```
int  *ip;
```

ip is declared to be a pointer variable of int type.

```
float  *fp;
```

fp is declared to be a pointer variable of float type.

DO YOU KNOW

A pointer always stores an address of a variable.

Program 5.1. To illustrate the concept of pointers and pointer operators [&,*]

```
#include  <stdio.h>
#include  <conio.h>

void main()
```

DO YOU KNOW

A pointer along with a structure is used to implement a linked list

```
{
        int   i  =  10,   *ip ;
        float  f  =  3.4,   *fp ;
        char  c  =  'a',  *cp ;

        clrscr() ;

        printf (" i  =  %d  \n",  i ) ;
        printf ("  f  =  %f  \n" ,  f ) ;
        printf ("  c =  %c \n" ,   c ) ;

        ip  =  &i ;
        printf (" \n Address of i  =  %u \n", ip) ;
        printf (" Value of i  =  %d \n", ip) ;

        fp  =  &f ;

        printf (" \n Address of f  =  %u \n", fp) ;
        printf (" Value of f  =  %f \n", fp) ;


        cp  =  &c ;

        printf (" \n Address of c  =  %u \n", cp) ;
        printf (" Value of c  =  %c \n", cp) ;
        getch() ;
    }
```

INPUT-OUTPUT

```
i  =  10
f  =  3.400000
c  =  a

Address of i =  65524
Value of  i  =  10

Address of f = 65520
Value of f  =  3.400000

Address of c = 65519
Value of c  =  a
```

EXPLANATION

In Program 5.1, i, f and c are declared to be variables of int, float and char type, respectively. ip, fp and cp are declared to be pointer variables of type int, float and char, respectively.

The initial values of i, f and c are displayed. The pointer variable ip is assigned the address of I using the statement ip = &I; the address of i and its value are displayed through the pointer variable. The same thing is repeated fro char and float variables.

5.3. POINTER ARITHMETIC

Following are operations that can be performed over pointers :

1. Adding an integer value to a pointer
2. Subtracting an integer value from a pointer
3. Comparing two pointers if they point the elements of the same array
4. Subtracting one pointer from another pointer if both point to the same array
5. Assigning one pointer to another pointer provided both are of same type.

But the following are not possible :

1. Addition of two pointers
2. Subtraction of one pointer from another pointer when they do not point to the same array.
3. Multiplication of two pointers
4. Division of one pointer by another pointer

Suppose p is a pointer variable to integer type. The pointer variable p can be subjected to the following operations :

1. We can increment it using increment operator ++.

   Suppose the address store in p initially is 100. After the statement p++; is executed, the content in p gets incremented by 2, the size of int type. Thus, it becomes 102.

2. we can decrement it using decrement operator –

   Suppose the address store in p initially is 100. After the statement p—; is executed, the content in p gets decremented by 2, the size of int type. Thus, it becomes 98.

   In general, if a pointer variable is incremented using ++, it gets incremented by the size of its data type, *i.e.*, a char pointer gets incremented by 1, the size of char type. A float pointer gets incremented by 4, the size of float data type. In the case of decrement operator, a pointer variable gets decremented by the size of its data type.

3. An integer value can be added to it, *i.e.*,

   ```
   p = p + integer value
   ```

   The content of p will now get incremented by the product of integer value and the size of int type.

4. An integer value can be subtracted from it, *i.e.*,

   ```
   p = p - integer value
   ```

   The content of p will now get decremented by the product of integer value and the size of int type.

Program 5.2. To illustrate the concept of pointer arithmetic..................................

```
#include  <stdio.h>
#include  <conio.h>

void main()
 {
      int    i  , *ip ;

      clrscr() ;

      ip  =  &i ;
      printf ("ip   =  %u \n",  ip ) ;
      i++ ;
      printf ("After  ip++    ip  =  %u \n",  ip ) ;
       i-- ;
      printf ("After  ip--     ip  =   %u \n",  ip ) ;
      ip  =  ip+2 ;
      printf ("After  ip  =  ip+2  ip  =  %u \n",  ip ) ;
      ip  =  ip-2 ;
      printf ("After  ip  =  ip-2  ip  =  %u \n",  ip ) ;
      getch() ;
 }
```

INPUT-OUTPUT

```
ip    =   65524
After  ip++      ip  =  65526
After  ip--      ip  =  65524
After  ip  =  ip+2  ip  =  65528
After  ip  =  ip-2  ip  =   65524
```

EXPLANATION

In Program 5.2, i is declared to be a variable of int type and ip, a pointer to int type. Initially, the address of i is assigned to ip and it is displayed. The pointer variable ip is incremented using ++ operator. As can be seen, the value in ip gets incremented by two, the size of int type. The pointer variable ip is then decremented using--operator. Now, the value in ip is decremented two, the size of int data type. Then, an integer constant 2 is added to ip. As a result, the value of ip gets increment by 4, *i.e.*, 2*size of int type. On the execution of the statement ip = ip-2; the value of ip gets decremented by 4, *i.e.*, 2 * sizeof(int).

5.4. POINTERS AND ARRAYS

We are now acquainted with both arrays and pointers. In our earlier discussions, we treated them as separate entities. But in C, there is a close relationship between them so we will discuss them together. Elements of an array are accessed through pointers internally. Let us now try to unravel the bonding between them.

5.4.1. Pointers and One-dimensional Arrays

Let us first consider relationship between one-dimensional arrays and pointers. Suppose a is a one-dimensional array of int type and of size 10, which is declared as follows :

```
int  a[10];
```

We know that a block of memory consisting of 10 contiguous locations, gets allocated and all the locations share common name a and are distinguishable by subscript values [0-9].

Here, the array name, a gives the base address of the array. That is, the address of the first element a[10] of the array. So a being equivalent to &a[0] can be regarded as a pointer to integer but in the capacity of a constant pointer. That is, a cannot be incremented or decremented.

Since a is the address of a[0], *a is the element stored in the first location. Since a is a constant pointer, it cannot be incremented to point to the next location. But the expression (a+1) gives the address of the second element a[1] of the array. *(a+1) gives the elements itself stored at a[1]. Similarly, (a+2) is the address of a[2] and *(a+2) is the value at a[2].

$$
\begin{aligned}
*a &= *(a + 0) = a[0] \\
&\ *(a + 1) = a[1] \\
&\ *(a + 2) = a[2] \\
&\ *(a + 3) = a[3] \\
&\ *(a + 4) = a[4] \\
&\ \quad \vdots \\
&\ *(a + i) = a[i]
\end{aligned}
$$

In general, (a+i) gives the address of ith element in the array a and *(a+i) is the element stored in the ith location of the array.

Program 5.3. To illustrate processing one-dimensional array using pointer ....

```
#include  <stdio.h>
#include  <conio.h>
void main()
 {
      int  a[10], n, i ;
      clrscr() ;
      printf ("Enter no. of elements \n") ;
      scanf ("%d", &n) ;

      printf ("\n Enter %d elements \n", n) ;
      for ( i = 0;  i < n ;  i++ )
            scanf ( "%d",  a+i ) ;

      printf ("n The list of elements \n") ;
      for ( i = 0;  i < n ;  i++ )
           printf ("%d", *(a + i) ) ;
```

```
        getch();
    }
```

INPUT-OUTPUT

```
Enter no. of elements
5
Enter 5 elements
1  2  3  4  5
The list of elements
1  2  3  4  5
```

EXPLANATION

a is declared to be an array of int type of size 10. n and i are declared to be variables of int type. The variable n is to accept the number of elements to be accepted into the array a, the value of which should lie within 1 and 10 and the variable I is to select each location of the array. We know that the array name a gives the address of the first location of the array and the expression a+i gives the address of ith location of the array. The following segment of Program 12.3 enables us to accept n values into the array.

```
for (i   =   0;   i < n;   i++)
scanf("%d",   a  +  i);
```

Note that the second argument a+i is equivalent to &a[i], *(a+0) gives the value at the first location of the array and in general, *(a+i) gives the value at ith location of the array a. The following segment of Program 12.3 displays the n values stored in the array.

```
for (i   =   0;   i < n;   i++)
printf("%d   \n",   *(a + i));
```

Note that the second argument *(a+i) is equivalent to a[i].

5.4.2. Pointers and Two-dimensional Arrays

Suppose a is an array of two dimensions of int type, which is declared as follows :

```
int  a[2]  [3]
```

We know that a 2-d array is essentially an array of 1-d array. So each element in a 2-d array happens to be a 1-d array.

Here also, the array name a gives its base address, *i.e.*, the address of its first element. First element in this case is the first 1-d array. So a is the address of its first 1-d array. Thus, a is a pointer to its first 1-d array. As we have seen earlier, since 1-d array itself is a pointer, a can now be regarded as a pointer to pointer to int type. Note the two levels of indirections.

(a + 0) is the address of first 1-d array　　　*(a + 0) is the first 1-d array
(a + 1) is the address of second 1-d array　　　*(a + 1) is the second 1-d array
(a + 2) is the address of third 1-d array　　　*(a + 2) is the third 1-d array
:
:
(a + n-1) is the address of (nth) 1-d array　　　*(a +n-1) is the (nth) 1-d array

Now, *(a + i), (i - 1)th 1-d array of a is similar to any 1-d array, say b, which is declared as follows :

```
int  b[10];
```

Here, we know that the array name b gives the address of its first element. On the similar lines,
*(a + 0)
= *(a + 0) + 0 give the address of the first element in the first 1-d array of a
*(a + 0) + 1 give the address of the second element in the first 1-d array of a
*(a + 0) + 2 give the address of the third element in the first 1-d array of a and so on
((a + 0) + 0 give the first element in the first 1-d array a[0][0]
((a + 0) + 1 give the second element in the first 1-d array a[0][1]
((a + 0) + 2 give the third element in the first 1-d array a[0][2]

Similarly,
((a + 1) + 0 give the first element in the second 1-d array a[1][0]
((a + 1) + 1 give the second element in the second 1-d array a[1][1]
((a + 1) + 2 give the third element in the second 1-d array a[1][2]

In general, a[i][j] can be written as *(*(a + i) + j)

Program 5.4. To illustrate processing 2-d arrays pointer_notations ....

```
#include  <stdio.h>
#include  <conio.h>

void main()
 {
     int  a[10] [10], m, n, i, j ;
     clrscr() ;
     printf ("Enter the order of the matrix  a  \n") ;
     scanf ("%d%d",  &m,  &n ) ;
     printf ("\n Enter elements of the matrix  a  of order
                                          %d * %d \n", m, n );
     for ( i = 0;  i< m ;  i++ )
     for ( j = 0;  j < n  ;  j++ )
            scanf ( "%d",   *(a + i) + j ) ;

     printf (" Matrix  a  \n") ;
     for ( i = 0; i <m ;  i++ )
     {
          for ( j = 0; j< n ;  j++ )
              printf ("%4d ", *(*(a + i) + j )) ;
          printf (" \n") ;
     }

     getch();
 }
```

INPUT-OUTPUT

```
Enter the order of the matrix  a
2  2

Enter elements of the matrix  a  of order  2  *  2
1  2  3  4

Matrix  a
1   2
3   4
```

EXPLANATION

a is declared to be an array of two dimensions and of size 10 * 10. m, n, i and j are declared to be variables of int type. The variables m and n are to collect the number of rows and the number of columns of the 2-d array. The variables i and j are to select rows and columns of the array, respectively. The segment of the program responsible for accepting the elements into the array a is :

```
for(i   =  0;  i   <  m;   i++)
for(i   =  0;  i   <  m;   i++)
    scanf("%d",   *(a   + i)   +   j);
```

Note that the second argument to scanf(), *(a + i) + j is same as &a[i][j].

The segment of the program responsible for displaying the elements of the array a is :

```
for(i = 0;  i   <  m;  i++)
{
    for (j   =   0;  j   <   n;   j++)
        printf("%4d",   *(*(a   +   i) + j));
    printf("\n");
}
```

Note that the second argument to printf(), *(*(a + i) + j) is same as a[i][j].

5.5. POINTERS AND STRINGS

In the previous section, we learnt that pointers and arrays are close associates. They are so close to each other that arrays are processed using pointers internally. Strings also being arrays enjoy the same kind of relation with pointers. We know that a string is a sequence of characters terminated by a null character '\0'. The null character marks the end of the string. We also know that the name of an array gives the base address of the array, that is, the address of the first element of the array. this is true even in the case of strings. Consider the string variable char s[20]; here s gives the base address of the array, that is, the address of the first character in the string variable and hence can be regarded as a pointer to character. Since each string is terminated by a null character, it is enough for us to know the starting address of a string to be able to access the entire string. In view of these observations, we conclude that a string can be denoted by a pointer to character.

```
char   s[20];
```

```
char  *cp;
cp  =  s;
```

s is declared to be an array of char and it can accommodate a string. To be precise, s represents a string.

cp is declared to be a pointer to char type. Assignment of s to cp is perfectly valid since s is also a pointer to char type. There is type compatibility. Now, cp also represents the string in s.

Let us now consider a string constant "abcd". We know that a string constant is a sequence of character enclosed within a pair of double quotes. The string constant is stored in some part of memory and it requires five bytes of space. Here also the address of the first character represents the entire string. The value of the string constant thus is the address of the first character 'a'. So it can be assigned to a pointer to char type. If cp is a pointer to char type, cp = "abcd" is perfectly valid and cp now represents the entire string "abcd".

Program 5.5. To illustrate pointer to strings

```
#include  <stdio.h>
#include  <conio.h>

void main()
 {
      char  str [20],  *cp ;

      clrscr()  ;
      printf (" Enter a string \n")  ;
      scanf ("%s",  str) ;
      cp  =  str ;
      printf (" string in str  =  %s \n",  cp) ;

      cp  =  "abcd" ;
      printf ("%s  \n", cp ) ;
      getch();

 }
```

INPUT-OUTPUT

```
Enter  a  string
Program
string  in  str  =  Program
abcd
```

EXPLANATION

In Program 5.5, str is declared to be a string variable of size 20. Maximum 20 characters can be stored in it. cp is declared to be a pointer to char type. A string is accepted into the variable str using the statement scanf(%s", str); the pointer variable cp, being a pointer to char type is then assigned str. It is perfectly correct since str, the array name, gives the base address of the array and hence is appointer to char type. Pointer assignment is permissible. After the assignment of str to cp, cp will also represent the string in str. It is confimed by the statement printf("%s", cp), which displays the string in str.cp is then assigned a string constant "abcd" and the string constant is displayed by the statement printf("%s", cp);

Thus, a pointer to char type can represent a string and it can be made to represent different strings at different points of time. But an array of char type can represent only one string throughout the program. It is because of the fact that the name of a string variable is a constant pointer and it cannot be assigned any other address.

In the case of an array of char type, a string can be accepted into it through console. But in the case of a pointer to char type, we can only assign a string but cannot accept a string through the console.

Comparison between an array of char and a pointer to char

S.No.	Array of char	Pointer to char
1.	We can initialize a string during declaration char str[20] = "abcd"	We can initialize a string during declaration char *cp[20] = "abcd"
2.	A string variable or a string constant cannot be assigned.	A string variable or a string constant can be assigned.
3.	The number of bytes allocated fro a string variable is determined by the size of the array.	The number of bytes allocated fro a string variable is determined by the number of characters within the string.
4.	An array of char refers to the same string (storage area).	A pointer to char can represent different string (storage area) at different points of time.
5.	A string can be accepted through the console char str[20]; *scanf("%s", str);* is valid.	A string cannot be accepted through the console. char *cp; *scanf("%s", cp);* is not valid since cp contains garbage value.

5.5.1. Array of Pointers to Strings

We know that a pointer variable to char type can represent a string. If we are to deal with more than one string then we can use an array of pointers to char type to represent the strings. An example of the declaration of an array of pointers to char type is given as follows :

```
char   *ptr[10];
```

Here, ptr[0], ptr[1],...., ptr[9] are all pointers to char type and each of them can denote a string.

Program 5.6. To illustrate array of pointers to strings ....

```
#include  <stdio.h>
#include  <conio.h>

void main()
 {
    char  *names[5] = {"Nishu", "Harsha", "Shobha",
                                "Devaraj", "Asha"};
    int  i ;
```

```
        clrscr() ;
        printf (" The list of five Names \n") ;
         for ( i = 0;  i < 5 ;  i++ )
               printf ("%s \n", names[i] ) ;
         getch();
    }
```

INPUT-OUTPUT

```
The  list  of  five  Names
Nishu
Harsha
Shobha
Devaraj
Asha
```

EXPLANATION

names is declared to be an array of pointers to char type and of size 5. The array of pointers is initialized with five strings. names[0] denotes the string "Nishu"; names[1] denotes the string "Harsha", and so on. i is declared to be a variable of int type. It is to select each string in the list of strings. A for loop is used to display the strings represented by the pointers of the array names.

5.6. POINTERS AND STRUCTURES

The concept of pointers can be used in combination with structures in two ways : (*a*) Since the address of a structure variable can be retrieved using & operator, we can think of pointers to structures. (*b*) Having pointers to different data types as members of structures.

5.6.1. Pointers to Structures

Similar to pointers to basic type like int, float and char, we can have pointers to structures also. The pointer variables should be declared to be the corresponding type. We use & operator to retrieve the address of a structure variable.

> **DO YOU KNOW**
>
> A pointer can grow as well as shrink as & when required.

t is declared to be variable of struct temp type. tp is declared to be a pointer to struct temp type. The statement tp = &t; assigns the address of t to tp. The pointer variable tp is now said to point tot. the members of t are accessed through tp with the help of operator -> known as structure-pointer operator. -> expects the pointer variable to its left and member name to its right. tp -> i, tp -> f and tp -> c represent t.i, t.f and t.c, respectively.

Program 5.7. To illustrate pointers to structures ....

```
#include  <stdio.h>
#include  <conio.h>
struct  temp
 {
     int    i ;
```

```
        float  f ;
        char  c ;
    } ;

    void main()
    {
        struct  temp  t  =  {123, 34.56, 'p' }, *tp ;
        clrscr() ;

        tp  =  &t ;
        printf (" t.i = %d \n", tp->i ) ;
        printf (" t.f = %f \n", tp->f ) ;
        printf (" t.c = %c \n", tp->c ) ;
        getch();
    }
```

INPUT-OUTPUT

```
t.i  =  123
t.f  =  34.560000
t.c =   p
```

EXPLANATION

The structure variable t is initialized with values 123, 34.56 and the character constant 'p'. The memory locations allocated fro t will be filled with the given values as follows :

123	34.560000	p
t.i	t.f	t.c

The pointer variable tp is then assigned the address of t. The members of t are then accessed by means of tp with the use of the operator ->.

5.6.2. Structures Containing Pointers

```
struct temp
{
    int  i;
    int  *ip;
};
```

The structure definition includes ip, a point6er to int type as one of its members.

Program 5.8. To illustrate structure containing_pointers

```
#include  <stdio.h>
#include  <conio.h>
```

```
struct  temp
 {
      int    i ;
      int    *ip ;
 } ;
void main()
 {
      struct  temp  t ;
      int     a =  10 ;

      clrscr() ;
      t.i  =  a ;
      t.ip = &a ;
      printf (" Value of a = %d \n", t.i ) ;
      printf (" Address of a = %u \n", t.ip ) ;
      getch();
 }
```

INPUT-OUTPUT

```
Value  of  a  =  10
Address  of  a  =  65520
```

EXPLANATION

The definition of the structure struct temp includes two members i and ip. The member i is of type int and the member ip is a pointer to int type. In the main(), t is declared to be a variable of struct temp type. The storage for the variable t is as follows :

t.i	t.ip

t.i is a variable of int type whereas t.ip is a pointer variable to int type. a is declared to be a variable of int type and it is initialized with the value 10. The value of a is assigned to t.i and the address of a is assigned to t.ip, a pointer to int type. The value of a and its address are displayed through t.i and t.ip, respectively.

A structure containing a pointer to the same type is referred to as self-referential structure.

```
struct  list
{
    int  n;
    struct list *p;
};
```

DO YOU KNOW

Self referential structure is used to store the data in memory using either linked list or trees or graphs.

5.7. POINTERS AND FUNCTIONS

Here, we explore the possibility of the following :

- Passing pointers as arguments to functions

- Returning a pointer from a function
- Pointer to a function
- Passing one function as an argument to another function

5.7.1. Passing Pointers as Arguments to Functions

In our earlier programs which involve functions with arguments, we passed the values of actual arguments. The mechanism of calling a function by passing values is called call by value. We now discuss passing pointers as arguments to functions. The mechanism of calling a function by passing pointers is called call by reference. If a function is to be passed a pointer, it has to be reflected in the argument-list of the function prototype and the header of the function as function_name(type *ptr);.

Program 5.9. To illustrate call by value and call by reference ....

```
#include  <stdio.h>
#include  <conio.h>

void  inc (int) ;
void  incr (int*) ;

void main()
 {
      int    a  =  10,  b  =  10 ;

      clrscr() ;
      printf (" Call By Value \n") ;
      printf ("a = %d \n",  a ) ;
      inc (a) ;  printf ("After calling inc() \n" ) ;
      printf ("a = %d \n",  a ) ;
      printf (" Call By Reference \n") ;
      printf ("b = %d \n",  b ) ;
      incr (&b) ;     printf ("After calling incr() \n" ) ;
      printf ("b = %d \n",  b ) ;
      getch() ;
 }

Void inc (int  x)
{
      x++ ;
      printf ("x = %d \n",  x ) ;
}

Void incr (int   *p)
{
      (*p)++ ;
}
```

INPUT-OUTPUT

```
Call  By  Value
a   =  10
x   =  11
After calling inc()
a   =  10

Call By Reference
b   =  10
After calling incr()
b   =  11
```

EXPLANATION

The function inc() is defined with a formal argument x of type int. The purpose of the function is to increment the value passed to it by one. Incr() is another function defined with one formal argument p, which is a pointer to int type. So the function incr() takes the address of a variable of int type and the purpose of the function is to increment the value pointed to by the address.

In the main(), a and b are declared to be variables of int type and are initialized with the value 10. To begin with, the value of a is displayed to be 10. The function inc() is invoked by passing a as the actual parameter. As a result, the value of a is copied to x, the formal parameter of the function and within the function, x is incremented by one, it becomes 11 and is displayed. After the exit of the inc(), the control is transferred back to the main(). In the main(), the value of a is once again displayed. It turns out to be 10 only. The function inc() has not been able to change the value of a, which belongs to main(). It has acted on x, its formal parameter (a copy of a).

We now understand the working of incr(). The initial value of b in main() is displayed. Then a call is made to incr() by passing the address of the variable b to it as incr(&b). Now, the address of b is copied to p, the formal parameter of incr(). Within the function incr(), the value at address in p is incremented by one, *i.e.*, (*p)++ which turns out to be the variable b of main() itself. After the function is exited, the value of b is redisplayed and it is seen to be 11. So, the function incr() could change the value of the variable b, which belongs to main().

Program 5.10. To illustrate call by value and call by reference ....

```
#include  <stdio.h>
#include  <conio.h>

void  swap (int, int) ;
void  swapr (int*, int*) ;

void main()
 {
      int    a  =  5,  b  =  6 ;
      int    c  =  7,  d  =  8 ;
      clrscr() ;
```

```
        printf (" Call By Value \n" ) ;
        printf ("a = %d b = %d \n",  a , b ) ;
        swap(a, b) ;  printf ("After calling swap() \n");
        printf ("a = %d b = %d \n",  a, b ) ;

        printf (" Call By Reference \n" ) ;
        printf ("c = %d d = %d \n", c, d ) ;
         swapr(&c, &d) ; printf ("After calling swapr() \n");
        printf ("c = %d d = %d \n", c, d ) ;
        getch() ;
   }

void swap(int x, int y) ;
{
        int  t ;

        t  =  x ;
        x  =  y ;
        y  =  t ;

         printf ("x = %d y = %d \n",  x, y ) ;
}

void swapr(int    *p, int *q)
{
            int  t ;

           t   =  *p ;
            *p   =  *q ;
            *q   =  t ;
}
```

INPUT-OUTPUT

```
Call  By  Value
a   =  5   b  =  6


After calling swap()
a   =  5   b  =  6
```

```
Call By Reference
c   =  7    d  =  8

After calling swapr()
c   =  8    d  =  7
```

EXPLANATION

The function swap() is defined with two formal arguments x and y of int type. The purpose of the function is to interchange the values taken by the formal arguments x and y. Within the in function, the following statements accomplish interchanging of values of x and y.

```
t  =  x;
x  =  y;
y  =  t;
```

t is a local variable of the function, which acts as a temporary variable used to interchange the values.

The function swapr() is defined with two formal arguments p and q which are pointers to int type. The purpose of this function is to interchange the values of the variable pointed to by the pointers p and q. Within the function, the following statements accomplish interchanging of values of the location pointed to by p and q.

t = *p;	The value at address in p is moved to t.
*p = *q;	The value at address in q is moved to the location whose address is in p.
*q = t;	The value of t (the value at address in p) is moved to the location whose address is in q.

t is a local variable of the function, which acts as a temporary variable. It is used to interchange the value.

In the main(), the local variables of int type a, b and c and d are initialized with the values 5,6,7 and 8, respectively. The variables a and b are to be passed as arguments to swap(). Before calling swap(), the values of a and b are displayed to be 5 and 6, respectively. Then the function swap() is called by passing a and b as its arguments as swap(a, b);. The values of a and b are copied to x and y, the formal arguments of the function, respectively. The statements of the function interchange the values of x and y and they are displayed to be 6 and 5, respectively. After the exit of the function swap(), the main() function regains the control. The values of a and b are once again displayed in main()and they are seen to be 5 and 6 only, *i.e.*, the values of a and b are not interchanged by swap(). The swap() has acted on x and y, the copies of a and b, respectively not on a and b themselves.

Let us now understand the working of swapr(). The values of c and d of main() are displayed befoe calling swapr(). They are seen to be 7 and 8, respectively. The function swapr() is then called by passing the addresses of c and d as the arguments. The addresses of c and d are taken by the formal arguments of swapr(), p and q respectively. Within the function, the values at addresses in p and q are interchanged. Even within the swapr() also, locations of c and d are the targets. After the exit of the swapr(), control is gained by main(). In main(), the values of c and d are once again displayed. They are seen to be 8 and 7, respectively, *i.e.*, the function swapr() has been able to interchange the values of c and d themselves.

Difference between call by value and call by reference

S.No.	Call by value	Call by reference
1.	Only values are passed as arguments.	Addresses are passed as arguments.
2.	Constants can be passed as actual arguments.	Constants cannot be passed as actual arguments. Exception to this are string constants.
3.	Functions cannot return more than one value.	The effect of changing the values of the variables belonging to the calling program is equivalent to returning more than one value.

5.7.2. Returning a Pointer from a Function

Under some circumstance, we require functions to return a pointer. The fact that a function returns a pointer should be indicated by its declaration and definition.

The syntax of declaration of a function returning a pointer is as follows :

SYNTAX

```
data-type   *function_name(arguments);
```

Note the presence of * before function_name. It is to indicate that the function returns a pointer to data-type.

Example :

```
int   *sum(int,   int);
```

The function sum() is declared to indicate that it requires two parameters of int type and it returns a pointer to int type.

Program 5.11. To illustrate a function returning a pointer

```
#include   <stdio.h>
#include   <conio.h>
int      *sum (int, int) ;

void main()
  {
     int      a,   b,   *s ;

     clrscr() ;

     printf (" Enter two numbers \n" ) ;
     scanf ("%d%d",    &a,    &b ) ;
     printf ("a = %d b = %d",    a,  b ) ;
     s   =   sum (a,   b) ;
     printf ("s = %d \n", *s) ;
     getch() ;
  }
```

```
int   *sum (int a,  int b) ;
{
        int  s ;
        s  =  a + b ;
        return  (&s) ;
}
```

INPUT-OUTPUT

```
Enter   two   numbers
4   5
a   =  4   b   =  5    s   =  9
```

EXPLANATION

The function sum() is defined with two arguments of type int and it is made to return a pointer to int type. The purpose of the function is to find the sum of the values passed to it and return the address of the variable, which stores the sum of the values.

In the main(), values are accepted into the variables a and b. A call is then made to sum() by passing a and b as its actual arguments as

```
s   =   sum(a,   b);
```

Note that s is declared to be a pointer to int type. The value pointed to by s, the sum of a and b, is then displayed.

5.7.3. Pointers to Functions

In the case of arrays, we know that name of an array gives its base address(address of the first location) and thus acts as a pointer to the first location of the array. This is true even in the case of functions. Name of a function also gives its starting address. Thus, we can use a pointer to invoke a function before which the pointer has to be declared appropriately and assigned the address of the function.

The syntax of declaring a pointer to a function is as follows :

SYNTAX

```
data-type (*variable_name)(argument_types);
```

variable_name is declared to be a pointer to a function, which return a value of type data-type.

Example :

```
int    (*fnp)   (int,    int);
```

fnp is declared to be a pointer to a function, which takes two arguments of int type and returns a value of type int. fnp can now be assigned the address of a function returning a value of int type.

```
int    sum(int,   int);
fnp    =   sum;
```

fnp now points to the function sum().

Program 5.12. To illustrate pointers to functions ....

```
#include   <stdio.h>
#include   <conio.h>
```

```
int    sum (int, int) ;

void main()
 {
        int     a,   b,   s,   (*fnp)    (int, int);
        clrscr() ;
        printf  (" Enter two numbers \n" ) ;
        scanf  ("%d%d",    &a,    &b ) ;
        printf  ("a = %d b = %d \n",    a, b ) ;
        fnp    =    sum ;
        s   =   (*fnp) (a,   b) ;
        printf  ("sum = %d \n", s) ;

        getch() ;
 }
int    sum (int a,    int b) ;
{
        int   s ;
        s   =   a + b ;
        return   s ;
}
```

INPUT-OUTPUT

```
Enter two numbers
4   6
a    =  4    b  =  6

sum    =   10
```

EXPLANATION

The function sum() is defiend with two arguments of type int and it is made to return value of int type. The purpose of the function is to find the sum of the values passed to it and return the sum of the values.

In the main(), a, b and s are declared to be variables of int type. fnp is declared to be a pointer to a function which returns a value of int type. Values are accepted int the variable a and b. The variable fnp is assigned the address of the function sum() with the statement

```
fnp   =   sum;
```

Then, a call is made to sum() throught eh pointer fnp with the statement

```
s   =   (*fnp)   (a,   b);
```

The variable s collects the value returned by the function and it is displayed.

5.7.4. Passing One Function as an Argument to Another Function

We know that the name of a function always gives the starting address of the function and thus it is a pointer to the function. We also knows that we can pass pointers to data types as arguments to functions. These facts lead us to think of the possibility of passing a pointer to a function to another function. This is certainly possible. The following program illustrate this.

Program 5.13. To illustrate passing one function as an argument to another function

```
#include   <stdio.h>
#include   <conio.h>

int   less than (int, int) ;
int   greaterthan (int, int) ;
void sort (int a[ ], int n, int (*fnp)  (int, int)) ;
void main()
 {
        int    a[10],  i,  n ;

        clrscr() ;
        printf (" Enter no. of values \n" ) ;
        scanf ("%d",  &n ) ;
        printf ("Enter %d Value \n",  n ) ;
        for ( i = 0;  i < n ;  i++ )
        scanf ("%d",  &a[ i ] ) ;

        printf (" Unsorted List \n") ;
        for ( i = 0;  i < n ;  i++ )
        printf ("%d \n", a[ i ] ) ;

        sort (a, n, less than) ;
        printf (" sorted in decreasing order \n") ;
        for ( i = 0;  i < n ;  i++ )
        printf ("%d \n", a[ i ] ) ;

        sort (a, n, greaterthan) ;
        printf (" sorted in increasing order \n") ;
        for ( i = 0;  i < n ;  i++ )
        printf ("%d \n", a[ i ] ) ;

        getch() ;
 }
int   less than (int  a,  int  b) ;
{
        return  (a < b ) ;
}

int   greaterthan (int  a,  int  b) ;
{
        return  (a < b ) ;
}

void sort (int a[ ], int n, int (*fnp)  (int, int)) ;
{
        int   i,  j,  t ;
```

```
        for ( i = 0;  i < n - 1 ;  i++ )
        for ( j = i + 1;  j < n ;  j++ )
        if (fnp (a[ i ],  a[ j ] ) )
        {
            t  =  a[ i ] ;
          a[ i ]  =  a[ j ] ;
          a[ j ]  =   t ;
        }
}
```

INPUT-OUTPUT

```
Enter no. of values
5
Enter 5 Values
1 3 4 2 6

Unsorted List
1 3 4 2 6

Sorted in decreasing order
6 4 3 2 1

Sorted in increasing order
1 2 3 4 6
```

EXPLANATION

The function less than() is defined with two formal arguments a and b of int type and it is made to return a value of int type. The purpose of this function is to collect two numbers into a and b and return 1 of a is found to be less than b otherwise 0.

The function greaterthan() is also defined with two formal arguments a and b of int type and it is made to return a value of int type. The purpose of this function is to collect two numbers into a and b and return 1 if a is found to be greater than b otherwise 0.

The function sort() is defined with three formal arguments a, n and fnp. The argument a is a pointer to integer type; n is of int type and fnp is a pointer to a function which returns an integer value. The purpose of the function is to take an array of integers and sort the element of the array in the order determined by the function pointed to by fnp.

In the main(), a is declared to be an array of int type and of size 10. Maximum 10 elements can be accepted into the array. i and n are declared to be variables of int type. The variable n is to collect the number of elements to be accepted into the array a and the variable i is to select each location of the array.

After accepting a value for the variable n, n values are accepted into the array a and they are displayed. The list is then sorted in the decreasing order with the call sort(a, n, less than).

Within the function sort(), the function less than() is invoked to compare two numbers which returns true when the first number is less than the second number being compared. Thus the list is sorted in the decreasing order. The sorted list is then displayed.

The list is then sorted in the increasing order with the call sort(a, n, greaterthan). Within the function sort(), this time, the function greaterthan() is invoked to compare two numbers which returns true when the first number is greater than the second number being compared. Thus the list is sorted in the increasing order. The sorted list is once again displayed.

A list of n numbers is accepted into the array a. The unsorted list is displayed. The function sort() is invoked by passing the array a, n and less than as the actual arguments (note that less than is the address of the function less than(), which is mapped onto the third argument fnp of sort()) as sort(a, n, less than); the function call sorts the element in the array a in the increasing order. (Note that this time, greaterthan, the address of the function greaterthan() is passed and it is mapped onto fnp, the third parameter of the function sort()).

5.8. POINTERS TO POINTERS

We now know that a pointer is a special type of variable and it can collect the address of another variable. Once a pointer is assigned the address of a plain variable, the value of the plain variable can then be accessed indirectly through the pointer with the help of * (value at address) operator. Since the pointer is also a variable, it is also allocated memory space of size two bytes and its address also is retrievable. The address of the pointer variable itself can also be stored in another appropriately declared variable. The variable which can store the address of a pointer variable itself is termed as pointer to pointer. Note that a pointer to pointer adds a further level of direction. The syntax of declaring a pointer to pointer is as follows :

SYNTAX

```
data-type **variable;
```

Note the usage of two *'s before the pointer variable name. They are to indicate that the variable is a pointer to a pointer to a variable of type data-type.

Consider the following declarations :

```
int   a  =  10,   *ap,   **app;
```

Here, a is plain integer variable. The variable ap is a pointer to int type and the variable app is declared to be a pointer to pointer to int type (Note that two *'s are used). Now ap can be assigned the address of the variable a. The variable app can be assigned the address of ap. The following statements accomplish these assignments :

```
ap    =  &a;
app    =   &ap;
```

To access the value of a through ap, we use *ap

To access the value of a through app, we use **ap.

Program 5.14. To illustrate a pointer to a pointer ....

```
#include   <stdio.h>
#include   <conio.h>
void main()
 {
     int  a  =  10,  *ap,  **app ;

     clrscr() ;
     ap   =  &a ;
     app   =   &ap ;

     printf (" Address of a   =   %u \n", ap) ;
     printf (" Address of ap   =  %u \n", app) ;
     printf (" value of a through ap   =  %d \n", *ap) ;
```

```
        printf (" value of a through app  = %d \n", **app) ;
        getch() ;
    }
```

INPUT-OUTPUT

```
Address  of  a  =  65524
Address  of  app  =  65522
Value  of  through  ap  =  10
Value  of  through  app  =  10
```

In Program 5.14, we have seen that there are two levels of indirection to access the value of the variable a. This was accomplished through a pointer to pointer. Theoretically, the levels on indirection can be further extended. There is no constraint imposed on this. Further levels of indirection like three levels, four levels are rarely used because of the following fact : Higher the level of indirection, Lesser will be the degree of comprehensibility.

5.9. DYNAMIC MEMORY ALLOCATION

We know that the variable are name memory locations and they are to hold data to be manipulated by the programs. So far, in all our programs, memory locations for the variables were allocated during compilation time itself. For instance, the declarations int a; float f; and char c; direct the compiler to allocate two bytes, four bytes and one byte of memory to hold integer value, float value and char value, respectively. So the quantum of memory is determined by the compiler itself depending on the type of variables. This type of memory allocation is termed as static memory allocation. Consider the declaration of arrays in C.

```
int   a[10];
```

declares a to be an array of int type and size 10. The compiler thus allocates 10 contiguous memory locations as follows.

a[0]	a[1]	a[2]	a[3]	a[4]	a[5]	a[6]	a[7]	a[8]	a[9]

Memory locations are allocated for the array statically. Here we identify two shortcomings. First'y, if we need to deal with more than ten values of int type during retime of the program, we cannot increase the size of the array during runtime. Secondary, many a time, not all the locations are used, when this is the case, we cannot decrease the size of the array. As a result, memory goes wasted. This is where the concept of dynamic memory allocation comes into picture. As the name itself indicates, memory can be allocated dynamically, *i.e.*, during runtime of the programs. While allocating, we can specify the size of the memory block to be allocated and also we can releases the unwanted memory blocks for later use. The standard C library has the following built-in functions, which help in dealing with the dynamic memory allocation.

1. malloc()
2. calloc()
3. realloc()
4. free()

The declarations for the functions are available in the header file malloc.h

5.9.1. malloc()

The malloc() allocates a block of memory of requested size and returns a pointer to the first byte of the block. The syntax of its usage is as follows :

```
ptr  =  (data_type*)   malloc(byte_size);
```

where byte_size is the number of bytes to be allocated and ptr is a pointer of type data-type, which collects the address of the first byte of the memory block allocated. The memory block allocated will have garbage values. If the allocation is not successful, the function returns NULL a value.

5.9.2. free()

The free() is to release a block of memory, which is previously allocated so that it can be reused for some other purpose. The syntax of its usage is as follows :

```
free(ptr);
```

where ptr is a pointer, which points to the firs byte of the memory block to be released.

Program 5.15 : To illustrate malloc(): Creation of an array dynamically

```
#include  <stdio.h>
#include  <conio.h>
#include  <malloc.h>

void main()
 {
      int     *a,  n,  i ;

      clrscr() ;
      printf (" Enter no. of elements \n") ;
      scanf ("%d",   &n ) ;
       a = malloc (n * sizeof (int) );
                                 /* allocate memory for n integers */
      printf (" Enter %d  elements \n",  n ) ;
      for ( i = 0;  i < n ;  i++ )
      scanf ("%d",   a + i ) ;
      printf (" The list of elements") ;
      for ( i = 0;  i < n ;  i++ )
      printf (" %d \n", *(a + i) ) ;
      free (a) ;     /* Release the allocated memory  */
      getch() ;
 }
```

INPUT-OUTPUT

```
Enter  no.  of  elements
5

Enter  5  elements
1  2  3  4  5

The  list  of  elements     1  2  3  4  5
```

EXPLANATION

In Program 5.15, the variable a is declared to be a pointer to int type. n and i are declared to be plain variables of int type. The purpose of the program is to create an array dynamically by specifying the size of the array during runtime. The pointer variable a is to collect the address of the first byte of the memory block allocated by the malloc(). The variable n is to accept the size of the array during runtime. The variable i is used as the array index.

After accepting the size of the array, the required amount of memory space is allocated with the statement.

```
a = malloc(n * sizeof(int));
```

On successful execution of the statement, a block of memory large enough to accommodate n integers is allocated in the memory heap and the starting address of the block is assigned to the pointer variable a. Now, the array is filled with integer values and they are displayed. Note that the constant pointer expression *(a + i) has been used to access each element of the array. Then the statement free(a) releases the block of the memory pointed to by the variable a.

5.9.3. calloc()

The calloc() allocates multiple blocks of memory, all of which are of same size and returns a pointer to the first byte of the firs block. The syntax of its usage is as follows :

```
ptr = (data_type*) calloc(n, block_size);
```

where n is the number of blocks of memory to be allocated. Block_size is the number of bytes in each block and ptr is a pointer of type data-type, which collects the address of the first bytes of the first memory block allocated. The memory block allocated will be filled with zeros. If the allocation is not successful, the function returns NULL value.

Program 5.16. To illustrate calloc(): Creation of an array dynamically

```
#include <stdio.h>
#include <conio.h>
#include <malloc.h>
void main()
 {
     int *a, n, i ;

     clrscr() ;
     printf ("Enter no. of elements \n") ;
     scanf ("%d", &n ) ;
     a = calloc (n, 2 ); /* allocate memory for n integers */
     printf ("Enter %d elements \n", n ) ;
     for ( i = 0; i < n ; i++ )
     scanf ("%d", a + i ) ;
     printf ("The list of elements \n") ;
     for ( i = 0; i < n ; i++ )
     printf ("%d \n", *(a + i) ) ;
     free(a);      /* Release the allocated memory */
     getch();
 }
```

INPUT-OUTPUT

```
Enter no. of elements
5
```

```
Enter   5   elements
4  5  6  7  8

The  list  of  elements    4  5  6  7  8
```

EXPLANATION

In Program 5.16, the variable a is declared to be a pointer to int type. n and i are declared to be plain variables of int type. The purpose of the program is to create an array dynamically by specifying the size of the array during runtime. The pointer variable a is to collect the starting address of the first block of the memory blocks allocated by the calloc(). The variable n is to accept the size of the array during runtime. The variable I is used as the array index.

After accepting the size of the array, the required amount of memory space is allocated with the statement.

```
a   =   calloc(n,    sizeof(int));
```

On successful execution of the statement, n blocks of memory, all of equal size (sizeof(int)) are allocated in the memory heap and the starting address of the first block is assigned of the pointer variable a. Now, the array is filled with integer values and they are displayed. Note that the constant pointer expression *(a + i) has been used to access each element of the array. Then the statement free(a) releases the block of memory pointed to by the variable a.

5.9.4. realloc()

We now know that malloc() and calloc() are used to dynamically allocate memory space required to store data. Under some circumstances, we may need to alter the size of the memory block allocated by either malloc() or calloc(), during runtime itself. This is accomplished by realloc(). The syntax of its usage is as follows :

```
new_ptr  =  (data-type*)  realloc(old_ptr,  new_size);
```

Where old_ptr points to the block of memory previously allocated by malloc() or calloc(). new_size is the size of the block to be reallocated. On successful execution, the function returns a pointer to the first byte of the reallocated memory block. If the function does into execute successfully, it returns NULL value.

Here, new_size may be smaller or larger than the earlier size of the memory block. The position of the old memory block and that of the reallocated block may be same or different. If they are different, the data in the old block will be safely shifted to the new block.

Program 5.17. To illustrate realloc() and free()

```
#include  <stdio.h>
#include  <conio.h>
#include  <string.h>
#include  <conio.h>

void main()
 {
     char *str ;
     /* allocate memory for string */
     clrscr() ;
```

```
    str  =  (char  * )  malloc (7) ;

    /* copy "Mysore" into  string */
    strcpy (str,  "Mysore") ;
    printf ("String  %s  is at address %p \n",  str,  str) ;

    str  =  (char  * )  realloc (str,  10) ;
    strcpy (str,  "Bangalore") ;
    printf ("String  %s  is at address %p \n",  str,  str) ;

    /*  free  memory  */
    free(str);
    getch() ;
}
```

INPUT-OUTPUT

```
String  Myscore  is  at  address  07EC
String  Bangalore  is  at  address  07F8
```

EXPLANATION

ptr is declared to be pointer variable to char type. It can collect the address of a character. The statement ptr = malloc(7), on its successful execution, allocates a block of memory fo size seven bytes in the memory heap and assigns the starting address of the block to the variable ptr. The string constant "Mysore" is then copied to the memory block. Since the length of the string constant "Mysore" is 6, the memory block allocated was of size 7 bytes (6 + Null character). Later, in order to write the string "Bangalore", the size of the block has been altered by the use of realloc() with the statement str = realloc(str, 10). Note that the new size is 10 bytes (length of "Bangalore" + Null character). Then the string constant "Bangalore" has been copied to the new memory block.

SUMMARY

1. Pointers are a powerful concept in C and add to its strength. Mastery over pointers empowers a C programmer to deal with system level jargons.
2. C provides two special operators known as pointer operators. They are & and *. & stands for 'Address of' and it is used to retrieve the address of a variable. * stands for 'value at address' and it is used to access the value at a location by means of its address.
3. The concept of pointers can be used in combination with structures in two ways : (*a*) Since the address of a structure variable can be retrieved using & operator, we can think of pointers to structures. (*b*) Having pointers to different data types as members of structures.
4. To allocate the memory to the variables dynamically we will use the functions viz malloc, calloc, relloc, free
5. The malloc() allocates a block of memory of requested size and returns a pointer to the first byte of the lbock.
6. The free() is to release a block of memory, which is previously allocated so that it can be reused for some other purpose.
7. The calloc() allocates multiple blocks of memory, all of which are of same size and returns a pointer to the first byte of the firs block.

8. We now know that malloc() and calloc() are used to dynamically allocate memory space required to store data. Under some circumstances, we may need to alter the size of the memory block allocated by either malloc() or calloc(), during runtime itself. This is accomplished by realloc().

EXERCISES

1. What is Call by Value and Call by Reference. Demonstrate with an example.
2. What is the output of the following code:

```
main()
{
int a[] = { 11, 14, 23, 31, 14};
int *ptr[] = {a, a+1, a+2, a+3, a+4};
int **ptr1 = ptr;
for (i=0;i<4;i++)
{
if(i<3)
{
printf("\n %d",**ptr1);
continue;
printf("\n %d",*ptr);
}
else
break;
printf("\n %d",ptr);
}
printf("\n %d %d %d",ptr, *ptr, **ptr);
}
```

3. int a[50];

```
int *ptr;
ptr = a;
```

To access the 7th element of an array *(*ptr+6) is the correct statement (T/F)

4. Add the missing statement for the following program to print 80,

```
main()
{
int j, *ptr;
*ptr = 80;
printf("\n %d",j);
}
```

5. Define "pointer in C" and demonstration with example program?
6. Demonstrate the relationship between arrays & pointers with suitable program.
7. How the pointers are passed to function? Explain through an example.
8. what is a pointer? How is a pointer initialized ?
9. write a program to reverse the character of a string using pointers.
10. explain what will be the output of following program :

```
#include < stdio.h>
main ()
{
```

```
int a, b, * ptr ;
a =9;
ptr= & a;
b= * ptr ;
a++;
}
```

What is value of a, b, * ptr?

11. Explain the following :

(*i*) int * p;

(*ii*) int ** p;

(*iii*) int * p();

(*iv*) a =& b;

(*v*) Arithmetic operations on pointers.

12. What is pointer ? How is a pointer initialized ?

13. Write a program using pointers to need in a array of integers and print its elements in reverse order ?

14. Differentiate between a pointer to constant & a constant pointer ?

15. what do you mean by dynamic memory allocation? Allocate a 2 dimemsional integer array of size MXN dynamically.

16. Interpret the meaning of 'P' in the following declaration

(*i*) int (*P (char *a)) [10];

(*ii*) int (* P (char *a[]));

(*iii*) int (*P [10]) (char a);

(*iv*) int *(*P)(char *a[]);

17. a[i] and i[a] are equivalent. Justify.

18. Differentiate between static memory allocation and dynamic memory allocation.

19. Differentiate between malloc() and calloc().

20. What is a self-referential structure ?

❑❑❑

CHAPTER 6

LINKED LIST

6.1. INTRODUCTION

A linked list is a dynamic data structure of elements which may be integer, float, string or a structure. The individual elements of linked list are called node and they are declared as structure in C. A linked list is a self-referential data type because it contains a pointer or link to another data of the same type. Linked lists permit insertion and removal of nodes at any point in the list in constant time, but do not allow random access. The list is called linked list as the nodes are linked to each other with the help of pointers. It is called dynamic as we may add or delete during program execution. Again for each node memory is llocated dynamically malloc or calloc.

Several different types of linked list exist: singly-linked lists, doubly-linked list and circularly-linked lists, header linked list, unrolled linked list, skip lists etc.

6.2. DYNAMIC MEMORY ALLOCATION

When we declare variables, arrays, structure (in definition) etc. the memory for them is allocated at the compile time. this allocation is **static allocation.** When you declare the array say **int a[5]** we give the size of array *i.e.* number of elements. We just cannot take input for *x* from user during program execution and declare the array by writing **int a[x].** Memory for array in the above declaration is allocated at compile time so compiler requires the size of the array. The size once fixed cannot be changed during the program execution. This may sometimes results in wastage of memory when size specified in the beginning is more then actually used in the program.

When you write your C program, the program's data and code is stored in the RAM. All the local and global variables used in the program are stored in a fixed area of memory allocated to that program.

This area allocated is fixed and remain constant during the execution of the program. The local variable are always stored on the stack and are destroyed automatically as soon as control goes out of scope. Global and static variables are known as load time variables as they come into local variables are efficiently handled by C compiler. But the programmer must know the amount of space required for every program.

The technique through which a program can obtain space in the RAM during the execution of the program and not during compilation is called dynamic memory allocation. In this method space for the variables (array, strings, structure etc.) is allotted from a free memory region which is known as **heap.** This area is not fixed in nature and keeps changing as the memory is allocated and deallocated from this region. The entire runtime view of memory for a program is given below.

Dynamic memory allocation is the allocation of memory at run time *i.e.* when program executes then required memory is asked from the user and memory is allocated with the help of certain inbuilt functions which we explain shortly. There are four such functions which we can use for allocation, reallocation and for freeing the memory. The header file for the functions is **alloc. h/stdlib.h**

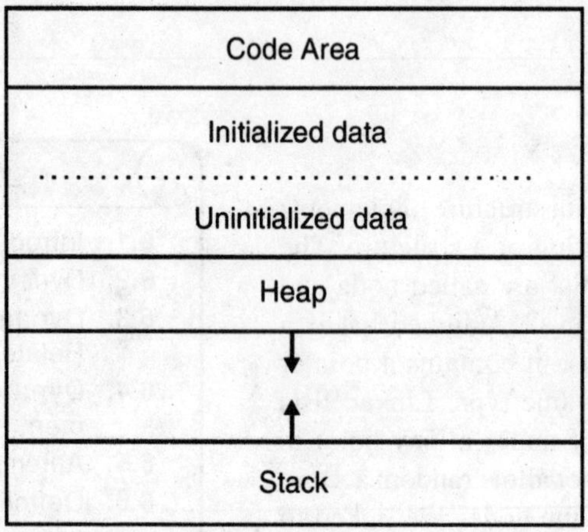

Fig. 6.1. Run time memory area of a C program.

6.3. DYNAMIC MEMORY ALLOCATION RELATED FUNCTIONS

1. void *malloc(size_t size);

This function allocates requested block of memory specified by the size which is of type **size_t. size_t** is type of **unsigned integer.** The function returns a pointer to the first byte of allocated memory of type **void** so you can typecast it to of any type of pointer say **int, char** or **float.**

2. void *calloc(size_t n_elem, size_t size);

This function allocates a block of **n_elem * size** bytes and initializes them to **0.** We use **calloc** to allocate memory for array of **n_elem** each of size **size.** Similar to **malloc** a pointer is returned which points for first byte of allocated memory.

3. void * realloc (void*block, size_t size);

This function is used to modify the previously allocated block of memory. The function **realloc** allocates a new memory block of size **size** and returns a pointer to the first byte of newly allocated block.

4. void free(void*block);

free() de-allocates *i.e.* releases the memory allocated previously by a call to **malloc, calloc** or **realloc.**

All the four functions are declared in the header file **alloc.h,** so this file must be included in all the programs which make use of any of the above mentioned function..

To give you a feel of dynamic memory allocation and de-allocation we write a very simple program that allocates memory dynamically for variables.

Program 6.1. /*Program demo of malloc and free */..

```
#include<alloc.h>
main ( )
{
    char * ch;
    int *p;
    float *f;
    clrscr ();
    ch =  (char*)malloc  (sizeof(char));
    p = (int*) malloc (sizeof (int));
    f = (float*)malloc(size of (float));
    printf  ("Enter a character\n");
    *ch = getchar();
    printf("Enter an integer\n");
    scanf("%d", p);
    printf ("Enter an float\n");
    scanf ("%f", f);
    printf(" U entered \n%c %d %f\n", *ch, *p, *f);
    free (ch);
    free (p);
    free (f);
    getch ();
|
```

OUTPUT

```
Enter a character
A
Enter an integer
15
Enter a float
2. 34
U entered
A 15 2. 34
```

In the program, we have allocated memory for an **integer, a char and a float** variable dynamically. According to the type we have type-casted the **malloc** function. The **sizeof** function has been used to determine the size of each type which in turbo c turns out to be 1, 2 and 4 respectively. In the end we de-allocate the memory through **free** function.

This was a very trivial example which simply demonstrated how to use **malloc** and **free** function. Usually you won't be allocating memory for simply a variable in this manner.

6.4. DYNAMIC STORAGE MANAGEMENT

In case of static storage management scheme, the net amount of memory requital Ini various data for a program is allocated before the starting of the execution of the program. Once memory is allocated, it neither can be extended nor can be returned to memory bank for the use of other programs at the same time. Dynamic storage management scheme allows the user to allocate and deallocate as per the necessity during the execution of programs. This dynamic memory management scheme is suitable in multiprogramming as well as single user environment where generally more than one program reside in the memory and then memory requirement can be known only during their execution. Operating systems (OS) generally provides the service of dynamic memory management.

Consider a small memory of 1024 bytes. Suppose a request is made for three blocks of storage of 350, 125 and 250 bytes. Let us assume that these blocks are allocated sequentially as shown in Fig. 6.2.

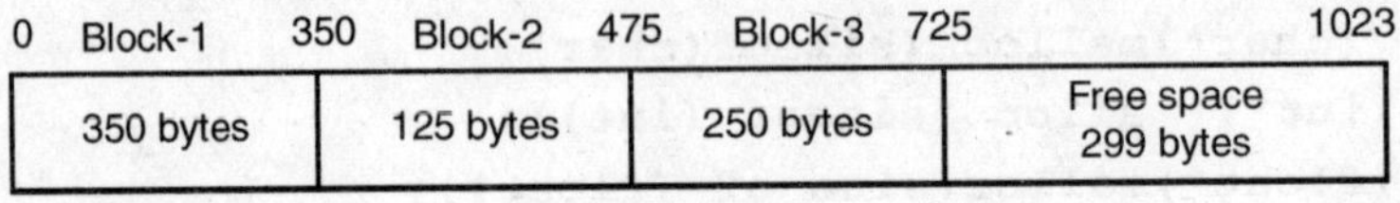

Fig. 6.2.

Now let us suppose that the second block of 125 bytes is set free. Although there are 424 bytes of free space available, a request of 400 bytes could not be satisfied since the free space is divided into noncontiguous blocks.

If block-3 is set to be free, it is not desirable to retain three free blocks of 125, 250 and 299 bytes. Rather the block can be combined into a single large block of 674 bytes so that large memory requests can be further satisfied. After combination they appear in two fragments of that of size 350 bytes and 674 bytes respectively which is shown in Fig. 6.3.

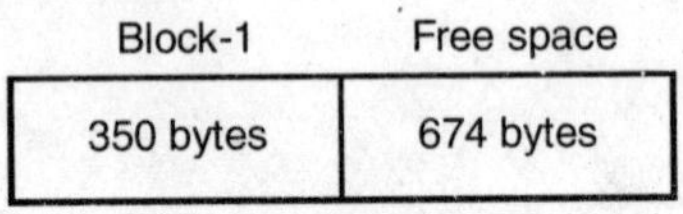

Fig. 6.3.

6.5. AUTOMATIC LIST MANAGEMENT

The idea behind incorporating automatic list management into a programming system is that the programmer should not have to decide when a node should be allocated or freed. Instead, the system will automatically allocate nodes that are being created and make available for reuse any nodes that are no longer accessible.

There are two principal methods used in automatic list management :

(*i*) reference count method and

(*ii*) Garbage collection method

6.5.1. Reference Count Method

Under this method each node has an additional count field that keeps a count (called the reference count) of the number of pointers point to that node. Each time a pointer is set to point that node, the reference count in that node is increased by 1. Each time some pointer that had been pointing to a node is changed, the reference count in that node is decreased by 1. When the reference count in any node becomes 0, that node can be returned to the available list of free nodes.

The reference count method has two disadvantages:

- the additional space required in each node for the count
- the first node of a circular list will never be reduced to 0.

6.5.2. Garbage Collection

The other method of detecting and reclaiming free nodes is called *garbage collection,* Under this method, nodes no longer in use remain allocated and undetected until all available storage has been allocated. A subsequent request for allocation can not be satisfied until nodes that has been allocated but are no longer in use are recovered. When a request is made for additional nodes and there are none available, a system routine called the garbage collector is called. This routine searches through all of the nodes in the system, identifies those that are no longer accessible from an external pointer, and restores the inaccessible nodes to the available pool. The request for additional nodes is then fulfilled with some of the reclaimed nodes and the system continues processing user requests for more space. When available space is used up again garbage collector is called once more. Garbage collection is usually done in rwo phases :

(1) ***Marking phase :*** Involves marking all nodes that are accessible from an external pointer.

(2) ***Collection phase :*** Involves proceeding sequentially through memory and freeing all nodes that have not been marked.

6.5.2.1. Marking Phase

In order to be able to carry out marking, we need a mark field in each node. It will be assumed that this mark field can be changed at any time by marking algorithm. Marking algorithms mark all directly accessible nodes *(i.e.* nodes accessible through program variables referred to as pointer variables) and also all indirectly accessible nodes *(i.e.* nodes accessible through link fields of nodes in accessible list). It is easy to mark all directly accessible nodes. The indirectly accessible nodes are marked by systematically examining all nodes reachable from these directly accessible nodes.

It is sometimes inconvenient to reserve one field in each node solely for the purpose of marking. In that case a separate area in memory can be reserved to hold a long array of mark bits, one bit for each node that may be allocated.

One aspect of garbage collection is that it must run when there is very little space available. This means that auxiliary table and stacks needed by the garbage collector must be kept to a minimum since there is little space available for them. An alternative is to reserve a specific percentage of memory for the exclusive use of the garbage collector. However, this effectively reduces the amount of memory available to the user and means that the garbage collector will be called more frequently.

Whenever the garbage collector is called, all user processing comes to a halt while the algorithm examines all allocated nodes in memory. For this reason it is desirable that the garbage collector be called as infrequently as possible. For real time applications, in which a computer must respond to a user request within a specific short time span, garbage collection has generally been considered an unsatisfactory method of storage management. The garbage collector must be called before all space has been exhausted so that user processing can continue in whatever space is left, while the garbage collector recovers additional space.

It is possible that at the time the garbage collection program is called, users are actually using almost all the nodes that are allocated. Thus almost all the nodes are accessible and garbage collector recovers very little additional space. After the system runs fora short time, it will again be out of space; the garbage collector will again be called only to recover very few additional nodes, and the vicious cycle starts again. This phenomenon, in which system storage management routines such as garbage collection are executing almost all the time, is called *thrashing*.

Clearly, thrashing is a situation to be avoided. One drastic solution is to impose the following condition. If the garbage collector is run and does not recover a specific percentage of the total space, the user's space is then recovered and made available to other users.

6.5.2.2. Collection and Compaction

Once the memory locations of a given system have been marked appropriately, the collection phase may begin. The purpose of this phase is to return all those locations that were previously garbage (not used by any program but unavailable to any user) to available memory. It is easy to pass through memory sequentially by examining each node and return unmarked nodes to available storage.

After the collection algorithm has completed, all unused nodes are on the available list, and all nodes that are in use by programs have their mark fields turned off (for the next call to garbage collection).

Although at this point (following the marking and collection phases of the system) all nodes that are not in use are on the available list, the memory of the system may not be in an optimal state for future use. This is because the interleaving of the occupied nodes with available nodes may make much of the memory on the available list unusable.

For example, memory is often required in blocks (group of contiguous nodes) rather than as single discrete nodes one at a time. The memory request by a compiler for space in which to store an array would require the allocation of such a block. If, for example, all the odd locations in memory were occupied and all the even locations were on the available list, a request for even an array of size 2 could not be honored, despite the fact that half of memory is on the available list. Although this example is probably not very realistic, there certainly exist situations in which a request for a contiguous block of memory could not be honored, despite the fact that sufficient memory does indeed exist.

There are several approaches to this problem. Some methods allocate and free portions memory in blocks (groups of contiguous nodes) rather than in units of individual nodes. This guarantees that when a block of storage is freed (retuned to the available pool), a block will be available of subsequent allocation requests.

However, even if storage is maintained as units of individual nodes rather than as block, it is still possible to provide the user with blocks of contiguous storage. The process of moving all used (marked) nodes to one end of memory and all the available memory to the other end is called *compaction,* and an algorithm that performs such a process is called a *compaction* (or *compacting} algorithm.*

6.6. DEFINITION AND RELATED CONCEPTS

List is a finite set of zero or more elements. The elements of a list may be *atoms* or *lists.* An **atom** is assumed to be distinguishable from a list. Lists are classified into two types – linear and generalized. **Linear list** contains atom elements only, and are sequentially ordered. **Generalized list** could contain both list and atom elements.

6.6.1. Linear List

Linear ordering of atoms is the essential propertyof linear list. The following is the notation for this type of list :

$$L = (i_1, i_2, i_3, ..., i_n)$$

Several **operations** could be done with a linear list. Insertion may be done at any position. Similarly, any element may be deleted from any position. The following are the operations that can be done on linear lists :

- Initializing (setting the list to NULL)
- Determining if a list is empty (checking if L != NULL)
- Finding the length (getting the number of elements)
- Accessing the ith element, 0 i $n - 1$
- Updating the ith element
- Deleting the ith element
- Inserting a new element
- Combining two or more lists into a single list
- Splitting a list into two or more lists
- Duplicating a list
- Erasing a list
- Searching for a value
- Sorting the list.

6.6.2. Generalized List

Generalized list could contain list and atom elements. It has depth and length. Generalized list is also known as **list structure**, or simply **list**. The following is an example:

$L = ((a, b, (c, ())), d, (), (e, (), (f, (g, (a))), d))$

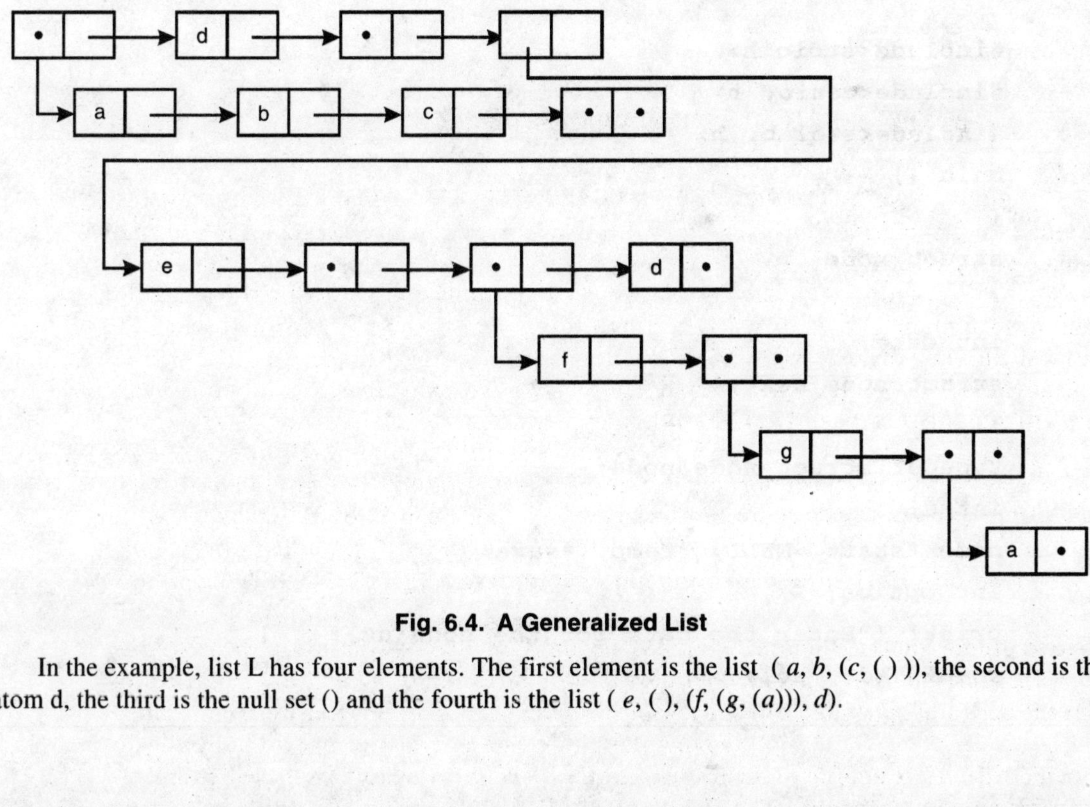

Fig. 6.4. A Generalized List

In the example, list L has four elements. The first element is the list $(a, b, (c, ()))$, the second is the atom d, the third is the null set () and the fourth is the list $(e, (), (f, (g, (a))), d)$.

6.7. SINGLY LINKED LIST

The simplest kind of linked list is a singly-linked list (SLL) whIch has one link per node. This link points to the next node in the list, or to a null value or empty list if it is the final node. The structure of a node in a SLL is given as in C:

```
struct node
{
int data;
struct node*next;
};
```

Here the structure node contains one integer element data and a pointer of type struct node next to point to the next node. If no such node is present the next will have NULL.

The above declared structure is called self-referential structure as in the declaration it contains a pointer to itself.

When we declare array, the array name represents the base address and we can access all its elements if we know its base address. But in the linked list the different nodes are created dynamically and their addresses may not be contiguous that's why we have a struct node*next in the declaration of node. In this next pointer we store the address of next node. For the same reason we must have the starting address of the linked list *i.e.* address of the first node of the linked list. The starting node in a SLL is pointed by a pointer of node type *i.e.* known as start pointer. We discuss more about the SLL but first we see a small program for creation and display of data in a SLL.

Program 6.2. /*program Demo of simple linked list without any user defined function*/

```
#include<stdio.h>
#include<conio. h>
#include<stdlib. h>
main ()
{
struct node
{
int data;
struct node*next;
};
typedef struct node node;
int d;
node *start =NULL, *temp, *save;
int option;
printf ("Enter the data for the node\n");
scanf("%d", &d);
```

```
start=(node*)malloc(sizeof (node));
start->data =d;
start->next= NULL;
save = start;
printf("Do you want to enter more data\n");

printf ("Press 1 for yes, 0 for No\n");
scanf("%d", &option)
while (option = = 1)

{
printf ("Enter the data for the node\n");
scanf("%d", &d);
temp (node*)malloc(sizeof (node));
temp->data = d;
temp->next NULL;
save->next = temp;
save = temp;
printf("Do you want to enter more data\n");
printf ("Press 1 for yes, 0 for No\n");
scanf("%d", &option)

}
printf ("The linked list is\n");
while(start ! = NULL)
{
printf("%d->", start->data);
start =start->next;
}
getch ( );
}
```

OUTPUT

```
Enter the data for the node
   10
Do you want to enter more data
```

```
Press 1 for yes, 0 for No
   1
Enter the data for the node
   20
Do you want to enter more data
Press 1 for yes, 0 for No
   1
Enter the data for the node
30
Do you want to enter more data
Press 1 for yes, 0 for No
1
Enter the data for the node
40
Do you want to enter more data
Press 1 for yes, 0 for No
0
```

The linked list is

```
10->20->30->40->
```

In the program the structure of the node is defined as discussed earlier. We declare a pointer start of type struct node to point to the first node of linked list. We have to be cautious about start as if we loose this address we won't be able to track the linked list.

Initially we allocate memory for the first node and start point to it. We take input from user in d and assign d to data member of node and NULL to the next member field as this is our first node and there is no node following this node. We also take an additional pointer save that points to this node. Every time a new node is added at the end of this node, save will start pointing to this new node so that we can form a linked list by following the next pointer of the node.

DO YOU KNOW

ADI SHAMIR (BORN 1952) was born in Tel Aviv, Israel. His undergraduate degree is from Tel Aviv University (1972) and his Ph.D. is from the Weizmann Institute of Science (1977). Shamir was a research assistant at the University of Warwick and an assistant professor at M.I.T. He is currently a professor in the Applied Mathematics Department at the Weizmann Institute and leads a group studying computer security. Shamir's contributions to cryptography, besides the RSA cryptosystem, include cracking knapsack cryptosystem, cryptanalysis of the Data encryption Standard (DES), and the design of many cryptographic protocols.

Next allocation of node depends upon user response. Next time when user presses 1, this is stored in the option variable. The data for the next node is taken from the user. The new node is pointed by the pointer temp. The statement save->next = temp links last node of the linked list to the new node. Next statement save = temp causes temp becomes save. This process of addition of node continues as user gives her response in the form of 1. The complete process is shown in the Fig. 6.1.

Note in the whole process of creation of linked list, start pointer keep pointing to the first node of the linked list and does not change. When the option is not equal to 1, we assume that user is not interested in adding more nodes to the linked list. Now for displaying the linked list, we start with start pointer within the while loop and displays the data portion of the node, start then points to the next node of the linked list and so on. This continues till start does not become NULL, that denotes end of the linked list.

(*i*) First node is created

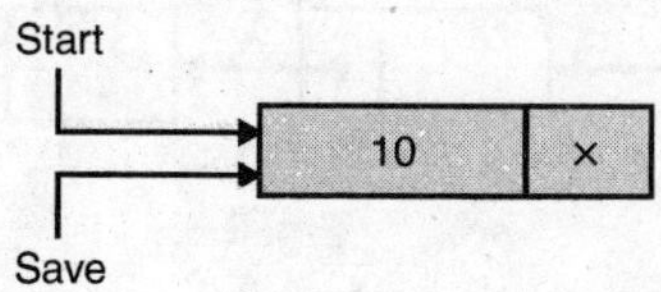

(*ii*) Second node is created at the end of first node

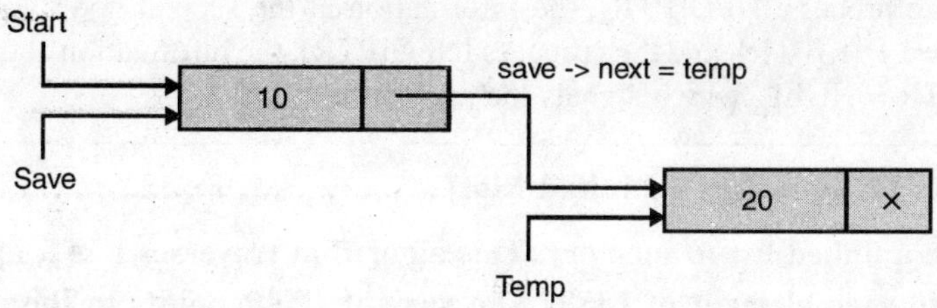

(*a*)

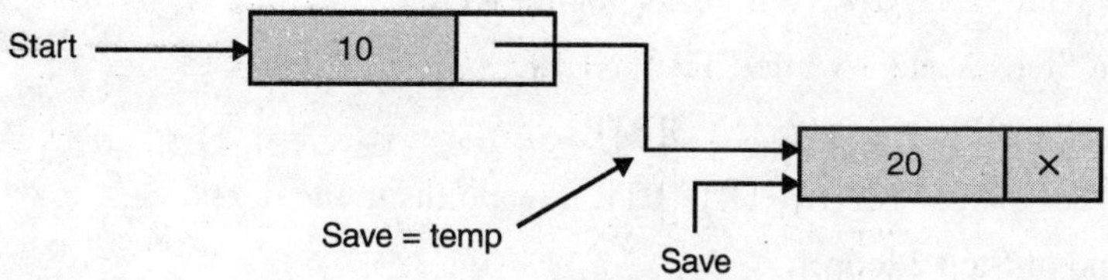

(*b*)

Fig. 6.5. Creation of linked list.

Algorithm for Traversing a Linked List

Let LIST be a linked list in memory stored in linear arrays INFO and LINK with START pointing to the first element and NULL indicating the end of LIST. Suppose we want to traverse LIST in order to process each node exactly once. This section presents an algorithm that does so and then uses the algorithm in some applications.

Out traversing algorithm uses a pointer variable PTR which points to the node that is currently being processed. According, LINK[PTR] points to the next node to be processed. Thus the assignment.

```
PTR:= LINK[PTR],
```

moves the pointer to the next node in the list as shown below in Fig. 6.6.

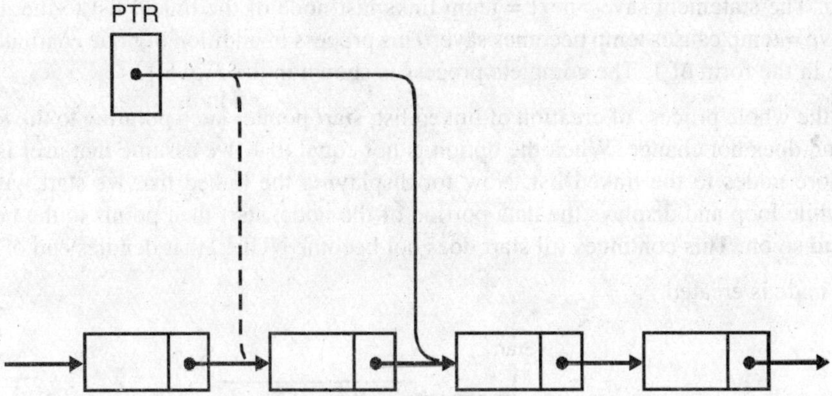

Fig. 6.6.

The details of the algorithm are as follows. Initialize PTR or START. Then process INFO[PTR], the information at the first node. Update PTR by the assignment PTR:= LINK[PTR], so that PTR points to the second node. Then process INFO[PTR], the information at the second node. Again update PTR by the assignment PTR:= LINK[PTR], and then process INFO[PTR], the information at the third node. And so on. Continue until PTR = NULL, which signals the end of the list.

Algorithm 6.1. (Traversing a Linked List) ....

Let LIST be a linked list in memory. This algorithm traverses LIST, applying an operation PROCESS to each element of LIST. The variable PTR points to the node currently being processed.

1. Set PTR := START. [Initializes pointer PTR.]
2. Repeat Steps 3 and 4 while PTR ≠ NULL.
3. Apply PROCESS to INFO[PTR].
4. Set PTR := LINK[PTR]. [PTR now points to the next node.]
   [End of Step 2 loop.]
5. Exit

Caution : As with linear arrays, the operation PROCESS in Algorith 6.1 may use certain variables which must be initialized before PROCESS is applied to any of the elements in LIST. Consequently, the algorithm may be prreceded by such an initialization step.

6.7.1. List Representations

A way to represent a list is to organize the elements one after another in a sequential structure like an array. Another way to implement it is to chain nodes containing the elements of the list using linked representation.

6.7.1.1. Sequential Representation of Singly-Linked Linear List

In sequential representation, the elements are stored *contiguously*. There is a pointer to the last item in the list. The following shows a list using sequential representation:

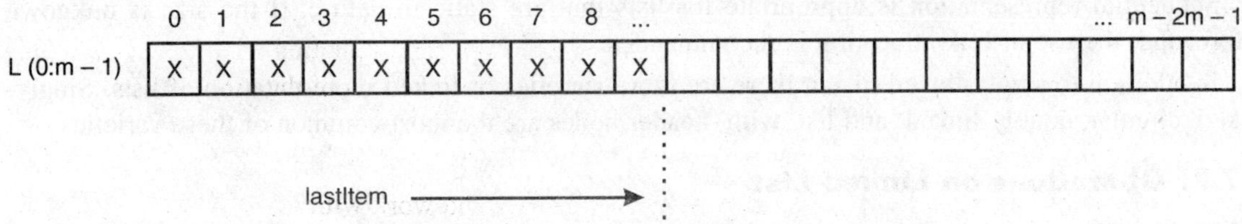

Fig. 6.7. Sequential Representation of List.

With this representation, it would take $O(1)$ time to access and update the ith element in the list. By keeping track of the last item, it would take $O(1)$ time to tell if the list is empty and to find the length of the list. It has a condition, though, that the first element should always be stored at the first index $L(0)$. In that case, insertion and deletion would require shifting of elements to ensure that the list satisfies this condition. In the worst case, this would entail shifting *all* the elements in the array, resulting to time complexity of $O(n)$ for insertion and deletion, for n elements. In combining two lists, a larger array is required if the combined size will not fit in any of the two lists. That would entail copying all the elements of the two lists into the new list. Duplicating a list would require traversing the entire list, hence, a time complexity of $O(n)$. Searching for a particular value would take $O(1)$ time if the element is the first one in the list; however, worst case is when it is the last, where traversal of the entire list would be necessary. In such a case, the time complexity is $O(n)$.

Sequential allocation, being static in nature, is a disadvantage for lists with unpredictable size, *i.e.*, the size is not known at the time of initialization, and with a lot of insertions and deletions, would eventually need to grow or shrink. Copying the overflowed list into a larger array and discarding the old one would do, but it would be a waste of time. In such cases, it would be better to use linked representation.

6.7.1.2. Linked Representation of Singly-Linked Linear List

A chain of linked nodes could be used to represent a list. The following is an illustration :

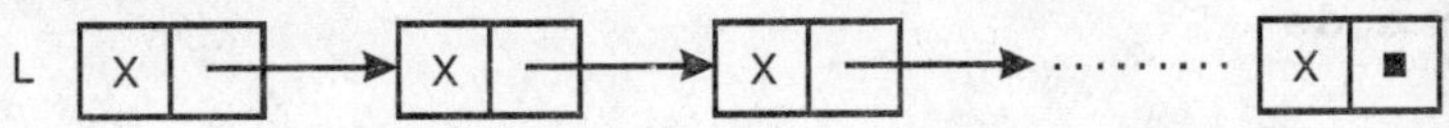

Fig. 6.8. Linked Representation of List.

To access the ith element with linked allocation, the list has to be traversed from the first element up to the ith element. The worst case is when $i = n$, where n is the number of elements. Hence, the time complexity of accessing the ith element is $O(n)$. Similarly, finding the length would entail traversing the entire list, so a time complexity of $O(n)$. If insertion is done at the beginning of the list, it would take $O(1)$ time. However, with deletion and update, search has to be performed, resulting to a time complexity of $O(n)$. To tell if the list is empty, it would take constant time, just like in sequential representation. To copy a list, every node is copied as the original list is traversed.

The following table summarizes the time complexities of the operations on lists with the two types of allocation :

Operation	*Sequential Representation*	*Linked Representation*
Determining if a list is empty	$O(1)$	$O(1)$
Finding the length	$O(1)$	$O(n)$
Accessing the ith element	$O(1)$	$O(n)$
Updating the ith element	$O(1)$	$O(n)$
Deleting the ith element	$O(n)$	$O(n)$
Inserting a new element	$O(n)$	$O(1)$

Sequential representation is appropriate for lists that are static in nature. If the size is unknown beforehand, the use of link allocation is recommended.

In addition to singly-linked linear, there are more varieties of linked representation of lists. Singly-linked circular, doubly-linked and list with header nodes are the most common of these varieties.

6.7.2. Operations on Linked List

At this point you must have understood how to create a SLL in C and work with it. In the following section we discuss various other operations that can be performed over an SLL. These operations are not limited to SLL and can be performed over any other type of linked list. Some of them are:

(*i*) Creation of linked list
(*ii*) Displaying a linked list
(*iii*) Sorting a linked list
(*iv*) Searching an element in linked list
(*v*) Concatenation of two linked list
(*vi*) Merging of two linked list
(*vii*) Splitting a linked list
(*viii*) Adding a node in beginning, at end and in between
(*ix*) Deletion of first, last or any intermediate node
(*x*) Reversing a linked list
(*xi*) Counting number of nodes in linked list

Before doing any of the above operation we simply tell you one more thing that data for a node in linked list can be any things *i.e.* in the above program if you want to create a linked list of strings say names of persons then you can declare the structure as :

```
struct node
{
char name[15];
struct node*next;
};
```

Similarly you can also have a structure book, item, employee, person with member fields as desired by you as data field of the node. Try to rewrite the previous program with these data fields.

6.7.3. Adding a Node at The End and Display

To understand adding the node at the beginning of a linked list and after displaying the elements consider the code given below. We are giving you this program only as it is the first program with user defined functions. In the subsequent discussion of various operations we will give you only the functions for the said operation.

Program 6.3. /* Program creating Linked list adding node at the end and displaying using functions */ ....

```
#include <stdio. h>
#include <conio. h>
```

```
#include  <stdlib. h>
struct node
{
int data;
struct node * next;
};
trpedef struct node node;
void append (node **, int);
void display (node *);
node*alloc()
{
return (node*)malloc(sizeof(node));
}
void main()
{
node *start;
start =NULL;
append (&start, 1);
append (&start, 3);
append (&start, 5);
printf ("Linked list is \n");
display (start);
append (&start, 6);
append (&start, 11);
printf ("After adding more, Linked list is \n");
display(start);
}
void append (node **s, int num)
{
node *temp, *ptr;
if (*S = = NULL)
{
temp =alloc( );
temp - > data = num;
temp  - >next = NULL;
*S=temp;
}
else
{
```

```
temp=*S;
while (temp - > next != NULL)
temp = temp -> next;
ptr = alloc();
ptr - > data = num;
ptr -> next = NULL;
temp - > next = ptr;
}
}
void display (node *S)
{
while (S!= NULL)
{
printf ("%d->", S -> data);
S = S -> next;
}
printf ("\n");
}
```

OUTPUT

```
Linked list is
   1- >3-> 5->
after adding more, Linked list is
   1- > 3-> 5-> 6-> 11->
```

In the program we have two functions append and display. The prototype of append is:

```
void append (node**, int);
```

The above declaration tells the compiler that append function will take first argument as address of the pointer of type node and second argument as integer.

The first argument will always be &start, as start is our main pointer to the linked list. The reason for having address of the pointer instead of address of a structure node variable is that passing address of start let us do the changes on the actual pointer and that changes are consistent with the original pointer *i.e.* actually reflected back outside the function.

At places where we do not pass the address of start and simply pass the start pointer as in the display function, it means we are not interested in modifying the contents of the linked list but simply want to access the elements without doing any changes in the linked list.

In the main start is NULL, and for the first time we call function append as

```
append(&start, 1)
```

Address of start is passed to the pointer variable S and inside the function append, *S is equal to start. Now whatever changes we do using *S that will actually be occurring onto the original linked list.

Initially as start is NULL, if condition will be satisfied and first node will be created pointed by *S (start in turn), For the allocation of memory to every new node, we have written the function alloc that allocates memory for the new node and return a pointer to the newly created node.

When append is called second time with data 3, if condition fails and controls goes to else part. The logic for adding the node at the end of the linked list is traverse the entire linked list using a temporary pointer and reach up to the end of the linked list. When end of the linked list is met then add the node at the end. This is done as follows:

The temp pointer points to the beginning of the linked list. Inside the while we check while next of temp is not NULL, and inside the body we move pointer temp to point to the next of the temp. As we have added just one node in our linked list, condition will be false. Next memory for the new node is allocated pointed by pointer ptr. The data and next fields are filled for this node. This new node added at the end of the linked list by writing the statement

```
temp->next = ptr;
```

For all the remaining calls to append, this process is repeated. For displaying the linked we simply pass the start pointer. The coding is similar to one as given in the earlier program. The complete process of adding the node is shown in the Fig. 6.4.

(*i*) append is called first time with data 1

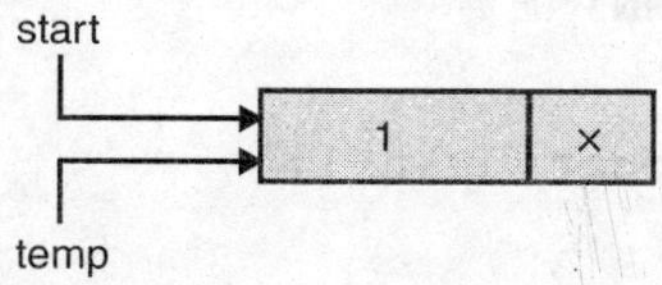

(*ii*) append is called second time with data 3

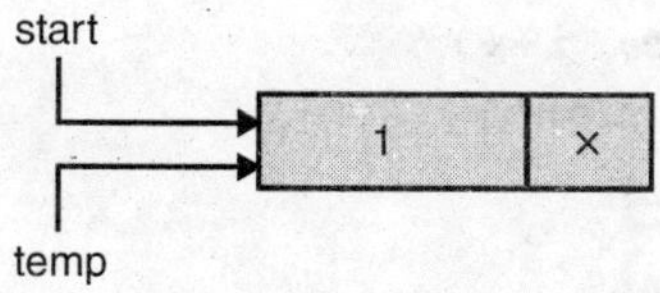

(as while loop fails we have the following fig.)

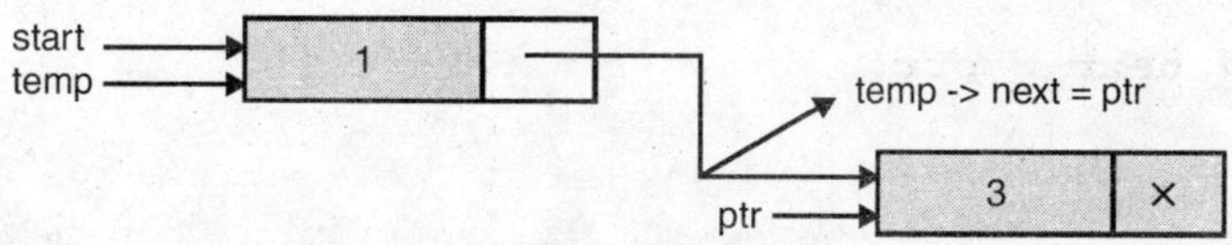

(*iii*) append is called with data 5

(In the else part, when while loop fails, temp points to last no.)

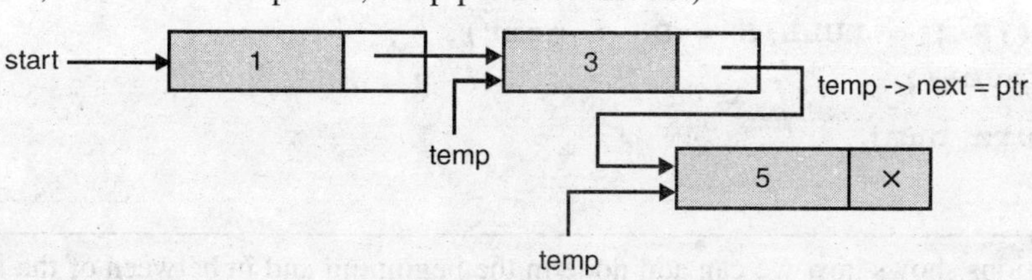

Fig. 6.9. Creation of linked list with append function.

6.7.4. Insertion, Addition at Beginning and Counting Number of Nodes

```
void addatbeg (node **S, int num)
{
node *temp;
temp = alloc();
temp -> data = num;
temp -> next =*S;
*S = temp;
}
void insert (node *S, int loc, int num)
{
node *temp, *ptr;
int i, NN;
temp = S;
NN = count(S);
if (loc<1 || loc>NN)
{
printf("Invalid location\n");
exit (0) ;
}
for (i = 1; i <loc; i++)
    temp = temp -> next;
ptr = alloc();
ptr - > data = num;
ptr - > next = temp -> next;
temp -> next = ptr;
}

int count (node * S)
{
int num = 0;
for(;S ! = NULL;S = S -> next)
    num++;
return num;
}
```

The functions shows how we can add node in the beginning and in between of the linked list. It also has one function count that returns number of nodes present in the linked list.

For, adding node at the beginning of the linked list, the logic is simple.

(*i*) Allocate a new node, say **temp (temp = alloc( ))**

(*ii*) Let next of temp point to start node **(temp->next = *S)**

(*iii*) Make start point to **temp. (*S = temp)**

This is as shown in the figure given below:

(*i*) addatbeg called first time with data 5

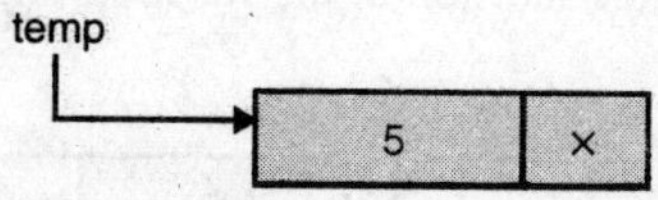

(As first time start is NULL so temp->next will be equal to temp-> next = NULL)

Next *S = temp causes the following figure :

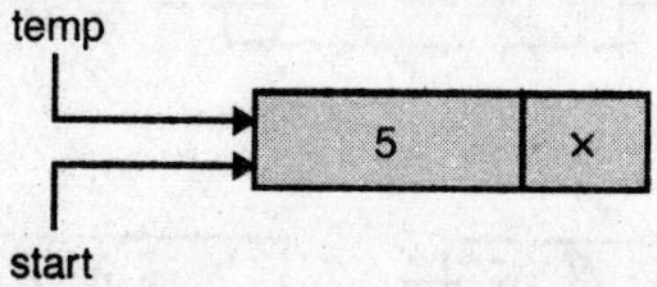

(*ii*) addatbeg called with data 3

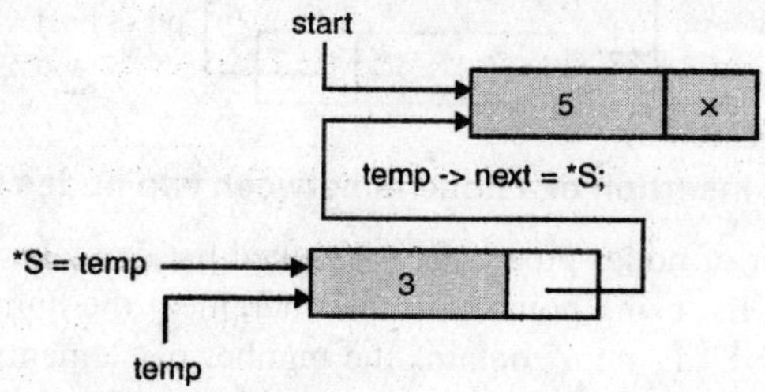

(After *S = temp, start point to newly added node)

(*iii*) addatbeg called with 1

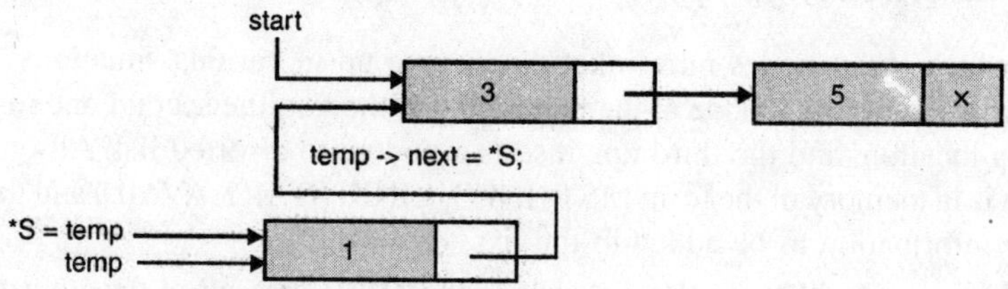

(After *S = temp; start point to newly created node)

Fig. 6.10. Adding node at the beginning of linked list.

For inserting the new node in between any nodes in the linked, we have written insert function. The function takes three arguments: first is pointer to the linked list, second is the location of the node after which new node is to be inserted and third is data for the new node. Using a temporary pointer temp that points to start of the linked list and a for loop we reach up to the node after which new node is to be inserted. Now we allocate the memory for the new node using alloc. This node is pointed by pointer ptr.

Insertion of the new node after the temp node is done follows :

(*i*) next of ptr points to next of temp (ptr->next = temp->next)

(*ii*) next of temp points to ptr (temp->next= ptr)

(*i*) Initially linked list

(*ii*) Insert called with location 1 and data 2

(As i = 1 and loc is also 1, so condition in the for loop fails, temp points to the first node.)

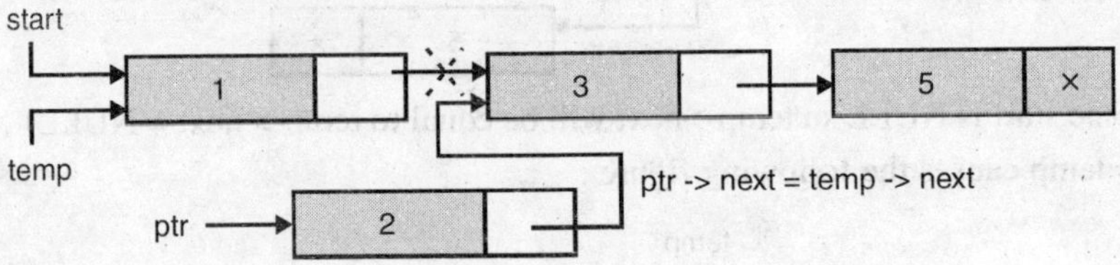

(Figure after temp-> = ptr ; is show below)

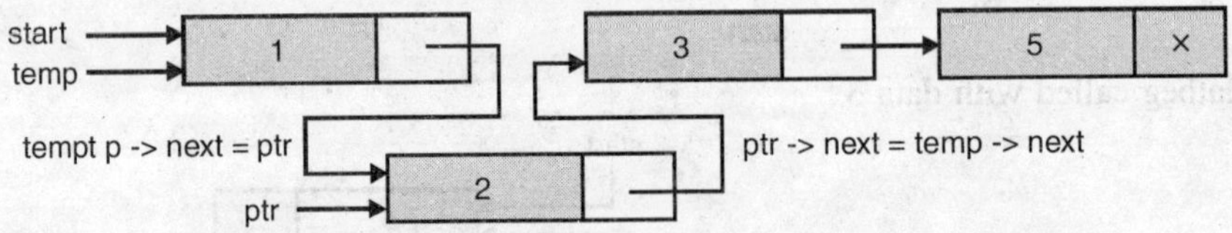

Fig. 6.11. Insertion of a node in between two nodes in linked list

For counting the number of nodes present in the linked list, we take a variable num, initialized to 0. Then we traverse the linked list using pointer ptr and increment the num. This continues till S does not become NULL. When S is NULL, num contains the number of elements in the linked list. The same is returned back from num.

Algorithm For Insertion Algorithms

Algorithms which insert nodes into linked lists come up in various situations. We discuss three of them here. The first one inserts a node at the beginning of the list, the second one inserts a node after the node with a given location, and the third one inserts a node into a sorted list. All our algorithms assume that the linked list is in memory in the form LIST (INFO, LINK, START, AVAIL) and that the variable ITEM contains the new information to be added to the list.

Since our insertion algorithms will use a node in the AVAIL list, all of the algorithms will include the following steps :

(*a*) Checking to see if space is available in the AVAIL list. If not, that is, if AVAIL = NULL, then the algorithm will print the message OVERFLOW.

(*b*) Removing the first node from the AVAIL list. Using the variable NEW to keep track of the location of the new node, this step can be implemented by the pair of assignments (in this order)

```
NEW := AVAIL,    AVAIL := LINK[AVAIL]
```

(*c*) Copying new information into the new node. In other words,

```
INFO[NEW] := ITEM
```

The schematic diagram of the latter two steps is shown in Fig. 6.12.

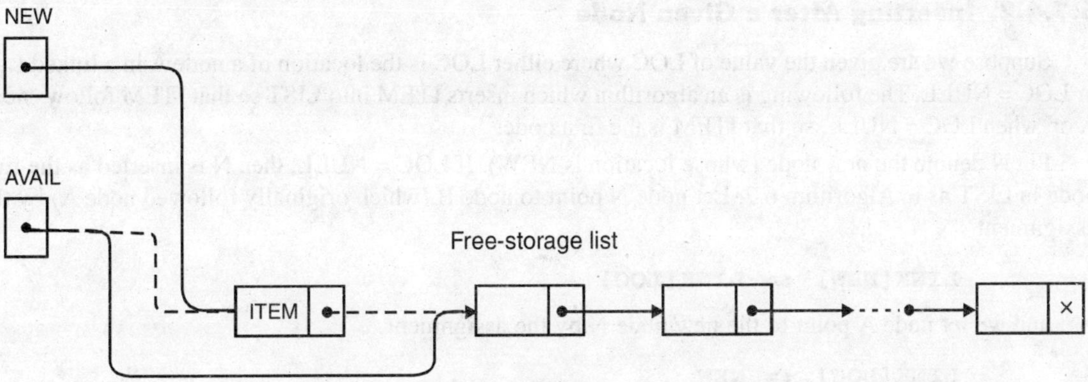

Fig. 6.12.

6.7.4.1. Inserting at the Beginning of a List

Suppose our linked list is not necessarily sorted and there is no reason to insert a new node in any special place in the list. Then the easiest place to insert the node is at the beginning of the list. An algorithm that does so follows :

Algorithm 6.2. INSFIRST(INFO, LINK, START, AVAIL, ITEM) ....

This algorithm inserts ITEM as the first node in the list.

1. [OVERFLOW?] If AVAIL = NULL, then: Write: OVERFLOW, and Exit
2. [Remove first node from AVAIL list.]
   Set NEW:= AVAIL and AVAIL:= LINK[AVAIL].
3. Set INFO[NEW]:= ITEM. [Copies new data into new node.]
4. Set LINK[NEW] := START. [New node now points to original first node.]
5. Set START := NEW. [Changes START so it points to the new node.]
6. Exit

Steps 1 to 3 have already been discussed, and the schematic diagram of Steps 2 and 3 appears in Fig. 6.12. The schematic diagram of Steps 4 and 5 appears in Fig. 6.13.

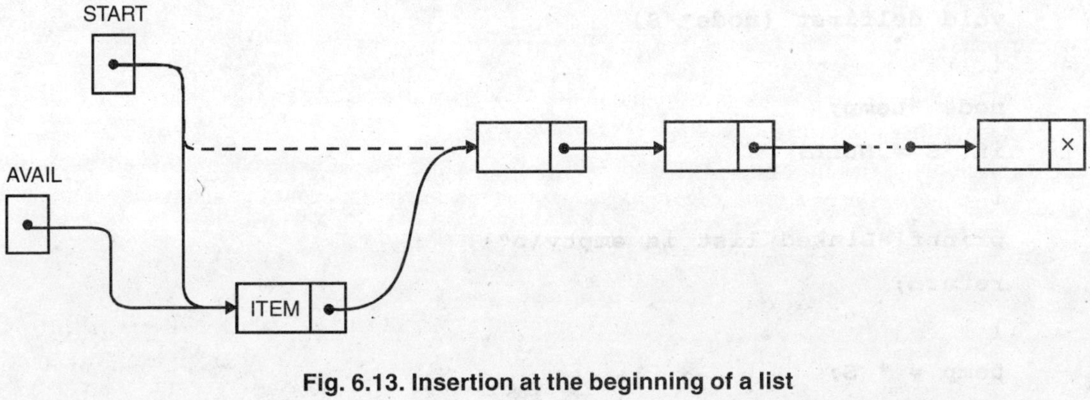

Fig. 6.13. Insertion at the beginning of a list

6.7.4.2. Inserting After a Given Node

Suppose we are given the value of LOC where either LOC is the location of a node A in a linked LISt or LOC = NULL. The following is an algorithm which inserts ITEM into LIST so that ITEM follows node A or, when LOC = NULL, so that ITEM is the first node.

Let N denote the new node (whose location is NEW). If LOC = NULL, then N is inserted as the first node in LIST as in Algorithm 6.2. Let node N point to node B (which originally followed node A) by the assignment

```
LINK[NEW]  := LINK[LOC]
```

and we let node A point to the new node N by the assignment

```
LINK[LOC]  := NEW
```

A formal statement of the algorithm follows.

Algorithm 6.3. INSLOC(INFO, LINK, START, AVAIL, LOC, ITEM)..............................

This algorithm inserts ITEM so that ITEM follows the node with location LOC or inserts ITEM as the first node when LOC = NULL.

1. [OVERFLOW?] If AVAIL = NULL, then: Write: OVERFLOW, and Exit
2. [Remove first node from AVAIL list.]

   Set NEW := AVAIL and AVAIL := LINK [AVAIL]
3. Set INFO[NEW] := ITEM. [Copies new data into new node.]
4. Set LINK[NEW] := START and START := NEW.

   Else: [Insert after node with location LOC.]

   Set LINK[NEW] := LINK[LOC] and LINK[LOC] := NEW.

   [End of If structure.]
5. Exit

6.7.5. Deletion of a Node In Linked List

```
void delfirst (node**S)
{
node *temp;
if(*S = NULL)
{
printf("Linked list is empty\n");
return;
}
temp = * S;
```

```
* S = (*S)->next;
free (temp);
}
void del(node*S, int d)
{
node*ptr, *save;
ptr = S;
while(ptr->data!=d)
{
save = ptr;
ptr = ptr->next;
if(ptr = = NULL)
{
printf("node with specified data does not exist\n");
return;
}

}
save->next = ptr->next;
free (ptr);
}
void dellast (node**S)

{
node *temp = *s;
node *save;
while(temp->next ! = NULL)
{
save = temp;
temp = temp->next;
}
free (temp);
save->next = NULL;
}
```

The function shows how we can perform deletion of a node in a linked. For deletion there may be three different cases:

(*i*) Deletion of first node

(*ii*) Deletion of last node

(*iii*) Deletion of any intermediate node

For each of the above case, a separate function is written. During deletion process the node is removed from the linked list and memory allocated to it will be de-allocated. Deletion of first node of a linked list is very simple. All we have to do is take a temporary pointer temp that point to the starting node. Let start pointer to its next node and free the temp. That's it! This is shown below:

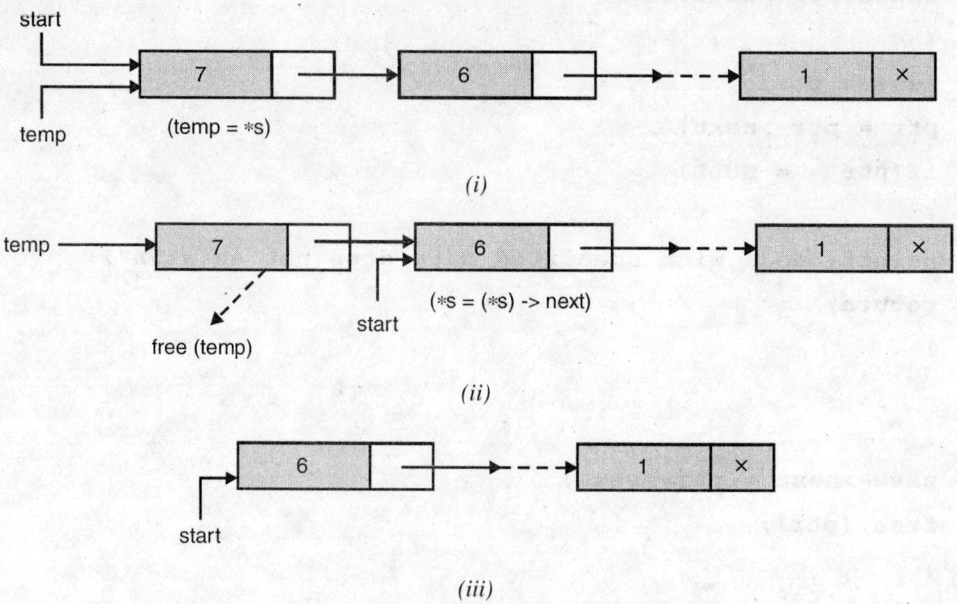

Fig. 6.14. Deletion of first node.

For deletion of last node, we have to reach up to the last node. For that we have to traverse the entire linked list by taking a temporary pointer and while loop. During this process we also keep addresses of the previous node in another pointer called save. The loop runs till next of temp is not equal to NULL. After the loop we simply free the temp and let next of save point to NULL.

(*i*) Initial

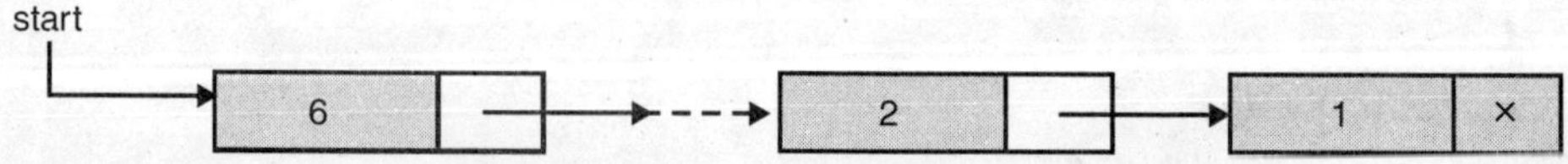

(*ii*) After while loop ends

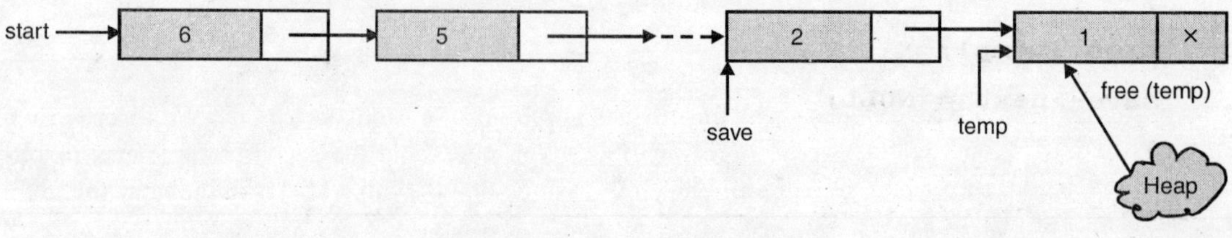

(*iii*) After statements free (temp); save -> next = NULL

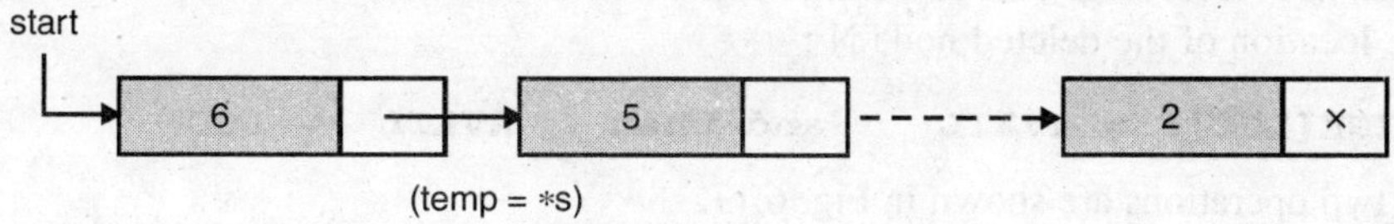

Fig. 6.15. Deletion of last node.

For deletion of any intermediate node, there are two ways: either we delete the node at specified location or delete the node with specified data. We have shown the second approach in the program. To achieve this we asked the user, data for the node to be deleted. We take a temporary pointer ptr and traverse the whole linked list. During traversing we keep the addresses of the previous node in a pointer called save. We also compare data of the node pointed by ptr is equal to the data of the node we want to delete. If this matches we come out from loop. In case this does not matches, it means data does not reside in any of the node and pointer ptr becomes NULL. In this case we return from the function by displaying appropriate message.

If the match has indeed occurred, it means ptr points to the node to be deleted and save points to the node previous to the node pointed by ptr. The node is deleted from the linked list by following statement:

```
save->next = ptr->next;
```

The memory of the node is de-allocated by the statement free(ptr). This memory is brought back to heap.

(*i*) Initial linked list

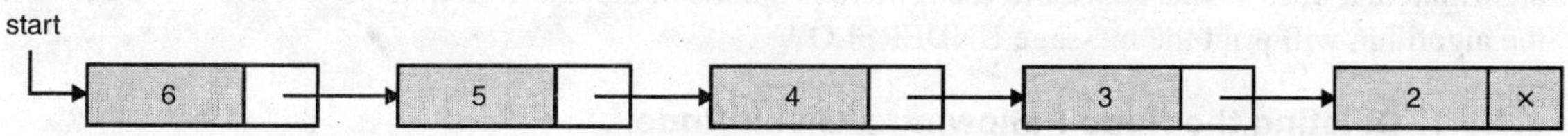

(*ii*) function del called with d = 3 (when the specified node is found, then ptr points to it and save points to node previous to prt.

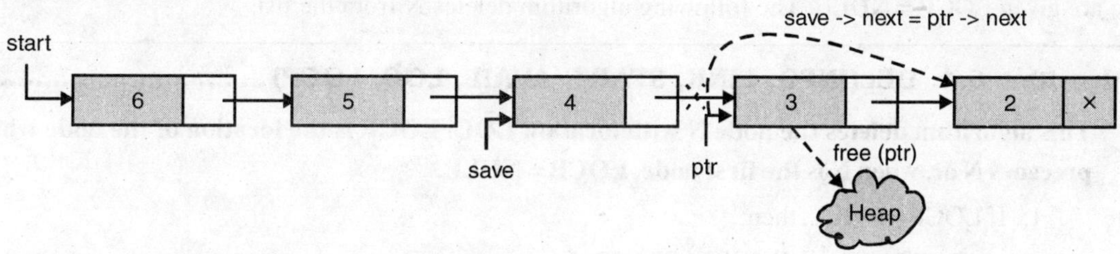

(*iii*) After the statement save -> next = ptr -> next

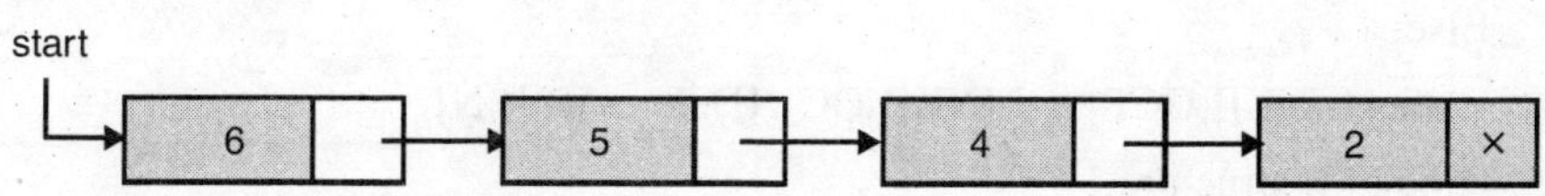

Fig. 6.16. Deletion of intermediate node.

Algorithm for Deletion Algorithms

Algorithms which delete nodes from linked lists come up in various situations. We discuss two of them here. The first one deletes the node following a given node, and the second one deletes the node with a given ITEM of information. All our algorithms assume that the linked list is in memory in the form LIST (INFO, LINK, START, AVAIL).

All of our deletion algorithms will return the memory space of the deleted node N to the beginning of the AVAIL list. Accordingly, all of our algorithms will include the following pair of assignments, where LOC is the location of the deleted node N :

```
LINK[LOC]: = AVAIL    and then    AVAIL := LOC
```

These two operations are shown in Fig. 6.17.

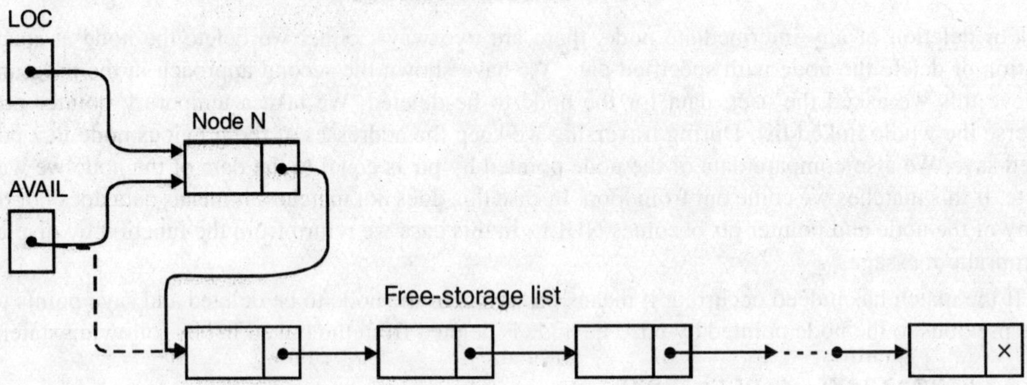

Fig. 6.17. LINK[LOC] := AVAIL and AVAIL := LOC.

Some of our algorithms may want to delete either the first node or the last node from the list. An algorithm that does so must check to see if there is a node in the list. If not, *i.e.*, if START = NULL, then the algorithm will print the message UNDERFLOW.

6.7.5.1. Deleting the Node Following a Given Node

Let LIST be a linked list in memory. Suppose we are given the location LOC of a node N in LIST. Furthermore, suppose we are given the location LOCP of the node preceding N or, when N is the first node, we are given LOCP = NULL. The following algorithm deletes N from the list.

Algorithm 6.4. DEL(INFO, LINK, START, AVAIL, LOC, LOCP) ....

This algorithm deletes the node N with location LOC. LOCP is the location of the node which precedes N or, when N is the first node, LOCP = NULL.

1. If LOCP = NULL, then:

   Set START := LINK[START]. [Deletes first node]

   Else:

   Set LINK[LOCP] := LINK[LOC]. [Deletes node N]

   [End of If structure.]
2. [Return deleted node to the AVAIL list.]

   Set LINK[LOC] := AVAIL and AVAIL := LOC.
3. Exit

Figure 6.18 is the schematic diagram of the assignment

```
START := LINK[START]
```

which effectively deletes the first node from the list. This covers the case when N is the first node.

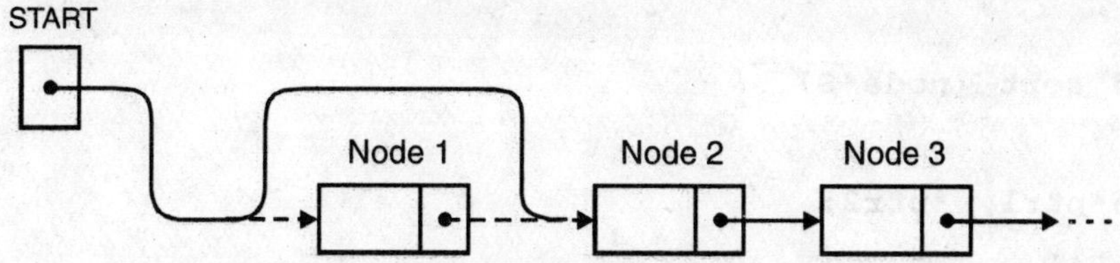

Fig. 6.18. START := LINK[START].

Figure 6.13 is the schematic diagram of the assignment

```
LINK[LOCP] := LINK[LOC]
```

which effectively deletes the node N when N is not the first node.

The simplicity of the algorithm comes from the fact that we are already given the location LOCP of the node which precedes node N. In many applications, we must first find LOCP.

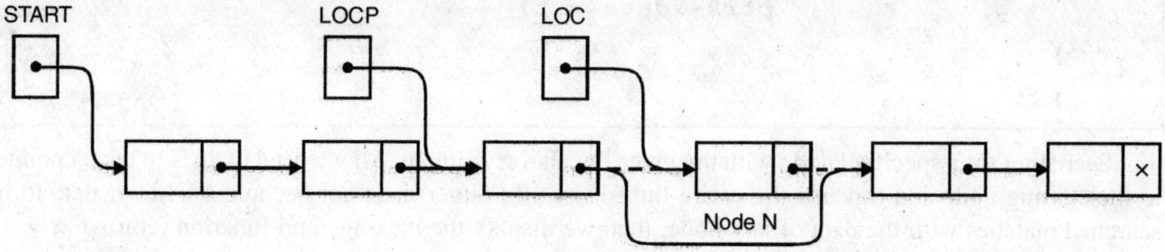

Fig. 6.19. LINK[LOCP] := LINK[LOC].

6.7.6. Searching and Sorting in a Linked List

```
void search(node*S, int d)
{
node*ptr= S;
int pos = 1;
while(ptr!= NULL)
{
if(ptr->data =  = d)
{
printf('Element exist in node no. %d\n", pos);
return;
}
ptr = ptr->next;
pos++;
```

```
}
printf ("Element does not exist in linked list\n");
}
void sort (node*S)
{
node*ptrl, *ptr2;
int t;
for(ptrl = S;ptrl;ptrl ptrl->next)
        for (ptr2 = ptrl->next;ptr2;ptr2 = ptr2->next)
                if(ptr->data   >  ptr2->data)
                {
                        t = ptrl->data;
                        ptrl->data = ptr2->data;
                ptr2->data = t;
}
}
```

Searching for a specified node with the given data is very simple. All we need to do is to take a pointer to the starting node and traverse the entire linked list till pointer does not become NULL. If data to be searched matches with the data of any node, then we display the message and function returns.

Sorting of linked list is similar to the sorting of array as given in the chapter "Arrays". Instead of i we have ptrl and instead of j we have ptr2. The pointer ptrl is initialized to pointer S (*i.e.* start pointer) and ptr2 is initialized to ptrl->next. For the increment of ptrl and ptr2, we have statements ptrl = ptrl->next and ptr2 = ptr2->next. Rest is simple to understand.

Algorithm for Searching in a Linked List

List LIST be a linked list in memory, Suppose a specific ITEM of information is given. this section discusses two searching algorithms for finding the location LOC of the node where ITEM first appears in LIST. The first algorithm does not assume that the data in LIST are sorted, whereas the second algorithm does assume that LIST is sorted.

If ITEM is actually a key value and we are searching through a file for the record containing ITEM, then ITEM can appear only once in LIST.

6.7.6.1. LIST is Unsorted

Suppose the data in LIST are not necessarily sorted. Then one searches for ITEM in LIST by travering through the list using a pointer variable PTR and comparing ITEM with the contents INFO[PTR] of each node, one by one, of LIST. Before we update the pointer PTR by

PTR := LINK[PTR]

we require two tests. First we have to check to see whether we have reached the end of the list; *i.e.*, first we check to see whether

```
PTR = NULL
```

If not, then we check to see whether

```
INFO[PTR] = ITEM
```

The two tests cannot be performed at the same time, since INFO[PTR] is not defined when PTR = NULL. Accordingly, we use the first test to control the execution of a loop, and we let the second test take place inside the lopo. The algorithm follows.

Algorithm 6.5. SEARCH(INFO, LINK, START, ITEM, LOC) ....

LIST is a linked list in memory. This algorithm finds the location LOC of the node where ITEM first appears in LIST, or sets LOC = NULL.

1. set PTR := START
2. repeat Step 3 while PTR ≠ NULL:
3. if ITEM = INFO[PTR], then:
   set LOC:= PTR, and Exit.
   else:
   set PTR := LINK[PTR]. [PTR now points to the next node.]
   [end of If structure.]
   [End of step 2 loop.]
4. [Search is unsuccessful.] Set LOC := NULL.
5. Exit.

The complexity of this algorithm is the same as that of the linear search algorithm for linear arrays. That is, the worst-case running time is proportional to the number n of elements in LIST, and the average-case running time is approximately proportional to *n*/2 (with the condition that ITEM appears once in LIST but with equal probability in any node of LIST).

6.7.6.2. LIST is Sorted

Suppose the data in LIST are sorted. Again we search for ITEM in LIST by traversing the list using a pointer variable PTR and comparing ITEM with the contents INFO[PTR] of each node, one by one, of LIST. Now, however, we can stop once ITEM exceeds INFO[PTR].

The complexity of this algorithm is still the same as that of other linear search algorithms; that is, the worst-case running time is proportional to the number *n* of elements in LIST, and the average-case running time is approximately proportional to *n*/2.

Algorithm 6.6. SRCHSL (INFO, LINK, START, ITEM, LOC)..

LIST is a sorted list in memory. This algorithm finds the location LOC of the node where ITEM first appears in LIST, or sets LOC = NULL.

1. Set PTR := START
2. Repeat Step 3 while PTR ≠ NULL:
3. If ITEM < INFO[PTR], then:
   Set PTR:= LINK[PTR]. [PTR now points to next node.]
   else if ITEM = INFO[PTR], then:
   Set LOC := PTR, and Exit. [Search is successful.]

Else:

Set LOC := NULL, and Exit. [ITEM now exceeds INFO[PTR].]

[End of If structure.]

]End of Step 2 loop.]

4. Set LOC:= NULL.
5. Exit

Recall that with a sorted linear array we can apply a binary search whose running time a proportional to $\log_2 n$. On the other hand, a *binary search algorithm cannot be applied to a sorted linked list, since there is no way of indexing the middle element in the list.* This property is one of the main drawbacks in using a linked list as a data structure.

6.7.7. Reversing Linked List

```
void reverse (node*S)

{
node*ptr1, *ptr2;
int t;
for (ptr1= S;ptr1;ptr1 = ptr1->next)
        for(ptr2 = ptr1->next;ptr2;ptr2 ptr2->next)
        {
            t = ptr1 -> data;
              ptr1->data = ptr2->data;
            ptr2->data = t;
        }
}
```

Logic for reversing a linked list is similar to reversing an array. Instead of i and j we have ptrl and ptr2. For the increment of i we have ptrl = ptrl->next and for the increment of j we have ptr2 = ptr2->next. The main point to note here is that data othe different nodes are swapped and nodes itself are not swapped *i.e.* positions of the node have not been changed. In the next version of the same operation we use another concept for reversing the linked list.

6.7.8. Copying One Liked List To Another

```
void copy(node*L1, node**L2)
{
node * save;
if(L1 = = NULL)
{
printf ("Linked list is empty, cannot copy\n");
return;
}
(*L2) = alloc ( );
(*L2)->data = L1 ->data;
(*L2)->next = NULL;
save = *L2;
```

```
    L1 = L1->next;
    while (L1 ! = NULL)
    {
    save ->next = alloc ( );
    save =save->next;
    save->data = L1->data;
    save->next = NULL;
    L1 = L1->next;
    }
    }
```

For copying one linked list to another we have written function copy that takes two arguments. First argument denotes the source linked list and is of node* type, second argument denotes the destination linked list that is of node** type. The first node of the source linked list is copied outside the while loop. A pointer save points to it that will be used inside the while loop. Inside the while loop, the next of save points to newly allocated node, this is caused by the following statement:

```
    save->next = alloc();
```

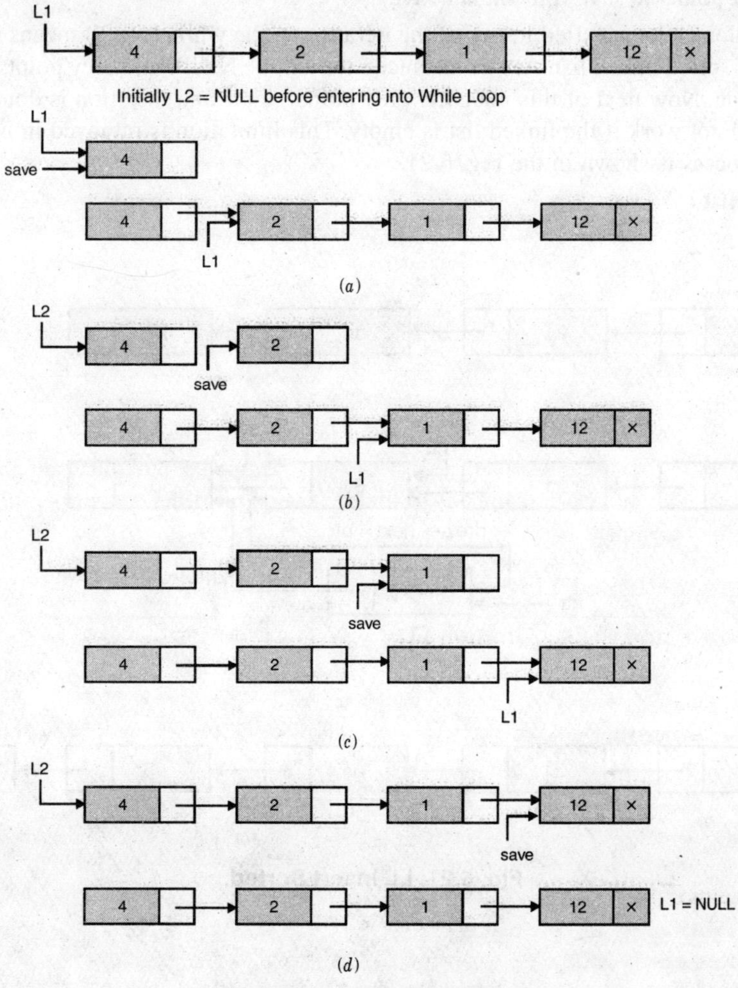

Fig. 6.20. Linked list Copy

Next statements save = save->next causes save to point to this newly allocated node. In the next statement we simply copy the data of the current node of the source linked list and let next of save be equal to NULL. The source linked list points to the next node to be copied in the next iteration of the while loop. This continues till source linked list does not become NULL.

6.7.9. Inserting Element into Sorted (Ascending Order) Linked List

For the insertion into sorted linked list we have written the insert function. Note prior using this function we assume that -we have a linked list in which elements are added in ascending order. The insert function is called with the address of start pointer and data, to be inserted. A new node is allocated pointed by ptr, data and next fields are filled. We take a temporary pointer save that points to beginning of the linked list, for traversal purpose. Now if the node is to be added before the start node, if condition checks that, If the condition turns out to be true, node is added and function returns. Otherwise the whole linked list is traversed using the pointer save and while loop. For inserting the node in between the linked list, we have written if condition. For the first iteration of the while loop if condition will be false, An additional pointer prev saves the save pointer while save points to its next node. If the if condition is satisfied, the usual process of insertion take place, *i.e.*

(*i*) next of prev points to ptr (prev->next = ptr)

(*ii*) next of ptr points to save (ptr->next =save)

If the if condition is not satisfied in any of the iteration of the while loop, it means new node is to be added at the end of sorted linked list. At this instance save will be NULL and prev points to previous node of save *i.e.* last node. Now next of this node *i.e.* prev points to ptr and insertion is done. Note one thing insert function will not work if the linked list is empty. This limitation is removed in next program. The whole insertion process is shown in the Fig. 6.21.

Initial linked list :

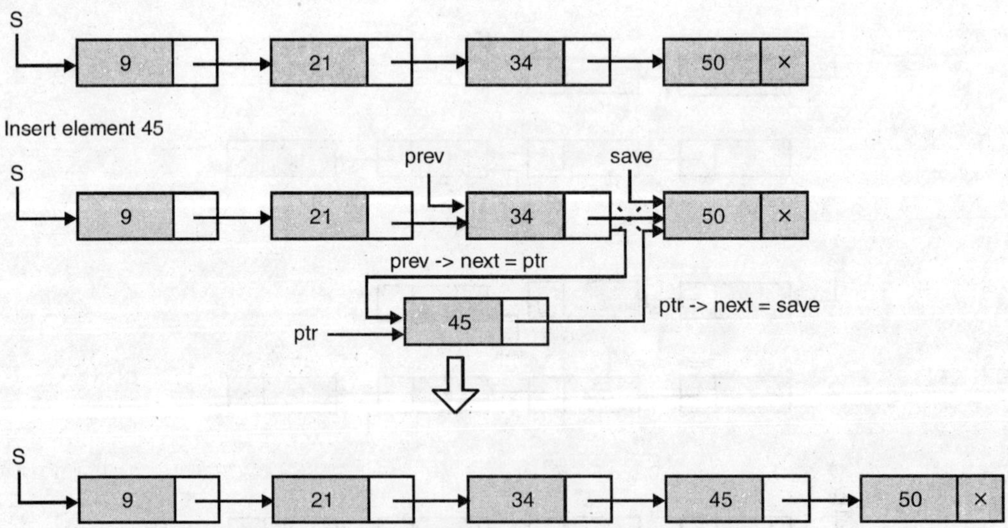

Fig. 6.21. LL Insert Sorted.

Algorithm For Inserting into a Sorted Linked List

Suppose ITEM is to be inserted into a sorted linked LIST. Then ITEM must be inserted between nodes A and B so that

INFO(A) < ITEM ≤ INFO(B)

The following is a procedure which find the location LOC of node A, that is, which finds the location LOC of the last node in LIST whose vlaue is less than ITEM.

Traverse the list, using a pointer variable PTR and comparing ITEM with INFO[PTR] at each node. While traversing, keep track of the location of the preceding node by using a pointer variable SAVE, as shown in Fig. 6.17. Thus SAVE and PTR are updated by the assignments

SAVE := PTR and PTR := LINK[PTR]

The traversing continues as long as INFO[PTR] > ITEM, or in other words, the traversing stops as soon as ITEM £ INFO [PTR]. Then PTR points to node B, so SAVE will contain the location of the node A.

The formal statement of our procedure follows. The cases where the list is empty or where ITEM < INFO[START], so LOC = NULL, are treated separately, since they do not involve the variable SAVE.

Algorithm 6.7. FINDA(INFO, LINK, START, ITEM, LOC)

This procedure finds the location LOC of the last node in a sorted list such that INFO[LOC] < ITEM, or sets LOC = NULL.

1. [List empty?] If START = NULL, then: Set LOC: = NULL, and Return
2. [Special case?] If ITEM < INFO[START], then: Set LOC: = NULL, and Return.
3. Set SAVE := START and PTR := LINK[START]. [Initializes pointers.]
4. Repeat Steps 5 and 6 whlile PTR ≠ NULL
5. If ITEM < INFO[PTR], then :
   Set LOC := SAVE, and Return
   [End of If structure.]
6. Set SAVE := PTR and PTR := LINK[PTR]. [Updates pointers.]
   [End of Step 4 loop.]
7. Set LOC := SAVE.
8. Return.

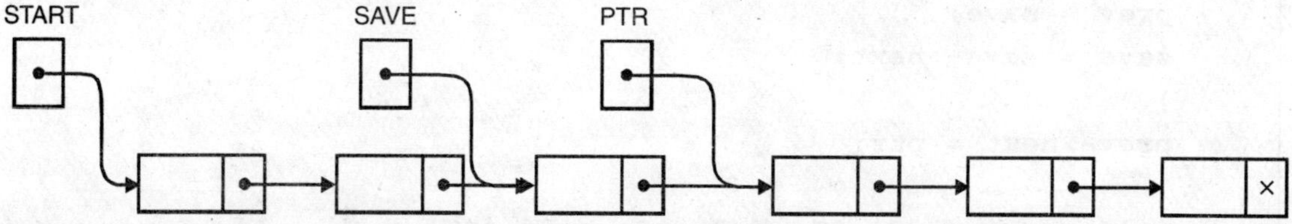

Fig. 6.22.

Now we have all the components to present an algorithm which inserts ITEM into a linked list. The simplicity of the algorithm comes from using the previous two procedures.

Algorithm 6.8. INSSRT (INFO, LINK, START, AVAIL, ITEM) ....

This algorithm inserts ITEM into a sorted linked list.

1. [Use Procedure 6.7 to find the location of the node preceding ITEM.]
   Call FINDA(INFO, LINK, START, ITEM, LOC).
2. [Use Algorithm 6.3 to insert ITEM after thenode with location LOC.]
   Call INSLOC (INFO, LINK, START, AVAIl, LOC, ITEM).
3. Exit

6.7.10. Merging of Two Linked List

```
void insert(node**S, int d)
{
node*ptr, *prev;
node*save =*S;
ptr =alloc( );
 ptr->data = d;
ptr->next = NULL;
if((*S)->data    >    d | | (*S) = = NULL)
{
*S = ptr;
(*S)->next = save;
return;
}
while(save! = NULL)
{
if(save->data    >    d)
{
prev->next = ptr;
ptr->next = save;
return;
}
prev = save;
save = save->next;
}
prev->next = ptr;
}
void merge (node *F, node *S, node **T)
{
node *TL = NULL;
if(F=NULL && S = = NULL)
return;
```

```
while(F! = NULL && S! = NULL)
{
if (*T = = NULL)
{
*T = alloc ( );
TL = *T;
}
else
{
TL->next = alloc ( );
TL = TL -> next;
}
if (F->data    <    S->data).
{
TL->data = F->data;
F = F->next;
}
else if (S->data   <   F->data)
{
TL->data = S->data;
S = S->next;
}
else
{
TL->data = S->data;
F = F->next;
S = S->next;
}
}
while (F != NULL)
{
TL->next = alloc
TL = TL->next;
TL->data =F->data;
F = F->next;
}
while(S != NULL)
{
TL->next = alloc();
TL = TL->next;
TL->data =S->data;
```

```
    S = S->next;
    }
    TL->next = NULL;
    }
```

Merging is an operation in which we combine elements of two arrays, linked list etc into third structure in sorted order. For that we require that the two source linked list should already be sorted. For this purpose we have written just one insert function for the addition of nodes to the linked regardless of position of the new node, be it in the beginning, end or in between.

For the merging of two linked list we have written merge function. The first two argument to this function represent the two source linked list that are to be merged and third argument is the third linked list (initially empty) for holding the final merged linked list.

In the merge function, if both the source linked lists are empty we return. The while loop runs till one of the source linked list does not become NULL. If *TL is empty *i.e.* first node to be added to destination linked list, we allocate the memory and *TL points to it, else next of *TL points to newly allocated node and finally *TL points to this new node.

Next using else-if ladder we check data of which source linked list is smaller. If F->data is less then S->data then F->data is assigned to data of new node and F points to its next node else if S->data is smaller then F->data then data of S is assigned to data of new node and S points to its next node. In case F->data and S-data is equal, data of any node is assigned to new node and both F and S points to their next node.

When either S or F becomes NULL, control comes out from the while loop. At moment we have no idea whether S is NULL or F is NULL. For this reason we run two while loop, one for each F and S. The first while loop assigns all pending nodes from the first linked to the target linked list till F becomes NULL. Second while does the same for second linked list S. Either of the two while loop wil run only when number of elements in both the linked lists is not same.

The whole operation of merging is illustrated in the Fig. 6.18.

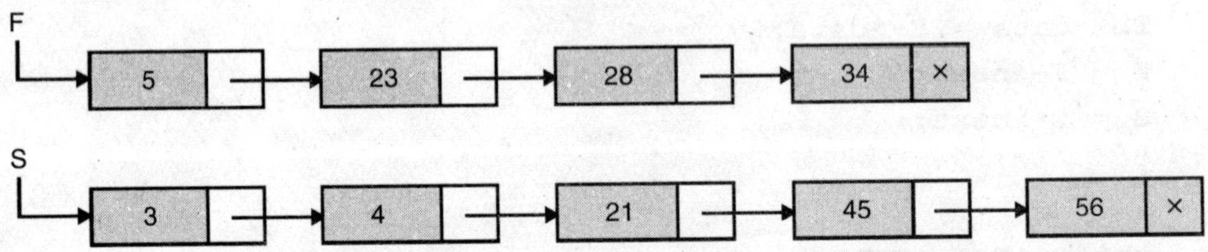

Initial source linked list

(*i*) F -> data > S -> data (from initial Fig.)

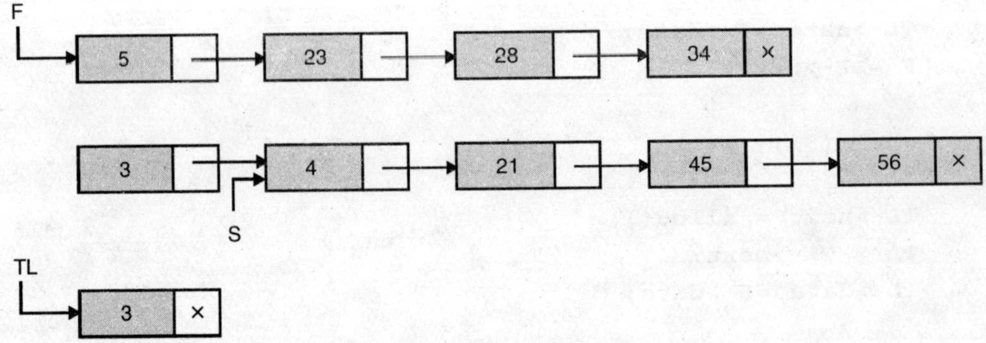

(*ii*) F -> data > S -> data (from previous figure)

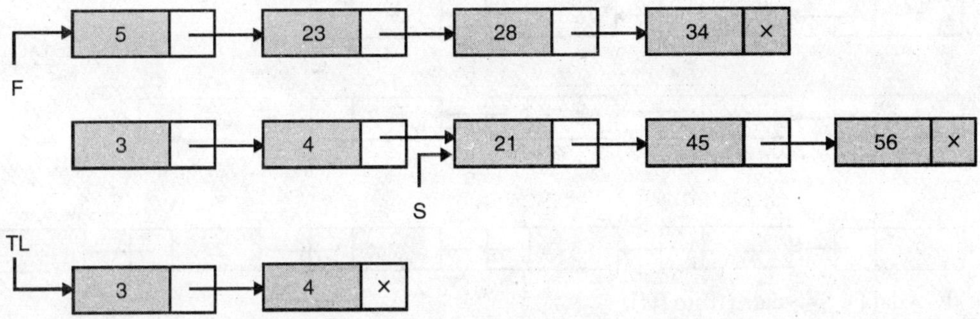

(*iii*) F -> data < S -> data (from (*ii*))

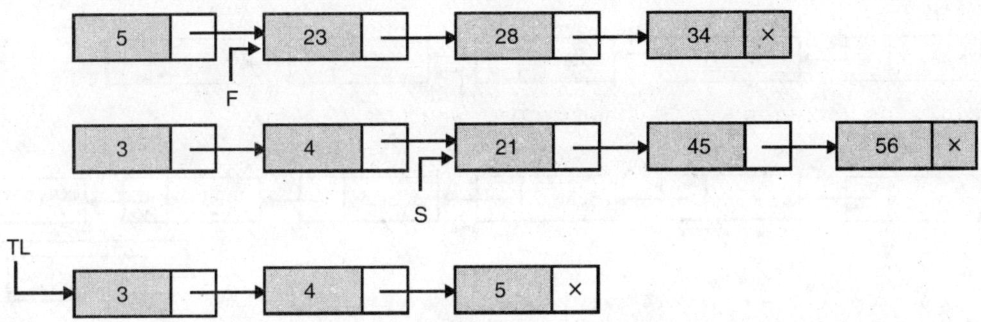

(*iv*) F -> data > S -> data (from (*iii*))

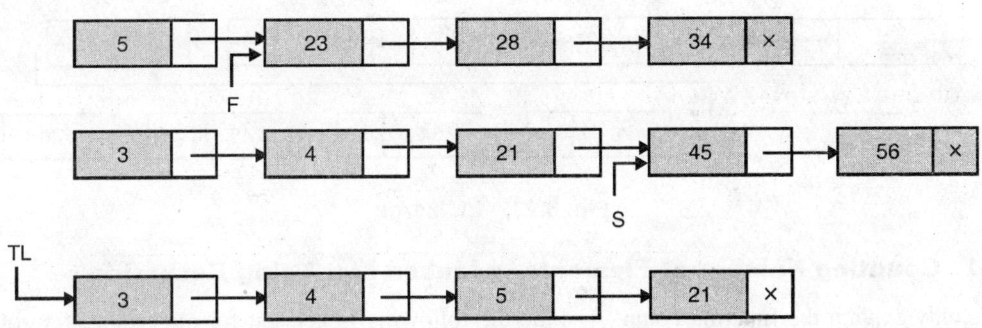

(*v*) F -> data < S -> data (from (*iv*))

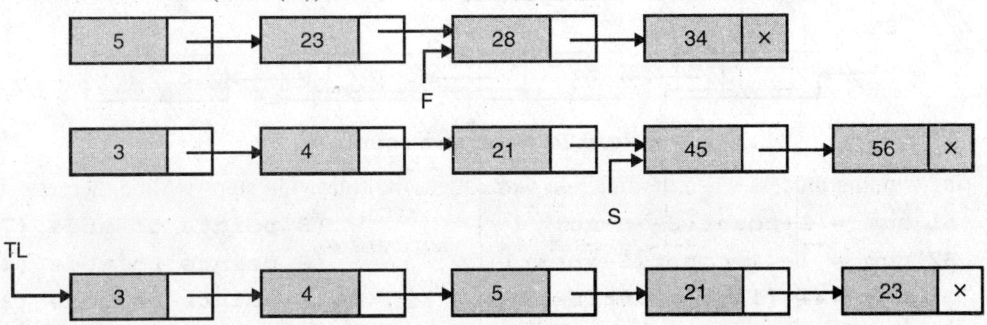

(*vi*) F -> data < S -> data (from (*v*))

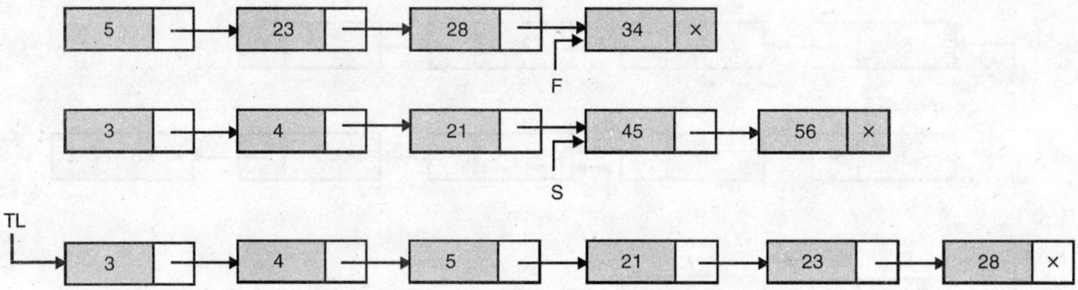

(*vii*) F -> data < S -> data (from (*vi*))

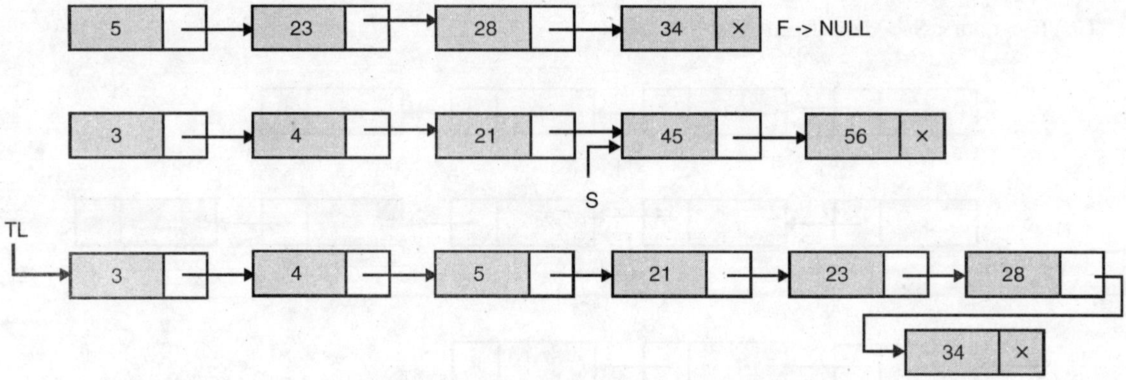

(*viii*) F is NULL so all remaining nodes from second linked are assigned to linked list pointed by TL.

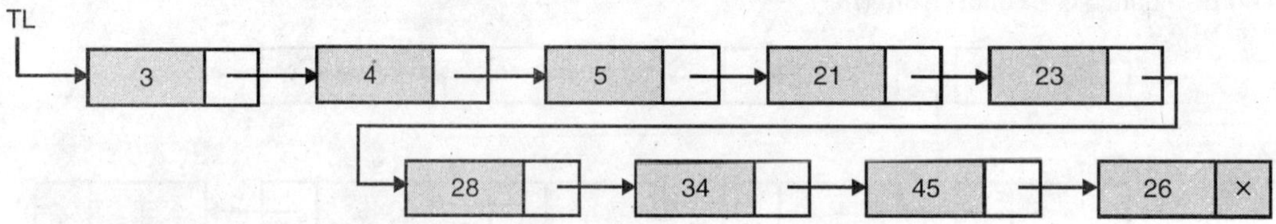

Fig. 6.23. LL Merge

6.7.11. Counting Number of Elements in Linked List Using Recursion

We only explain the function count. Assume the following linked list for our explanation of count function.

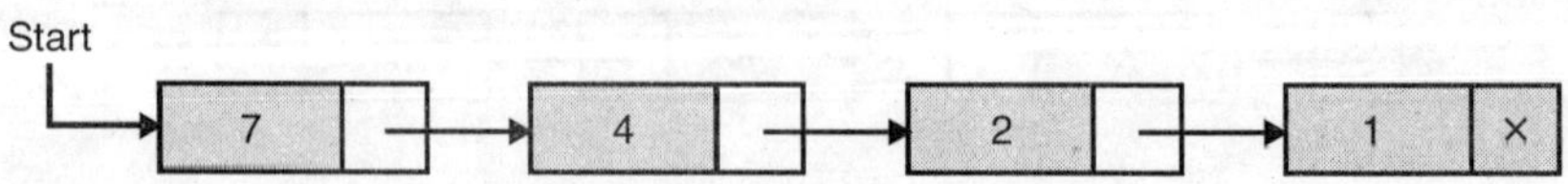

Fig. 6.24. Recursive count.

During winding process when the function was called, the following steps will be there:

```
S1 num = 1+count(S->next)                  (S points to node (7))
S2 num = 1+(1+count(S->next))              (S points to node (4))
S3 num = 1+(1+(1+count(S->next)))          (S points to node (2))
S4 num = 1+(1+(1+(1+count(S->next))))      (S points to node (1))
```

Now S->next is NULL *i.e.* S is now NULL so function returns and unwinding takes place.

Expression in the innermost parenthesis in S4 will become 1+0 *i.e.* 1 this, 1 will return to S3, then expression will become 1+1 *i.e.* 2, 2 will return to S2 then expression will become 1+2 *i.e.* 3, 3 will return to S1, then expression will become 1+3 *i.e.* 4. At this step unwinding will stop and final value of num will be 4, This num will be returned from the function.

6.8. DOUBLY LINKED LIST

A doubly linked list (DLL) is similar to singly linked list that we have seen earlier in all the programs done earlier. A DLL has an additional pointer that point to the node previous to it. Thus structure of a DLL is given as:

```
struct dnode
{
int data;
struct dnode*prev;
struct dnode*next;
};
```

Here the data field and next pointer field is same as belong to singly linked list and prev pointer field stores the address of the previous node. Apart from one start pointer that points the starting node of the DLL, a DLL has one more pointer by the name last that points to the last node of the linked list. Following figure shows DLL :

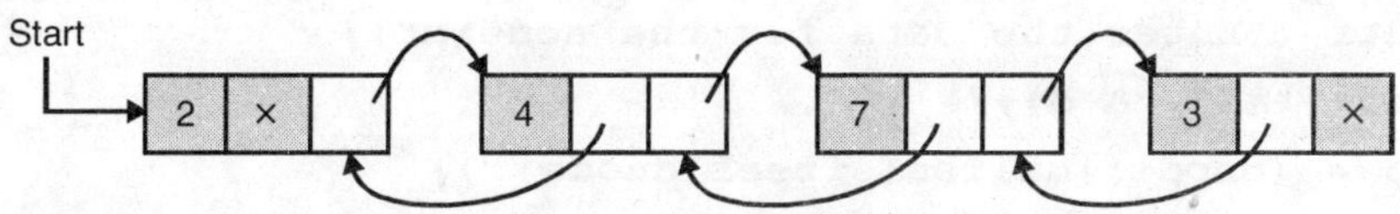

Fig. 6.25. A Doubly linked list.

The program given below demonstrates DLL creation and display without using any user defined function.

Program 6.4. /* Program Demo of doubly linked list without any user defined function*/

```
#include<stdio. h>
#include<conio. h>
#include<stdlib. h>
void main()
{
struct dnode
{
int data;
struct dnode*prev;
struct dnode*next;
};
```

```
typedef struct dnode dnode;
int d;
dnode *start, *last, *temp;
int option;
start = last =NULL;
clrscr( );
printf ("Enter the data for the node\n");
scanf("%d", &d);
start = (dnode*)malloc (sizeof (dnode));
start->data =d;
start->next = NULL;
start->prev = NULL;
last = start;
printf ("Do you want to enter more data\n");
printf ("Press 1 for yes, 0 for No\n");
scanf("%d", &option);
while (option = =1)
{
printf ("Enter the data for the node\n");
scanf ("%d", & d);/
temp = (dnode*)malloc(sizeof(dnode) );
temp->data =d;
temp->next = NULL;
last->next =temp;
temp->prev = last;
last = temp;
 printf("Do you want to enter more data\n");
printf ("Press 1 for yes, 0 for No\n");
scanf("%d", &option);
}
printf("\nLinked list using next pointer\n");
while(start !=NULL)
{
printf("%d->", start->data);
start  = start->next;
}
printf("\nLinked list using prev pointer\n");
while(last ! = NULL)
{
```

```
printf("%d->", last->data);
last = last->prev;
}
getch ( );
}
```

OUTPUT

```
Enter the data for the node
1
Do you want to enter more data
Press 1 for yes, 0 for No
1
Enter the data for the node
2
Do you want to enter more data
Press 1 for yes, 0 for No
1
Enter the data for the node
3
Do you want to enter more data
Press 1 for yes, 0 for No
1
Enter the data for the node
4
Do you want to enter more data
Press 1 for yes, 0 for No
0
Linked List using next pointer
1-> 2-> 3-> 4->
Linked list using prev pointer
4-> 3- >2 ->1 ->
```

Using the start pointer and next field the DLL can be traversed in forward direction and using the last pointer and prev field the DLL can be traversed in backward direction. For the start pointer prev field is NULL and for the last pointer next field will be NULL.

The first node is created outside the while loop and start and last points to as it is the only node present in the DLL. All successive nodes will be added at the end of the DLL. The creation of next node depends upon user response. If the option is 1, new node is created and temp points to it. Addition of this new node the end of DLL is very simple:

(*i*) next of last points to temp (last->next = temp)

(*ii*) prev of temp points to last (temp->prev =last)

(*iii*) next of temp is NULL (temp->next= NULL)

(*iv*) temp becomes last (last = temp)

The whole creation process is shown in the Fig. 6.26.

(*i*) data 1

(*ii*) data 2

(*iii*) data 3

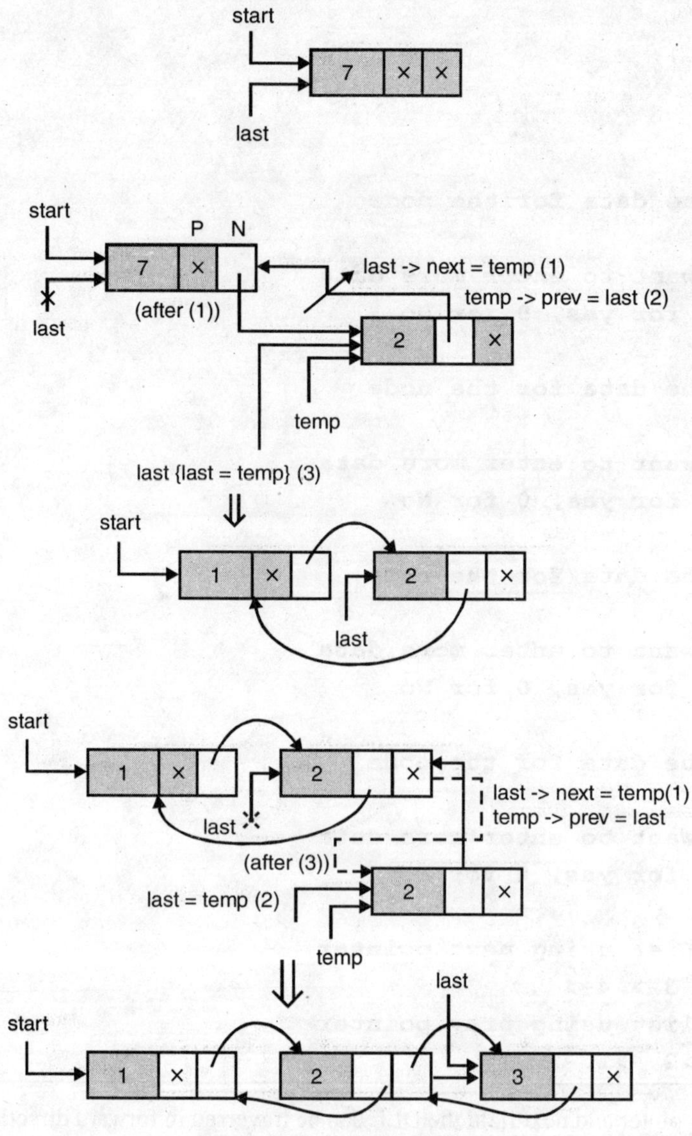

Fig. 6.26. DLL insertion.

For displaying the DLL, we have traversed it using both start pointer and next field and last pointer and prev field. This is simple to understand.

6.8.1. Introduction

Doubly-linked lists are formed from nodes that have pointers to both left and right neighbors in the list. The following figure shows the node structure for a doubly-linked list.

Fig. 6.27. Doubly-Linked List Node Structure.

The following is an example of a doubly-linked list :

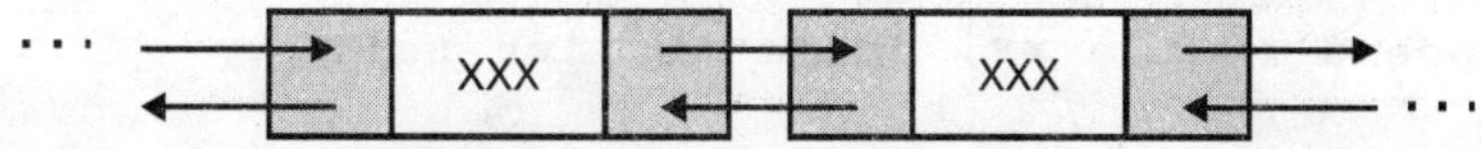

Fig. 6.28. Doubly-Linked List.

With doubly-linked list, each node has two pointer fields - LLINK and RLINK that points to left and right neighbors respectively. The list can be traversed in either direction. A node **i** can be deleted knowing only the information about node **i**. Also, a node **j** could be inserted either before or after a node **i** by knowing the information about **i**. The following figure illustrates deletion of node **i**:

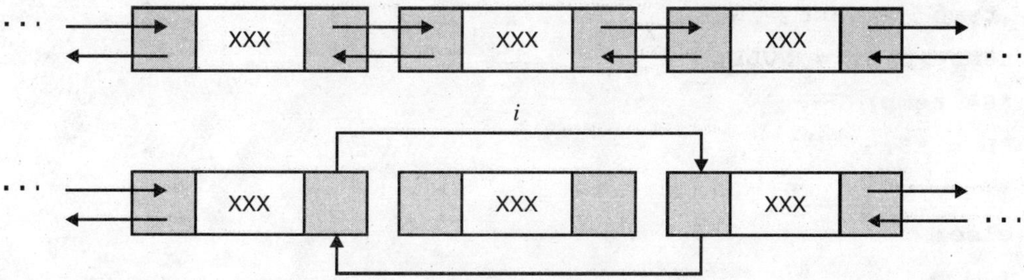

Fig. 6.29. Deletion from a Doubly-Linked List.

A doubly-linked list could be constituted in the following ways :

- Doubly-linked linear list
- Doubly-linked circular list
- Doubly-linked circular list with list head

6.8.1.1. Properties of Doubly-Linked List

LLINK(L) = RLINK(L) = L means list L is empty. We can delete any node, say node, in L in O(1) time, knowing only the address. We can insert a new node, say node , to the left (or right) of any node, say node, in O(1) time, by knowing only since it only needs pointer reassignments, as illustrated below :

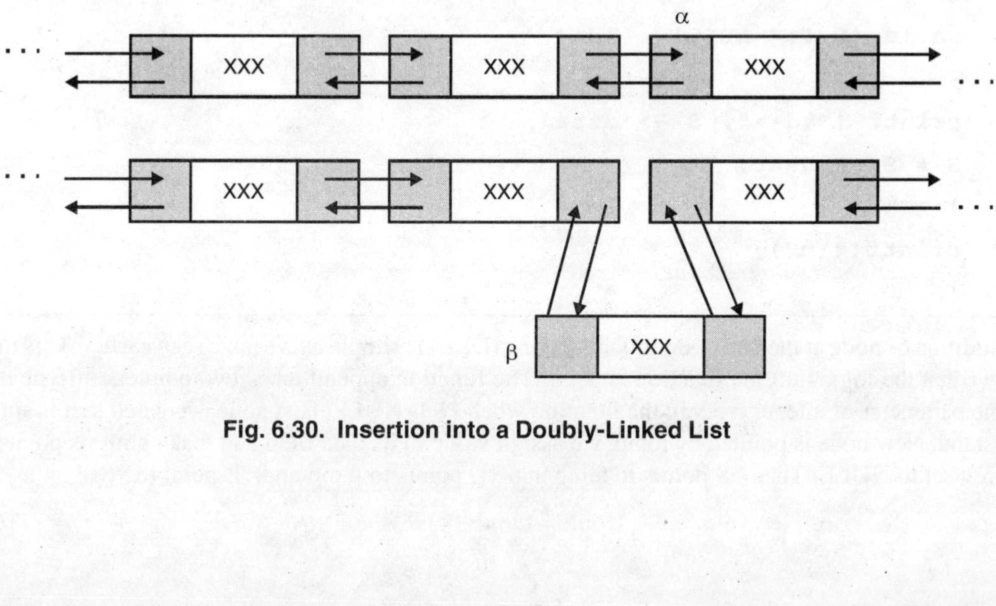

Fig. 6.30. Insertion into a Doubly-Linked List

6.8.2. Function Append and Display

```
void append (dnode **, dnode**L, int num)
{
dnode *temp;
if (*S = =  NULL)
{
temp= alloc ( );
temp -> data =num;
 temp -> next = NULL;
temp->prev = NULL;
*S= temp;
*L = *S;
}
else
{
temp = alloc ( );
temp -> data =num;
temp -> next =NULL;
temp -> prev = *L;
 (*L)->next =temp;
*L= temp;
}
}
void display (dnode *S)
{
while (S ! = NULL)
{
printf ("%d->", S -> data);
S = S -> next;
}
printf("\n");
}
```

Addition of node at the end of doubly linked list (DLL) is simple as we have seen earlier. This time we have written the logic into the function append. The function append takes two pointers of type node** and one parameter of integer type. In the function when *S is NULL, first node is created that is simple to understand. New node is pointed by temp, we assign value to its data field and make both its pointer next and prev set to NULL. Then *S points to temp and *L points to temp and *L point to *S.

In the successive call to function append else part executes. In the else part temp is added at the end of the DLL as :

(*i*) prev of temp points to *L (temp->prev = *L)

(*ii*) next of *L points to temp (*L->next = temp)

(*iii*) *L points to temp (*L= temp)

Follow the figure for insertion in DLL.

6.8.3. Adding New Node at the Beginning

```
void addatbeg (dnode **S, dnode**L, int num)
{
dnode *temp;
temp = alloc( );
temp -> data= num;
if (*S = = NULL && *L = = NULL)
{
temp -> next =NULL;
temp -> prev = NULL;
*S   = temp;
*L = *S;
}
else
{
temp = alloc ( );
temp -> data = num;
temp -> next =*S;
temp->prev = NULL;
(*S) -> prev = temp;
*S = temp;
}
}
```

Algorithm for Inserting in Doubly LL

Suppose we are given the locations LOCA and LOCB of adjacent nodes A and B in LIST, and suppose we want to insert a given ITEM of information between nodes A and B. As with a one-way list, first we remove the first node N from the AVAIl list, using the variable NEW to keep track of its location, and then we copy the data ITEM into the node N; that is, we set :

NEW := AVAIL, AVAIL := FORW[AVAIL], INFO[NEW] := ITEM

Now, as shown in Fig. 6.22, the node N with contents ITEM is inserted into the list by changing the following four pointers :

FORW[LOCA] :=NEW, FORW[NEW] := LOCB

BACK[LOCB] := NEW, BACK[NEW] := LOCA

The formal statement of our algorithm follows :

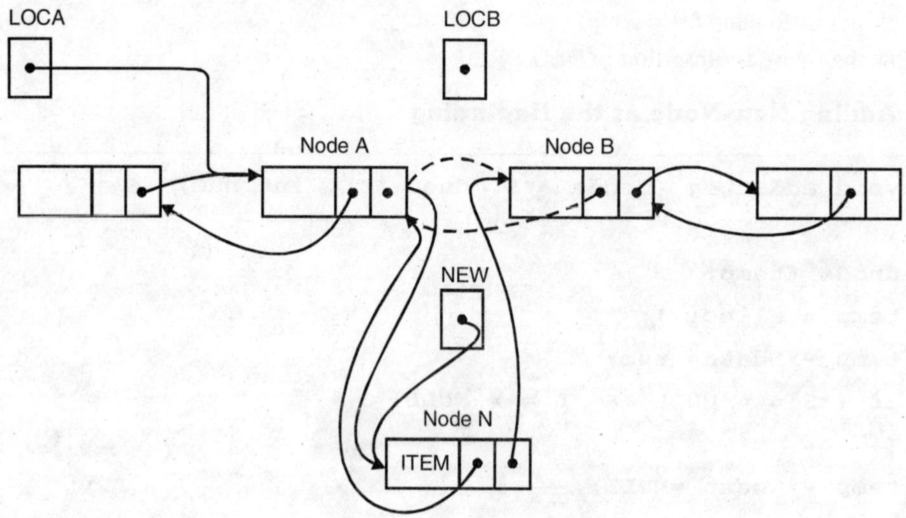

Fig. 6.31. Inserting node N.

Algorithm 6.9. INSTWL(INFO, FORW, BACK, START, AVAIL, LOCA, LOCB, ITEM)

1. [OVERFLOW?] [If AVAIL = NULL, then: Write: OVERFLOW, and Exit.
2. [Remove node from AVAIl list and copy new data into node.]
   Set NEW := AVAIL, AVAIL := FORW[AVAIL], INFO[NEW] := ITEM.
3. [Insert node into list.]
   Set FORW[LOCA] := NEW, FORW[NEW] := LOCB,
   BACK[LOCB] := NEW< BACK [NEW]: = LOCA
4. Exit

Algorithm 6.10 assumes that LIST contains a header node. Hence LOCA or LOCB may point to the header node, in which case N will be inserted as the first node or the last node. If LIST does not contain a header node, then we must consider the case that LOCA = NULL and N is inserted as the first node in the list, and the case that LOCB = NULL and N is inserted as the last node in the list.

6.8.4. Insertion and Count in Doubly Linked List

```
void insert (dnode **S, int loc, int num)
{
dnode *temp *ptr;
int i, NN;
temp = *S;
NN = count(*S);
```

```
if (loc<1 | | loc> = NN)
{
printf ("Invalid location\n");
exit (0) ;
}
for (i = 1; i <loc; i++)
temp =temp -> next;
ptr = alloc ( );
ptr -> data = num;
ptr -> next = temp -> next;
temp->next->prev =ptr;
ptr->prev =temp;
temp -> next =ptr;
}
int count (dnode * S)
{
int num = 0;
for(;S = NULL;S = S -> next)
        num++;
return num;
}
```

Initial DLL

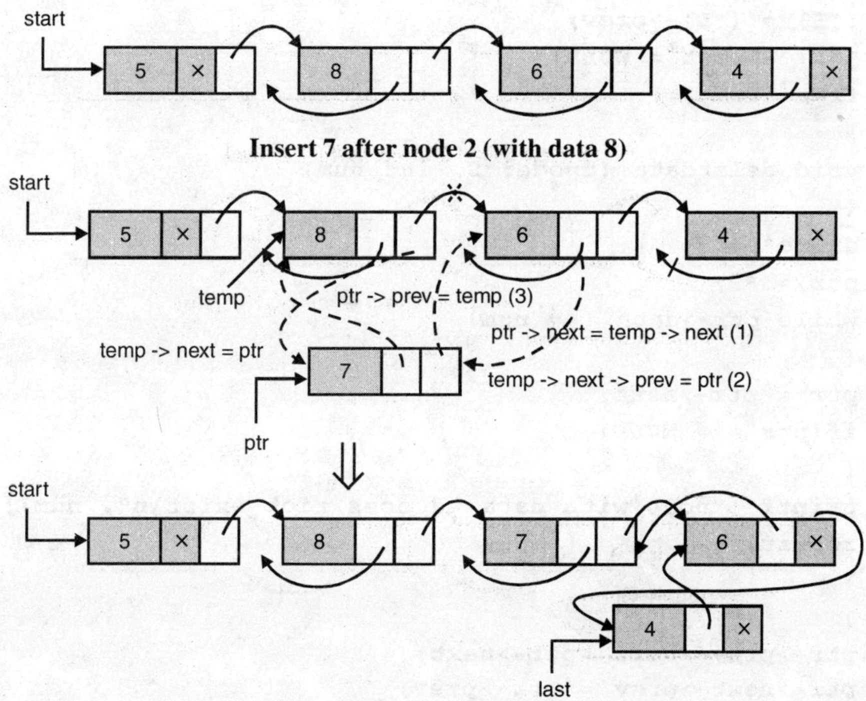

Fig. 6.32. Insertion of new node in between in DLL.

We have seen insertion in singly linked list. We simply show the pointer movements that causes insertion of new node in the doubly linked list. The new node to be inserted is pointed by pointer ptr. The node after which new node is to be inserted is pointed by temp. The insertion is done as

(*i*) next of ptr points to next of temp (ptr->next = temp->next)

(*ii*) prev of next of temp points to ptr (temp->next->prev = ptr)

(*iii*) prev of ptr points to temp (ptr->prev = temp)

(*iv*) next of temp points to ptr (temp->next = ptr)

This is shown in the Fig. 6.32.

6.8.5. Deletion in Doubly Linked List

```
void delfirst (dnode**S)
{
dnode* temp;
temp = *S;
(*S) = (*S)->next;
(*S)->prev = NULL;
free (temp);
}
void dellast (dnode**L)
{
dnode*temp;
temp =*L;
(*L) = (*L)->prev;
(*L) ->next = NULL;
free (temp);
}
void delatdata (dnode**S, int num)
{
dnode*ptr;
ptr = *S;
while(ptr->data ! = num)
{
ptr = ptr->next;
if(ptr = = NULL)
{
printf ("node with data %d does riot exist\n", num);
return;
}
}
ptr->prev->next =ptr->next;
ptr->next->prev = ptr->prev;
free (ptr);
}
```

For the deletion we have written three different functions: one deletes the first node, another deletes the last node and last one deletes any intermediate node. For deletion of first node of the doubly linked list (DLL) we simply have to write the following steps:

(*i*) pointer temp points to *S (temp = *S)

(*ii*) *S points to next of *S (*S = (*S)->next)

(*iii*) set prev of *S to NULL ((*S)->prev = NULL)

(*iv*) free the temp pointer (free(temp))

For deletion of last node of the doubly linked list (DLL) we do not have to traverse the DLL to reach up to the last node, as we know that in a DLL we have the last pointer that points to the last node, so we simply have to write the following steps for the deletion of last node in a DLL:

(*i*) pointer temp points to L (temp = *L)

(*ii*) *L points to prev of *L (*L = (*L)->prev)

(*iii*) set next of *L to NULL ((*L)->next =NULL)

(*iv*) free the temp pointer (free(temp))

For deletion of any intermediate node with the given data, we have to re up to the node using a temporary pointer that we wish to delete. During this traversing if the pointer becomes NULL it means desired node is not present in the DLL. If the node is found then it is pointed by the pointer ptr. Deletion of this node is of two step process:

(*i*) next of prev of ptr points to next of ptr (ptr->prev->next = ptr ->next)

(*ii*) prev of next of ptr points to prev of ptr (ptr->next->prev= ptr->prev)

Finally we can free the ptr by the statement free(ptr);

All three deletion operations are shown in the following figure:

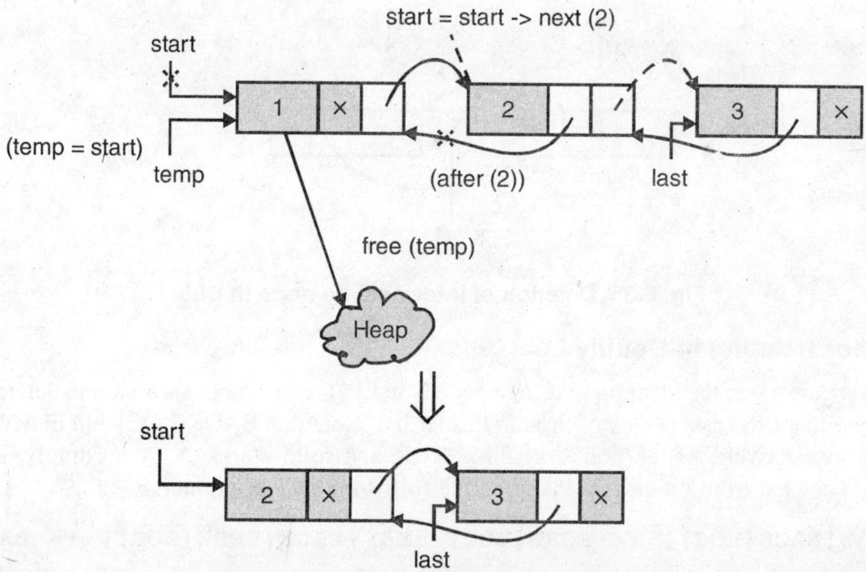

Fig. 6.33. Deletion of first node in DLL.

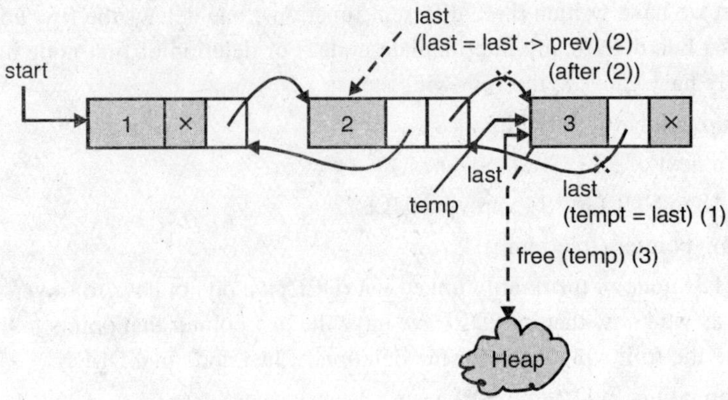

Fig. 6.34. Deletion of last node in DLL.

Deletion of node with data = 2

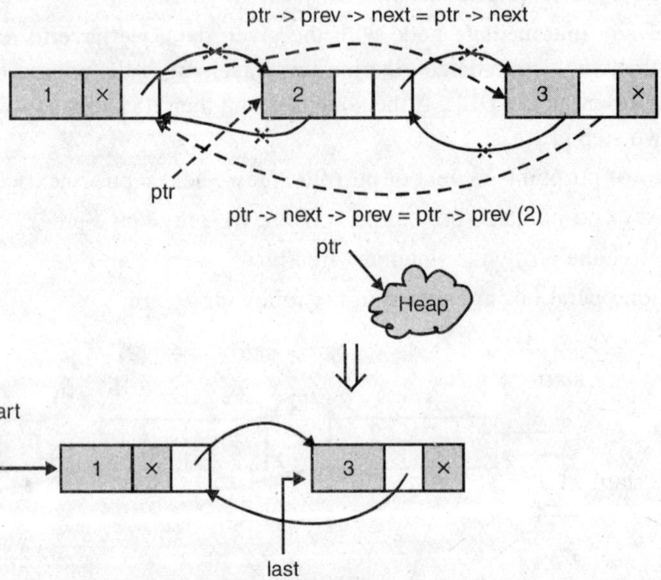

Fig. 6.35. Deletion of intermediate node in DLL.

Algorithm for Deleting in Doubly LL

Suppose we are given the location LOC of a node N in LIST, and suppose we want to delete N from the list. We assume that LIST is a two-way circular header list. Note that BACK[LOC] and FORW{LOC] are the locations, respectively, of the nodes which precede and follow node N. Accordingly, as shown in Fig. 6.35, N is deleted from the list by changing the following pair of pointers :

```
FORW[BACK[LOC]] := FORW[LOC] and  BACK[FORW[LOC]] := BACK[LOC]
```

The deleted node N is then returned to the AVAIl list by the assignments :

```
FORW[LOC] := AVAIl and   AVAIL := LOC
```

The formal statement of the algorithm follows :

Algorithm 6.10. DELTWL(INFO, FORW, BACK, START, AVAIL, LOC) ....

1. [Delete node]
   Set FORW[BACK][LOC]] := FORW[LOC] and
   BACK[FORW[LOC]] : = BACK[LOC]
   [Return node to AVAIL list.]
2. Set FORW[LOC] := AVAIL and AVAIL := LOC
3. Exit

Here we see one main advantage of a two-way list : If the data were organized as a one-way list, then, in order to delete N, we would have to traverse the one-way list to find the location of the node preceding N.

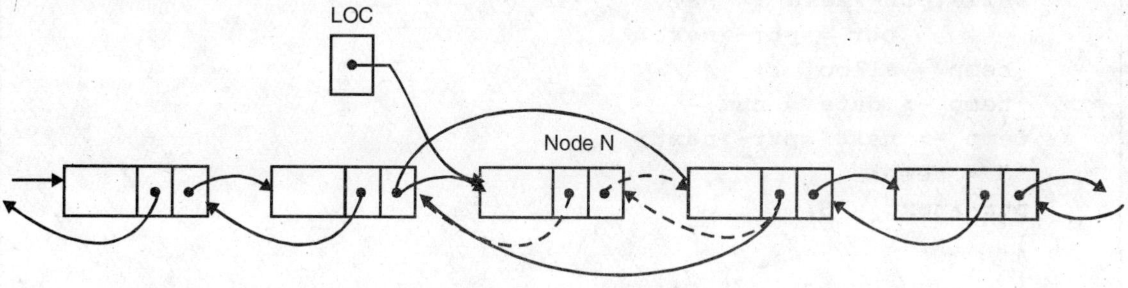

Fig. 6.36. Deleting node N.

6.9. CIRCULAR LINKED LIST (CLL)

A circular linked list is the linked list in which last node always point to the first node. In other way we can say that first node and final node are linked together. A CLL can be there for SLL and DLL. To traverse a circular linked list, you begin at any node and follow the list in either direction until you return to the original node. Viewed in this way CLL can be seen as having no beginning or end. But in general we will always have the start pointer that will point to the first node of the CLL when CLL was created. This type of list is most useful for management of buffers where data is consumed at regular intervals and in cases where you have one object in a list and wish to see all other objects in the list. The pointer pointing to the whole list is usually called the end pointer sometimes as there is no start node in the CLL.

6.9.1. Singly CLL

Singly CLL is similar to ordinary SLL but the next pointer of the last node points to the first node. As in a SLL, new nodes can only be efficiently inserted after a node when we already have a pointer to that node. For this reason, it's usual to keep a reference to only the last element in a singly CLL as this allows quick insertion at the beginning, and also allows access to the first node through the last node's next pointer. The function given can be used for adding a node at the end and beginning. Note instead of keeping a pointer to last node we keep a pointer to start node only. Try the before said logic yourself.

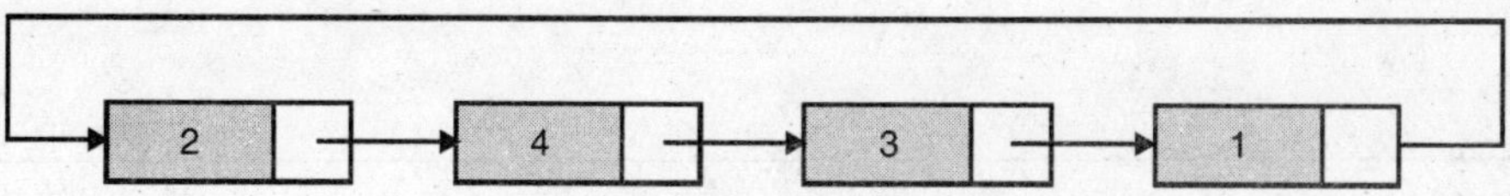

Fig. 6.37. Singly circular linked list.

```
void addatbeg(node**S, int num)
{
node *temp, *ptr;
if(*S = = NULL)
{
(*S) = alloc( );
(*S)->data=num;
(*S)->next = (*S);
}
else
{
ptr = *S;
while(ptr->next != *S)
        ptr = ptr->next;
 temp = alloc( );
 temp -> data = num;
temp -> next =ptr->next;
*S = temp;
ptr->next =*S;
}
}
void addatend (node **S, int num)
{
      node *temp, *ptr;
if (*S= = NULL)
{
temp = alloc( );
temp -> data = num;
temp -> next = temp;
*S = temp ;
}
else
}
temp = *S;
while (temp -> next != *S)
            temp = temp -> next;
 ptr = alloc( );
ptr -> data = num;
 temp -> next= ptr;
 ptr -> next =(*S);
}
}
```

To perform addition in the beginning and end of the circular linked list we have written two separate functions: addatbeg and addatend.

Addition of a node in the beginning is very simple and we have seen it earlier in simple linked list. When the circular linked list is empty and the node to be added is the first node then we simply make next of start point back to start. In case circular linked list is already created and node to be added as the first node, then we have to have a pointer to the last node so that last node can point to the newly created first node. For that we traverse the entire linked list using a temporary pointer ptr till next of ptr does not point to starting node. When the loop ends, next of ptr points to start. After that we create a new node pointed by temp. Now is added in the beginning of circular linked list as follows:

(*i*) next of temp points to next of ptr (temp->next = ptr->next)

(*ii*) start (*S) points to temp (*S = temp)

(*iii*) next of ptr points to start (*S) (ptr->next = *S)

DO YOU KNOW

LEONARD ADLEMAN (BORN 1945) was born in San Francisco, California. He received a B.S. in mathematics (1968) and his Ph.D. in computer science (1976) from the University of California, Berkeley. Adleman was a member of the mathematics faculty at M.I.T. from 1976 until 1980, where he was a co-inventor of the RSA cryptosystem, and in 1980 he took a position in the computer science department at the University of Southern California (USC). He was appointed to a chaired position at USC in 1985. Adleman has worked on computer security, computational complexity, immunology, and molecular biology. He invented the term "computer virus". Adleman's recent work on DNA computing has sparked great interest. He was a technical advisor for the movie *Sneakers*, in which computer security played an important role.

For adding a node at the end of a circular linked list, we traverse the entire circular linked list and reach up to the end of the linked list. The last node is pointed by the pointer temp. Now new node is created and is pointed by ptr.This node is added to the circular linked list as:

(*i*) next of temp points to ptr (temp->next = ptr)

(*ii*) next of ptr points to start (*S) (ptr->next =*S)

Both types of addition are shown in figure given below:

(*i*) addatend called with data 1

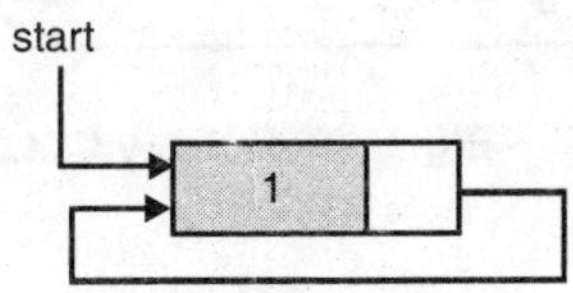

(*ii*) Addatend called with data 3

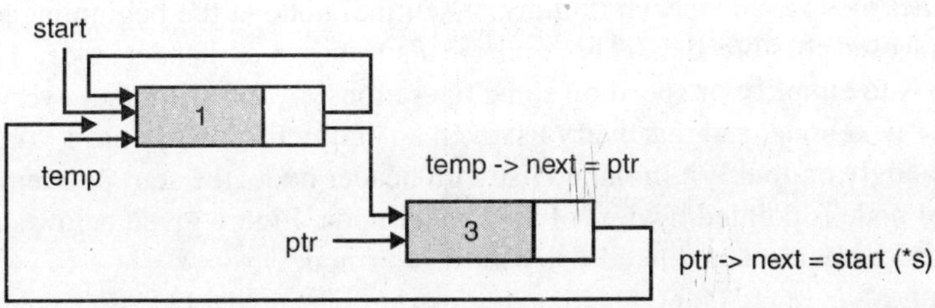

(*iii*) After addition of data 5 at the end.

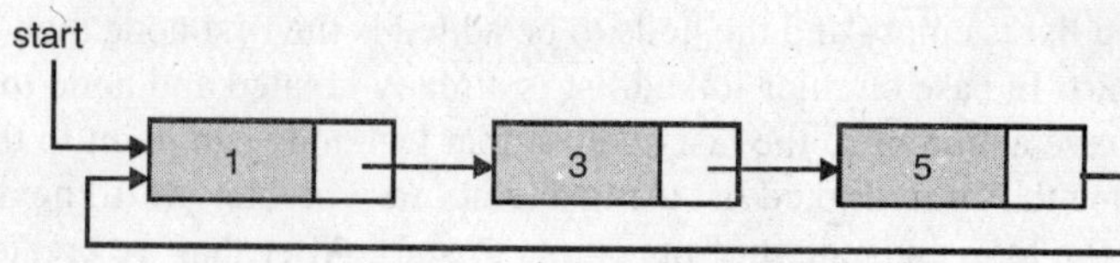

(*iv*) addatbeg called with data 6

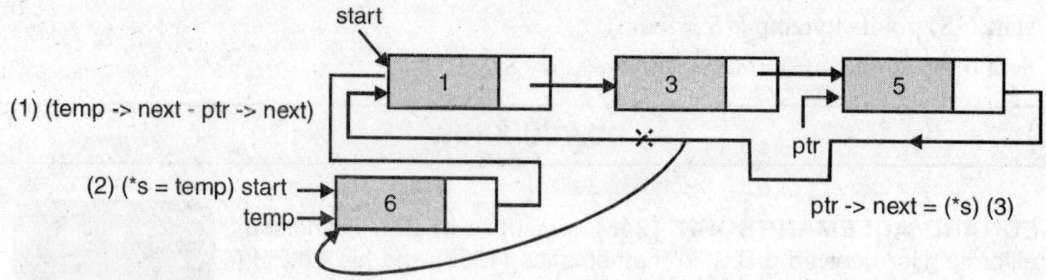

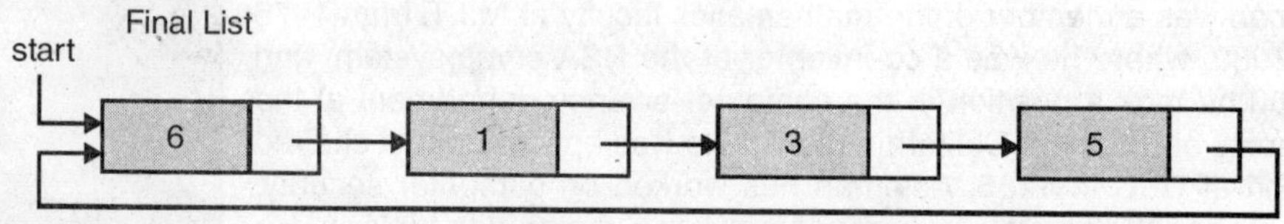

Fig. 6.38. Addition of mode in the circulat linked list.

6.9.2. Doubly CLL

In a doubly CLL, each node has two links, similarly to a DL as discussed earlier, except that the previous pointer of the first node points to the last node and the next pointer of the last node points to first node. As in a DLL insertion and removal can be done at any point with access to any nearby node same is the case with doubly CLL.

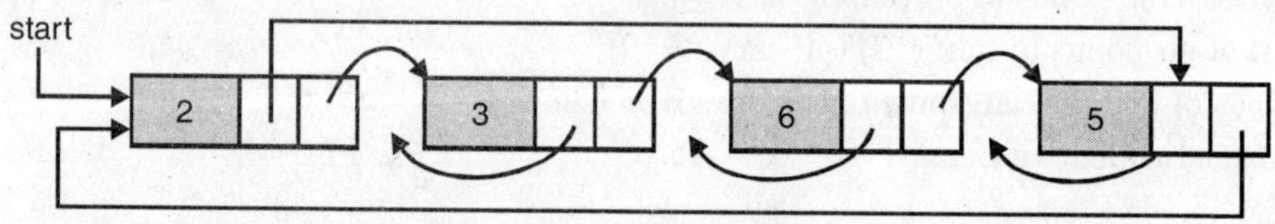

Fig. 6.39. Doubly CLL.

6.10. HEADER LINKED LIST

Linked lists sometimes have a special dummy or sentinel node at the beginning and/or at the end of the list, which is not used to store data. This node is also known as header node. The purpose of this special header node is to simplify or speed up some operations, by ensuring that every data node always has a previous and/or next node, and that every list even an empty list always has a "first" and "last" node (depend upon list is singly or doubly). In linked list with header node, the start pointer always point to the header node and first node is pointed by next of the header node. Figure given below shows singly, singly circular, doubly and doubly circular linked list with header node.

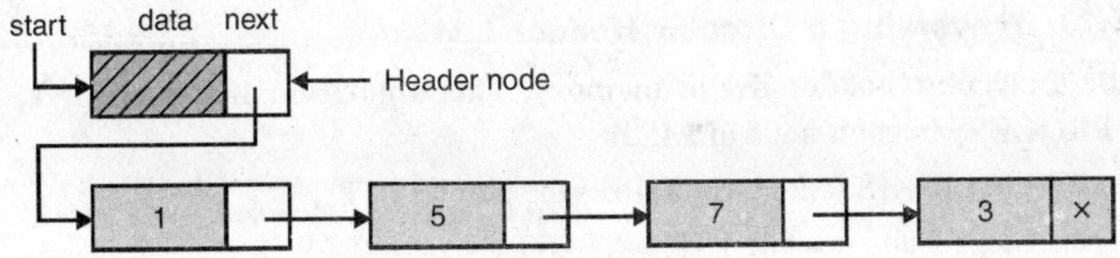

(a) Singly header linked list

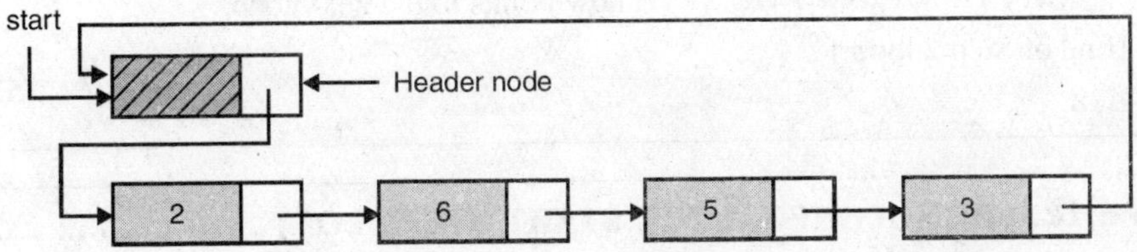

(b) Single circular header linked list

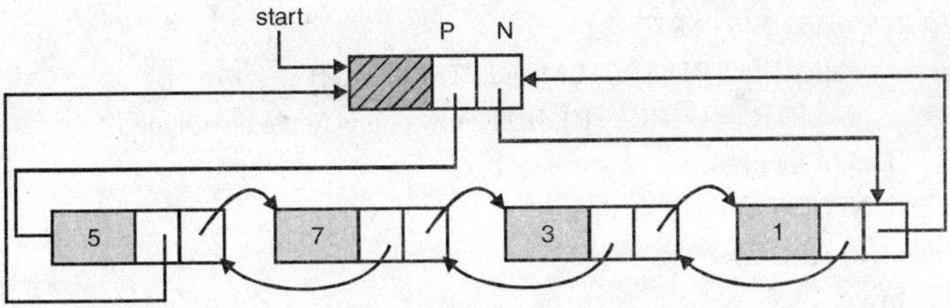

(c) Doubly circular header linked list

P pointer point to first node and N pointer point to last node

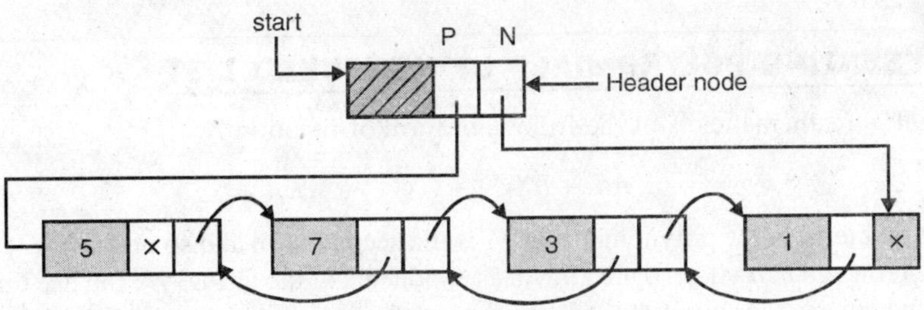

(d) Doubly header linked list

Fig. 6.40. Header linked list.

Circular header lists are frequently used instead of ordinary linked lists because many operations are much easier to state and implement using header lists. This comes from the following two properties of circular header lists :

(1) The null pointer is not used, and hence all pointers contain valid addresses.

(2) Every (ordinary) node has a predecessor, so the first node may not require a special case.

Algorithm 6.11. Traversing a Circular Header List ....

Let LISt be a circular header list in memory. This algorithm traverse LIST, applying an operation PROCESS to each node of LIST.

1. Set PTR := LINK[START]. [Initializes the pointer PTR.]
2. Repeat Steps 3 and 4 while PTR ≠ START :
3. Apply PROCESS to INFO[PTR].
4. Set PTR := LINK[PTR]. [PTR now points to the next node.

   [End of Step 2 loop.]
5. Exit

Algorithm 6.12. SRCHHL(INFO, LINK, START, ITEM, LOC) ....

LIST is a circular header list in memory. This algorithm finds the location LOC of the node where ITEM first appears in LIST or sets LOC = NULL.

1. Set PTR := LINK[START]
2. Repeat while INFO[PTR] ≠ ITEM and PTR ≠ START :

   Set PTR := LINK[PTR]. [PTR now points to the next node.]

   [End of loop].
3. If INFO[PTR] = ITEM, then :

   Set LOC:= PTR

   Else:

   Set LOC:=NULL.

   [End of If structure]
4. Exit

6.11. REPRESENTING POLYNOMIALS USING LINKED LIST

A polynomial in mathematics is of the following form of function X.

$$AX^n + BX^{n-1} + CX^{n-2} + \ \ldots\ldots$$

where AX^n is the first term in the polynomial, BX^{n-1} is the second term and so on. A, B, C etc are known as coefficient of the term and n, $n-1$.... 0 are known as exponents of the terms. We can use linked list to store the polynomials where each term will be stored in one node and all nodes (terms) will be linked together using pointer. The structure of the node will be as follows :

```
struct pnode
{
int coff;
int exp;
struct pnode * next;
};
```

Here the coff represent coefficient and exp represent exponent for a term of the polynomial. The pointer next is same we have used earlier.

Assume our first polynomial is :

$$1X^5 + 3X^2 + 5X^1 + 6X^0$$

And second polynomial is

$$6X^4 + 2X^3 + 6X^2 + 4X^1 + 2X^0$$

Then both the polynomials are maintained as linked list as shown below :

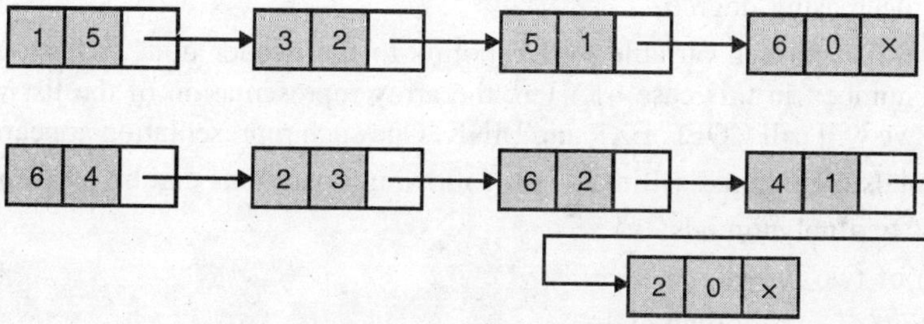

Fig. 6.41. Linked list representation of polynomial.

Polynomials

Header linked lists are frequently used for maintaining polynomials in memory. The header node plays an important part in this representation, since it is needed to represent the zero polynomial. This representation of polynomials will be presented in the context of a specific example.

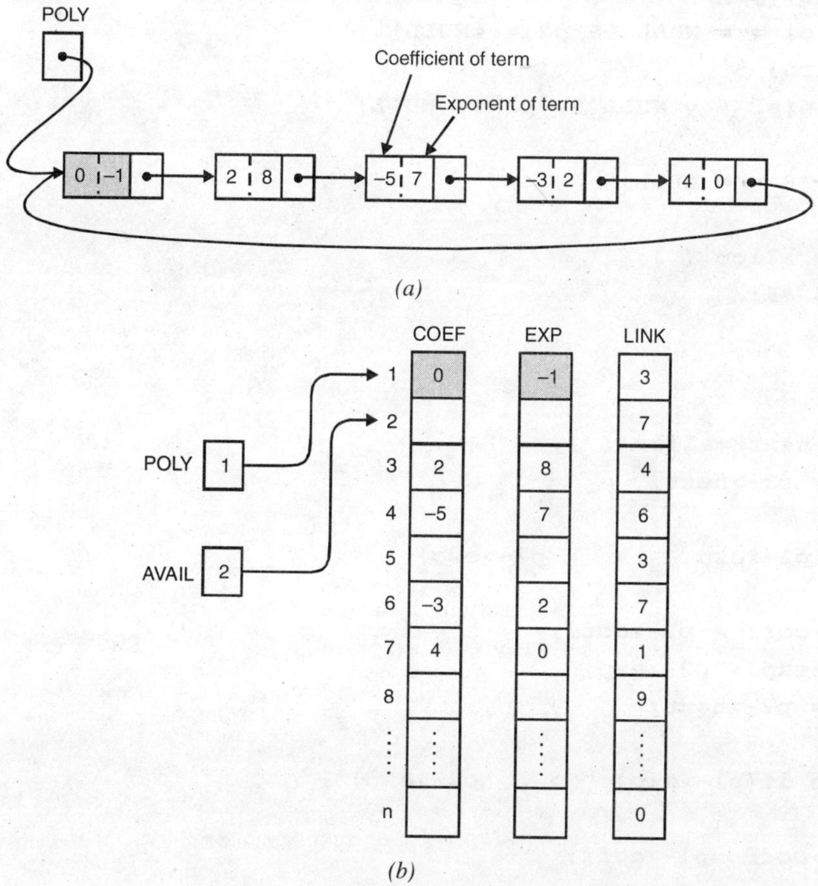

Fig. 6.42. p(x) = $2x^8 - 5x^7 - 3x^2 + 4$.

Let $p(x)$ denote the following polynomial in one variable (containing four nonzero terms) :

$$p(x) = 2x^8 - 5x^7 - 3x^2 + 4$$

Then $p(x)$ may be represented by the header list pictured in Fig. 6.42 (*a*) where each node corresponds to a nonzero term of $p(x)$. Specifically, the information part of the node is divided into two fields representing respectively, the coefficient and the exponent of the corresponding term, and the nodes are linked according to decreasing degree.

Observe that the list pointer variable POLY points to the header node, whose exponent field is assigned a negative number, in this case –1. Here the array representation of the list will require three linear arrays, which we will call COEF, EXP and LINK. One such representation appears in Fig. 6.42 (*b*).

On the polynomials maintained as linked list, following operations can be performed:

(*i*) Addition of two polynomials

(*ii*) Subtraction of two polynomials

(*iii*) Multiplication of two polynomials

We discuss addition and multiplication of two polynomials as subtraction is similar to addition.

6.11.1. Addition of Two Polynomials

```
void addpoly (pnode *p1, pnode *p2, pnode**R)
{
pnode *p3;
if (p1 = = NULL && p2 = =NULL)
return;
while(p1 ! = NULL && p2 != NULL)
{
if (*R = = NULL)
{
*R = alloc ( );
p3 = *R;
}
else
{
p3->next =alloc ( );
p3 = p3->next;
}
if (p1->exp    <    p2->exp)
{
p3->coff = p2->coff;
p3->exp = p2->exp;
p2 = p2->next;
}
else if(p1->exp    >    p2->exp)
{
p3->coff =p1->coff;
p3->exp = p1->exp;
```

```
p1 = p1->next;
}
else if (p1->exp = = p2->exp)
{
p3->coff = p1->coff + p2->coff;
p3->exp = p1->exp;
p1 = p1->next;
p2 = p2->next;
}
while (p1 ! = NULL)
{
if (*R = = NULL)
{
*R = alloc ( );
p3 = *R;
}
else
{
p3->next = alloc ( );
p3 = p3->next;
}
p3->coff = p1->coff;
p3->exp =p1->exp;
p1 = p1->next;
}
while (p2 ! = NULL)
{
if (*R = = NULL)
{
*R = alloc ( );
p3 = * R;
}
Else
{
p3 -> next = alloc( );
p3 = p3 -> next;
}
p3->coff = p2->coff;
p3->exp = p2->exp;
p2 = p2->next;
}
p3 ->next = NULL;
}
```

For the addition of two polynomials we have to cheek each term of the two polynomials, if the exponents of the two terms match, we add their corresponding coefficients and make one term. If the exponent of first polynomial's term is greater then exponent of second polynomial's term then it is added as it to the resultant polynomial and pointer to first polynomial points to its next term. In case reverse is true, second polynomial's term is added to the resultant polynomial and pointer to second polynomial points to its next term. This whole logic is written inside the first while loop. The loop runs till one of the linked list does not become NULL.

The main pointer to third linked list (resultant or target) is denoted by *R and an alternate pointer is by p3. Initially when the pointer *R3 is NULL, allocation for the first node is done and p3 points to *R. On the successive iteration of while loop, next of p3 points to new node and finally p3 points to new node. This is because of following statements:

```
p3->next = alloc; p3 = p3->next;
```

In each iteration of while loop, the else-if ladder checks exponent of which term of two polynomials is greater, lesser or equal and corresponding action is taken as discussed above.

When the control comes out from the loop, we are not sure whether p1 is NULL or p2 is NULL and if initially p1 or p2 was NULL so control did not enter into the body of the while loop at all. For this reason we allocate memory for the new node using the same if-else logic as done in the main while loop in both the while loop. Both the while loops are similar to the each other with the difference that first while loop runs for first polynomial and second runs for second polynomial. In both the loops, remaining and all the terms are assigned to the resultant polynomials by exactly copying all the terms from the concerned polynomial.

(*a*)

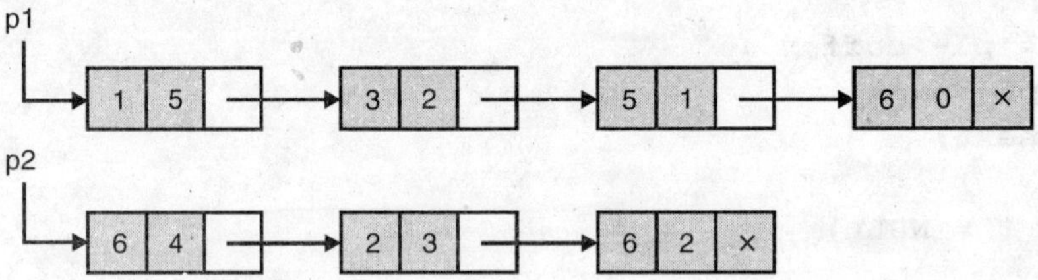

(*b*) p1 -> exp > p2 -> exp

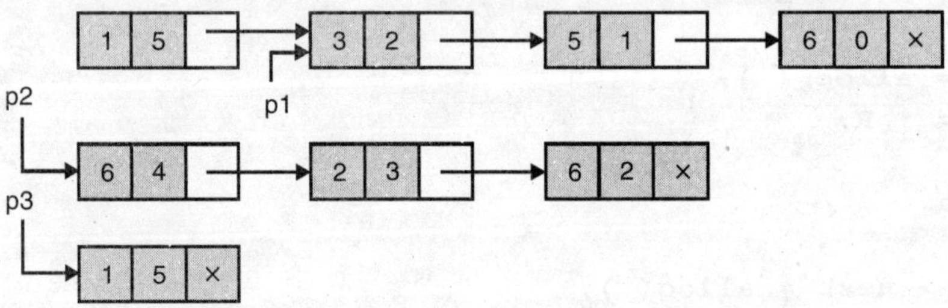

(*c*) p1 -> exp < p2 -> exp (from (*b*)).

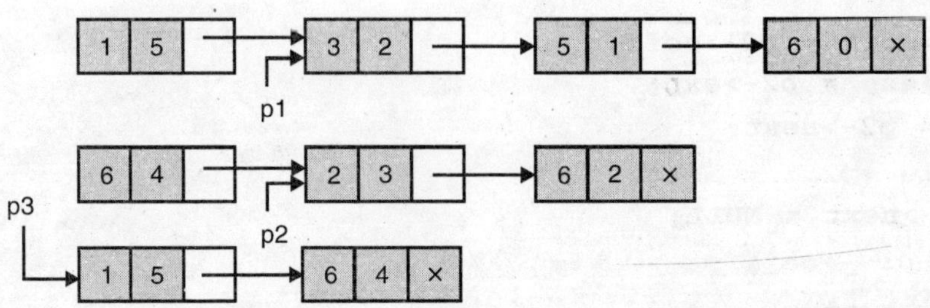

(*d*) p2 -> exp > p1 -> exp

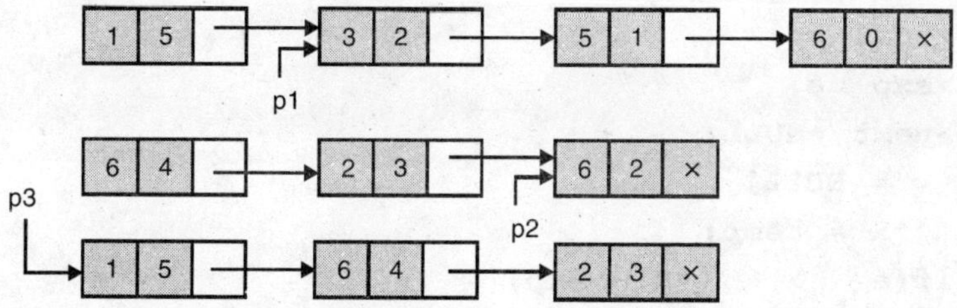

(*e*) p1 -> exp = = p2 -> exp (add the coefficient)

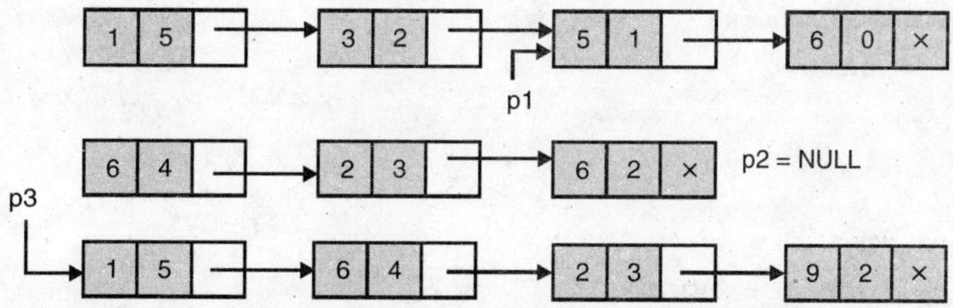

(f) As p2 = NULL (from (e)) all remaining nodes from p1 are assigned to p3

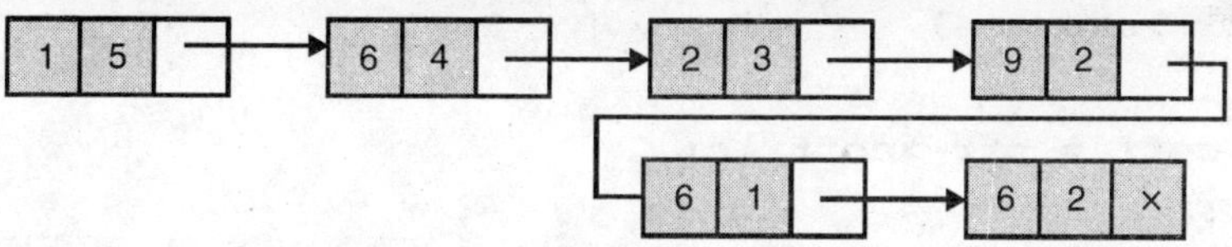

Fig. 6.43. Addition of two polynomial maintained as linked list.

6.11.2. Multiplication of Two Polynomials

```
void multipoly(pnode *p1, pnode *p2, pnode**R)
{
pnode *save= p2, *temp, *ptr;
int c, e;
if(p1 = =NULL &&p2 = =NULL)
return;
else if(p1 = =NULL)
*R = p2;
        else if(p2 = = NULL)
            *R = p1;
else
{
while (p1 ! = NULL)
{
while (p2 ! = NULL)
{
c = p1 ->coff*p2->coff;
e =p1->exp+p2->exp;
```

```
temp = alloc( );
temp->coff =c;
temp->exp =e;
temp->next =NULL;
if(*R = = NULL)
        *R = temp;
else if(e   >   (*R)->exp)
{
temp->next = *R;
(*R) = temp;
}
else
{
ptr = *R;
while (ptr ! = NULL)
{
if(ptr->exp = e)
{
ptr->coff = ptr->coff +c;
break;
}
else if (ptr->exp  > e && (ptr->next->exp  <  e | |  ptr->next = NULL))
{
temp->next =ptr->next;
ptr->next = temp;
break;
}
ptr = ptr -> next;
}
}
p2 =p2->next;
}
p2 = save;
p1 = p1 -> next;
}
}
}
```

For multiplication of two polynomials we have to multiply each term of the first polynomial with each term of the other polynomial. In the multipoly function we first check whether any of the source linked list (maintained as polynomial, represented by p1 and p2) is NULL. If any of them *i.e.* p1 or p2 is NULL, the other one is assigned to *R (target polynomial storing the multiplication of two polynomials). In case both the polynomials are not NULL, the else part executes. In the else part, the outer while loop runs till p1! =NULL. For every term (node) of this p1 polynomial, inner while loop runs. The inner while loop runs for every term (node) of second polynomial *i.e.* p2. In the inner while we multiply the two coefficients, and sum the exponents of the nodes pointed by p1 and p2. This is stored in two variables c and e respectively.

A new node temp is created and its member fields are filled, Now if *R is NULL, then *R points to temp else if e is greater then the exponent of the starting node i, e. *R then new node is added in the beginning of the target linked list. If this is not the case then we find a suitable place for the new node to be inserted or appended. For this we take a temporary pointer ptr and traverse the entire linked list till ptr! = NULL. If the exponent of the ptr is equal to e we add the corresponding coefficient fields, else if the e lies between exponent of ptr and exponent of next of ptr then node is added in between. In case exponent of ptr is greater then e and next of ptr is equal to NULL, then node is added at the end of target linked list.

When all the terms of the second polynomial have been multiplied with the first term of the first polynomial, then inner while loop ends as p2 points to NULL. To make p2 again point to original second polynomial, note we have stored p2 into the temporary pointer save in the beginning of the function. Now save is assigned back to p2, p1 points to its next and this process repeats.

6.12. UNROLLED LINKED LIST

An unrolled linked list is a variation on the linked list which stores multiple elements in each node. An unrolled linked list is known as cache sensitive structure. In C a typical node for an unrolled linked list will look like as:

```
#define MAX 40
struct node
{
int count;
int arr [MAX];
struct node*next;
};
```

Each node holds up to a certain maximum number of elements, typically just large enough so that the node fills a single cache line or a small multiple thereof (Cache memory is a very fast memory attached to the CPU. Number of frequently used references constitutes one cache line. Frequently used references can be put into the cache so that main memory is accessed later in case a reference is not found in cache memory). A position in the list is indicated by both a reference to the node and a position in the elements array. It's also possible to include a previous pointer for an unrolled doubly-linked linked list. Assuming maximum 10 elements per node, unrolled linked list is shown in the figure given below:

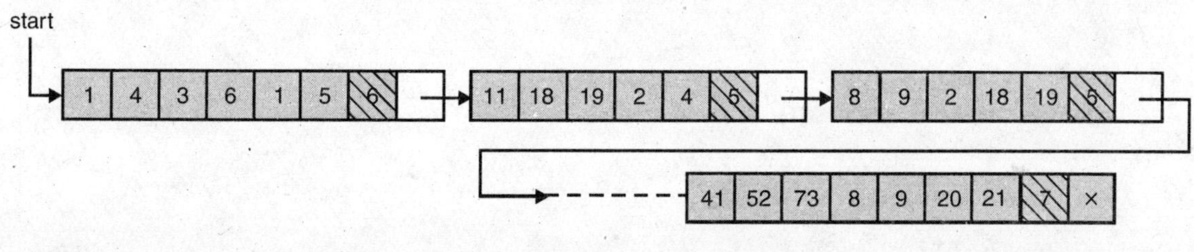

Fig. 6.44. Unrolled linked list.

Insertion

If there is space left in the array of the node in which we wish to insert the value, we simply insert it. To insert a new element, we simply find the node the element should be in and insert the element into the array arr, incrementing count. If the array already contains MAX values, we create a new node, insert it after the current one, and move half the elements to that node, creating room for the new value.

Deletion

For deletion we simply remove the value from the array. If the number of elements in the array drops below N/2, we take elements from a neighbolng array to fill it back up. If the neighboring array also has N/2 values, then we merge it with the neighboring array instead.

Performance

The primary advantage of unrolled linked lists is their cache behavior. Due to this cache behavior unrolled linked lists also perform sequential traversal much more rapidly. Another advantage of unrolled linked lists is decreased storage requirements. Another advantage of unrolled linked lists is that they perform a number of operations, typically associated with arrays, much more quickly than ordinary linked lists. For example, when indexing into unrolled linked list, we can progress a node at a time rather than an element at a time, reducing the indexing time to O(N/MAX) instead of O(N) (N =total number of elements). One disadvantage of unrolled linked list is the overhead per node for references and element counts.

6.13. APPLICATIONS OF LINKED LIST

1. Linked lists are used as a building block for many other data structures, such as stacks, queues, trees, graphs etc.
2. The "data" field of a node can be another linked list. By this device, one can construct many linked data structures with lists; this concept was first implemented in the Lisp programming language, where linked lists are a primary data structure, and is now a common feaure of the functional programming style. Thus it is an application of linked list in functional programming.
3. Sometimes, linked lists are used to implement associative arrays.
4. Linked lists are used for the implementation of polynomials as discussed above.
5. In the solution of Josephus problem, circular linked list is used.
6. For the implementation of sparse matrix (see next chapter) linked list is used.

SOLVED PROBLEMS

1. Write C rotine to find frequency of specific element passed as argument.

Ans.

```
    int FindFreq ( node* ptr, int num)
{
int f = 0;
while (ptr! = NULL)
{
if (ptr -> data = = num)
        f ++;
ptr = ptr -> next;
}
return f;
}
```

2. Write C routine to count number of nonzero elements in list.

Ans.

```
int countNZ (node*ptr)
{
int f=0;
while (ptr!=NULL)
{
if (ptr->data ! = 0)
     f++;
ptr=ptr->next;
}
return f;
}
```

3. Write C routine to add a value to every node in the linked list supplied as argument.

Ans.

```
void IncrK(node*ptr,int K)
{
while (ptr! =NULL)
{
ptr->data=ptr->data+K
ptr=ptr->next;
}
}
```

4. Write a C routine that deletes the last node in the linked list.

Ans.

```
void delLast (node**S)
{
node*ptr=*S;
node* save;
if ( (*S) ->next = =NULL)
{
ptr=*S;
*S=NULL;
free (ptr) ;
return;
}
while (ptr->next ! =NULL)
{
save=ptr;
ptr=ptr->next;
```

```
}
save->next = NULL;
free(ptr);
}
```

If condition when there is just one element in the linked list; code following if causes save points to second last element of the linked list and ptr points to last node.

5. Write a C routine that removes the first node in a linked list and put it at the end of linked list.

Ans.

```
void FirstAtLast (node**S)
{
node*ptr*S;
node*save, * first;
first=*S;
(*S) = (*S) ->next;
while (ptr !=NULL)
{
save=ptr;
ptr=ptr->next;
}
save->next=first;
first->next=NULL;
}
```

Note in the above we have assumed that linked list contains at least two nodes.

6. Write a C routine that swaps k and k+l th node using pointer movements.

Ans.

```
void Swap(node**S,int k)
{
node*ptr=*S;
node* save, *pnext;
int i;
ptr = *S;
for(i=1 ;i<k; i++)
{
save=ptr;
ptr=ptr->next;
}
pnext=ptr->next;
save->next=pnext;
ptr->next=pnext->next;
pnext->next=ptr;
}
```

It is assumed that linked list contains sufficient number of nodes so that above routine can be applied over it. Pointer ptr points to kth node, pnext points to k+1 th node, and save points to k-1 th node. The main pointer mo ;ements written after the for loop and are easy to understand.

7. What will be the declaration of structure if instead of linked list, array is used for the implementation of polynomial ?

Ans. The declaration will be:

```
struct Pnode
{
int coff[MAX], expo[MAX};
int noterms;
};
```

where value of MAX may be chosen appropriately as per the maximum number of terms in polynomial assumed in a particular context.

8. A method of reversing the linked list we have seen in this chapter? Can you think of any other method that prints the list in reverse ?

Ans.

```
void reverse (node * R)
{
node* save, *ptr;
 ptr = NULL;
while (ptr ! =R)
{
save=R;
while (save->next != ptr)
          save=save->next;
printf("%d",save->data);
ptr=save;
}
}
```

We leave it for you to understand how the routine works.

9. WAP to create, Traverse a linked list ?

Ans.

```
#define NULL '\0'
   struct Node
   {
      char info;
      struct Node *link;
   };

   void main()
```

```
{
    char num, ch;
    struct Node *start, *new, *temp;
    start = NULL;
    do
    {
        printf(" Enter the info part of a node ");
         scanf("%c",&num);

        if ( start == NULL)
        {
            start = (struct Node *) malloc (sizeof (struct Node));
            start -> info = num;
            start -> link = NULL;
        }
        else
        {
            temp = start;
            while ( temp -> link != NULL)
            {
                temp = temp  ->  link;
            }
            new =  (struct Node *) malloc (sizeof (struct Node));
            r  ->  info  =  num;
            r  ->  link  =  NULL;
            temp  ->  link  =  r;
        }
        printf(" Do you want to insert more nodes (Y / N) ");
         scanf("%c",&ch);
    } while ( ch == 'y'  || ch == 'Y');
  for(  temp=start ; temp != NULL ; temp = temp -> link)
    {
        printf(" %c ->>",temp  ->info);
    }
    getch();
}
```

10. **WAP to count the total number of nodes in a linked list?**

Ans.

```
STRUCT Node
    {
        char info;
```

```
    STRUCT Node * link;
  }i;

  LOGIC :
  int k;
for( k = 1 , i=start ; i -> link != NULL ; k++, i = i -> link);
  printf("total number of nodes = %d",k);
```

Semicolon used after the for loop means that there is no body (statements) of the for loop.

11. WAP to traverse a given linked list?

Ans.

```
Traverse_list(node * i)
  {
  for(  i=start ; i != NULL ; i = i -> link)
     {
        printf(" %c ->>",i->info);
     }
  }
```

12. WAP to insert a node at a particular location in a given linked list ?

Ans.

```
Insert_list(node * i, int LOC)
  {
  node  *w, *j;
  int k;
for(  k=1, j=start ; k <= LOC ; w = j, k++, j = j -> link);
     w -> link = i;
     i -> link = j;
  }
```

13. WAP to search a node at a particular location in a given linked list ?

Ans.

```
search_list(char  i)
  {
  node   *j;
  int k;
for(  k=1, j=start ; j != NULL;  k++, j = j -> link)
{
  if ( j -> info = = i)
  {
     printf("the node is at location =%d",j);
     Break;
  }
```

```
    }
    if ( j = = NULL)
    {
        printf("the node is not in the list");
            }
    }
```

14. WAP to delete a node from a particular location from a given linked list?

Ans.

```
delete_list( int LOC)
    {
    node   *w, *j;
    int k;
for(  k=1, j=start ; k <= LOC ; w = j, k++, j = j -> link);
        w -> link = j -> link;
        free ( j );  / /  delete ( j );
    }
```

15. Find the number NUM of elements in a linked list.

Ans. The following procedure finds the number NUM of elements in a linked list

COUNT (INFO, LINK, START, NUM)

1. Set NUM :\ =. [Initialize counter.]
2. Set PTR := START. [Initializes pointer.]
3. Repeat Steps 4 and 5 while PTR ≠ NULL.
4. Set NUM := NUM + 1. [Increases NUM by 1.]
5. Set PTR := LINK[PTR]. [Updates pointer.]

   [End of Step 3 loop.]
6. Return.

16. Consider a poly a polynomial $p(x, y, z)$ in variables x, y and z. Unless otherwise stated, the terms in $p(x, y, z)$ will be ordered *lexicographically*. That is, first we order the terms according to decreasing degrees in x; those with the same degree in x, we order according to decreasing degrees in y; those with the same degres in x and y we order according to decreasing degrees in z. Suppose

$$p(x, y, z) = 8x^2y^2z - 6yz^8 + 3x^3yz + 2xy^7z - 5x^2y^3 - 4xy^7z^3$$

(a) Rewrite the polynomial so that the terms are ordered.

(b) Suppose the terms are stored in the order shown in the problem statement in the linear arrays COEF, XEXP, YEXP and ZEXP, with the HEAD node first. Assign values to LINK so that the linked list contains the ordered sequence of terms.

Ans. (a) Note that $3x^3yz$ comes first, since it has the highest degree in x. Note that $8x^2y^2z$ and $-5x^2y^3$ both have the same degree in x but $-5x^2y^3$ comes before $8x^2y^2z$, since its degree in y is higher. And so on. Finally we have

$$p(x, y, z) = \quad 3x^2yz - 5x^2y^3 + 8x^2y^2z - 4xy^7z^3 + 2xy7z - 6yz^8$$

(b) Figure 6.37 shows the desired data structure.

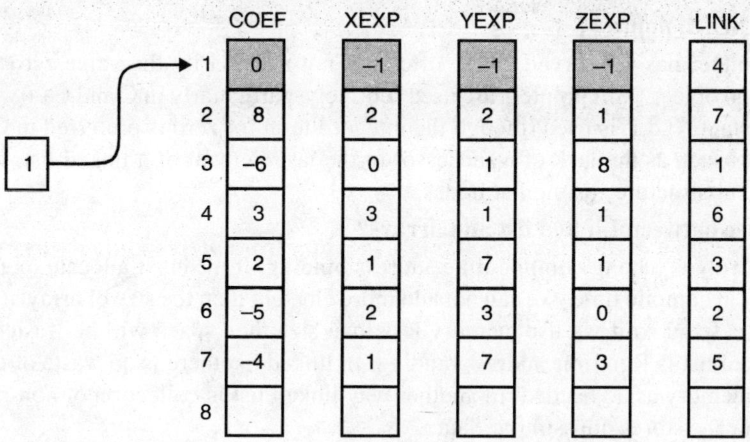

Fig. 6.37

17. Difference between Static & Dynamic Memory Allocation?

Ans.

Static Memory Allocation	Dynamic Memory Allocation
Static memory allocation is before run time, but the values of variables may be changed at run time.	Dynamic memory allocation is at runtime
Static memory allocation saves running time, but can't be possible in all cases.	Dynamic Allocation doesn't save running time
The static memory allocation stores it's data in the "data segment" of the memory.	Dynamic memory allocation stores it's memory on heap,
Static memory must be declared at compile-time.	Dynamic memory can be declared at run-time using the new and delete operators (or malloc and free in C),

18. Under what circumstances would you prefer to use a doubly linked list against a singly linked list?

Ans. Advantages of Doubly Linked list compared to Singly Linked list:

1. Doubly linked list offers easy implementation of many operations, whereas singly linked list requires more information for the same operation like for the deletion of a node in a singly linked list, we definitely need the pointer ptr to node previous to it while in case of doubly linked list, we just need the pointer ptr to the node being deleted as shown below:

```
ptr->right->left=ptr->left;
ptr->left->right=ptr->right;
free(ptr);
```

2. A doubly linked list can be traversed in both directions (forward and backward). A singly linked list can only be traversed in one direction.

Disadvantage of Doubly Linked list :

But then some tradeoffs that is doubly linked list occupy more space and often more operations are required for the similar tasks as compared to singly linked lists for example while inserting or deleting it needs more space (to store extra pointers)..

19. What is NULL pointer?

Ans. A null pointer has a reserved value, often but not necessarily the value zero, indicating that it refers to no object. Null pointers are used routinely, particularly in C and C++ where the compile-time constant NULL is used (though the integer literal for zero is preferred in C++), to represent conditions such as the lack of a successor to the last element of a linked list, while maintaining a consistent structure for the list nodes.

20. Difference between Linked list and arrays?

Ans.
1. An Array is having compile time address binding. It means it allocate memory space for an array at compile time. We can not add more element than the size of array at run time. In array size is fixed so if we use memory less than size then space will be wasted. In another way Linked list is Run time address binding. In linked list there is no waste of memory. User can use memory as he needed. In another way linked list is collection of non-similar data types. So it cans stores non-similar data.
2. In array we can insert or delete an element and run time memory allocation is not possible in the arrays we can add memory and delete memory allocations in arrays it is not possible to store large amount of data where it is possible with linked list.
3. Array is a collection of similar entities like integers, float etc. Its syntax is int arr[] that is array is of int datatype. They store the variables in contiguous memory locations. The advantage of linked list lies in its flexibility in assigning memory locations. The element stored in linked list need not to be in contiguous memory locations. Hence memory can be utilized efficiently.
4. Memory utilization is very high in linked list i.e. when users want another node then he will create and the nodes in the list are not contiguous. So the memory is greatly utilized.
5. Inserting or deleting nodes from a linked list is very easy at any place of the linked list.
6. Computer memory is divided into blocks. Each work use these blocks in non-contiguous manner. But array elements are stored in continuous memory allocation. This will adversely affect unuse of the memory blocks. Array searches for bigger blocks it needed, So that many small blocks of memory is unused .The main advantage of linked list is it can use non-continuous memory blocks, Because of it has the relation between the nodes, and this is main merit over array.
7. Data access is very faster in arrays
8. Linked List has an extra Overhead in its each node to store the pointer to the next node.
9. Insertion/Deletion of values in middle of array is not possible.
10. Array can be access randomly and it can even access the middle element by just the array name with the subscripted, a [5].The elements are allocated a contagious memory. Whereas in linked list more space is required for the pointer and the information. Accessing elements in linked list is sequential.
11. The Main advantage of array data structure is which is used to hold like kind of data. In other words arrays hold (save) similar kind of data items and array items stored in contiguous locations in memory. On the other hand link list may hold unlike kind of data (group of data) and items stored are not contiguous in linked list.
12. Arrays are of types one, two, multi dimensional but link list have types single and doubly which further has types circular, header, circular header link list.

21. What are the advantages of circular linked list over linear linked list?

Ans. It's not that one is better than the other. They are used in different circumstances. A linear linked list is used like an array, with the added benefits of random insertion/removal of elements, etc. A circular linked list is often used as a buffer where one portion of the program produces data and another consumes it, such as in communications.

22. List Advantages & Disadvantages of Doubly Linked, Circular Lists with a Header Node?

Ans. The primary advantage of a doubly linked list is that given a node in the list, one can navigate easily in either direction. This can be very useful, for example, if the list is storing strings, where the strings are lines in a text file (e.g., a text editor). One might store the "current line" that the user is on with a pointer to the appropriate node; if the user moves the cursor to the next or previous line, a single pointer operation can restore the current line to its proper value. Or, if the user moves back 10 lines, for example, one can perform 10 pointer operations (follow the chain) to get to the right line. For either of these operations, if the list is singly linked, one must start at the head of the list and traverse until the proper point is reached. This can be very inefficient for large lists.

The primary disadvantage of doubly linked lists are that (1) each node requires an extra pointer, requiring more space, and (2) the insertion or deletion of a node takes a bit longer (more pointer operations).

The primary advantage of a header node is that insertion and deletion require no special cases. The primary disadvantage of using a header node is that it becomes unclear what the header node stores. Is it a regular list node? Or is it a special node? If it's a regular node, what data should it store? For a list of strings, for example, one often wants to provide a function to return a string from a given node. If the list is empty, what is returned? It can't be NULL, because the function has a return type of String. It should be a string that can't possibly ever actually be used. Which string is that?

23. If you are using C language to implement the heterogeneous linked list, what pointer type will you use?

Ans. The heterogeneous linked list contains different data types in its nodes and we need a link, pointer to connect them. It is not possible to use ordinary pointers for this. So we go for void pointer. Void pointer is capable of storing pointer to any type as it is a generic pointer type.

24. List out few of the applications that make use of Multilinked Structures?

Ans. 1. Sparse matrix, 2. Index generation.

25. Whether Linked List is linear or Non-linear data structure?

Ans. According to Access strategies Linked list is a linear one.

According to Storage Linked List is a Non-linear one.

26. Write a Program to Split a given linked list into two parts from the given position from which to split?

Ans. /* Structure for a linked list*/

```
    struct list{
        int num;
        struct list* next; /* pointer to next node in list*/
    };
```

/* Function to split a linked list into two parts, given the position from which to start splitting*/

```
void split(struct list* head, struct list** dest, int
        position)
        {
            int offset = 0;
            struct list* temp = NULL;
```

```
        struct list* newStart = NULL;
        while((offset < position -1) && (head->next !=
    NULL))
        {
            head = head->next;
            offset++;
        }
        temp = head;
        newStart = head->next;
        while(head->next != NULL)
        {
            *dest = head;
            head = head->next;
            (*dest) = (*dest)->next;
        }
        *dest = newStart;
        temp->next = NULL;
    }
void insert(struct list **head,int value)
    {
        struct list* temp = NULL;
        temp = *head;
        if(NULL == *head)
        {
            *head = (struct list*) malloc(sizeof(struct
    list));
            (*head)->num = value;
            (*head)->next= NULL;
        }else
        {
            while(temp->next != NULL)
            {
                temp = temp->next;
            }
            temp->next = (struct list*)
    malloc(sizeof(struct list));
            temp->next->num = value;
            temp->next->next = NULL;
        }
    }
```

/* Function to traverse the linked list given the starting node pointer */

```
void traverse(struct list* node)
    {
        while(node != NULL)
        {
            printf("%d->",node->num);
            node = node->next;
```

```
        }
        printf("NULL\n");
    }
```

/* Function to find the length of linked list using recursion, given the starting node pointer*/

```
int length(struct list* node)
    {
        if(node->next == NULL)
            return 1;
        else
            return (1 + length(node->next));
    }
```

/* Function to use the linked list functions*/

```
int main()
    {
        struct list* node = NULL;
        struct list* new = NULL;
        int i;
        for(i=0; i < 10; i++)
        {
            insert(&node,i);
        }
        split(node,&new,6);
        traverse(node);
        traverse(new);
        printf("Old List length [%d]",length(node));
        printf("New List Length [%d]",length(new));
    }
```

27. Write a function to find the length of a given linked list?

Ans. /* Function to find the length of linked list using recursion, given the starting node pointer*/

```
int length(struct list* node)
    {
        if(node->next == NULL)
            return 1;
        else
            return (1 + length(node->next));
    }
```

28. Write a function to find middle of a given linked list?

Ans. /* Function to find the middle node for the linked list, given the starting node pointer*/

```
struct list* middle(struct list* head)
    {
        struct list* far = NULL;
        if(!head->next)
            return far=head;
        else{
            printf("List not empty\n");
```

```
            far = head->next;
            while((head->next) && (far->next))
            {
                head = head->next;
                if(far->next->next)
                    far = far->next->next;
                else
                    break;
            }
        }
        return head;
    }
```

29. Write a function to reverse a given linked list using Recursion ?

Ans. **/* Function to print data in reverse order for a linked list, given the starting node pointer*/**

```
void reverseprint(struct list* node)
    {
                    if(NULL == node)
                            return;
                    reverseprint(node->next);
                       printf("%d",node->num);
    }
```

30. Write a function to remove duplicates from a given linked list?

Ans. **/* Function to remove duplicates from a linked list, given the head pointer for the list*/**

```
void removeDuplicates(struct list** head)
    {
        struct list* temp = NULL;
        struct list* dup = NULL;
        temp = *head;
        while(temp->next != NULL)
        {
            if(temp->num == temp->next->num)
            {
                dup = temp->next;
                if(temp->next)
                    temp->next = temp->next->next;
                else
                    temp->next = NULL;
                FREE(dup);
                temp = *head;
                continue;
            }
            temp = temp->next;
        }
    }
```

31. Given a polynomial

$$P(x) = 2x^4y^3z^2 + 4x^5yz + 10x^2y^2 + 5x^3z^2 + 10z^3 + 5y^3 + 11y^2z^3$$

(*a*) Arrange the above polynomial lexicographically.

(*b*) Represent it using linked list.

Ans. (*a*) Lexicographically means that we have to arrange the polynomial according to decreasing power of exponent *x*, *y* and *z* as given below :

$$P(x) = 4x^5yz + 2x^4y^3z^2 + 5x^3z^2 + 10x^2y^2 + 5y^3 + 11y^2z^3 + 10z^3$$

(*b*)

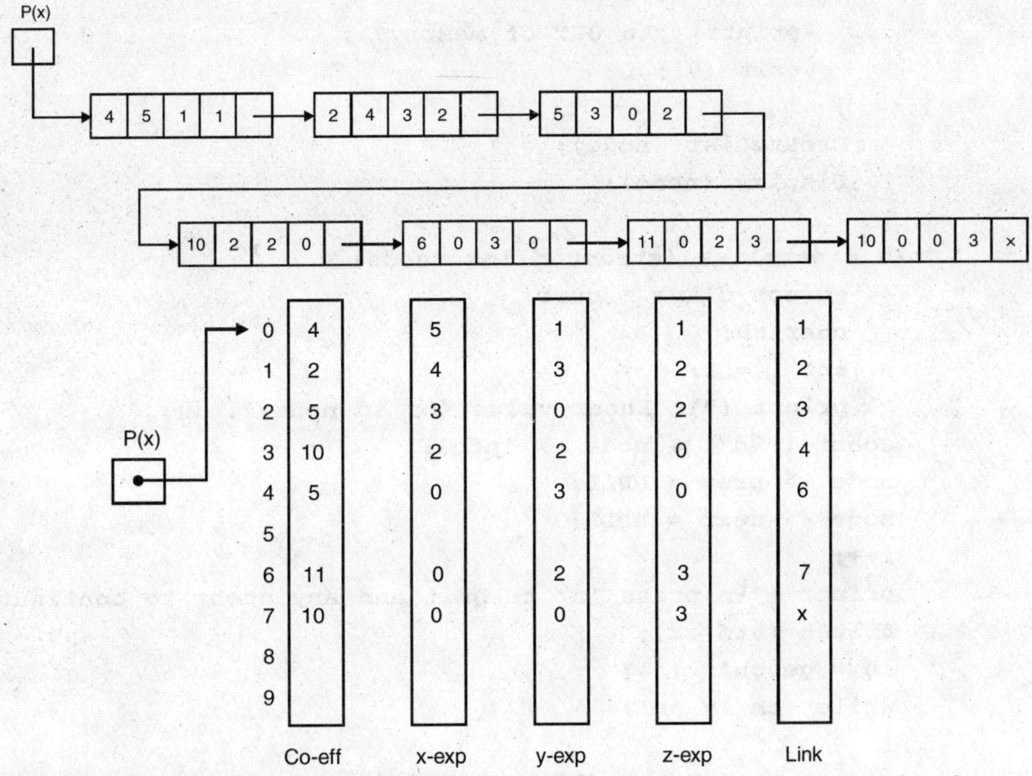

32. Write a program to create a doubly linked list and traverse it in both the directions ?

Ans.

```
#include<stdio.h>
   #include<malloc.h>
   struct dlink
   {  int info;
      struct dlink *prev;
      struct dlink *next;
   };

   void createList (struct dlink *);
   void Display (struct dlink *);
```

```
void main ( )
{
    struct dlink * node;
    clrscr ( );

    node = (struct dlink *) malloc
        (size of (struct dlink));
    if (node = = NULL)
    {
        printf ("\n OUT of memory");
        exit (0);
    }
    CreateList (node);
    Display (node);
}
void createlist (struct dlink *node)
{   struct dlink * curr;
    char ch;
    int i = 1;
    printf ("\n Enter value for %d node:", i);
  scanf ("%d", & node -> info);
  node -> prev = NULL;
  node -> next = NULL;
  i++;
  printf ("\n press 'n' to quit and any otehr to continue :");
  fflush (std in);
  ch = getchar ( );
  while (ch != 'n')
  {
    curr = (struct dlink *) malloc (size of (struct dlink));
    if (curr = = NULL)
    {
        printf ("\n OUT OF MEMORY");
        exit (0);
    }
  node → next = curr;
  curr → prev = node;
  node = node → next;
  printf ("\n Enter value for %d node:", i);
  scanf ("%d", & node → info);
  node → next = NULL:
```

```
        printf ("\n press 'n' to quit:");
        fflush (stdin);
        ch = getchar( );
        i++
        }
    }

    void Display (struct dlink * node)
    {
        struct dlink * temp;
        printf ("\n values of nodes in the list in forward
              direction \n");

        while (node! = NULL)
        {  temp = node;
           printf ("%d", node → info);
           node = node → next;
        }
        printf ("\n\n values of nodes in the list in backward
            direction \n\n");
        node = temp;
        while (node != NULL)
        {
            printf ("%d", node → info);
            node = node → prev;
        }
    }
```

33. Write a single 'C' statement to perform the following task. Each time restrart with the original list as given below.

(*a*) *P* should point to the first node in the list.

(*b*) Print the value of the next node pointed by *P*.

(*c*) *q* should point to the second node in the list.

(*d*) Delete the node which is next to the node pointed by *q*.

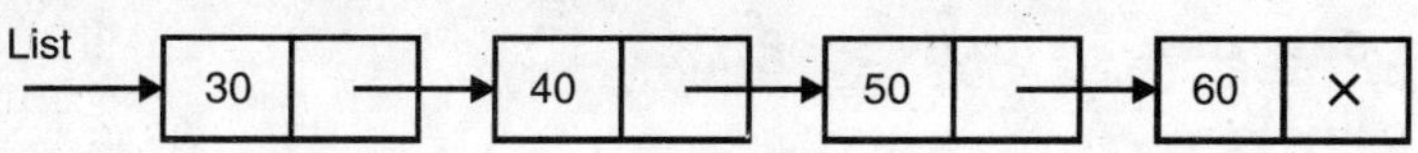

Ans. (*a*) P = List

(*b*) printf ("%d", P → next);

(*c*) q = list → next

(*d*) q → next = *q* → next → next;

34. How can we search for data in a linked list ?

Ans. Unfortunately, the only way to search a linked list is with a linear search, because the only way the member's of a linked list can be accessed is sequentially. Sometimes its quicker to take the data from a linked list and store it in a different data structure so that searches can be more efficient.

35. How can we reverse a linked list without any 'C' pointers ?

Ans. One way is to reverse the data in the nodes without changing the pointers themselves. One can also create a new linked list which is the reverse of the original linked list. Please note that you still uses the "next" pointer field to traverse through linked list.

36. Write down the procedure for doubly linked list containing *n* nodes, where *n* > 0 :

(1) Delete the ith node.

(2) Insert a new node at 5th position (assume *n* > 4)

Ans. (1) Assuming a doubly linked list with n number of nodes, let $n = 4$.

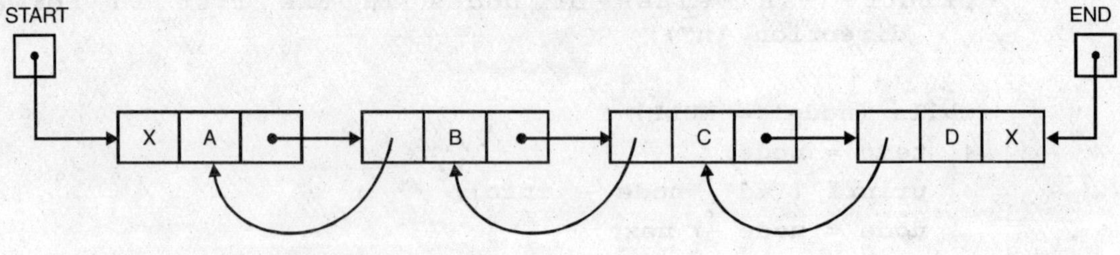

Delete (node * START, node * i)

```
(1)    set P = START
(2)    Repeat while ((P → next)! = 1)
   (a) Set P = (P → next)
(3)    set (i → next) → prev = P
(4)    Set P → next = i → next
(5)    Exit
```

(2) Insert (node * START, node * new)

```
(a)    Set P = START, Set i = 1
(b)    Repeat while (i ≠ 4) do
(a)    Set P = (P → ne × t)
(b)    Set i = i + 1
(c)    Set (P → next) → Prev = new
(d)    Set new → next = P → next
(e)    Set P → next = new
(f)    Set new → prev = P
(g)    Exit
```

37. Write a function/algorithm to insert a node 'q' which is in between nodes 'p' and 'r' such that 'p' and 'r' are consecutive nodes ?

Ans.

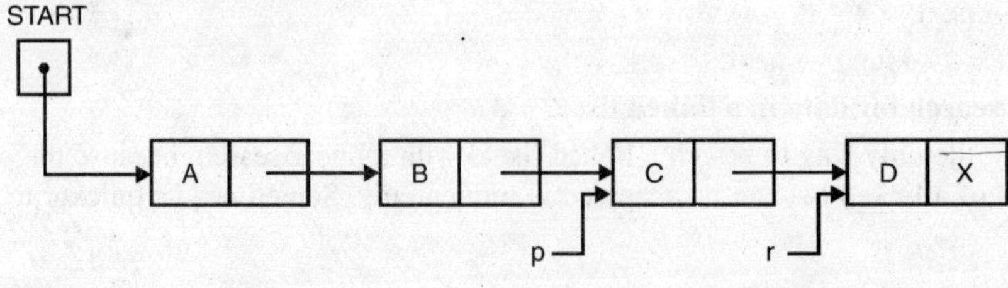

A new node which is to be inserted in between 'p' and 'r' is

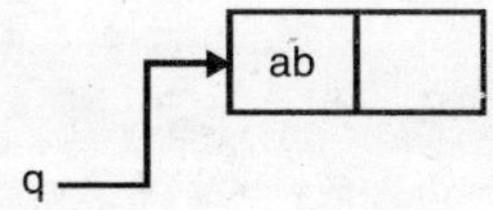

where p, q and r are the three pointers points to the nodes of the linked list.

Insert (node * P, node * r, node * q)

```
(1)    Set P → next = q
(2)    Set q → next = r
(3)    Exit
```

SUMMARY

1. A linked list is a dynamic data structure of elements which may be integer, float, string or a structure.
2. The simplest kind of linked list is a singly-linked list (SLL) whIch has one link per node. This link points to the next node in the list, or to a null value or empty list if it is the final node.
3. Logic for reversing a linked list is similar to reversing an array.
4. Merging is an operation in which we combine elements of two arrays, linked list etc into third structure in sorted order. For that we require that the two source linked list should already be sorted
5. A doubly linked list (DLL) is similar to singly linked list that we have seen earlier in all the programs done earlier. A DLL has an additional pointer that point to the node previous to it.
6. A circular linked list is the linked list in which last node always point to the first node. In other way we can say that first node and final node are linked together. A CLL can be there for SLL and DLL.
7. Singly CLL is similar to ordinary SLL but the next pointer of the last node points to the first node. As in a SLL, new nodes can only be efficiently inserted after a node when we already have a pointer to that node.
8. In a doubly CLL, each node has two links, similarly to a DL as discussed earlier, except that the previous pointer of the first node points to the last node and the next pointer of the last node points to first node.
9. A polynomial in mathematics is of the following form of function *X*.
$$AX^n + BX^{n-1} + CX^{n-2} + \ldots\ldots\ldots$$
where AX^n is the first term in the polynomial, BX^{n-1} is the second term and so on.
10. An unrolled linked list is a variation on the linked list which stores multiple elements in each node. An unrolled linked list is known as cache sensitive structure.
11. Linked lists are used as a building block for many other data structures, such as stacks, queues, trees, graphs etc.

EXERCISES

1. What do you understand by Linked List?
2. Is linked list is linear or non-linear data structure? Justify
3. What is self-referential structure in C?
4. What possible operations can be performed over a linked list?
5. How doubly linked list differs from singly linled list?
6. What are the advantages of a circular linked list over non circular singly linked list?
7. What do you understand by header and sentinel in a linked list?

8. Show figure with header node for
   (*i*) Singly linked list
   (*ii*) Singly circular linked list
   (*iii*) Doubly linked list
   (*iv*) Doubly circular linked list
9. What is unrolled linked list? What is the basic concept behind this type of linked list?
10. What are the advantages & disadvantages of linked list over arrays?
11. Write all possible disadvantages of linked list.
12. Discuss the operation of searching in a linked list.
13. What are the various applications of linked list?
14. What is the significance of header in a linked list? How can a polynomial be represented in a linked list?
15. Write algorim/C code for single linked list for the followings:
   (*i*) Count the number of nodes in a given linked list
   (*ii*) Delete the kth node from linked list
   (*iii*) Create a new list by reversing the list
   (*iv*) Concatenates two lists
   (*v*) Split a list into even and odd list
16. Perform the same operations as given in problem 15 over a doubly linked list.
17. Discuss the structure of a node in unrolled linked list.
18. Discuss the case of reversing a doubly linked list with change of pointers.
19. Consider the following operations;
   (*i*) Removing last node in singly linked list
   (*ii*) Removing last node in doubly linked list
   (*iii*) Removing first node in singly linked lint
   (*iv*) Removing first node in doubly Linked list

   What is the time complexity in each case, considering number of elements in both types of linked list is *N*.
20. How can end of list be checked in circular linked list?
21. Assuming a linked list is given in increasing order of integers. How can a binary search be applied over this linked list to search for a specific element?
22. Write a C routine that removes duplicate elements in a linked list.
23. Write a C routine that removes all occurrences of a given data node from linked list.
24. Modify the routine in problem 23 to remove only first and last occurrence separately in linked list.

❑❑❑

CHAPTER 7

Sparse Matrix

7.1. INTRODUCTION

A sparse matrix is a matrix where majority of its elements are zero and very few non-zero elements exist. The zero elements occupies unnecessary memory space, therefore we need only to store the values of nonzero members, and rows and columns in which member occurs. The figure given below shows some of the examples sparse matrices,

$$\begin{bmatrix} 0 & 2 & 3 & 0 & 0 \\ 0 & 2 & 1 & 0 & 0 \\ 0 & 0 & 0 & 0 & 5 \\ 0 & 3 & 0 & 0 & 0 \end{bmatrix} \quad \begin{bmatrix} 2 & 0 & 0 & 0 & 0 & 0 \\ 0 & 0 & 0 & 6 & 8 & 0 \\ 0 & 0 & 0 & 2 & 0 & 0 \\ 0 & 1 & 4 & 4 & 0 & 4 \end{bmatrix} \quad \begin{bmatrix} 0 & 0 & 0 & 0 & 0 \\ 0 & 2 & 4 & 6 & 0 \\ 0 & 3 & 0 & 4 & 0 \\ 0 & 0 & 3 & 0 & 3 \\ 0 & 0 & 0 & 0 & 0 \end{bmatrix} \quad \begin{bmatrix} 0 & 0 & 0 & 0 \\ 0 & 4 & 0 & 0 \\ 3 & 0 & 0 & 1 \\ 1 & 0 & 3 & 0 \end{bmatrix}$$

size 4 × 5 *size 4 × 6* *size 5 × 5* *size 4 × 4*

Fig. 7.1 (a). Sparse matrices.

Sometimes zero elements are not shown in sparse matrices and only nonzero elements are shown. This is shown for the figure shown above:

$$\begin{bmatrix} - & 2 & 3 & - & - \\ - & 2 & 1 & - & - \\ - & - & - & - & 5 \\ - & 3 & - & - & - \end{bmatrix} \quad \begin{bmatrix} 2 & - & - & - & - & - \\ - & - & - & 6 & 8 & - \\ - & - & - & 2 & - & - \\ - & 1 & 4 & 4 & - & 4 \end{bmatrix} \quad \begin{bmatrix} - & - & - & - & - \\ - & 2 & 4 & 6 & - \\ - & 3 & - & 4 & - \\ - & - & 3 & - & 3 \\ - & - & - & - & - \end{bmatrix} \quad \begin{bmatrix} - & - & - & - \\ - & 4 & - & - \\ 3 & - & - & 1 \\ 1 & - & 3 & - \end{bmatrix}$$

size 4 × 5 *size 4 × 6* *size 5 × 5* *size 4 × 4*

Fig. 7.1 (b). Sparse matrices.

7.2. REPRESENTATION OF SPARSE MATRIX

As we know that sparse matrix is a matrix where high proportion of zero elemets are present. Presence of zero elements serves no purpose, instead we are interested in only nonzero entries. We, therefore need a way to represçnt spare matrices in terms of its nonzero elements only. For that, in general two types representation for sparse matrix exist:

(*i*) Using three column form

(*ii*) Using linked list.

We study both the forms one by one.

7.2.1. Representation of Sparse Matrix Using Three Column Form

For representing sparse matrix, we have 3-column rows. The columns of the first row stores: number of rows, number of columns ,and total number of non-zero elements in order. The next rows for this representation consist of row number, column number and non-zero element at that row-column intersection. For example a tuple of the form :

2 1 8

tells us that 3rd row and 2nd column (indexing starts at 0)has a non zero value 8

For the following sparse matrix of 4 by 4,

$$\begin{bmatrix} 0 & 2 & 0 & 0 \\ 4 & 0 & 3 & 0 \\ 0 & 1 & 0 & 1 \\ 0 & 2 & 0 & 0 \end{bmatrix}$$

The three column rows form will be :

```
4   4   6
0   1   2
1   0   4
2   1   1
2   3   1
3   1   2
```

Program 7.1. /* The program given below converts a sparse matrix into 3-column row form */ ....

```
#include<stdio. h>
#include<conio. h>
#include<stdlib. h>
#define ROW 3
#define COL 3
struct sparse
{
int*pr;
int*pc;
int*pv;
};
```

```
typedef struct sparse sp;

void main()
{
int arr[ROW] [COL];
void input(int [ ] [COL]);
void show(int [][COL]);
int getNZ(int[][COL]);
void convert(int [][COL], sp*);
void disp_tuple (sp);
sp ptr;
int count;
input (arr);
printf("\nOriginal matrix is \n");

show(arr);
        count= getNZ(arr);
ptr. pr = (int*)malloc(sizeof(int)*(count+1));
ptr. pc = (int*)malloc(sizeof(int)*(count+1));
ptr. pv =(int*)malloc(sizeof(int)*(count+1));
ptr. pr[0] =ROW;
ptr. pc[0] =COL;
ptr. pv[0] =count;
convert (arr, &ptr);
printf("'\nThe tuple form of sparse matrix is\n");
disp_ptuple (ptr);
}
void input(int p[] [COL])
{
int i, j;
for(i = 0;i<ROW;i++)
{
for(j = 0;j<COL;j++)

{
printf("\nEnter arr[%d] [%d]element:" i, j);
scanf("%d", &p[i][j]);
}
}

}
void show(int p[] [COL])
```

```
{
int i, j;
for(i =0;i<ROW;i++)
{
for(j =0;j<COL;j++)
{
        printf("%3d", p[i][j]);
}
printf (\n");
}
}
int getNZ(int p[] [COL])

{
int c =0, i, j;
for(i = 0;i<ROW;i++)
    for(j = 0;j<COL;j++)
        if(p[i] [j] ! = 0)
                c++;
return c;
}
void convert(int p[] [COL], sp*ptr)

{
int i, j, k = 1;
for(i =0;i<ROW;i++)
    for(j =0;j<COL;j++)
            if(p[i] [j]! = 0)
            {
                ptr->pr[k] = i;
                ptr->pc[k] = j;
                ptr->pv[k] = p[i] [j];
              k++;
        }
}
void disp_tuple(sp ptr)
{
int i;
printf("ROW\tCOL\tVALUE\n");
for(i =0;i<ptr. pv[0]+1;i++)
printf("%d\t%d\t%d\n", ptr. pr[i], ptr. pc[i], ptr. Pv[i]);
}
```

OUTPUT ....

```
Enter arr[0] [0]element: 1
Enter arr[0] [1]element: 0
Enter arr[0] [2]element: 9
Enter arr[1] [0] element: 0
Enter arr[1] [1]element: 5
Enter arr[1] [2]element: 0

Enter arr[2] [0]element: 0
Enter arr[2] [1]element: 3
Enter arr[2] [2]element: 6
Original matrix is
   1  0  9
   0  5  0
   0  3  6
The tuple form of sparse matrix is
ROW COL VALUE
   3  3   5
   0  0   1
   0  2   9
   1  1   5
   2  1   3
   2  2   6
```

In the program we have written a structure sparse that has three data members: pr, pc, pv. All three are integer pointers. Pointer pr acts as 1-D array for all the row elements (*i.e.* stores all row numbers where the element is non zero); Pointer pc acts as 1-D array for all the column elements (*i.e.* stores all column number where the element is non zero); Pointer pv acts as 1-D array for all the non zero elements at the corresponding row and column. For example for the give sparse matrix representation :

3	4	7
0	0	1
0	1	2
0	3	8
1	2	6
1	3	3
2	1	9
2	2	4

pr[0] is 3, pc[0] is 4 and pv[7]. The entries pr[0] and pc[0] together tells size of the matrix from this representation is created was 3 by 4. The entry pv[0] tells us that total number of nonzero elements in the matrix are 7. Next the index 1, entry pr[1], pc[1], pv[1] tells us that first nonzero element in the matrix is row number 0, column number 0 and value of the element is 1.

Note that number of elements in all three arrays pr, pc and pv are 1+number of nonzero elements. In the above case number of nonzero elements are 7 so size of all three arrays is 7+1 *i.e.* 8.

In the program we take input into the original 2-D array arr using the function input. For displaying the original matrix we have written the function show. The function getNZ gives us number of nonzero elements in the main matrix arr. The function is of main interest to us is convert.

DO YOU KNOW

FIBONACCI (1170-1250) (short for *filius Bonacci*, or son of Bonacci") was also known as Leonardo of Pisa. He was born in the Italian commercial center of Pisa. Fibonacci was a merchant who traveled extensively throughout the Mideast, where he came into contact with Arabian mathematics. In his book *Liber Abaci*, Fibonacci introduced the European world to Arabic notation for numerals and algorithms for arithmetic. It was in his book, his famous rabbit problem. Fibonacci also wrote books on geometry and trigonometry and on Diophantine equations, which involve finding integer solutions to equations.

In the convert function we pass the input matrix and a pointer to structure sparse. Before calling the convert function we have called getNZ function to find out number of nonzero elements, This is stored in the variable count. Using count we allocate memory for the three pointer pr, pc and pv. After allocation we set the first entry into three arrays pr0o], pc[0], pv[0] as the number of rows, number of columns and number of non-zero elements.

```
ptr. pr = (int*)malloc(sizeof(int)*(count+1));
ptr. PC = (int*)malloc(sizeof(int)*(count+1));
ptr. pv = (int*)malloc(sizeof(int)*(count+1));
ptr. pr[0] =ROW;
ptr. pc[0] = COL;
ptr. pv[0] = count;
```

Here ptr is a variable of type structure sparse.

The body of the convert function is simple. We take a counter variable k and initialized it to 1. We run a nested for loop and check for the non zero element in the main matrix. Wherever the element is nonzero the if condition is satisfied and corresponding row, column and value is written into the kth entry in array pr, pc and pv. The code is repeated here once again for a quick reference

```
if(p[i] [j] != 0)
{
ptr->pr[k]= i;
ptr->pc[k]= j;
ptr->pv[k]= p[i] [j];
k++;
}
```

For displaying the sparse matrix in 3-column form, we have written the function disp_tuple. Inside the function we pass a parameter of sparse type. Inside the body, using a for loop we display every tuple of the sparse matrix as:

```
for(i = 0;i<ptr. pv[0]+1;i++)
      printf("%d\t%d\t%d\n", ptr. pr[i], ptr. pc[i], ptr. pv[i]);
```

The program presented above suffers from one limitation where original 2-D array as sparse matrix and three column form both exist simultaneously and we talk about saving space. To save the space we need to create 2-D array dynamically and after converting it into 3 column row form we can deallocate the 2-D array. For that the code will be:

```
int **arr;
int j, ROW, COL;
void input(int**, int, int);
printf ("Enter number of rows\n");
scanf("%d", &ROW);
printf ("Enter number of columns\n");
scant ("%d", &COL) ;
arr = (int**)malloc(sizeof(int)* ROW);
        for(j = 0;j<ROW;j++)
                 arr[j] = (int*)malloc(sizeof(int)* COL);
```

The following code can be put into the main function. The function input takes input as elements into the 2-D array. This is given below:

```
void input(int **p, int ROW, int COL)
{
int i, j;
for(i = 0;i<ROW;i++)
{
for(j = 0;j<COL;j++)
{
printf("\nEnter arr[%d] [%d]element: ", i, j);
scanf("%d", &p[i] [j]);
}
}
}
```

The function can be called from main as

```
input(arr, ROW, COL);
```

Note one more thing after the conversion from main sparse matrix to 3-column tuple form representation has been done we free all the memory occupied by the main sparse matrix. Add the following code at the end of convert function.

```
for(i = 0;i<ROW;i++)
       free(p[i]);
free (p);
```

The other way is for saving space is not creating a 2-D array at all. We can simply ask from the user three piece of information: the row number, column number and nonzero value at that row-column. This information can directly be stored into the 3-column tuple. We leave this to you to write program using this logic.

1. Getting Transpose of Sparse Matrix

Obtaining transpose of a matrix is similar to we find transpose of a 2-D array. In the tuple form representation we simply interchange values of all elements of pr array to pc array. For that we have written the function transpose. The logic is very simple but it has one problem. The tuples in the transpose matrix will not be in proper sequence in the order of rows. Only the code for transpose is presented below. Rest of the code is similar we have seen earlier.

```
void transpose(sp p1, sp*p2)
{
p2->pr = (int*)malloc(sizeof(int)*(p1. pv[0]+1));
p2->pc =(int*)malloc(sizeof(int)*(p1. pv[0]+1));
p2->pv = (int*)malloc(sizeof(int)*(p1. pv[0]+1));
int i;
for(i = 0;i<p1. pv[0]+1;i++)
{
p2->pr[i] = p1. pc[i];
p2->pc[i] = p1. pr[i];
p2->pv[i] = p1. pv[i];
}
}
```

2. Addition of Two Sparse Matrices

For the addition of two sparse matrices presented in 3-column tuple form, we have written the following code. The logic is very simple. We consider each tuple of both the matrices and compare their row numbers. For that we keep two control variable i and j denoting each current tuple in both the matrices. The index k is for tuple in the destination matrix. If both the rows are same then their corresponding columns are compared. Depending upon which column is greater i or j is incremented else both are incremented. In case rows are not same, the complete tuple with smaller row value is assigned to destination matrix. See the code given below and go through explanation that follows after the code.

```
void add(sp sm1, sp sm2, sp*sm3)
{
int r1, r2, c1, c2, v1, v2;
int i = 1, j = 1, k = 1;
int count = 1;
int max, m1, m2;

m1 = sm1. pv[0];
m2 = sm2. pv[0];
```

```
max = m1 + m2;
sm3->pr =  (int*)malloc(sizeof(int)*(max+1));
sm3->pc = (int*)malloc(sizeof(int)*(max+1));
sm3->pv = (int*)malloc(sizeof(int)*(max+1));
while (count < = max)

{
r1 = sm1. pr[i];
r2 = sm2. pr[j];
c1 = sm1. pc[i];
c2 = sm2. pc[j];
v1 = sm1. pvIi];
v2 = sm2. pv[j];
if(i>m1 | | j>m2)
break;
if (r1 = = r2)
{
if (c1 = = c2)
{
sm3->pr[k] = r1;
sm3->pc[k] = c1;
sm3->pv[k] = v1+v2;
i++;

j++;
k++;
}
else if (c1 < c2)
{
sm3->pr[k] = r1;
sm3->pc[k] = c1;
sm3->pv[k] = v1;
i++;
k++;
}
else

{
sm3->pr[k] = r2;
sm3->pc[k] = c2;
sm3->pv[k] = v2;
j ++;
k++;
}
count++;
```

```
}
else if (r1 < r2)
{
sm3->pr[k] = r1;
sm3->pc[k] = c1;
sm3->pv[k] = v1;
i++;
count++;
k++;
}

else
{
sm3->pr[k] = r2;
sm3->pc[k] = c2;
sm3->pv[k] = v2;
j++;
count++;
k++;
}
}
while(i < = m1)
{
sm3->pr[k] = sm1. pr[i];
sm3->pc[k] = sm1. pc[i];
sm3->pv [k] = sm1. pv [i];
i++;
k++;
}

while(j  <= m2)
{
sm3->pr[k]= sm2. pr[j];
sm3->pc[k] = sm2. pc[j];
sm3->pv[k] =sm2. pv[j];
j ++;
k++;

}
k--;
sm3->pr[0] = ROW;
sm3->pc[0] = COL;
sm3->pv[0] = k;
}
```

For the addition of two sparse matrices we have written the function add. The function takes three arguments: first two are parameters of sparse type and denote source matrices, third argument is of sparse* type and denote the destination matrix.

In the function add we have 6 variables r1, r2, c1, c2, v1, and v2. All these six variables are for representing the values of rows, columns and values at row-column for the two source matrices. For handling one row at a time from first matrix we have counter variable i, similarly for handling second matrix we ha counter variable j and for destination matrix we have counter variable k, All these variables are initialized to 1 as at index 0 we have number of row, columns and total number of nonzero elements stored. Number of nonzero elements of first and second matrix is stored in ml and m2 respectively. Sum of these two is stored in variable max.

For the destination matrix sm3, memory is allocated for pr, pc and pv using this max variable. The main while loop runs till count < = max. In the beginning of the while loop, r1, r2, c1, c2, v1 and v2 are initialized with the first row of both the sparse matrices represented by sm1 and sm2.

7.2.2. Linked List Representation of Sparse Matrix

For representing a sparse matrix using linked list we require declaration of three structures: **headnode, colnode and rownode.**

(*i*) The structure headnode that contains 4 members: nrows for containing the number of rows, ncols for number of columns, nvals for number nonzero elements and head pointer of rownode type for pointing to first row of the sparse matrix maintained as linked list.

(*ii*) The structure rownode that contains 4 members: row is for row number, a pointer down of rownode type for pointing to next row and a pointer right of co1rode type for pointing to first node of linked list, maintained for all nonzero columns for that row.

(*iii*) The structure colnode that contains 3 members: first is the column number col, second is the value val and third is the next pointer of type colnode that points to next node for this linked list.

All the structures are declared below:

```
struct headnode
{
int nrows;
int ncols;
int nvals;
struct rownode*head;
};
struct rownode
{
int row;
struct rownode*down;
struct colnode*right;
};
struct colnode
{
int col;
int val;
struct colnode*next;
};
```

To understand the usage of above structures, consider the following sparse matrix :

$$\begin{bmatrix} 0 & 3 & 0 & 2 & 5 \\ 0 & 4 & 0 & 0 & 0 \\ 0 & 3 & 4 & 0 & 0 \\ 0 & 1 & 0 & 0 & 3 \end{bmatrix}$$

Sprase matrix of size 4 × 5

The corresponding linked list representation is shown below :

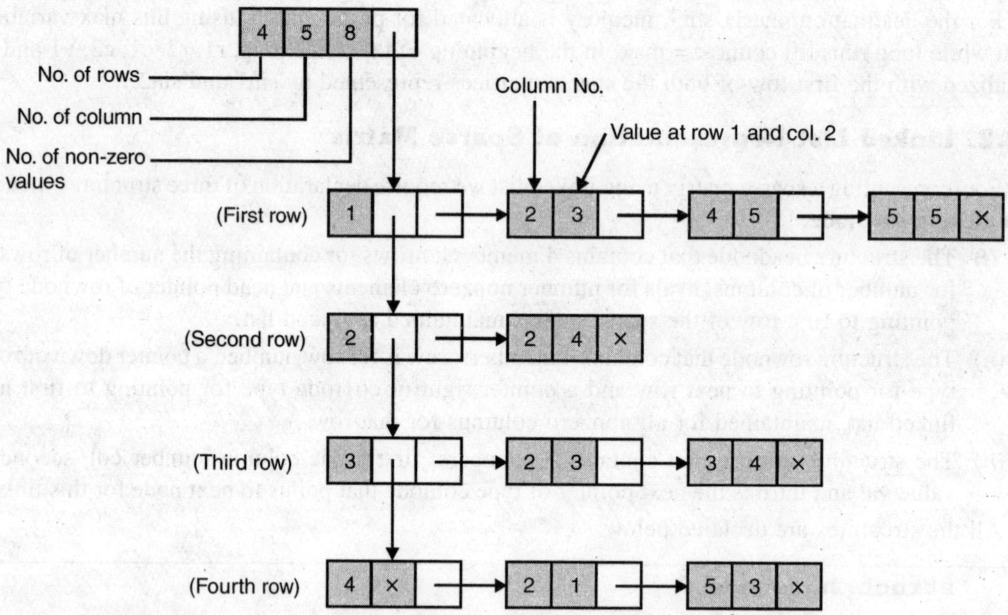

Fig. 7.2. Linked list representation of the matrix shown in previous figure.

The complete program for converting a sparse matrix into its equivalent linked list repesentation is given below:

Program 7.2. /*Program Converting a sparse matrix into its equivalent liked list representation*/ ....

```
#include<stdio. h>
#include<conio. h>
#include<stdlib. h>
struct headnode
{
int nrows;
jut ncols;
int nvals;
struct rownode*head;
```

```
};
struct rownode
{
int row;
struct rownode*down;
struct colnode*right;
};

struct colnode
{
int cal;
int val;
struct colnode*next;
};

typedef struct headnode Hnode;
typedef struct rownode Rnode;
typedef struct colnode Cnode;
Rnode* rnalloc()
{
return (Rnode*)malloc(sizeof(Rnode));
}
Cnode* cnalloc()
{
return (Cnode*)malloc(sizeof(Cnode));
}
void main()
{
int **arr;
int j, ROW, COL;
void input(int**, int, int);
void show(int int, int);
void init(Hnode**, int, int);
void convert(int **, int, Hnode**);
void display(Hnode*);
Hnode*SP;
SP = NULL;
printf ("Enter number of rows\n");
scanf("%d", &ROW);
printf ("Enter number of colurnns\n");
scanf("%d", &COL);
```

```
arr = (int**)malloc(sizeof(int)* ROW);
for(j =0;j<ROW;j++)
arr[j] = (int*)malloc(sizeof(int)* COL);
input(arr, ROW, COL);
printf("\nOriginal matrix is \n");
show(arr, ROW, COL);
init(&SP, ROW, COL);

convert(arr, ROW, COL, &SP);
printf("\nThe tuple form of sparse matrix is\n");
display(SP);
}
void input(int **p, int ROW, int COL)
{

int i, j;
for(i = 0;i<ROW;i++)
{
for(j =0;j<COL;j++)
{

printf("\nEnter arr[%d] {%d] element: ", i,j);
scanf("%d", &p[i] [j]);
}
}

}
void show(int **p, int ROW, int COL)
{
int i, j;
for(i = 0;i<ROW;i++)
{
for(j = 0;j<COL;j++)
                printf("%3d", p[i] [j]);
printf ("\n");
}

}
void init(Hnode**SP, int R, int C)
{
(*SP) = (Hnode*)malloc(sizeof(Hnode));
(*Sp)->nrows = R;
```

```
(*Sp)->ncols = C;
(*Sp)->nvals = 0;
(*Sp)->head = NULL;
}

void convert(int **p, int ROW, int COL, Hnode**SP)
{
int i,j;

Rnode *rptr *rtemp *rsave;
Criode *cptr, *ctemp;
for(i = 0;i<ROW;i++)
for(j = 0;j<COL;j++)
if(p[i] [j]! = 0)
{

(*SP) ->nvals++;
cptr =cnalloc ();
cptr->col =j;
cptr->val = p[i] [j];
cptr->next = NULL;
if((*Sp)->head = = NULL)
{
rptr = rnalloc ();
rptr->down NULL;
rptr->row = i;
rptr->right = cptr;
(*Sp)->head = rptr;
}
else
{
rsave = NULL;
rtemp = (*Sp)->head;
while(rtemp! = NULL && rtemp->row  <   i)
{
rsave = rtemp;
rtemp = rtemp->down;
}
if(rtemp->row = = i)
{
ctemp = rtemp->right;
while(ctemp->next != NULL)
```

```
ctemp = ctemp->next;
ctemp->next = cptr;
}

else if(rtemp = =NULL)
{
rptr = rnalloc();
rptr->down = NULL;
rptr->right = cptr;
rptr->row = i;
rsave->down = rptr;
}
}
}

}
void display(Hnode*SP)
{
Rnode*rptr;
Cnode*cptr;
rptr = SP->head;
printf ("%d\t%d\t%d\n", SP->nrows, SP->ncols,sp->nvals);
while(rptr ! = NULL)
{
cptr = rptr->right;
while(cptr! = NULL)
{
printf("%d\t%d\t%d ->", rptr->row, cptr->col, cptr ->val);
cptr = cptr->next;
}
rptr = rptr->down;
printf("\n");
}
}
```

OUTPUT ....

```
Enter number of rows
   3
Enter number of columns
   3
Enter arr[0] [0] element: 1
Enter arr[0] [1]element: 2
```

```
Enter arr[0] [2] element: 0
Enter arr[1] [0] element: 0
Enter arr[1] [1]element: 6
Enter arr[1] [2]element: 7
Enter arr[2] [0]element: 0
Enter arr[2] [1] element: 9
Enter arr[2] [2]element: 0
Original matrix is
        1    2    0
        0    6    7
        0    9    0
The tuple form of sparse matrix is
        3    3    5
        0    0    1
        0    1    2
        1    1    6
        1    2    7
        2    1    9
```

In the program each of the above structure as discussed earlier is type defined as: headnode as Hnode, rownode as Rnode, and colnode as Cnode.

In the main we have a pointer SP of Hnode type. The function init allocates memory for the node pointed by SP and initializes its various data members. The function convert is the crux of the program and we will discuss it first. In the function we run a nested for loop for all the elements of the sparse matrix arr. For all nonzero elements if condition will be satisfied. We consider the same matrix for illustration purpose as shown in the output of the program. Consider the first time if condition will be true (at row 0 and col 0) nvals be incremented by 1:

```
(*Sp)->nvals++;
```

A column node is created and pointed by cptr. Its various fields are filled. first if condition will be true as (*SP)->head is NULL. A new row is created and pointed by rptr; right of rptr points to cptr and down will be NULL and head (*SP) will point to rptr.

For the second non zero element (row 0, col 1) again column node will be created and cptr will point to it, The if condition will be false and else part will executes. In the else part rtemp pointer will point to head of (*SP). Now in the else part it will be checked whether the nonzero element belong to any existing linked list (for each of the row, we have one linked list) or new linked list will be created. Here the row number matches with i. Now addition of new node is similar to adding a new node at the end of the linked list.

For the third nonzero element (row 1, col 1) in the else part the while loop, causes rsave to point to previous node in the linked list of rows and rtemp will be NULL. The last last else-if condition will be true and new row node will be created. Whose next point to cptr and down will be NULL.

For displaying the linked list form of sparse matrix, we start with the first row and display all nonzero elements in that row by following the next pointer field. This is done in the inner while loop. Next we use down pointer for advancing to the next row.

7.3. SPECIAL TYPES OF SPARSE MATRICES

The following symmetric matrices (rows and columns are same) are some special forms of sparse matrices.

1. Lower Triangular matrix
2. Upper Triangular matrix
3, Diagonal matrix
4. Tridiagonal matrix

A square matrix is said to be lower triangular when all non zero elements (few may be zero) of matrix resides at and below the main diagonal and all elements above main diagonal are zero. For example the matrices shown below (wherc X is any non zero element) is lower triangular matrix :

$$\begin{bmatrix} \times & 0 & 0 & 0 \\ \times & \times & 0 & 0 \\ \times & \times & \times & 0 \\ \times & \times & \times & \times \end{bmatrix} \qquad \begin{bmatrix} \times & 0 & 0 & 0 & 0 \\ \times & \times & 0 & 0 & 0 \\ \times & \times & \times & 0 & 0 \\ \times & \times & \times & \times & 0 \\ \times & \times & \times & \times & \times \end{bmatrix}$$

Fig. 7.3. Lower triangular matrices.

A square matrix is said to be upper triangular when all non zero elements (few may be zero) of matrix resides at and above the main diagonal and all elements below main diagonal are zero. For example the matrix shown below (where X is any non zero element) is an upper triangular matrix:

$$\begin{bmatrix} 0 & \times & \times & \times \\ 0 & 0 & \times & \times \\ 0 & 0 & 0 & \times \\ 0 & 0 & 0 & 0 \end{bmatrix} \qquad \begin{bmatrix} \times & \times & \times & \times & \times \\ 0 & \times & \times & \times & \times \\ 0 & 0 & \times & \times & \times \\ 0 & 0 & 0 & \times & \times \\ 0 & 0 & 0 & 0 & \times \end{bmatrix}$$

Fig. 7.4. Upper triangular matrices.

A square matrix is said to be triadiagonal when all non zero elements (few may be zero) of matrix resides at the main diagonal and just above and below the main diagonal and all other elements are zero. For example the matrices shown below (where X is any non zero element) are triadiagonal matrix :

$$\begin{bmatrix} \times & \times & 0 & 0 \\ \times & \times & \times & 0 \\ 0 & \times & \times & \times \\ 0 & 0 & \times & \times \end{bmatrix} \qquad \begin{bmatrix} \times & \times & 0 & 0 & 0 \\ \times & \times & \times & 0 & 0 \\ 0 & \times & \times & \times & 0 \\ 0 & 0 & \times & \times & \times \\ 0 & 0 & 0 & \times & \times \end{bmatrix} \qquad \begin{bmatrix} \times & \times & 0 & 0 & 0 & 0 \\ \times & \times & \times & 0 & 0 & 0 \\ 0 & \times & \times & \times & 0 & 0 \\ 0 & 0 & \times & \times & \times & 0 \\ 0 & 0 & 0 & \times & \times & \times \\ 0 & 0 & 0 & 0 & \times & \times \end{bmatrix} \qquad \begin{bmatrix} \times & \times & 0 & 0 & 0 & 0 & 0 \\ \times & \times & \times & 0 & 0 & 0 & 0 \\ 0 & \times & \times & \times & 0 & 0 & 0 \\ 0 & 0 & \times & \times & \times & 0 & 0 \\ 0 & 0 & 0 & \times & \times & \times & 0 \\ 0 & 0 & 0 & 0 & \times & \times & \times \\ 0 & 0 & 0 & 0 & 0 & \times & \times \end{bmatrix}$$

Fig. 7.5. Tridiagonal matrices.

A square matrix is said to be diagonal when all non zero elements of matrix resides at the main diagonal For example the matrices shown below (where X is any non zero element) are a diagonal matrix:

$$\begin{bmatrix} \times & 0 & 0 \\ 0 & \times & 0 \\ 0 & 0 & \times \end{bmatrix} \qquad \begin{bmatrix} \times & 0 & 0 & 0 \\ 0 & \times & 0 & 0 \\ 0 & 0 & \times & 0 \\ 0 & 0 & 0 & \times \end{bmatrix} \qquad \begin{bmatrix} \times & 0 & 0 & 0 & 0 \\ 0 & \times & 0 & 0 & 0 \\ 0 & 0 & \times & 0 & 0 \\ 0 & 0 & 0 & \times & 0 \\ 0 & 0 & 0 & 0 & \times \end{bmatrix}$$

3×3 $\qquad$ 4×4 $\qquad$ 5×5

Fig. 7.6. Diagonal matrices.

Programs Based on Special Sparse Matrices

Program 7.3. /* Program To count sum of elements above and below main diagonal*/

```
#include<stdio. h>
#include<conio. h>
#define ROW 4
#define COL 4
void main()

{
int a[ROW] [COL];
int i, j, dsum = 0;
for(i =0;i<ROW;i++)
for(j = 0;j<COL;j++)
{
printf("\nEnter a[%d] [%d] element: ", i, j);

scanf("%d", &a[i] [i]);
}
}
printf("\n\nMatrix is\n");
for(i =0;i<ROW;i++)

{
for(j =0;j<COL;j++)
priritf("%2d", a[i] [j]);
print ("\n")
}

printf("\nElements above main diagonal are\n");
for(j = 1;j<ROW;j++)
```

```
for(i = 0;i<j;i++)
{

printf("%d ", a[i] [i]);
dsum = dsum+a[i] [j];
}

printf("\nSum of elements above main diagonal is %d\n", dsum)
dsum = 0;
printf("\nElements below main diagonal are\n");
for(i = 1;i<ROW;i++)
for(j = 0;j<i;j++)
printf("%d ", a[i] [j]);
dsum = dsum+a[i] [j];
}

printf("\nSum of elements below main diagonal is %d\n", dsum);

getch ();
}
```

OUTPUT ....

```
Enter a [0] [0] element: 2
Enter a [0] [1] element: 3
Enter a [0] [2] element: 4
Enter a [0] [3] element: 5
Enter a [1] [0] element: 6
Enter a [1] [1] element: 7
Enter a [1] [2] element: 8
Enter a [1] [3] element: 9
Enter a [2] [0] element: 1
Enter a [2] [1] element: 7
Enter a [2] [2] element: 9
Enter a [2] [3] element: 0
Enter a [3] [0] element: 3
Enter a [3] [1] element: 4
Enter a [3] [2] element: 5
Enter a [3] [3] element: 6
Matrix is
   3 6 7 8
   9 0 2 3
   5 7 1 0
   6 4 3 8
```

```
Elements above main diagonal are
   6 7 2 8 3 0
Sum of elements above main diagonal is 26
Elements below main diagonal are
   9 5 7 6 4 3
Sum of elements below main diagonal is 34
```

Note for all the elements that are above the main diagonal we have written a nested for loop as:

```
for(j = 1;j<ROW;j++)
   for(i = 0;i<j;i++)
```

The above for loops are written so that they displays elements from each column *i.e.* initially all elements from second column that are above the main diagonal are displayed, next from third column and so on.

In the first row there are 3 elements, second row 2 and in third row just 1. Our outer loop is for all the columns and inner loop is for all such rows. For j = 1 inner loop runs only once and element a[0] [1] is displayed. For j = 2, inner loop runs twice and elements a [0][2] and a[1][2] are displayed. For j = 3, inner loop runs thrice and elements a [0][3], a[1][3] and a[2][3] are displayed.

For all the elements that are below the main diagonal we have written nested for loop as:

```
for(i = 1;i<ROW;i++)
for(j = 0;j<i;j++)
```

The first row, starting from row 1 (the index) every row contains elements one more than the previous one. Have a look at matrix and the above for loop, it simple to understand.

Program 7.4. /* Program To check whether a matrix is lower triangular or not */....

```
#include<stdio. h>
#include<conio. h>
#define S 4
void main()
{
int a[S] [S];
int i, j, sum1= 0,  LT =1;
clrscr();
for(i = 0;i<S;i++)
{
for(j = 0;j<S;j++)
{
printf("\nEnter a[%d] [%d] element:", i, j);
scanf("%d", &a[i] [j]);
}
```

```
}
clrscr ( );
printf("\n\nMatrix is\n");
for(i = 0;i<S;i++)

{
for(j = 0;j<S;j++)
printf("%2d ", a[i] [j]);
printf("\n");
}
for(j= 1;j<S;j++)
for(i =0;i<j;i++)
if(a[i] [j])
{
LT = 0;
break;

}
if (!LT)
printf ("Matrix is not lower triangular\n");
else
printf ("Matrix is lower triangular\n");
}
```

OUTPUT ....

```
Enter a[0] [0] element: 1
Enter a[0] [1] element: 0
Enter a[0] [2] element: 0
Enter a[0] [3] element: 0
Enter a[1] [0] element: 2
Enter a[1] [1] element: 3
Enter a[1] [2] element: 0
Enter a[1] [3] element: 0
Enter a[2] [0] element: 5
Enter a[2] [1] element: 6
Enter a[2] [2] element: 7
Enter a[2] [3] element: 0
Enter a[3] [0] element: 4
Enter a[3] [1] element: 5
Enter a[3] [2] element: 6
Enter a[3] [3] element: 7
```

```
Matrix is
   1   0   0   0
   2   3   0   0
   5   6   7   0
   4   5   6   7
Matrix is lower triangular
```

In the previous program we have seen how to access the elements above the main diagonal in a matrix. In the program we simply check every element of the matrix that resides above the main diagonal to be zero. If any such element is nonzero then matrix is not lower triangular, so we set LT = 0 and come out from the loop. After the loop we check value of LT using if and display result accordingly.

Program 7.5. /* program to check Whether a matrix is upper triangular or not */....

```
#include<stdio.h>
#include<conio.h>
#define S 4
void main()
{

int a[S] [S];
int i, j, sum1 = 0, UT = 1;
// clrscr( );
for(i= 0;i<S;i++)
{
for(j = 0;j<S;j++)
{

printf("\nEnter a[%d] [%d] element:", i, j);
scanf("%d", &a[i] [j]);
}
}
printf ("\n\nMatrix is\n");
for(i = 0;i<S;i++)
{
for(j = 0;j<S;j++)
          printf("%2d", a[i] [j]);
printf("\n");
}
for(i = 0;i<S;i++)
        for(j = 1;j<i;j++)
                if(a[i][j])
              {
```

```
                UT = 0;
                break;
            }
    if( !UT)
    printf ("Matrix is not upper triangular\n");

    else
    printf ("Matrix is upper triangular\n");
    }
```

OUTPUT ....

```
    Enter a[0] [0] element: 1
    Enter a[0] [1] element: 2
    Enter a[0] [2] element: 3
    Enter a[0] [3] element: 4
    Enter a[1] [0] element: 0
    Enter a[1] [1] element: 3
    Enter a[1] [2] element: 4
    Enter a[1] [3] element: 5
    Enter a[2] [0] element: 0
    Enter a[2] [1] element: 0
    Enter a[2] [2] element: 6
    Enter a[2] [3] element: 8
    Enter a[3] [0] element: 0
    Enter a[3] [1] element: 0
    Enter a[3] [2] element: 0
    Enter a[3] [3] element: 8
    Matrix is
       1 2 3 4
       0 3 4 5
       0 0 6 8
       0 0 0 8
    Matrix is upper triangular
```

In the previous to previous program we have seen how to access the elements below the main diagonal in a matrix. In the program we simply check every element of the matrix that resides below the main diagonal to be zero. If any such element is nonzero then matrix is not upper triangular, so we set UT = 0 and come out from the loop. After the loop we check value of UT using if and display result accordingly.

Program 7.6. /* Program To check whether a matrix is tridiagonal or not*/.

```
#include<stdio. h>
#include<conio. h>
#define S 5
void main()
{
int a[S] [S] = {{1, 2, 0, 0, 0, }, {3, 4, 5, 0, 0}, {0, 4, 7,
8, 0}, {0, 0, 5, 8, 9}, {0, 0, 0, 8, 9} };
int i, j, L= 0, T, t = 0;
// clrscr();
printf("\n\nMatrix is\n");
for(i = 0;i<S;i++)
{
for(j = 0;j<S;j++)
printf("%2d", a[i][j]);
printf("\n");
}
/*Upper block of zeros*/
for(i = 0;i<S-2;i++)
for(j = i+2;j<S;j++)
        if( !a[i] [j])
         ++L;

/*Lower Block of zeros*/
for(i = 2;i<S;i++)
{

t++;
for(j = 0;j<t;j++)
        if(!a[i] [j])
        ++L;
}

/* At most 3*S-2 elements are nonzero*/
T = S*S-(3*S-2);
if(T = = L)
printf("Matrix is tridiagonal\n");
else
printf("Matrix is not tridiagonal\n");
}
```

OUTPUT

```
Matrix is
    1 2 0 0 0
    3 4 5 0 0
    0 4 7 8 0
    0 0 5 8 9
    0 0 0 8 9
Matrix is tridiagonal
```

In all the matrices shown above, starting from first row, there are S-2 rows where number of zero elements are S-2, S-2-1, S-2-2 and so on. For any matrix we first consider the S/2 rows in first half of the logic and remaining S/2 in second half of the logic. Not the last row of first half of the logic and first row of second half of the logic is same and in that row just one zero element is present (actually two zeros are present, but they are considered separately in both the halves). For the first half the logic is written as:

```
/*Upper block of zeros*/
for(i = 0;i<S-2;i++)
        for(j = i+2;j<S;j++)
              if( !a[i] [j])
                    ++L;
```

You can check first column in row i where zero element is present is at i+2. Using this loop we count number of zeros present in upper half. For counting number of zeros in the lower part of matrix, we have written the logic as:

```
for(i = 2;i<S;i++)
{
t++;
for(j=0;j<t;j++)


if (!a [i]  [j] )
++L;
}
```

The first row of the second half of matrix starts at row 2 and in every consecutive row, number of column increase by 1, so we have taken variable t. When both the code for upper and lower half finishes, L contains number of zeros present in both the halves, Now actual number of zeros present in a tridiagonal matrix can be calculated easily. For that we have this information that number of nonzero elements are 3*S-2 and total number of elements in the matrix are S*S, so total number of zero elements are S*S-(3*S-2). This value is compared with value L. If both matches then matrix are tridiagonal else matrix is not tridiagonal.

7.4. STORING SPECIAL SPARSE MATRICES IN MEMORY

All the elements of special types of sparse matrices can be conveniently stored in a 1-D array by choosing a appropriate addressing formula so that each element of these types of matrices map to a unique element in 1-D array. In this section we will see how to get that addressing formula for all special types of matrices discussed above.

Memory representation of some special types of matrices

(*i*) **Lower Triangular Matrix:** Consider the lower triangular matrix as give below:

$$\begin{bmatrix} a_{11} & & & \\ a_{21} & a_{22} & & \\ a_{31} & a_{32} & a_{33} & \\ \cdots & \cdots & \cdots & \\ a_{n1} & a_{n2} & a_{n3} & a_{nn} \end{bmatrix}$$

To write an access formula for the above matrix so that matrix can be stored in l-D array and given i, j, index in l-D array can be obtained, we proceed as follows :

Index K = No. of elements up to a_{ij} element

= total elements in first $(i - 1)$ rows + all the elements in the upto jth column

$= 1 + 2 + 3 + \ldots (i - 1) + j$

$= i(i - 1)/2 + j$

Now assuming array ARR as 1-D array for storing the elements of above Sparse matrix, we have

ARR [K] =MAT[i][j]

where $k = i(i - 1)/2 + j$.

For example MAT [3] [3] will be stored at index

K $= 3(3 - 1)/2 + 3 = (3 \times 2) / 2 + 3 = 6$th

Here it is assumed that array index start at 1.

(*ii*) **Upper triangular matrix :** Consider the following $n \times n$ upper triangular matrix

$$\begin{bmatrix} a_{11} & a_{12} & a_{13} & \cdots & a_{1n} \\ & a_{22} & a_{23} & \cdots & a_{2n} \\ & & a_{33} & \cdots & a_{3n} \\ & & & a_{44} & a_{4n} \\ & & & & \vdots \\ & & & & a_{nn} \end{bmatrix}$$

Similar to lower triangular matrix, we find an access formula for upper triangular matrix

Index

K = No of elements up to a_{ij} element

= Total no. of elements in first $(i - 1)$ rows + all elements up to i^{th} row and j^{th} column

$= n + (n - 1) + (n - 2) + (n - i + 2) + (j - i + 1)$

$= [(n + n + n + \ldots (i - 1 \text{ times})] - [(1 + 2 + 3 + \ldots + (i - 2) + (i - 1)] + j$

$= n \times (i - 1) - i(i - 1)/2 + j$

$= (i - 1) \times (n - i/2) + j$

So Arr $[K]$ = MAT $[i]$ $[j]$ $\{K = (i - 1)(n - i/2) + j$

Assuming $n = 5, i = 2, j = 3$

We have $K = (2 - 1)(5 - 2/2) + 3$

$= (1)(4) + 3 = 7$

i.e. array element MAT [2] [3] will be stored in 1-D array at 7th index.

(*iii*) **Tridiagonal matrix :**

$$\begin{bmatrix} a_{11} & a_{12} & & & & \\ a_{21} & a_{22} & a_{23} & & & \\ & a_{32} & a_{33} & a_{34} & & \\ & & a_{43} & a_{44} & a_{45} & \\ & & \vdots & \vdots & \vdots & \\ & & & a_{(n-1)(n-2)} & a_{(n-1)(n-1)} & a_{(n-1)n} \\ & & & & a_{n(n-1)} & a_{nn} \end{bmatrix}$$

Index

K = No. of elements upto a_{ij} element
= Total no. of elements up to a_{ij} element
= Total no. of elements in frst (i - 1) rows + all elements in ith row upto jth column
$= \{2 + [3 + 3 + ... \text{ upto } (i-2) \text{ terms}]\} + (j - i + 2)$
$= 2 + (i-2) \times 3 + j - (i-2)$
$= 2 + 2 \times (i-2) + j$
$= 2\,(i-1) + j$

SOLVED PROBLEMS

1. What is band matrix ?

Ans. A band matrix is a generalized form of Tridiagonal matrix where all non-zero elements must lie on the main diagonal or on one of the (A-1) diagonals below and above the main diagonal as shown below :

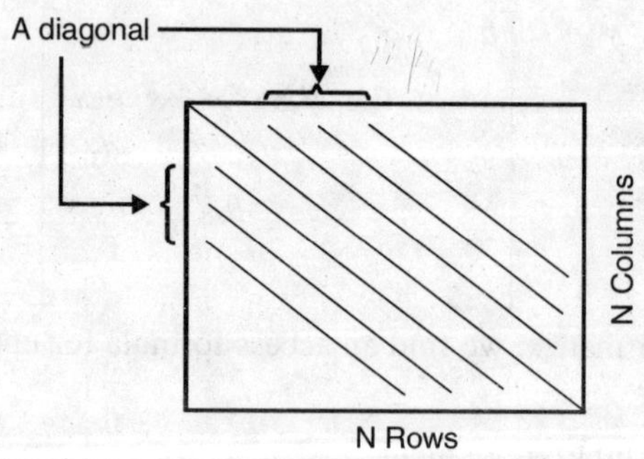

Fig. 7.7.

2. Consider the following sparse matrix shown below :

$$\begin{bmatrix} 2 & 0 & 0 & 3 & 0 & 0 \\ 0 & 6 & 8 & 0 & 0 & 0 \\ 5 & 0 & 0 & 2 & 0 & 0 \\ 0 & 1 & 0 & 4 & 0 & 4 \end{bmatrix}$$

Size 4 × 6

What is its 3-tuple and linked list representation?

Ans.

4	6	9
0	0	2
0	3	3
1	1	6
1	2	8
2	0	5
2	3	2
3	1	1
3	3	4
3	5	4

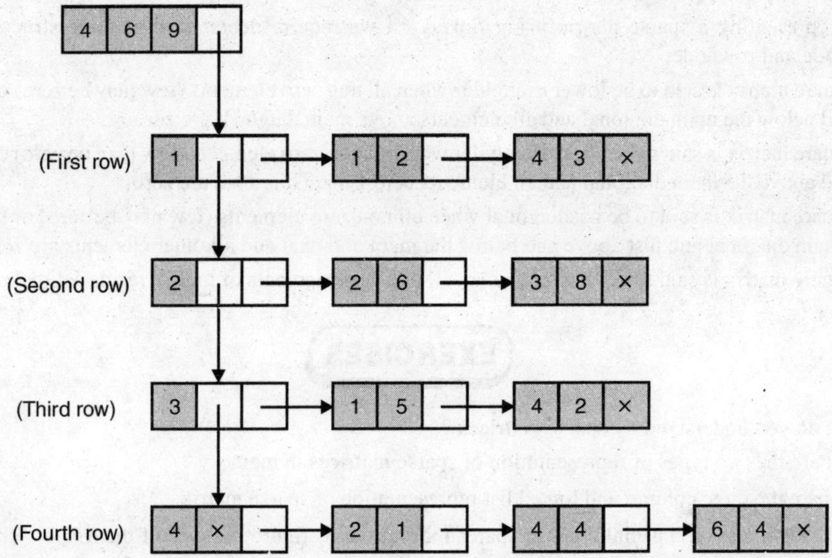

Fig. 7.8.

3. For the given sparse matrix obtain the array or three-column representation.

$$\begin{bmatrix} 15 & 0 & 0 & 22 & 0 & -15 \\ 0 & 11 & 3 & 0 & 0 & 0 \\ 0 & 0 & 0 & -6 & 0 & 0 \\ 0 & 0 & 0 & 0 & 0 & 0 \\ 91 & 0 & 0 & 0 & 0 & 0 \\ 0 & 0 & 28 & 0 & 0 & 0 \end{bmatrix}$$

Ans. The three column form will be :

6	6	8
0	0	15
0	3	22
0	6	–15
1	1	11

1	2	3
2	3	–6
4	0	91
5	2	28

SUMMARY

1. A sparse matrix is a matrix where majority of its elements are zero and very few non-zero elements exist. The zero elements occupies unnecessary memory space, therefore we need only to store the values of nonzero members, and rows and columns in which member occurs.
2. For representing a sparse matrix using linked list we require declaration of three structures: headnode, colnode and rownode.
3. A square matrix is said to be lower triangular when all non zero elements (few may be zero) of matrix resides at and below the main diagonal and all elements above main diagonal are zero
4. A square matrix is said to be upper triangular when all non zero elements (few may be zero) of matrix resides at and above the main diagonal and all elements below main diagonal are zero.
5. A square matrix is said to be triadiagonal when all non zero elements (few may be zero) of matrix resides at the main diagonal and just above and below the main diagonal and all other elements are zero.
6. A square matrix is said to be diagonal when all non zero elements of matrix resides at the main diagonal

EXERCISES

1. What do you understand by sparse matrices ?
2. What are the two types of representation of sparse matrices in memory ?
3. Compare the three column and linked list representation of sparse matrix.
4. What are the various special types of spares matrices draw figure for each of them?
5. How can we check whether a sparse matrix is
   (*a*) Upper triangular
   (*b*) Lower trigular
   (*c*) Tridiagonal
   (*d*) Diagonal
   Expain with proper logic.
6. For each of the special sparse matrix, generate an equivalent address formula so that elements of these matrices can be stored in 1-D array.
7. How can you perform multiplication of two sparse matrices represented as 3- column form? Write C routine for the same.

❑❑❑

CHAPTER 8

STACKS

8.1. INTRODUCTION

INSIDE THIS CHAPTER

A stack is a data structure in which last item inserted is taken out first. That's why they are known as LIFO (last in first out). LIFO constructions exist in the real world such as a stack of plates in a cafeteria, or a pile of newspapers/books in a store. Inserting an item in stack is termed as push and taking an item out from stack is termed as pop. Only one item can be pushed at a time which is added on top of previous item pushed. Similarly only one, the top most item can be popped back. For keeping track of top item we have an integer variable top, Push and Pop operations are done using this variable top. Note one thing Push and Pop terms are used only with the stacks and not anywhere else, whenever an item is pushed into the stack, the top is incremented by 1 and item is pushed onto the stack. Similarly whenever an item is deleted from stack, top is decremented by 1. To understand the stack diagrammatically follows the Fig. 8.1.

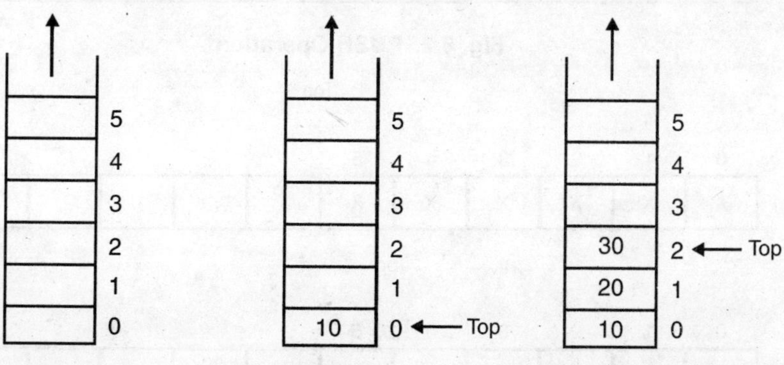

top = –1
stack is empty
(*i*)

10 pushed
top = 0
(*ii*)

20, 30 pushed
top = 2
(*iii*)

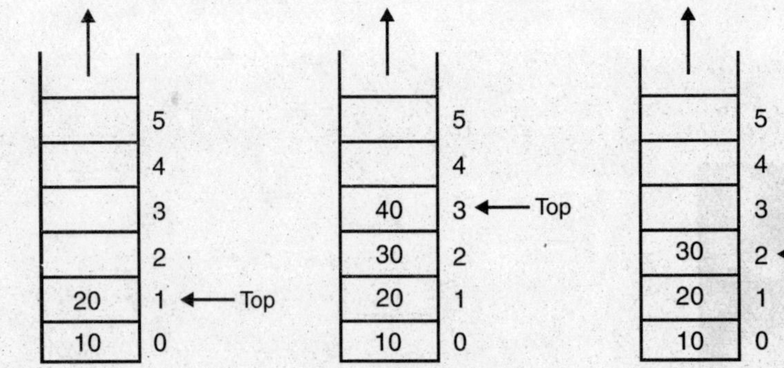

top item deleted
top = 1
(*iv*)

30, 40 pushed
top = 3
(*v*)

top item deleted
top = 2
(*vi*)

Fig. 8.1. Stack and its related operation (Push and Pop).

8.2. OPERATIONS

The following are the operations on a stack :

- Getting the size
- Checking if empty
- Getting the top element without deleting it from the stack
- Insertion of new element onto the stack (push)
- Deletion of the top element from the stack (pop)

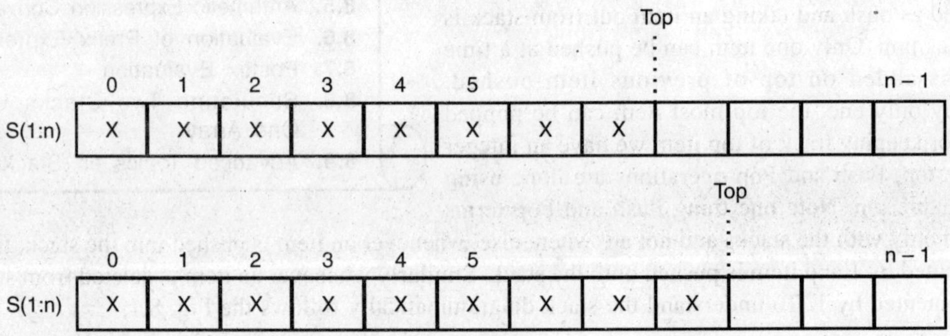

Fig. 8.2. PUSH Operation.

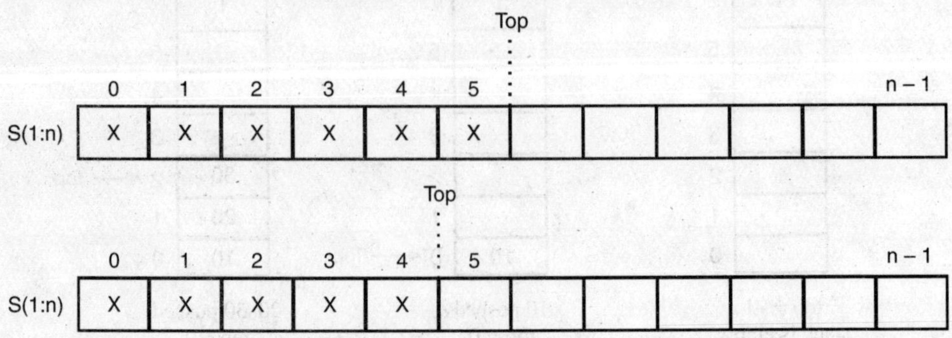

Fig. 8.3. POP Operation.

8.3. REPRESENTATION OF STACKS

In general, there are two types of representation of stacks in computer's memory. The one is using array and second is using linked list. We will see both one by one.

8.3.1. Array Representation of Stack

Sequential allocation of stack makes use of arrays, hence the size is static. The stack is **empty** if the top=–1 and **full** if top=n–1. Deletion from an empty stack causes an **underflow** while insertion onto a full stack causes an **overflow**. The following figure shows an example of the ADT stack :

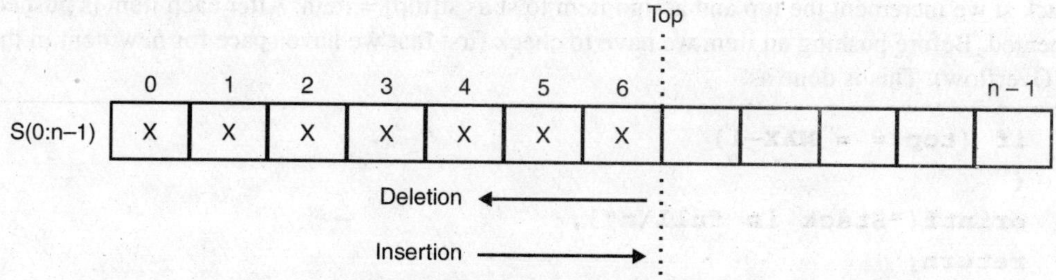

Fig. 8.4. Deletion and Insertion.

Array representation of stack is very simple. We simply take an array of a fixed size and an integer variable top. Elements are pushed and popped using the usual operation with array and top as discussed in the beginning of the chapter *i.e.* for pushing an item, top will be incremented by 1 and after popping an item from stack, top is decremented by 1. The main point to note in the array implementation of stack is that we cannot push more elements in the stack more than the size of the stack. That is if we try to push an item into the stack and stack was already full then this situation is known as "Overflow". Similarly if we try to remove an item (pop operation) from stack and stack is empty then this situation will be known as "Underflow". Both the situations can be handled effectively by the use of top variable. To understand the array representation of stack we have written the C code for the push and pop operation as shown below:

```
#define MAX 10
int top = -1;
int st[MAX];
void push(int item)
{
if (top = = MAX-1)
{
printf("Stack is full\n");
return;
}
st[++top] = item;
}
int pop( )
{
int d;
if (top = = -1)
{
printf("Stack is empty\n");
```

```
return -1;
}
d = st[top];
st[top--] = 0;
return d;
}
```

We have taken an array of size 10 by the name st which is our stack. When we want to push item into this stack st we increment the top and assign item to st as st[top] = item. After each item is pushed top is incremented. Before pushing an item we have to check first that we have space for new item in the stack (stack Overflow). This is done as

```
if (top = = MAX-1)
{
printf("Stack is full\n");
return;
}
```

If top = = 9, stack is full and function returns.

When an item is popped we take the value at st[top] and decrement the top by 1 as return st[top—]; Before popping an item we check stack is empty or not (stack Underflow). This is done as

```
if (top = = -1)
{
printf ("Stack i empty\n");
return -1;
}
```

If top is —1, stack is empty and function return the value -1.

The empty positions in array are denoted by 0. When an element is deleted, its position in the array is marked 0 to denote that element has been deleted at this position. The stack grows from lower index to higher index.

In the code given above, we have implemented a stack of integer elements. The elements of stack can be of any type, be it character, strings, pointers, structure or any other data type. For example to implement a stack of strings, the code can be written as:

```
#define MAX 10
int top = -1;
char st [MAX] [MAX]; /* 10 strings of maximum 10 characters*/
void push(char item[ ] )
{
if (top = = MAX-1)
{
printf("Stack is full\n");
return;
}
strcpy (st[++top], item);
```

```
}
char * pop( )
{
char d[10];
if (top = = -1)
{
printf("Stack is empty\n");
return  -1;
}
strcpy(d, st[top]);
st[top--] = 0;
return d;
}
```

Modified Representation Using Array Within Structure

The code we are going to present here is similar to the previous one with the difference that we have written one structure stack that represents stack data structure. It contains two data members: one is the array arr of integer type of size 10 for holding the elements for the stack and second is an integer variable top that points to the top element of the stack. The elements in the stack array and top are initialized as:

```
stack st= {{0}, -1};
```

This initializes all the array elements for st equal to 0 and top to -1. In the push and pop function address of this stack variable st is passed. Coding inside in functions is simple to understand as given earlier but here for accessing arr and top we are having a pointer of type stack in push and pop function.

```
#define MAX 10
typedef struct
{
int arr[MAX];
int top;
} stack;
void push(stack*, int);
int pop(stack*);
void push(stack *ptr, int d)
{
if(ptr->top = = MAX-1)
{
printf("Stack is full, while pushing %d\n", d);
return;
}
ptr->top++;
ptr->arr[ptr->top] = d;
}
int pop(stack*ptr)
{
int d;
```

```
if(ptr->top = = -1)
{
printf ("Stack is empty\n");
return -1;
}
d = ptr->arr[ptr->top];
ptr->arr[ptr->top) = 0;
ptr->top–;
return d;
}
```

Algorithm for Stack Implementation Using Arrays

Algorithm 8.1. PUSH (STACK, TOP, MAXSTK, ITEM) ....

This procedure pushes an ITEM onto a stack, MAXSTK is the maximum size of the stack.

1. [Stack already filled?]
   If TOP = MAXSTK, then: Print: OVERFLOW, and Return.
2. Set TOP := TOP + 1. [Increases TOP by 1.]
3. Set STACK[TOP] := ITEM. [Inserts ITEM in new TOP position.]
4. Return

Algorithm 8.2. POP(STACK, TOP, ITEM) ....

This procedure deltes the top element of STACK and assigns it to the variable ITEM.

1. [Stack has an item to be removed?]
   If TOP = 0, then: PRINT: UNDERFLOW, and Return.
2. Set ITEM: = STACK[TOP]. [Assigns TOP element to ITEM.]
3. Set TOP := TOP – 1. [Decreases TOP by 1.]
4. Return

8.3.2. Linked Representation of Stack

A linked list of stack nodes could be used to implement a stack. In linked representation, a node with the structure is :

```
struct Node{
int info;
Node *link;
}
```

The following figure shows a stack represented as a linked linear list:

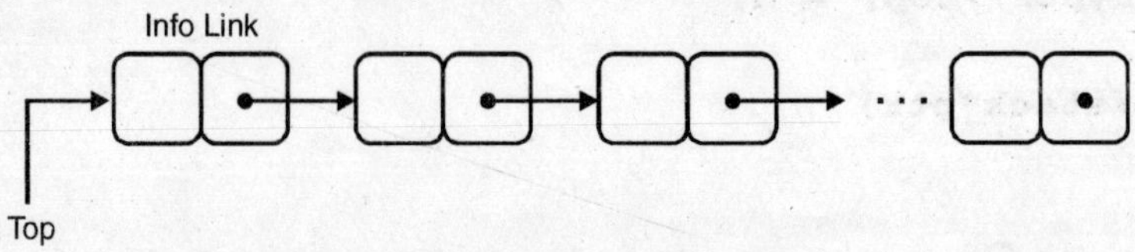

Fig. 8.5. Linked Representation.

We have studied linked list in the chapter 4. Linked representation of stack is nothing but a linked list of nodes where each node is an element of the stack. To point to the top most node added we have a pointer top that points to node recently added. Every new node is added in the beginning of the linked list and top point to it. The main advantage of this representation is that size of stack is not fixed so there is least chance of stack overflow (considering all allocation requests are fulfilled). The code for this is given below and figure following the code illustrates the working of code:

```
struct stack
{
int data;
struct stack * next;
};
typedef struct stack stack;
stack*alloc ( )
{
return (stack*) malloc (size of (stack));
}
void push(stack**T, int d)
{
stack *temp;
if(*T = = NULL)
{
(*T) = alloc ( );
(*T)->data = d;
(*T)->next = NULL;
}
else
{
temp = alloc( );
temp->data = d;
temp->next = (*T);
(*T) = temp;
}
}
int pop(stack **T)
{
stack*temp;
int d;
if(*T = NULL)
{
printf("Stack is empty\n");
```

```
    return -1;
    }
    d = (*T)->data;
    temp = *T;
    (*T) = (*T)->next;
    free (temp);
    return d;
    }
```

The coding is not hard to understand if you have gone through coding based on linked list. The function push takes two arguments: first is of node** type and second is an integer element that is to be pushed. The first parameter will always be &top (Assuming declaration stack*top = NULL exist before making first call to push). In the pop function we have just one argument to be passed of type node** i.e. &top. The function returns the data of the top node after deleting the node.

In the function push we first check if *T is NULL, if it is then we allocate the first node for the stack and assign d to its data field. In case *T is not NULL, we add node at the beginning of the stack. This is done as :

(*i*) next of temp points to *T (temp->next = *T)

(*ii*) *T points to temp (*T = temp)

In the pop function we store the data of the top node in a variable d by writing d = (*T)->data; then temp points to *T and (*T) points to its next node (*T = (*T)->next); the node temp is freed by writing free(temp) and in the end d is returned.

(*i*) 10 is pushed

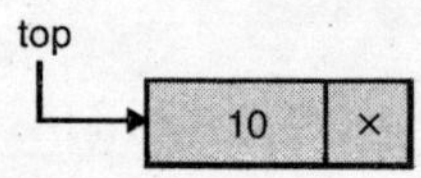

(*ii*) 20 is pushed

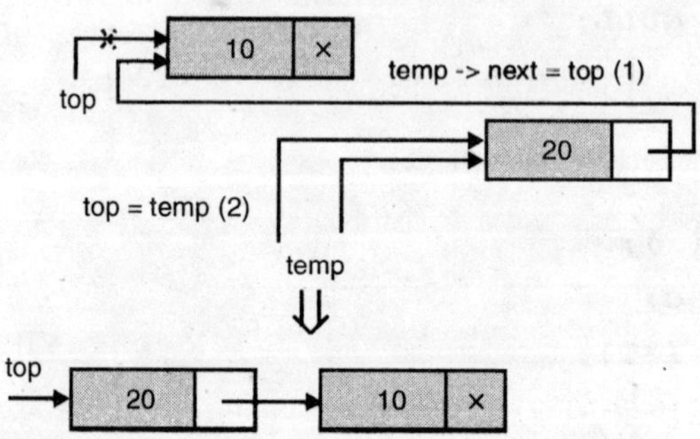

(*iii*) 30 is pushed

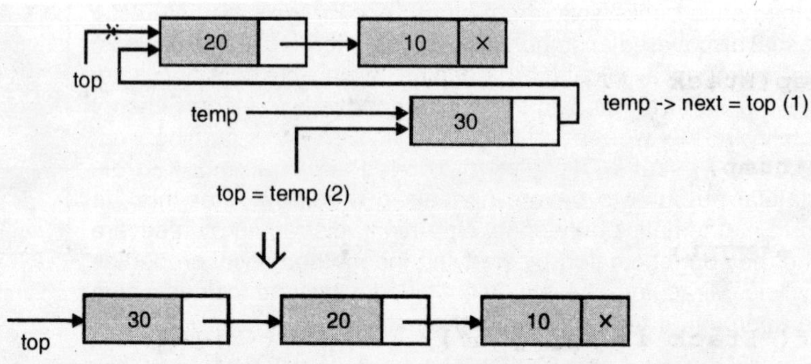

(*iv*) top element (node) is deleted

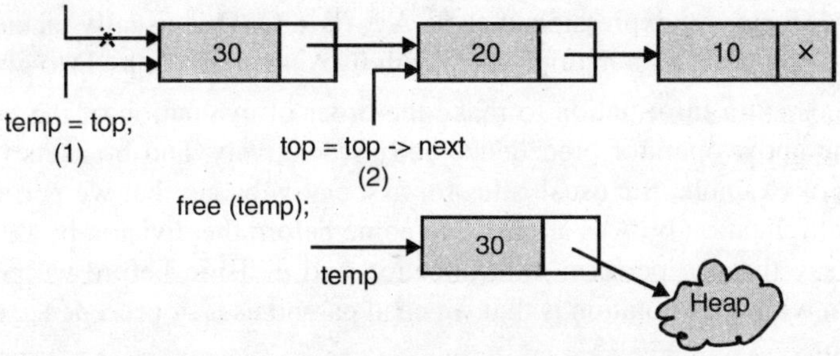

Fig. 8.6. Stack implemented using linked list.

8.4. STACK APPLICATIONS

There are number of applications of stack in computers. Some of them are:

(*i*) Conversion from infix to prefix /postfix and various other conversion

(*ii*) Use in function call and interrupt call

(*iii*) Use in recursion

(*iv*) Simulating memory buffers

(*v*) Evaluation in postfix / prefix expressions

(*vi*) Solving any problem where the problem characterizes stack

We will discuss some of the most useful applications in this section. We with application in conversion of arithmetic expression.

8.5. ARITHMETIC EXPRESSION CONVERSION

An arithmetic expression is combination of various constants, symbols (variables) and operators. We will discuss only the integer expressions in this context. Based on placement of operator with the operands there are 3 general notations of writing the arithmetic expressions. They are:

(*i*) Infix Notation

(*ii*) Prefix or Polish Notation

(*iii*) Postfix or Reverse Polish Notation

DO YOU KNOW

BLAISE PASCAL (1623-1662) exhibited his talents at an early age, although his father, who had made discoveries in analytic geometry, kept mathematics books away from him to encourage other interests. At 16 Pascal discovered an important result concerning conic sections. At 18 he designed a calculating machine, which he built and sold. Pascal, along with Fermat, laid the foundations for the modern theory of probability. In this work he made new discoveries concerning what is now called Pascal's Triangle. In 1654, Pascal abandoned his mathematical pursuits to devote himself to theology. After this, he returned to mathematics only once. One night, distracted by a severe toothache, he sought comfort by studying the mathematical properties of the cycloid. Miraculously, his pain subsided, which he took as a sign of divine approval of the study of mathematics.

In the infix notation the operator is written in between the operands. Infix notation is the usual way in which we write the expressions in simple arithmetic. The operators are written in between operands. For example **X+Y, 2*4+5** etc. An expression such as **A * (B + C)/D** is usually taken to mean something like: "First add B and C together, then multiply the result by A, then divide by D to give the final answer".

Infix notation needs extra information to make the order of evaluation of the operators clear: rules built into the language about operator precedence and associatively, and brackets ( ) to allow users to override these rules. For example, the usual rules for associatively say that we perform operations from left to right, so the multiplication by A is assumed to come before the division by D. Similarly, the usual rules for precedence say that we perform multiplication and division before we perform addition and subtraction. A problem with infix notation is that we need parentheses or precedence rules to handle more complicated expressions.

Prefix notation was introduced by the Polish logician Lukasiewicz, and is sometimes called "Polish notation". In prefix notation the operator is placed first followed by operands in order from left to right. For example to convert infix expression A*B into prefix it will be *AB. To convert the expression given above in prefix we perform the following steps :

= A * (+BC)/D

= (* A+BC)/D

= /*A+BCD

Although Prefix "operators are evaluated left-to-right", they use values to their right, and if these values themselves involve computations then this changes the order that the operators have to be evaluated in. In the example above, although the division is the first operator on the left, it acts on the result of the multiplication, and so the multiplication has to happen before the division (and similarly the addition has to happen before the multiplication).

For calling a function we write as sum (10, 20). The function name can be bought of as operator and its parameters are its operand. From this analogy the unction call can assumed to be written in prefix notation.

In postfix notation operators are put after the operands. The notation is also known are **Reverse Polish Notation.** In postfix notation X+Y can be written as Y+. The above expression given in infix as A * (B + C) I D can be converted to postfix as:

= A * (BC+)/D

= (ABC+*) / D

= ABC+*D/

The order of evaluation of operators is always left-to-right, and brackets cannot e used to change this order. Because the "+" is to the left of the "*" in the example above, the addition must be performed before the multiplication. Operators act on values immediately to the left of them. For example, the "+" above uses the "B" and "C". We can add (totally unnecessary) brackets to make this explicit:

((A (B C+),*) D/)

Thus, the "*" uses the two values immediately preceding: "A", and the result of the addition. Similarly, the "/" uses the result of the multiplication and the "D".

The advantages of prefix & postfix notations are

1. Parentheses are not required.
2. It can be used with any number of operands.
3. It is easy to translate prefix/postfix expressions

In the expression that we will be seeing, we assume the priority of the operator as shown in Table 8.1.

Table 8.1.

Operators	Priority
NOT	6
AND	5
OR	4
^ (exponention)	3
*, %, /	2
+, –	1

In any notation the conversion takes place by considering the priority of the different operators. The operator with the higher priority will be processed first. If the two operators with the same priority exist then processing will depend upon associatively of the operators i.e. either from left to right or right to left, The expression inside the parenthesis will be given highest priority. We will consider the associatively from left to right for all operators.

1. Conversion from Infix to Postfix

Example 8.1.

2*3/4 – 5 + 6 (Infix) = (23*) /4 – 5 + 6
= (23*4/) – 5 + 6
= (23*4/5–) + 6
= **23*4/5 – 6 +**

Example 8.2.

A* (B + C) – D/E^ (F*G) = A* (BC+) – D/E ^ (F*G)
= A*(BC+) – D/E ^ (FG*)
= A* (BC+) – D/(EFG*^)
= (ABC+*) D/ (EFG*^)
= (ABC+*) – (DEFG*^/)
= **ABC + *DEFG * ^/–**

Example 8.3.

A AND B + C OR D ^ E AND F/ D = (A B AND) + C OR D A E AND F/ D
= (A B AND) + C OR D ^ (E F AND) / D
= (A B AND) + (C D OR) ^ (E F AND) / D
= (A B AND) + (C D OR E F AND ^) / D
= (A B AND) + (C D OR E F AND ^ D /)
= **A B AND C D OR E F AND ^ D / +**

Example 8.4.

-A + C * D -E OR NOT F + 4 AND 5 = (-A) + C * D -E OR NOT F + 4 AND 5
= (-A) + C * D —E OR (NOT F) + 4 AND 5
= (—A) + C * D —E OR (NOT F) + (4 5 AND)
= (-A) + C * D -(E (NOT F) OR) + (4 5 AND)
= (—A) + (C D *)-(E (NOT F) OR) + (4 5 AND)

= ((-A) C D * +)—(E (NOT F) OR) + (4 5 AND)
= ((-A) C D * + E (NOT F) OR -) + (4 5 AND)
= (-A) CD*+E (NOTF) OR-45AND+

All the examples presented above are self explanatory. The main point to be noted in the last example is that unary operators are given higher priority then binary operators and they are written within the parenthesis as it. In all the above examples we have not used the stack. To show you the use of stack in the conversion, give you the algorithm for the same and then write the C code for the same.

Algorithm 8.3. INTOPOST (I, P) ....

1. Scan the input infix expression I from left to right, consider every operand and operator in the expression, taking it into some symbol say S.
2. If symbol S is an operand, add it to P (postfix expression being generated)
3. If a left parenthesis (is encountered push it onto stack.
4. If an operator say OP is encountered then
   (*i*) Pop from stack every operator and add to P, until stack does not become empty. Before adding it to P make sure the priority of the current operator in S is greater then equal to priority of the operator popped from stack.
5. If a right parenthesis) is encountered then
   (*i*) Pop from stack all operators and add to P till left parenthesis (is not encountered.
   (*ii*) Use parenthesis for matching purpose only, do not add it to P.
6. Repeat step 2 to 5 for the next symbol in infix expression.

In the algorithm given we have assumed the occurrence of sub expression enclosed within parenthesis, but to simplify the implementation process of the algorithm, we have developed two different implementations. The first implementation given below does not take into account use of parenthesis, and second does take into account.

```
#define MAX 30
char result[MAX], stack[MAX];
int op1, op2;
int top = -1;
char str[MAX];
void convert(char*s, char*t)
{
while (*s)
{
if(isdigit(*s)  ||  isalpha(*s))
{
*t = *S;
t++;
}
else if(*s = = '+'| | *s = = '-'| |*s = = '*'| | *s = = '^'| | *s=
      = '/')
{
```

```
if (top = = -1)
push(stack, &top, *s);
else
{
op1 = pop(stack, &top);
while (pr(op1) > = pr(*s))
{
*t= op1;
t++;
op1 = pop (stack, &top);
}
push (stack, &top,op1);
push (stack, &top,*s);
}
}
s++;
} /* end of while */
while(top != -1)
{
op1 = pop (stack, &top);
*t = op1;
t++;
}
*t = '\0';
}
int pr(char op)
{
if(op = = '^')
return 3;
else if(op = = '*' | | op = = '/')
return 2;
else if(op = = '+' | | op= = '-')
return 1;
else
return 0;
}
void push(char *st, int *sp, char item)
}
if(*sp = = MAX-1)
printf ("\nstack is full");
else
```

```
{
(*sp) ++;
st[*sp] = item;
}
}
char pop(char *st, int *sp)
{
int item;
if(*sp = = -1)
{
printf("\nstack is empty");
return -1;
}
else
{
item = st[*sp];
(*sp) - - ;
return item;
}
}
```

For the conversion process we have written function convert. The function takes two arguments of char* type. The first argument is the infix express stored in str and second is the result in which postfix expression will be stored.

We examine every character element of the source string pointed by the char pointer s for any digit, alphabet or any basic arithmetic operators. If the character element is any digit constant or alphabet then it is assigned to the target string pointed by t by writing *t =*s and t is incremented as t++. In case the character element is any operator (+,-,*, /,^) then if the stack is empty, it is pushed onto the stack, else the top element from the stack is popped up (that will also be an operator pushed earlier into the stack). Now the priority of the current operation is compared with the operator that has been popped up from stack (say op1). For that we have written a function pr that takes an operator and returns the priority of operator. Till the priority of op1 is greater than equal to the current operator, we assign that operator to the target string, increment it and again pop up next operator from the stack (till stack is not empty) and process repeats. The reason behind this is that a lower priority operator cannot be on top of higher priority operator, so when this situation arises we have to pop all the operators from the stack whose priorities are higher from the current operator and assign them to the target string. In case the priority of popped up operator is lower than the current operator then we pushed back the operator that was popped up and also push the current operator. These two statements are written outside the while loop in the else part. After the else part s is incremented and next character is examined.

When the input string exhausts, all the operators from the stack are popped up and assign to target string.

```
while(top != -1)
{
op1 = pop(stack, &top);
*t = op1;
t++;
}
```

Finally a null character is assigned to t to mark the end of string:

```
*t = '\0' ;
```

When function convert returns, the result string holds the converted postfix expression. The same is displayed to the user. For complete illustration see figure given below :

Steps	Stack	O/P string	R.I/P string	Remark
1.	—	—	2+ 3*4	Character is 2 assign to output string
2.	—	2	2+3*4	Character is + push onto stack
3.	+	2	2 +3*4	Character is 3 assign to output string
4.	+	23	2 + 3*4	Character is * push onto stack.
5.	+, *	23	2 + 3*4	Character is 4, assign to output string.
6.	+, *	234	2 + 3*4	Input string exhausts pop all from stack and assign to output string

Postfix expression is 234*+

(a) Converting infix expression into postfix, it is assumed that stack grows towards right.

Steps	Stack	O/P string	I/P string	Remark
1.	—	—	2 *3+4/5	Character is 2, assign to output string
2.	—	2	2* 3+4/5	Character is *, push onto stack
3.	*	2	2*3 +4/5	Character is 3, assign to output string
4.	*	23	2*3 + 4/5	Character is +, push onto stack but priority of * is higher than +, so * is popped first, assigned to output string, then + is pushed onto stack.
5.	+	23*	2*3+4/5	Character is 4, assign to output string
6.	+	23*4	2*3+4/5	Character is /, push onto stack.
7.	+, /	23*4	2*3+4/5	Character is 5, assign to output string
8.	+, /	23*45	2*3+4/5	Input string exhausts, pop all from stack and assign to output string

Postfix expression is 23*45/+

(b) Conversion from infix to postfix

Fig. 8.7.

The previous version of the function convert is less efficient in the case that it does not take, consider the expression written inside the parenthesis. For example 2*(3+4). The previous program will not work for this input. To solve the infix expressions that involve parenthesis we have written this program. Here regardless of priority of the any of the operators, expression inside the parenthesis has to be solved first. To incorporate this in our program we have added the following cod in the function:

```
else if (*s = = '(' )
push(stack, &top, *s);
else if(*s = = ')' )
{
op1 = pop(stack, &top);
```

```
while (op1!= ' (')
{
*t = op1;
t++;
op1 = pop(stack &top);
}
}
```

Whenever the input string contains current character as '(' it is pushed onto the stack. The expression inside parenthesis handled in the usual manner. When the input string contains current character as ')' then all operators from the stack are popped up till operator does not equal to '('. All popped operators are assigned to target string one by one as they are popped up from the stack. In between the target string pointer is also incremented.

Steps	Stack	O/P string	R.I/P string	Remark
1.	—	—	② *(3+4)/(5–2)	Character is 2, assign to output string
2.	—	2	2⊛ (3+4)/(5–2)	Character is *, push onto stack
3.	*	2	2 *(⃝3+4)/(5–2)	Character is (, push onto stack
4.	*, (	2	2*(③ +4)/(5–2)	Character is 3, assign to output string
5.	*, (	23	2*(3 ⊕4)/(5–2)	Character is +, push onto stack
6.	*, (, +	23	2*(3+④)/(5–2)	Character is 4, assign to output string
7.	*, (, +	234	2*(3 + 4⃝)/(5–2)	Character is), pop from stack till operator is not equal to (, only + is popped.
8.	*	234	2*(3+4)⃝(5–2)	Character is /, priority of * and is same, so pop * from stack, assign to o/p string and push / onto stack
9.	/	234+*	2*(3+4)/(⃝5–2)	Character is (, push onto stack
10.	/, (	234+*	2*(3+4)/(⑤–2)	Character is 5, assign of output string
11.	/, (	234+*5	2*(3+4)/(5⊖2)	Character is –, push onto stack
12.	1, (, –	234+*5	2*(3+4)/(5–②)	Character is 2, assign to output string
13.	/, (, –	234+*52	2*(3+4)/(5–2⃝)	Character is), pop from stack all operator till operator not equal to (, assign same to the output string, here just one operator – is there.
14.	/	234+*52–	2*(3+4)/(5–2)	Input string exhausts, pop all from stack and assign to output string

Fig 8.8. Conversion of infix to postfix, with expression contains sub expression

2. Conversion from Infix to Prefix

Example 8.5.

= 2+3*4–5

= 2+(*34)–5

= (+2*34)–5

= **–5+2*34**

Example 8.6.

= 2+3/5^6–(4*5)

= 2+3/5^6–(*45)

= 2+3/(^56)-(*45)

= 2+(/3^56)-(*45)

= (+2/3^56)-(*45)

= **-+2/3^56*45**

Example 8.7.

= A AND B+ C^D OR E AND F

= (AND A B)+ C ^D OR E AND F

= (AND A B)+ C ^ D OR (AND E F)

= (AND A B)+ C ^ (OR D AND E F)

= (AND A B)+ (^ C OR D AND B F)

= **+AND AB ^ C OR D AND E F**

The algorithm and C code for the conversion is exactly same as we have see earlier in conversion from infix to postflx. The difference here is that for the conversion of infix expression to prefix expression, **we first reverse the expression and then pass it to the convert function. The final output string is also reverse and then displayed.** This conversion is shown using an example given below:

```
ip string = 2 + 3 * 4, reversed i/p string = 4 * 3 + 2
```

Steps	Stack	O/P string	R.I/P string	Remark
1.	—	—	④*3+2	Character is 4, assign to output string
2.	—	4	4⊛3+2	Character is *, push onto stack
3.	*	4	4*③+2	Character is 3, assign to output string
4.	*,	43	4*3⊕2	Character is +, push onto stack, but priority of * is higher than +, so pop ^ first from stack, assign to output string and push + onto stack
5.	+	43*	4*3+②	Character is 2, assign to output string
6.	+	43*2	4*3+2	Input string exhausts, pop all from stack & assign to output string

stack is now empty. Final o/p string : 43*2+

On reversing the above o/p string +2*34 (Desired Prefix Expression)

Fig. 8.9. Conversion from infix to prefix expression.

Note one important thing in the program that we first check of right parenthesis ')' first and pushed onto stack. As soon as in the input string next character is '(' we pop all operators from the stack till the operator is not ')'. This is reverse from the C code for conversion from infix to postfix. This is shown below:

```
else if(*s = =')' )
push (stack, & top, *s);
```

```
else if(*s = ='(' )
{
op1 = pop(stack, &top);
while(op1 ! =')' )
{
*t = op1;
t++;
op1 = pop(stack, &top);
}
}
```

i/p string : A*(B + C) ^ (D - E), Reversed i/p string :) E - D (^) C + B * A

Steps	Stack	O/P string	R.I/P string	Remark
1.	—	—	ⓘ E–D(^)C+B(*A	Character is), push onto stack
2.	),	—	) Ⓔ –D(^)C+B(*A	Character is E, assign to output string
3.	),	E	) E⊖ D(^)C+B(*A	Character is –, push onto stack
4.	), –,	E	)E– Ⓓ(^)C+B(*A	Character is D, assign to output string
5.	), –,	ED	)E–D(^)C+B(*A	Character is (, pop from stack all operator till operator popped is not equal to), only – is popped, stack is not empty
6.	—	ED–	)E– D(^)C+B(*A	Character is ^, push onto stack
7.	^,	ED–	)E– D(^)C+B(*A	Character is), pushed onto stack
8.	^,)	ED–	)E–D(^) Ⓒ+B(*A	Character is C, assign to output string
9.	^,)	ED–C	)E–D(^)C ⊕B(*A	Character is +, push onto stack
10.	^,), +	ED–C	)E–D(^)C+ Ⓑ (*A	Character is B, assign to output string
11.	^,), +	ED–CB	)E–D(^)C+B (*A	Character is (, pop from stack till the operator is not equal to), only + is popped from stack & is assigned to output string
12.	^,	ED–CB+	E–D(^)C+ B(⊛A	Character is *, push onto stack, but * has the lower priority than ^, so ^ will be popped and assigned to output string, * will be then pushed to stack
13.	*	ED–CB+^	E–D(^)C+B(* Ⓐ	Character is A, assign to output string
14.	*	ED–CB+^A	)E–D(^)C+ B(*A	Input string exhaust pop all from stack and assign to output string
15.	—	ED–CB+^A*		

On reversing the output string : *A^+BC–DE

Desired prefix expression is : *A^+BC–DE

Fig. 8.10. Conversion from infix to prefix with parenthesis included in expression/

3. Conversion from Postfix to Prefix

Example 8.8.

= 23*415-6+ (Postfix)
= (*23)4/5-6+
= (/*234)5-6+

= (-/*2345)6+

= **+-/*23456 (Prefix)**

Example 8.9.

= ABC+*DEFG*^/- (Postfix)

= A(+BC)*DEFG*^/-

= (*A+BC)DEFG*^/-

= (*A+BC)DE(*FG)^/-

= (*A+BC)D(^E*FG)/-

= (*A+BC)(/D^E*FG)-

= **-*A+BC/D^E*FG (Prefix)**

Example 8.10.

= A B AND C D OR E F AND ^ D / + (Postfix)

= (AND A B) C D OR E F AND ^D/+

= (AND A B) (OR C D) E F AND ^D/+

= (AND A B)(OR C D) (AND E F) ^D /+

= (AND A B) (^ OR C D AND E F) D /+

= (AND A B) (/^OR C D AND E F D)+

= **+ AND A B / ^ OR C D AND E F D (Prefix)**

In all the examples shown above, the logic for conversion from postfix is simple. Scan the input expression left to right, whenever we have two operands and one operator in order, place operator before the two operands and get one new operands. For example if two operands and A and B and operator is + *i.e.* A B + then we get hew operand as (+AB). The operands are saved in stack until an operator in encountered.

4. Conversion From Prefix to Postfix

Example 8.11.

Prefix expression: +-/*23456

Reverse expression: 65432*/-+

= 65432*/-+

= 654(23*)/-+

= 65(23*4/)-+

= 6(23*4/5-)+

= **23*4/5-6+ (Postfix form)**

Example 8.12.

Prefix expression: -*A+BC/D^E*FG

Reverse expression: GF*E^D/CB+A*-

= GF*E^D/CB+A*-

= (FG*)E^D/CB+A*-

= (EFG*^)D/CB+A*-

= (DEFG*^/)CB+A*-

= (DEFG*^/)(BC+)A*-

= (DEFG*^/)(ABC+*)-

= **ABC+*DEFG*^/-**

Conversion from prefix to postfix is similar to conversion from postfix to prefix The difference here is only that we have to first reverse the prefix expression so that operators appear on the right side in the expression. Another point to not here is that first operand nearer to operator is written first. For example if two operand and A and B and operator is + i.e. A B + then we get a new operand as (+BA). The operands are saved in stack until an operator in encountered.

5. Conversion from Postfix to Infix

Example 8.13.

= 23*4/5–6+ (Postfix form)
= (2 * 3)4/5–6+
= (2*3/4–5)6+
= **2*3/4–5+6**

Example 8.14.

ABC+*DEFG*^/- = A(B+C)*DEFG*^/-
= (A*B+C)DEFG*^/-
= (A*B+C)DE(F*G)^/-
= (A*B+C)D(E^F*G)/-
= (A*B+C)(D/E^F*G)-
= **A*B+C-D/E^F*G**

Conversion from postfix to infix is similar to conversion from postfix to prefix or prefix to postfix. The difference here is only that we have to place the operator between the two operands. For example if two operands are A and B and operator is + *i.e.* A B + then we get a new operand as (A + B). The operands are saved in stack until an operator in encountered.

6. Conversion from Prefix to Infix

Example 8.15.

Prefix expression: +-/*23456

Reverse expression: 65432*/-+

= 65432*/-+
= 654(2*3)/-+
= 65(2*3/4)-+
= 6(2*3/4-5)+
= 2*3/4-5+6

Example 8.16.

Prefix expression: -*A+BC/D^E*FG

Reverse expression: GF*E^D/CB+A*-

= GF*E^D/CB+A*-
= (F*G)E^D/CB+A*-
= (E^F*G)D/CB+A*-
= (D/E^F*G)CB+A*-
= (D/E^F*G)(B+C)A*-
= (D/E^F*G)(A*B+C)-
= **A*B+C-D/E^F*G**

Conversion from prefix to infix is similar to conversion from postfix to infix with the difference is that first operand nearer to operator becomes the left operand and second operand become the right operand. For example if two operands and A and B and operator is + i.e. A B + then we get a new operand as (B + A). The operands are saved in stack until an operator in encountered.

8.6. EVALUATION OF PREFIX EXPRESSION

For evaluating an expression given in prefix notation, the following algorithm given.

It is assumed that prefix expression is stored in a given string.

1. Reverse the string containing given prefix expression
2. for each character C in a given string

```
{
if C is an operand
push C onto stack;
else         /* C is an operator */
{
pop item from stack, and store  in  Opr1;
pop item from stack, and store in Opr2;
result Opr1 C Opr2, using C as an operator;
push result onto stack;
}
}
```

For example consider the infix expression 4*3+5, its prefix equivalent expression is +*435

In the first step we reverse the string and get 534*+

Now we run for loop for each character in the string. Special symbol $ is used to denote empty stack.

This is shown in tabular form as given below

Step No.	Character scanned	Stack	Intermediate Result	Comments
1.	5	$, 5,		5 is operand, push onto stack
2.	3	$, 5, 3		3 is operand, push onto stack
3.	4	$, 5, 3, 4		4 is operand, push onto stack
4.	*	$, 5,	4 * 3	* is operator, pop two operands, and apply operator in between
5.	*	$, 5, 12		Result from previous step pushed onto stack
6.	+	$	12 + 5	+ is operator, pop two operands, and apply operator in between and push onto stack
7.		$, 17		string is terminated, answer is in stack.

Fig. 8.11. Evaluation of prefix expression.

8.7. POSTFIX EVALUATION

The algorithm for evaluating a postfix expression is similar to algorithm given for evaluating a prefix evaluation with a bit difference that string is not reversed and Opr2 and Opr1 are swapped. By doing these changes in the above we get algorithm for evaluating a postflx expression.

1. for each character C in a given string

```
{
if C is an operand
push C onto stack;
else      // C is an operator
{
pop item from stack, and store in Opr2;
pop item from stack, and store in Opr1;
result = Opr1 C Opr2, using C as an operator;
push result onto stack;
}
}
```

Consider again the infix expression **4*3+5**, its postfix equivalent is **43*5+.** We apply the algorithm on this expression.

}

Step No.	Character scanned	Stack	Intermediate Result	Comments
1.	4	$, 4,		4 is operand, push onto stack
2.	3	$, 4, 3		3 is operand, push onto stack
3.	*	$,	4*3	* is operator, pop two operands, and apply operator in between
4.	*	$, 12		result from previous step pushed onto stack
5.	5	$, 12, 5		
6.	+	$	12 + 5	+ is operator, pop two operands, and apply operator in between and push onto stack.
7.		$, 17		string is terminated, answer is in stack.

Fig. 8.12. Evaluation of postfix expression.

The C code for the above is given below :

```
# define MAX 30
 int stack[MAX/2];
 int top = -1;
void eval (char*exp)
{
int val1, val2, ans;
while (*exp)
{
if (*exp = = ' ' | | *exp = = '\t')
{
exp++;
```

```
continue;
}
else if (isdigit (*exp))
push (stack, &top, (*exp)-48);
else
{
val1 = pop (stack, &top);
val2 = pop (stack, &top);
switch (*exp)
{
case '+':
ans = val2 + val1;
break;
case '-':
ans = val2 - val1;
break;
case '/':
ans = val2 / val1;
break;
case '*':
ans = val2 * val1;
break;
case '%':
ans = val2 % val1;
break;
case '^':
ans = pow (val2, val1);
break;
default:
printf ("Unknown operator");
exit (0);
}
push (stack, &top, ans);
}
exp++;
}
}
void push(int *st, int *p, int item)
{
if(*sp = = MAX-1)
printf("\nstack is full");
else
```

```
{
(*sp) ++;
st[*sp] = item;
}
}
int pop(int *st, int *sp)
{
int item;
if(*sp = = -1)
{
printf("\nstack is empty");
return -1;
}
else
{
item = st [*sp];
(*sp)–;
return item;
}
```

The function eval takes any postfix expression as input and finds the value of expression. The function eval takes just one argument: a pointer of char type. Inside this function we pass the input string that contains postfix expression. Inside function we are having a while loop. The loop runs till *exp! ='\0'

Step	i/p string	Stack	Remark
1.	② 34*+52^–	—	Character is 2, pushed onto stack
2.	2③ 4*+52^–	2,	Character is 3, pushed onto stack
3.	23④ *+52^–	2, 3	Character is 4, pushed onto stack
4.	234⊛+52^–	2, 3, 4	Character is *, take out two operands from stack, apply * over then as 3*4 = 12, push 12 onto stack
5.	234*⊕ 52^–	2, 12	Character is +, take out 2 and 12 (i.e. top two operands, apply + as 2 + 12, push onto stack character is 5, push onto
6.	234*+⑤ 2^–	14	Character is 5, push onto stack
7.	234*+5② ^ –	14, 5	Character is 2, push onto stack
8.	234*+52^–	14, 5, 2	Character is ^, pop two operands from stack, apply the operate ^ as 5 ^ 2 = 25.
9.	234*+52^⊖	14, 25	Character is –, pop two operands from stack & apply the operation as 14 – 25 = –11, –11 is pushed onto stack
10.	234*+52^ –	–11	Input string exhausts, result of the expression is in stack

Fig. 8.13. Evaluation of postfix expression.

The first if condition checks that if the expression contains any white space characters (space, tab etc) then increments the pointer to move to next character i.e. skipping the whitespace characters. In the expression if any digit is encountered then it is stored onto the stack. Note as every element in the expression is a char, every digit will be treated as character. To convert the digit in character form into numeric digit, we subtract the ASCII value of '0' i.e. 48 from the character digit to make it a numeric digit. For example '2'-48 is equal to 50-48 (50 is the ASCII value of '2') = 2 (which is a numeric 2).

If the current character in the expression is any operator say op (as allowable in the expression). we pop up two values from the stack. The first popped up value is stored in val1 and second popped up value is stored in val2. Operation val2 op val1 is then performed. For checking the value of the op, we have written switch case statement. The calculated expression is written back onto the stack by pushing it. When the while loop terminates the entry in the stack contains the evaluated answer of the postfrx expression.

8.8. SIMULATING TWO STACKS USING ONE ARRAY

Given just one array we can simulate two stacks, Lets assume that array name is STACK and size of this array is N. We want size of both the stack be N/2. Then our first stack will start from 0 and end at N/2-1 and second stack will start from N-1 and end at N/2. Thus our first stack will grow from left to right and second stack will grow from right to left. The purpose of simulating two or even more stacks within single array is to minimize the overflow problem when adding a new element to the stack. We do not discuss the topic at length but simply give you one program that simulates two stack within one array. Note we have restricted the size of each stack as 5.

Program 8.1. /* Program to simulate two stacks using one array*/ ....

```
#include<stdio. h>
#include <stdlib. h>
#include<conio. h>
#define MAX 10
tyedef struct
{
int arr[MAX];
int top[2];
} stack;
void push (stack*, int, int);
int show (stack *, int);
void main( )
{
stack st = {{0}, {-1, MAX}};
int d, ch;
do
{
printf("1. Push in First\n");
printf("2. Push in Second\n");
printf ("3. Pop from First\n");
printf ("4. Pop from Second\n");
printf ("5. Display First\n");
printf("6. Display Second\n");
 printf ("7 Exit\n");
```

```
printf ("Enter your choice\n");
scanf ("%d", &ch):
switch(ch)
{
case 1:printf ("Enter data\n");
 scanf("%d", &d);
push(&st, d, 1);
break;
case 2:printf("Enter data\n");
scanf ("%d", &d) ;
push(&st, d, 2);
break;
case 3:d = pop(&st, 1);
printf("Top element = %d\n", d);
break;
case 4:d = pop(&st, 2);
printf("Top element %d\n", d);
break;
case 5: printf("\nFirst Stack is\n");
show(st, 1);
break;
case 6: printf("\nSecond Stack is\n");
show(st, 2);
break;
case 7: printf ("Bye Bye\n");
exit (0) ;
getch( );
}
}while(ch> = 1 && ch< = 7);
}
void push(stack *ptr, int d, int SN)
{
switch (SN)
{
case 1: if(ptr->top[SN-1] = = MAX/2-1)
{
printf("Stack is full, while pushing %d\n", d);
return;
}
ptr->top [SN-1] ++;
ptr->arr[ptr->top[SN-1]] = d;
break;
```

```
case 2: if(ptr->top[SN-1] = = MAX/2)
{
printf ("Stack is full, while pushing %d\n", d);
return;
}
ptr->top [SN-1]-;
ptr->arr[ptr->top[SN-1]] =d;
break;
}
}
int pop (stack*ptr, int SN)
{
int d;
switch (SN)
{
case 1: if(ptr->top[SN-1] = = -1)
{
printf("Stack is empty\n");
return -1;
d = ptr->arr[ptr->top[SN-1]];
ptr->arr[ptr->top[SN-1]] = 0;
ptr->top[SN-1]-;
break;
case 2: if(ptr->top[SN-1] = = MAX)
{
printf ("Stack is empty\n");
return -1;
}
d = ptr->arr[t->top[SN-1]];
ptr->arr[ptr->top[SN-1]] = 0;
ptr->top [SN-1] ++;
break;
}
return d;
}
void show(stack S, int SN)
{
int i;
switch (SN)
{
case 1: for(i = 0;i<MAX/2;i++)
printf("%d ", S. arr[i]);
```

```
break;
case 2: for(i =MAX-1;i >= MAX/2;i—)
printf("%d ", S. arr[i]);
break;
}
printf ("\n");
}
OUTPUT:
1. Push in First
2. Push in Second
3. Pop from First
4. Pop from Second
5. Display First
6. Display Second
7. Exit
Enter your choice
1
Enter data
10
1. Push in First
2. Push in Second
3. Pop from First
4. Pop from Second
5. Display First
6. Display Second
7. Exit
Enter your choice
1
Enter data
20
1. Push in First
2. Push in Second
3. Pop from First
4. Pop from Second
5. Display First
6. Display Second
7. Exit
Enter your choice
2
Enter data
90
```

```
1. Push in first
2. push in second
3. pop from first
4. Pop from Second
5.Display First
6.Display Second
7.Exit
Enter your choice
2
Enter data
80
1. push in first
2. push in second
3. pop from first
4. pop from second
5. Display First
6. Display second
7. Exit
Enter your choice
5
First Stack is
10 20 0 0 0
1. Push in First
2. Push in Second
3. Pop from First
4. Pop from Second
5. Display First
6. Display Second
7. Exit
Enter your choice
6
Second Stack is
90 80 0 0 0
1. Push in First
2. Push in Second
3. Pop from First
4. Pop from Second
5. Display First
6. Display Second
7. Exit
```

```
Enter your choice
3
Top element 20
1. Push in First
2. Push in Second
3. Pop from First
4. Pop from Second
5. Display First
6. Display Second
7. Exit
Enter your choice
4
Top element 80
1. Push in First
2. Push in Second
3. Pop from First
4. Pop from Second
5. Display First
6. Display Second
7. Exit
Enter your choice
7
Bye Bye
```

In the program we are simulating two stacks using just one array. For that in the structure stack we are having one array by the name top. The size of array is 2. Element top[0] represents top for the first stack and top[1] represents top for the second stack. The size of both the stacks is of 5. First stack consists of elements at position 0 to 4 and second stack consists of elements at position 9 to 5. The first Stack grows from left to right and second stack grows from right to left. Due to this our first stack works in the normal fashion while in the stack second top decrements during the push operation and increments during the pop operation. The first stack will be full when top[0] is equal to 4 and second stack will be full when top[1] is equal to 5. This is shown in the figure

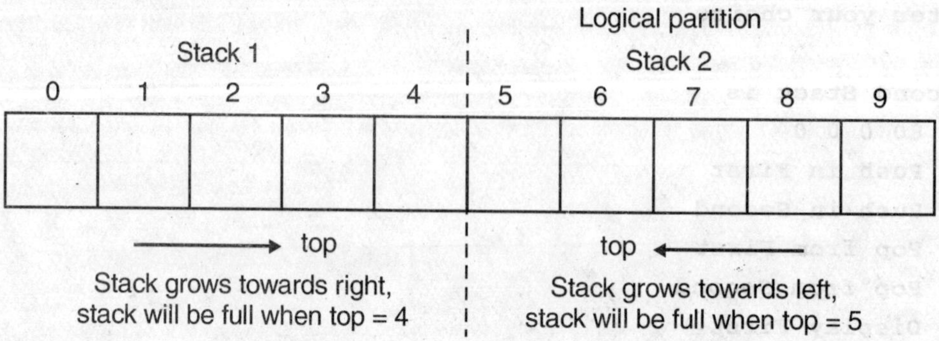

Fig. 8.10. Implementing two stacks using one array.

In the functions **pop, push and show,** one extra parameter SN has been passed. This is used to check onto which stack we want to perform any of the above three operations. The working of all three functions is simple to understand.

One important thing to note here is that we have set the size of both the stacks is 5 and they exist logically and not physically. One modification to above program can be done where we do not set the size of stack fixed, instead when either stacks crosses the boundary of other then we say overflow for stack has occurred. That is size of any of the stack will not be fixed and varies as long as there is space for an element in the array. Try it yourself how this can be done.

In the code given above we have simulated just two stacks within one array, but it is possible to implement as many stacks as per the application requirement.

8.9. ADVANCED TOPICS ON STACKS

8.9.1. Multiple Stacks using One-Dimensional Array

Two or more stacks may coexist in a common vector *S* of size *n*. This approach boasts of better memory utilization.

If two stacks share the same vector *S*, they grow toward each other with their bottoms anchored at opposite ends of *S*. The following figure shows the behavior of two stacks coexisting in a vector *S* :

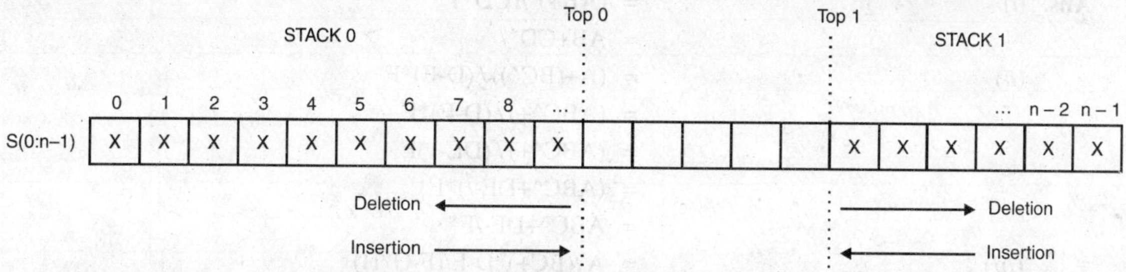

Fig. 8.11. Two Stacks Coexisting in a Vector

At initialization, stack 1's top is set to –1, that is, top1 = –1 and for stack2 it is top2 = *n*.

8.9.1.1. Three or More Stacks in a Vector S

If three or more stacks share the same vector, there is a need to keep track of several *tops* and *base* addresses. Base pointers defines the start of ***m*** stacks in a vector **S** with size ***n***. Notation of which is *B*(*i*):

$B[i]$	$= n/m * i - 1$	0<	$i < m$
$B[m]$	$= n - 1$		

B[*i*] points to the space one cell below the stack's first actual cell. To initialize the stack, tops are set to point to base addresses, *i.e.*,

```
T[i]  = B[i] ,     0<   i    <m
```

For example:

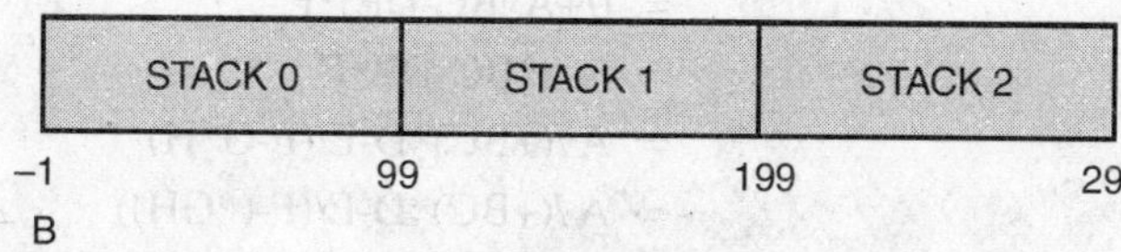

Fig. 8.12. Three Stacks in a Vector.

8.9.1.2. Three Possible States of a Stack

The following diagram shows the three possible states of a stack: empty, not full (but not empty) and full. Stack **i** is full if T[i] = B[i+1]. It is not full if T[i] < B[i+1] and empty if T[i] = B[i].

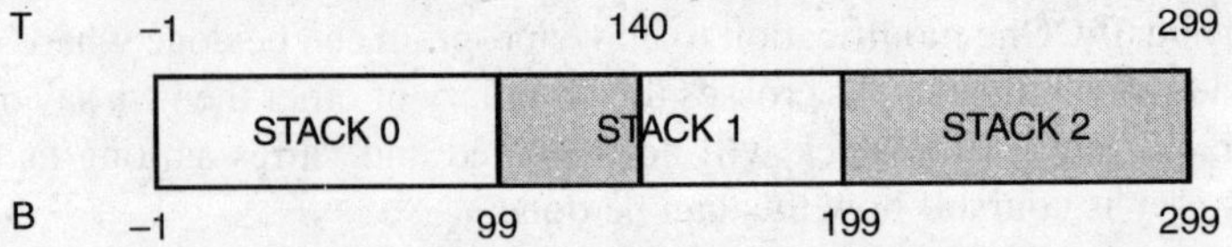

Fig. 8.13. Three States of Stack (empty, non-empty but not full, full)

SOLVED PROBLEMS

1. Convert the following infix expressions into equivalent postfix expressions

(*i*) (A+B)/(C*D)

(*ii*) (A+B^C)/(D-E) *F

(*iii*) A /(B+C)*D-E/(F-G*H)

Ans. (*i*) = (AB+) /(CD*)
= AB+CD*/

(*ii*) = (A+(BC^)) / (D-E)*F
= (ABC^+) / (D-E)*F
= (ABC^+) / (DE-)*F
= (ABC^+DE-/)*F
= ABC^+DE-/F*

(*iii*) = A /(BC+)*D-E/(F-G*H)
= A /(BC+)*D-E/(F-(GH*))
= A /(BC+)*D-E/(FGH*-)
= (ABC+/)*D-E/(FGH*-)
= (ABC+/D*)-E/(FGH*-)
= (ABC+/D*)-(EFGH*-/)
= ABC+/D*EFGH*-/-

2. Convert the same infix expressions given in question 1 to prefix form.

Ans. (*i*) = (+AB) / (C*D)
= (+AB) / (*CD)
= /+AB*CD

(*ii*) = (A+ (^BC)) / (D-E)*F
= (+A^BC) / (D-E)*F
= (+A^BC) / (-DE)*F
= (/+A^BC-DE)*F
= /+A^BC-DE*F

(*iii*) = A /(+BC)*D-E/(F-G*H)
= A /(+BC)*D-E/(F-(*GH))
= A /(+BC)*D-E/(-F*GH)

= (/A+BC)*D-E/(-F*GH)
= (*/A+BCD)-E/(-F*GH)
= (*/A+BCD)-(/E-F*GH)
= -*/A+BCD/E-F*GH

3. Consider the following postfix expression:

PostExp : 15, 8, 2,-,*, 3, 4, 5, +,*,-

Find its equivalent infix expression

Ans.
= 15, (8-2), *, 3, 4, 5, +,*,-
= 15, (8-2),*, 3, 4, 5, +,*,-
= 15*(8-2), 3, 4, 5, +,*,-
= 15*(8-2), 3, (4+5),*,-
= 15*(8-2), (3*(4+5)), -
= 15*(8-2)-(3*(4+5))

4. Write a recursive C routine that prints contents of stack in reverse order.

Ans.

```
void display_reverse (stack*T)
{
if (T->next != NULL)
display_reverse (T->next);
printf("%d->",T->data);
}
```

Assume the stack contents are 10, 20, 30, 40, and 50, with 50 on top of the stack. Now to display the elements in reverse i.e. from bottom to top we check whether next of T is not NULL, if it is then function display_reverse is called again with T->next as argument. This recursion continues till T->next is not NULL. During the winding process all return addresses are stored onto the stack. When winding stops (when T->next is NULL), unwinding starts and return addresses are popped from stack. Due to following statement printf("%d->",T->data); written after recursive call elements are displayed in reverse order.

5. Evaluate the following expression:

6,2,3,^ ,+,7,2,/,5,*,-,3,+

Ans. (*i*) 6, 8, +, 7,2,/,5,*,-,3,+

(*ii*) 14, 7,2,/,5,*,-,3,+

(*iii*) 14,3, 5*,-,3,+

(*iv*) 14, 15,-,3,+

(*v*) -1,3,+

(*vi*) 2

6. What are the notations used in Evaluation of Arithmetic Expressions using prefix and postfix forms?

Ans. Polish and Reverse Polish notations.

7. What is polish notation and reverse polish notation. Consider the following arithmetic expression P written in postfix :

P : 12, 7, 3, –, 1, 2, 1, 5, +, *, +

Translate P into its equivalent infix and prefix expression ?

Ans. Polish Notation : These notation is similar to prefix notation used in STACKS, in which operators comes before the operands as shown below :

+AB

Reverse Polish Notation : These expressions are similar to POSTFIX notation of STACK and here the operator comes after all its operands, the equivalent postfix expression is shown below :

AB+

POSTFIX EXP. **P**

P : 12, 7, 3, –, /, 2, 1, 5, +, *, *

(*a*) TO INFIX

Scan left to right and whenever an operator is visited then the operator should be inserted between two operands, which are just at left side of the operator visits.

(*i*) As in the above example, when first operator (–) encounters, then it is inserted between 7 and 3 as shown below :

12, 7 – 3, /, 2, 1, 5, +, *, *

(*ii*) Next operator (/) is inserted between 12 and (7 – 3).

12/(7 – 3), 2, 1, 5, +, *, *

(*iii*) Next operator (+) is encounted and it should be inserted in between 5 and 1 *i.e.*

12/(7 – 3), 2, 1 + 5, *, *

(*iv*) Next operator (*) is encountered and insert it between 2 and (1 + 5) *i.e.*

12/(7 – 3), 2 * (1 + 5), *

(*v*) The last operator (*) is encounter and it should insert in between

(12/(7 – 3)) and (2 * (1 + 5)) *i.e.*

(12/(7 – 3)) * (2 * (1 + 5))

The required *in expression.*

(*b*) TO PREFIX

Scan the POSTFIX EXPRESSION P. From left to right and whenever an operator is encountered, it should be inserted before the two operands, which are just at the left of operator encounted.

(*i*) In the above postfix expression, the first operator (–) is encountered and it will be inserted before 7 and 3 as shown below :

12, –73, /, 2, 1, 5, +, *, *

(*ii*) Next operator (/) is encounter and it will be inserted before 12 and (–73) *i.e.*

/ 12 (–73), 2, 1, 5, +, *, *

(*iii*) Next operator (+) is encounter, and it should be inserted before 1 and 5 *i.e.*

/ 12 (–73), 2, +15, *, *

(*iv*) Next operator (*) is encountered and it should be inserted before 2 and (+15)

/ 12 (–73), * 2 (+15), *

(*v*) Next operator (*) is encountered and it should be inserted before

/12 (–73) and * 2(+15)

$$* (/12 (-73)) (*2(+15))$$

The required **Prefix Expression.**

8. Convert the given infix expression into postfix using stacks :

$$Z : (A - B) * X + Y / (F - C * E) + D$$

Ans. To convert the above in fix. First attach or insert the left and right parenthesis at the start and end of the given infix expression as :

$$Z : ((A - B) * X + Y / (F - (*F) + D)$$

Now scan the infix left to right and use the following table :

	Element visit	STACK	EXP
Case I :		+	
	↑	+ ↑	
Case II :	+	↑	
		+	↑
Case III :		+	
	–	–	+

Element Scan	STACK	Expression (P)
(	(	
(	((	
A	((	A
–	((–	A
B	((–	A B
)	(	A B –
*	(*	A B –
×	(*	A B – X
+	(+	A B – X *
Y	(+	A B – X * Y
/	(+/	A B – X * Y
(	(+/(	A B – X * Y
F	(+/(	A B – X * Y F
–	(+/(–	A B – X * Y F
C	(+/(–	A B – X * Y F C
*	(+/(–*	A B – X * Y F C
E	(+/(–*	A B – X * Y F C E
)	(+/	A B – X * Y F C E * –
+	(+	A B – X * Y F C E * – / +
D	(+	A B – X * Y F C E * – / + D
)	–	A B – X * Y FCE * – / + D +

Hence the required POSTFIX EXPRESSION :

P : AB – X * Y F C E * – / + D +

9. Using the algorithm of postfix evaluation, evaluate the following :

P : (1, 2, 3, +, *, 3, 2, 1, –, +, *)

Each operand is of single digit ?

Ans. For **postfix evaluation :**

Element scan	Stack	Operation
(	(	
1	(1	
2	(1, 2	
3	(1, 2, 3	
+	(1, 5	2 + 3 = 5
*	(5	1 * 5 = 5
3	(5, 3	
2	(5, 3, 2	
1	(5, 3, 2, 1	
—	(5, 3, 1	(2 – 1) = 1
+	(5, 4	3 + 1
*	(20	5 * 4 = 20
)	–	

20 in STACK is popped and gets the required **Ans.**

10. Find the equivalent polish and reverse polish notation for the following infix expression :

(*i*) ((*A* + *B*) | *D*) ^ ((*E* – *F*) * *G*)

(*ii*) (*A* – *B*) | ((*D* + *E*) * *F*)

Ans. To convert from infix to prefix or postfix without using STACK follow the following steps :

(1) Scan the infix expression left to right

(2) Use BODMAS to solve the expression. For precedence of operator use the following table

OPERATOR According to decreasing precedence
$
–
*, /
+, –

(3) To convert from infix to prefix or polish notation

i.e. *A* + *B* ≡ +*AB*

Generally from infix to reverse polish notation *i.e.*

A + *B* ≡ *AB*+

(*i*) ((A + B)/D) ^ ((E – F) * G)

INFIX to PREFIX

((+AB)/D) ^ ((–EF) * G)

(/+ABD) ^ (*–EFG)

^/+ABD* – EFG

INFIX to POSTFIX

(AB+)/D) ^ ((EF–) * G)

(AB + D/) ^ (EF – G*)

AB + D/ EF – G * ^

(*ii*) (A – B) / ((D + E) * F)

INFIX to PREFIX

(–AB) / ((+DE) * F)

(–AB) / (*+DEF)

/–AB * + DEF

INFIX to POSTFIX

(AB –) / ((DE+) * F)

(AB–) / (DE + F*)

AB – DE + F */

Note :

Case I : To convert from infix to prefix or postfix expression.

Scan the infix expression from left to right according to BODMAS and whenever an operator is encountered it will be either shift toward left or right by one operand to get prefix and postfix respectively *i.e.*

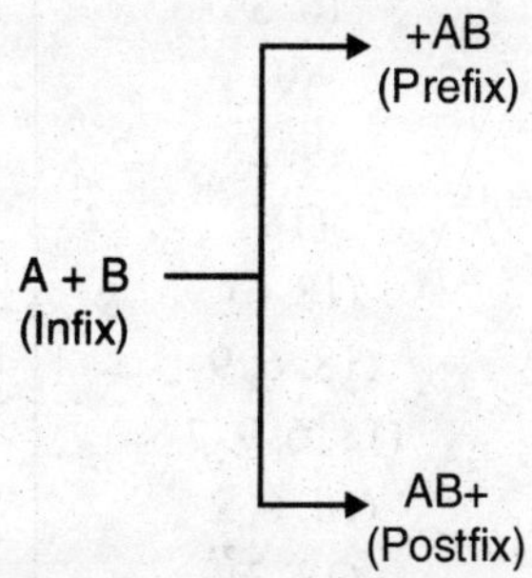

Case II : To convert from prefix to infix and postfix expression.

Scan the given prefix expression from right to left and whenever any operator encounted shift the operator either towards right by one or by two to get infix and postfix respectively, *i.e.*

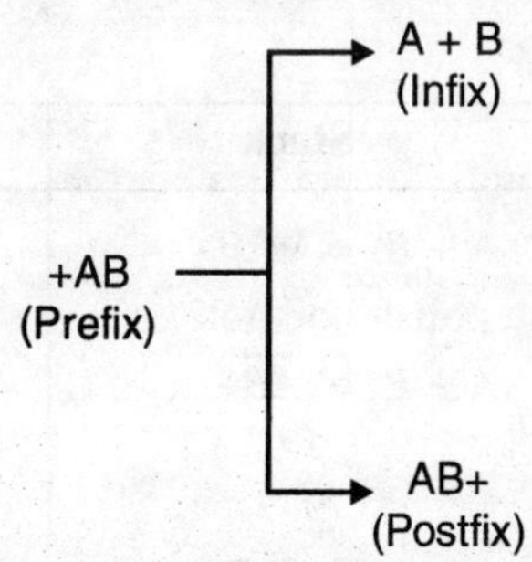

Case III : To convert from postfix to infix and prefix expression.

Scan the given postfix expression from left to right and whenever an operator encounteres, shifts the operator towards left by one or by two operands to get infix and prefix expression respectively *i.e.*

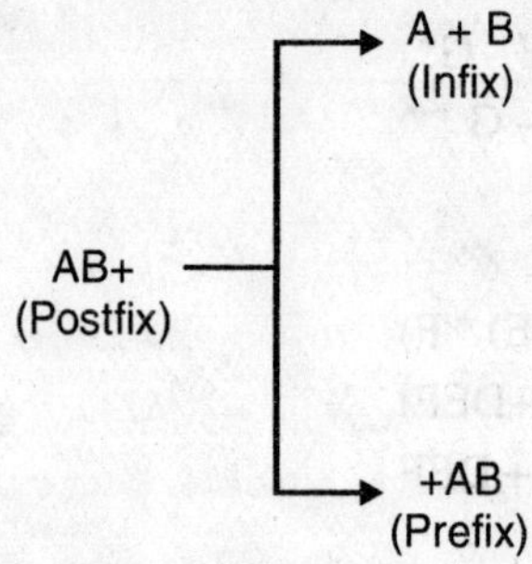

11. Evaluate the following expressions :

(*i*) 6, 3, +, 2, *, 6, 9, 7, –, /, –

(*ii*) –, *, 9, +, 2, 3, 1, 27, 3

Ans. (*i*) The given expression is in postfix notation. Hence its evaluation starts from left to right

Element scan	Stack	Operation
	(	
6	(6	
3	(6, 3	
+	(9	6 + 3 = 9
2	(9, 2	
*	(18	9 * 2 = 18
6	(18, 6)	
9	(18, 6, 9	
7	(18, 6, 9, 7	
–	(18, 6, 2	9 – 7 = 2
/	(18, 3)	6/2 = 3
–	(15	18 – 3 = 15

Whenever a right parenthesis is encounted, 15 is popped up.

(*ii*) The given expression is in prefix notation, hence it is a prefix evaluation, which starts from right to left.

Element scan	Stack	Operation
	(	
3	(3	
27	(3, 27	
/	(9	27/3 = 9
3	(9, 3	
2	(9, 3, 2	
+	(9, 5	2 + 3 = 5
9	(9, 5, 9	
*	(9, 45)	9 * 5 = 45
–	(36	45 – 9 = 36

and whenever a right parenthesis is encountered, the value 36 is popped up.

12. Convert the following infix expression to postfix using STACKS :

P : (A + (B * C – (D/E ^ F) * G))

Ans. First attach left and right parenthesis to the given infix expression :

Element scan	STACK	Expression
(	(	
A	(	A
+	(+	A
(	(+(	A
B	(+(	AB
*	(+(*	AB
C	(+(*	ABC
–	(+(–	ABC *
(	(+(–(	ABC *
D	(+(–(	ABC * D
/	(+(–(/	ABC * D
E	(+(–(/	ABC * DE
^	(+(–(/^	ABC * DE
F	(+(–(/^	ABC * DEF
)	(+(–	ABC * DEF ^ /
*	(+(–*	ABC * DEF ^ /
G	(+(–*	ABC * DEF ^ /G
)	(+	ABC*DEF^/G*–
)		ABC*DEF^/G*–+

Required Postfix expression

13. Convert the following infix expression into PREFIX notation using STACK ?

P : A + (B * C – (D/E ^ F) * G)

Ans. To convert it, first we have to reverse the given infix expression and attach left and right parenthesis to the generated expression as shown below :

P1 : ((G * (F ^ E/D) – C * B) + A)

Scan the expression P1, left to right and use following cases :

	Element Visit	Stack	Expression
Case I	↑	+ +↑	
Case II		↑	↑
Case III	+ –	+ + +–	

Element scan	STACK	Expression
(	(	
(	((	
G	((	G
*	((*	G
(	((*(	G
F	((*(	GF
^	((*(^	GF
E	((*(^	GFE
/	((*(/	GFE^
D	((*(/	GFE^D
)	((*	GFE^D/
–	((–	GFE^D/*
C	((–	GFE^D/*C
*	((–*	GFE^D/*C
B	((–*	GFE^D/*CB
)	(	GFE^D/*CB*–
+	(+	GFE^D/*CB*–
A	(+	GFE^D/*CB*–A
)	–	GFE^D/*CB*–A+

Reverse the generated expression and hence the required PREFIX EXPRESSION is

+ A – * B C * / D ^ E F G

14. How multi-stack can minimize the total number of overflows, explain with an example ?

Ans. In multi-stack, we can combine more than one stack together to increase the total STACK SIZE and if the SIZE is increased, then automatically the number of OVERFLOW decreases as shown below by an example :

Suppose, we have two stacks S_1 and S_2, S_1 have the MAXSIZE = 10 and elements pushed = 13 while S_2 have its MAXSIZE = 15 and elements PUSHED = 10

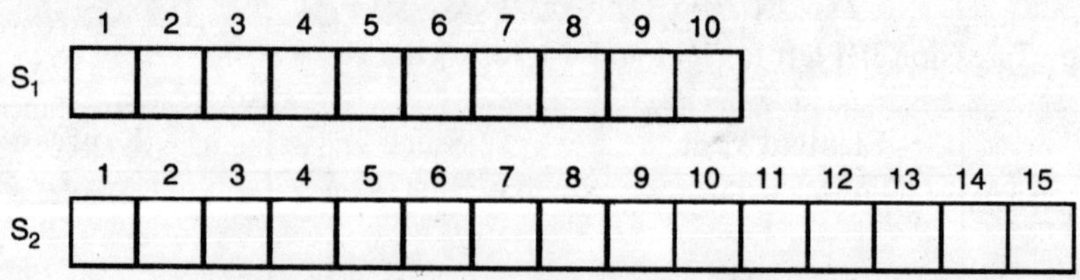

In stack S_1, there are 3 times overflow occurs while in stack S_2, the total of 5 locations remained empty.

To minimize the number of OVERFLOW, we will attach both the stack *i.e.* stack S_1 is from left to right while stock S_2 is from right to left and they are attached such that stack S_1 can use the memory of stack S_2 while stack S2 can use the memory of stack S1 and the total MAXSIZE = MAXSIZE(S1) + MAXSIZE(S2) as shown below :

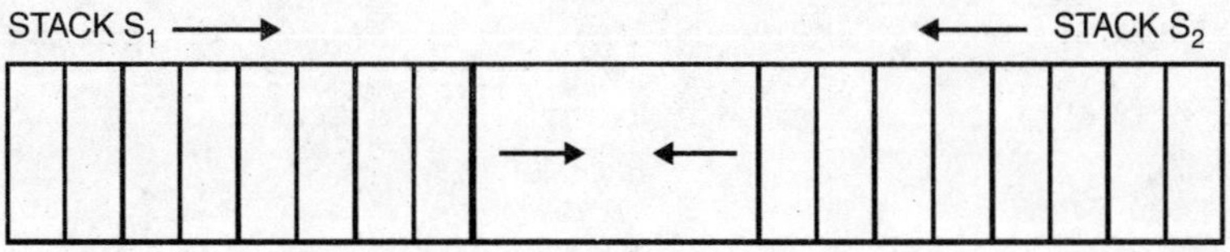

Total MAXSIZE = 25

Element PUSHED = 23

And number of OVERFLOWS are now nil.

SUMMARY

1. A stack is a data structure in which last item inserted is taken out first. That's why they are known as LIFO (last in first out).
2. Inserting an item in stack is termed as push and taking an item out from stack is termed as pop
3. An arithmetic expression is combination of various constants, symbols (variables) and operators.
4. Given just one array we can simulate two stacks, Lets assume that array name is STACK and size of this array is N. We want size of both the stack be N/2. Then our first stack will start from 0 and end at N/2-1 and second stack will start from N-1 and end at N/2. Thus our first stack will grow from left to right and second stack will grow from right to left. The purpose of simulating two or even more stacks within single array is to minimize the overflow problem when adding a new element to the stack.

EXERCISES

1. What is a stack? What are the basic operations that can be performed over the stack?
2. Is stack a linear or non-linear data structure?
3. What are various ways to represent a stack in memory? Discuss the relative merits and demerits of each.
4. When does stack underflow and overflow takes place in stack?
5. Do we need to check for stack overflow and underflow when stack is implemented using linked list?
6. Is it possible in C to create a stack of different types of elements? If possible then how?
7. What are the various ways to express an arithmetic expression?
8. Write equivalent infix expression for: +* AB/C-D
9. Write equivalent
   (*i*) A + (B ^ C) * D – (-E) (*ii*) AED ^ * BC* P/+
10. Convert the following postfix expression to infix: **ABCDE-+^+EF-**
11. What are the various applications of stacks?
12. Write a C program for implementing three stacks in an array of size N. Write just two functions for push and pop. Size of each stack is not fixed and will vary as push and pop operation take place.
13. Convert each of the following p (directly)
   (*i*) +-PQR (*ii*) + P-QR
   (*iii*) ++P-*^QRS/+TU *VWX (*iv*) +-^PQR*S**TUV
14. Convert the prefix expression given 13 to infix.
15. Convert each of the following postfix to prefix (directly)
   (*i*) PQ+R- (*ii*) PQR+-
   (*iii*) PQ-R+STU-+^ (*iv*) PQRST-+^*UV*-
16. Convert the postfix problem 15 to infix.
17. Using the C code as given in this chapter, evaluate the following postfix expression:
   (*i*) 1, 2, +, 3,-, 2, 1,+,3,^ (*ii*) 1, 2, 3, +, *, 3, 2,1,-,+,*

CHAPTER 9

QUEUES

9.1. INTRODUCTION

INSIDE THIS CHAPTER

The queue is a linear data structure where operations of insertion and deletions and performed at separate ends that are known as front and rear. Whenever a new item is added to the queue, rear pointer is used. During insertion operation rear is incremented by 1 and data is stored in the queue at that location indicated by rear. Insertion of queue always takes place using rear and this operation of adding new to the end of the queue is known as enque. If it is the first item inserted, then front also becomes 1. The front pointer is used when an item is deleted from queue. whenever an item is deleted from the queue front pointer is decremented by 1 and the deleted item is returned. If it was the last item in the queue front and rear both equal to 0. The queue is also known as FIFO(First In First Out) as items are added always at the end of queue (rear end) and are always deleted from front of the queue(beginning). In a FIFO data structure, the first element added to the queue will be the first one to be removed. This is equivalent to the requirement that whenever an element is added, all elements that were added before have to be removed before the new element can be invoked.

Queues occur naturally in situations where the rate at which clients' demand for services can exceed the rate at which these services can be supplied. For example, in a network where many computers share only a few printers, the print jobs may accumulate in a print queue. In an operating system with a GUI, applications and windows communicate using messages, which are placed in message queues until they can be handled.

9.2. REPRESENTATION OF QUEUE IN MEMORY

A queue can be represented in computer's memory using two ways: one is the array representation that is also known as static representation and second is using linked list that is also known as dynamic representation.

9.2.1. Array Representation of Queue

In the array representation of a linear queue, we take an array of fixed size and two variables front and rear. These two variables contain the current index at which insertion and deletion can be performed. The usual operation of queue is then used for insertion and deletion *i.e.* front for deletion and rear for insertion.

If the implementation uses sequential representation, a one-dimensional array/vector is used, hence the size is static. The queue is **empty** if the *front = rear* and **full** if *front=0* and *rear=n*. If the queue has data, **front** points to the actual front, while **rear** points to the cell after the actual rear. Deletion from an empty queue causes an *underflow* while insertion onto a full queue causes an *overflow*. The following figure shows an example of the ADT queue:

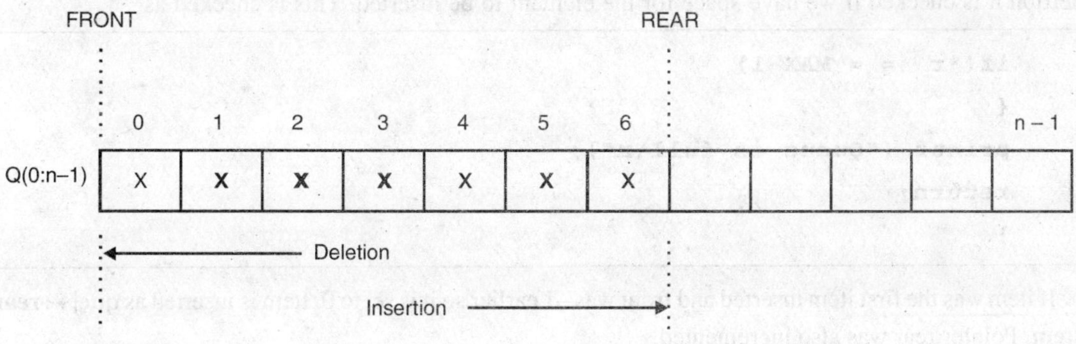

Fig. 9.1. Operations on a Queue.

To initialize, we set front and rear to 0 :

```
front = 0;
rear = 0;
```

To insert an item, say **x**, we do the following :

```
Q[rear] = item;
rear++;
```

and to delete an item, we do the following :

```
x = Q[front];
front++;
```

1. Insertion in Linear Queue

Consider the following C routine that does the insertion operation in linear queue. The function takes array maintained as queue, address of the front and rear variable and element to be inserted.

```
void enque(int que[ ], int*f, int*r, int ele)
{
if(*r  == MAX-1)

{
printf ("Queue is full\n");
return;
}
```

```
(*r) ++;
que[*r] = ele;
if(*f = = -1)
*f= 0;
}
```

Initially front and rear are initialized to –1. For insertion of item into the queue (implemented as array que) rear is incremented, for the insertion operation, we have written the function **enque**. Before insertion it is checked if we have space for the element to be inserted. This is checked as

```
if(*r  = = MAX-1)
{
printf ("Queue is full\n");
return;
}
```

If item was the first item inserted and front was –1 earlier so it is set to 0. Item is inserted as **que[++rear] = item.** Pointer rear was also incremented,

2. Deletion in Linear Queue

For the deletion in linear queue, we have written the following code. The function deque takes three parameters: array maintained as queue, address of front and rear variables,

```
int deque(int*que, int*f, int*r)
{
int d;
if(*f = = -1)
{
printf ("Queue is empty\n");
return -1;
}
d =que[*f];
que[*f]= 0;
if(*f = = *r)
          *f  = *r = -1;
else
        (*f) ++;
return d;
}
```

For the deletion of item from que, front pointer is used. During deletion the element in the que at the front is returned as que [front]. Again before deletion is checked whether que is empty or not as

```
if(*f= = -1)
{
printf ("Queue is empty\n");
return -1;
}
```

Now after deletion if front is equal to the rear (last item deleted) both are assigned value -1 else front is incremented.

3. Displaying the Linear Queue

For displaying the contents of the queue, we have written the function display. In the function starting from front, up to the rear all the elements are displayed.

```
void display(int*que, int*f, int*r)
{
int i;
for(i = *f ; i< = *r ; i++)
printf("%d", que[i]);
}
```

(*i*)

0	1	2	3	4	5	6	7	8	9
10	11	13	12	17	19	15			

Front = 0 Rear = 6

(*ii*) Front item 10 is deleted

0	1	2	3	4	5	6	7	8	9
0	11	13	12	17	19	15			

Front = 1 Rear = 6

(*iii*) Two items deleted 11, 13

0	1	2	3	4	5	6	7	8	9
0	0	0	12	17	19	15			

Front = 3 Rear = 6

(*iv*) New item 20 inserted

0	1	2	3	4	5	6	7	8	9
0	0	0	12	17	19	15	20		

Front = 3 Rear = 7

(*v*) Two new elements 25, 30 are inserted

0	1	2	3	4	5	6	7	8	9
0	0	0	12	17	19	15	20	25	30

Front = 3 Rear = 9

Next insertion causes queue overflow.

Fig. 9.2. Linear Queue insertion and deletion.

9.2.2. Linked List Representation of Linear Queue

A linear queue can be implemented using linked list. The structure of the node remains same but here we have two pointers front and rear. As stated earlier the front pointer points to the starting of the queue (first item inserted) and rear pointer points to the end of the queue (last item inserted). The structure of the node given below:

```
struct queue
{
int data;
struct queue*next;
};
typedef struct queue queue;
```

Linked allocation may also be used to represent a queue. It also makes use of nodes with fields INFO and LINK. The following figure shows a queue implemented as a linked list :

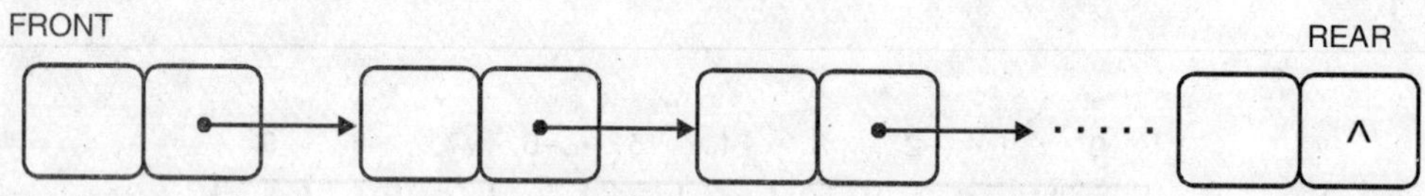

Fig. 9.3. Linked Representation of a Queue.

The queue is empty if front = null. In linked representation, since the queue grows dynamically, overflow will happen only when the program runs out of memory and dealing with that is beyond the scope of this topic.

1. Insertion

The code for the insertion is very simple.

```
temp = (queue*)malloc(sizeof(queue));
temp->data = d;
temp->next = NULL;
if (rear = =  NULL)
{
rear =  temp;
front = rear;
}
else
{
```

```
rear ->next = temp,
rear = temp;
}
```

DO YOU KNOW

PIERRE-SIMMON LAPLACE (1749-1827) Pierre-Simon Lap lace came from humble origins in north-mandy. In his childhood he was educated in a school run by the Benedictines. At 16 he entered the University of Caen intending to study theology. However, he soon realized his true interests were in mathematics. After completing his studies he was named a provisional professor of mathematics at the Paris Military School. Laplace is best known for his contribution to celestial mechanics, the study of the motion of heavenly bodies. His *Traite du Mecanique Celeste* is considered one of the greatest scientist works of the early nineteenth century. Laplace was one of the founder of probability theory and made many contributions to mathematical statistics. His work in this area is documented in his book *Theorie Analytique des Probabilitees*, in which he defined the probability of an event asa the ratio of the number of favourabe outcomes to thr total number of outcomes of an experiments. Lapalace is famous for his political flexibility. He was loyal, in sucession to the French Republlic, Napolean , king Louis XVIII. This permitted him to be productive before, during, and after the French Revolution.

In the insertion operation, temp points to the new node. If this is the first node to be inserted into the queue then front and rear both will be NULL and both now points to this new node. If rear is not NULL then insertion is similar to adding the node at the end of linked list. The next pointer of rear points to temp and rear becomes temp. For that we have written the statements:

```
rear->next = temp;
rear = temp;
```

2. Deletion

for deletion purpose, it is first checked whether front is NULL, if it in NULL, we display the message and return. In case the queue is not empty, deletion is done as:

temp pointer points to front and front pointer points to its next node. After displaying data for the node to be deleted, node is deleted by free (temp);

```
if(front = = NULL)
printf("Queue is empty\n");
else
{
temp = front;
front = front->next;
printf ("Front node deleted = %d\n", temp->data);
free (temp);
}
```

3. Display

For display purpose, front pointer is assigned to temporary pointer temp and data for all the nodes are displayed till temp does not become NULL.

```
if(front = = NULL)
printf("Queue is empty\n");
else
{
temp = front;
for(; temp! = NULL; temp = temp->next)
printf ("%d->", temp->data);
}
```

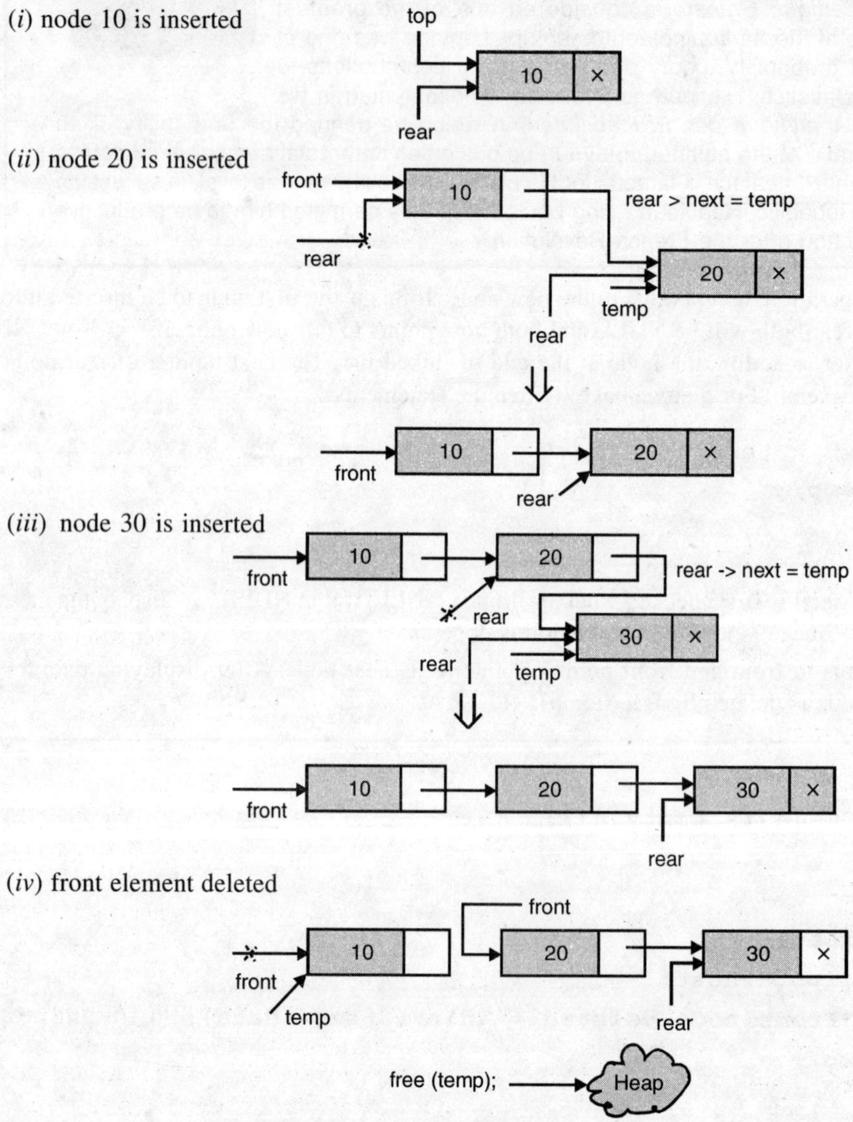

Fig. 9.4. Insertion and deletion in linear queue.

9.2.2.1. C Functions For Insertion, Deletion and Display

The structure of the node and function for allocation is defined as:

```
struct queue
{
int data;
struct queue*next;
};
typedef struct queue queue;
queue *alloc ( )
{
return (queue*) malloc (sizeof (queue) );
}
```

1. Insertion

For the insertion operation we have written the function enque that takes two arguments of queue** type which are &front and &rear and third one is the integer argument (data for the node to be added to queue). In the enque function *r denotes pointer rear and *f denotes pointer front. In the node to be inserted is the first node *r will be NULL. A new node is created and temp points to it. Member field's data and next field for the pointer temp are set. If *r is NULL, it points to temp and *f also points to *r. In case *r is not NULL, insertion is done as:

(*i*) next of *r points to temp ((*r)->next = temp)

(*ii*) *r points to temp (*r = temp)

```
void enque(queue**f, queue**r, int d)
{
queue*temp.
temp = alloc ( );
temp->data = d;
temp->next = NULL;
if (*r = = NULL)
{
*r = temp;
*f  =*r;
}
else
{
(*r)->next = temp;
*r = temp;
}
}
```

2. Deletion

For the deletion operation we have written the function deque that takes two pointers of queue** type which are &front and &rear. The function returns the data of the front node deleted. In the deque function *r denotes pointer rear and *f denotes pointer front. If the pointer *f is NULL, it means queue is empty and -l value is returned. In case pointer *f is not NULL, the node is deleted as:

(*i*) pointer temp points to *f (temp = *f)

(*ii*) *f points to its next node (*f = (*f)->next)

The data of temp is stored into the variable d and temp is freed. The data stored in d is returned.

```
int deque (queue**f, queue**r)
{
int d;
queue*temp;
if(*f  = = NULL)
{
printf ("Queue is empty\n");
d = -1;

}
else
{
temp = *f;
*f = (*f)->next;
if(*f = = NULL)
        *r = NULL;
d = temp->data;
free (temp);
}
return d;

}
```

3. Display

The C code for displaying is similar as given above and do not require much explanation.

```
void display (queue*f)
{
queue *temp;
if(f = = NULL)
{
printf("Queue is empty\n");
return;
}
printf("Queue is\n");
for(temp = f ; temp! = NULL ; temp = temp->next)
        printf ("%d->", temp->data);
}
```

Refer the Fig. 9.2 for more illustration.

9.2.2.2. Algorithm For Insertion, Deletion

Algorithm 9.1. QINSERT (QUEUE, N, FRONT, REAR, ITEM) ....

This algorithm inserts an element ITEM into a queue

1. If FRONT = 1 and REAR = N, or if FRONT = REAR + 1, then
   WRITE: OVERFLOW, and return
2. If FRONT = NULL, then:
   Set FRONT <- 1 and REAR <- 1
   Else if REAR = N, then:
   Set REAR <- 1;
   Else:
   Set REAR <- REAR + 1
3. Set QUEUE [REAR] <- ITEM
4. Return

Algorithm 9.2. QDELETE (QUEUE, N, FRONT, REAR, ITEM) ....

This algorithm deletes an element from the queue and assigns it to the variable ITEM

1. If FRONT = NULL, then
   WRITE: UNDERFLOW, and return
2. Set ITEM <- QUEUE [FRONT]
3. If FRONT = REAR, then:
   Set FRONT <- NULL and REAR <- NULL
   Else if FRONT = N, then:
   Set FRONT <- 1;
   Else:
   Set FRONT <- FRONT + 1
4. Return

9.3. CIRCULAR QUEUE

A circular queue is similar to the ordinary queue with the difference that queue is circular; that is pointer rear can point to beginning of the queue when it reaches at the end of the queue. The advantages of this type of queue is that empty location left due to deletion of elements using front can again be filled using rear pointer. We discuss just array representation of circular queue. Linked list presentation of circular queue is left for the user.

The code common to both insertion and deletion is written first,

```
#define MAX 5
int que[MAX] = {0};
int front, rear;
front = rear = -1;
```

A disadvantage of the previous sequential implementation is the need to move the elements, in the case of *rear = n* and *front > 0,* to make room for insertion. If queues are viewed as circular instead, there will be no need to perform such move. In a circular queue, the cells are considered arranged in a circle.

The *front* points to the actual element at the front of the queue while the *rear* points to the cell on the right of the actual rear element (clockwise). The following figure shows a circular queue:

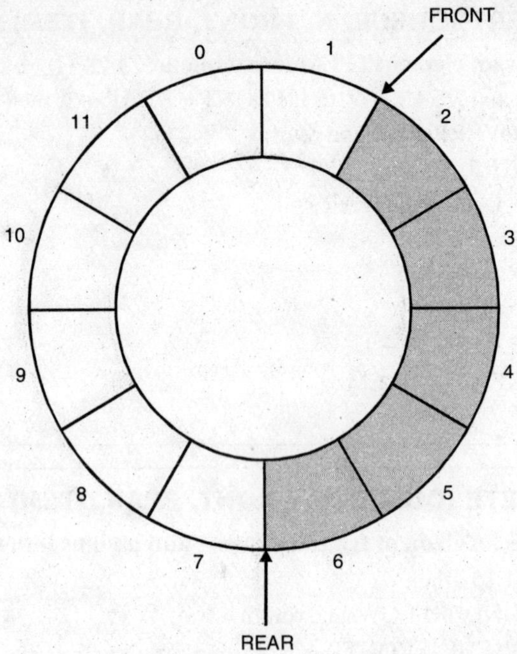

Fig. 9.5. Circular Queue.

To initialize a circular queue:

```
front = 0;
rear = 0;
```

To insert an item, say x:

```
Q[rear] = x;
rear = (rear + 1) mod n;
```

To delete:

```
x = Q[front];
front = (front + 1)     mod n;
```

We use the modulo function instead of performing an if test in incrementing rear and front. As insertions and deletions are done, the queue moves in a clockwise direction.If front catches up with rear, *i.e.*, if front = rear, then we get an **empty** queue. If rear catches up with front, a condition also indicated by front = rear, then all cells are in use and we get a **full** queue. In order to avoid having the same relation signify two different conditions, we will not allow rear to catch up with front by considering the queue as full when exactly one free cell remains. Thus a full queue is indicated by:

```
front == (rear + 1) mod n
```

1. Insertion in Circular Queue

For the insertion operation in the circular queue (CQ), the queue will be full if the front is at position 0 and rear is at MAX-1 or front is equal to rear +1. Again if currently rear is at position MAX-1 it will be at position 0 (queue is circular) else rear will be incremented by 1 and element will be inserted at this new position of rear. If function enque was called for the first time *f will be –1 so after the insertion of first element *f will also be 0.

```
void enque(int*que, int*f, int*r, int ele)
{
if( (*f = = 0 && *r = MAX-1) || (*f = = *r+1)
                printf("Queue is full\n");
else if(*r = MAX-1)
                *r = 0;
else
                 *r = *r+1;
que [*r] = ele;
if(*f = = -1)
       *f = 0;
}
```

2. Deletion in Circular Queue

For the deletion operation the data at front position is stored in the variable d and a 0 is stored at this position. Now if at present only one element is present in the queue, it means *f and *r will be same, first if condition is satisfied and both *f and *r are set to -1, If *f was at the last position in the array, it will be set to 0 (queue is circular) else *f will simply be incremented by 1.

```
int deque (int*que, int*f, int*r)
{
int d;
if(*f = = -1)
{
printf("Queue is empty\n");
return -1;
}
d = que [*f];
que[*f] = 0;
if(*f = = *r)
           *f =*r = -1;
else if(*f = = MAX-1)
          *f =0;
else
           (*f) ++;
return d;
}
```

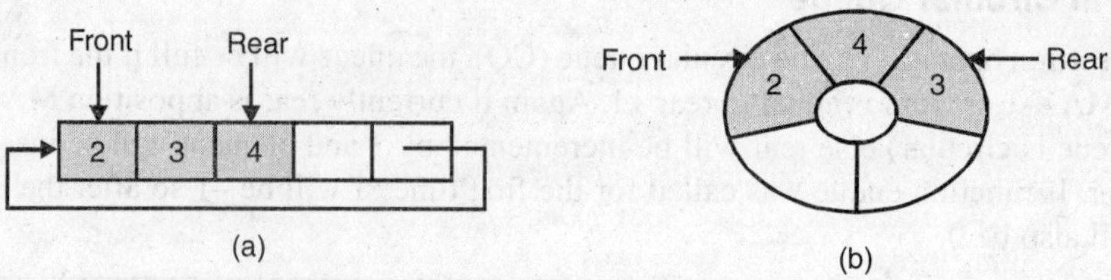

Fig. 9.6. Circular Queue.

9.4. PRIORITY QUEUE

A priority queue is a queue in each element is inserted or deleted on the basis of their priority. A higher priority element will be added first before any lower priority element. If priority of the two elements is same then they are added to the queue on first come first serve (FCFS) basis. The priority queue may be of two types: Ascending order priority queue and Descending order priority queue. In the ascending order priority queue a low priority number is given high priority and in descending order priority queue a higher number is given higher priority.

A priority queue can be represented using number of ways like using linked list, using 2-D array and using heaps. Here we will discuss the first two representation of priority queue.

A *priority queue* is a collection of elements such that each element has been assigned a priority and such that the order in which elements are deleted and processed comes from the following rules :

(1) An element of higher priority is processed before any element of lower priority.

(2) Two elements with the same priority are processed according to the order in which they were added to the queue.

A prototype of a priority queue is a timesharing system: programs of high priority are processed first, and programs with the same priority form a standard queue.

There are various ways of maintaining a priority queue in memory. We discuss two of them here : one uses a one-way list, and the other uses multiple queues. The ease or difficulty in adding elements to or deleting them from a priority queue clearly depends on the representation that one chooses.

9.4.1. Linked List Representation of Priority Queue

The structure of the node will be same as we have seen earlier but to provide the notion of priority in the node, the node includes one more data member prn of type integer. The structure of the node in a priority queue is given as :

```
struct pque
{
int data;
int prn;
struct pque*next;
};
```

This is type defined as : **typedef struct pque pque;**

Either the higher value of the prn termed as higher priority or higher value of pm termed as lower priority. Here we assume higher priority for the node having lower value of pm field. For example, priority of a node with prn 2 will be lower than a node with value of prn field is 1.

One Way List Representation of a Priority Queue

One way to maintain a priority queue in memory is by means of a one-way list, as follows :

(*a*) Each node in the list will contain three items of information : an information field INFO, a priority number PRN and a link number LINK.

(*b*) A node X precedes a node Y in the list (1) when X has higher priority than Y or (2) when both have the same priority but X was added to the list before Y. This means that the order in the one-way list corresponds to the order of the priority queue.

Priority numbers will operate in the usual way : the lower the priority number, the higher the priority.

Figure 9.7 shows a shcematic diagram of a priority queue with 7 elements. The diagram does not tell us whether BBB was added to the list before after DDD. On the other hand, the diagram does tell us that BBB was inserted before CCC, because BBB and CCC have the same priority numbre and BBB appears before CCC in the list. Figure 9.8 shows the way the priority queue may appear in memory using linear arrays INFO, PRT and LINK.

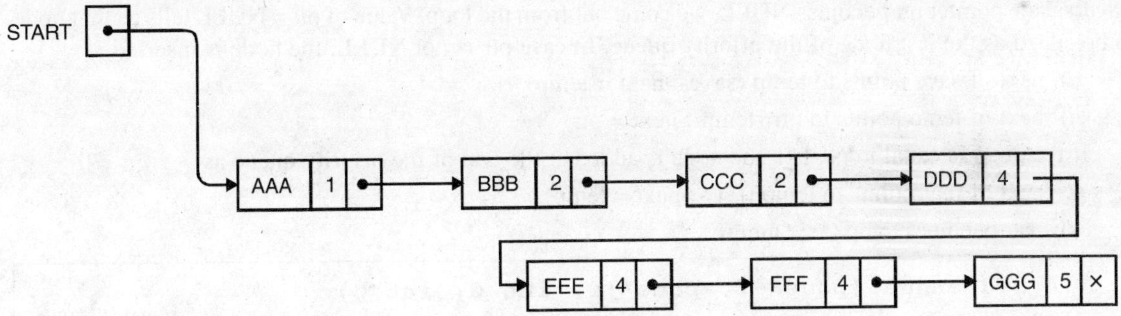

Fig. 9.7.

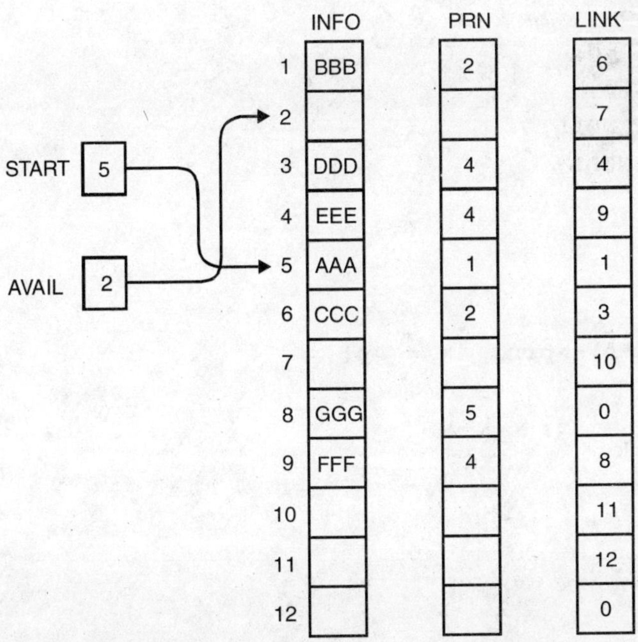

Fig. 9.8.

1. Insertion in Priority Queue

For the insertion operation we have written the function enque that takes four arguments: the first two arguments are of node** type and they always represent a pointer to front and rear pointer. The third and fourth arguments of integer type represent data to be inserted into the node and the priority of the node.

For insertion a new node temp is created and its various fields are filled. Now insertion is done on the basis of priority p of the new node, If the queue is empty then node will be added as the first node and both *r and *f will point to it. If the priority of the front node is greater then equal to the new node *i.e.* ((*f)->prn > = p)then node is added at the beginning of the queue as

```
temp->next = *f;
*f = temp;
```

If this is not the case then, the whole queue is traversed using the temporary pointer ptr. Using this pointer ptr we search for the node whose pm is less then equal to the p. The condition is written using while loop. The node previous to pointed by ptr is stored in the temporary pointer save. Now if during traversal the pointer ptr becomes NULL, we come out from the loop. Value of ptr = NULL tells us that node to be added as the last node of the priority queue. In case ptr is not NULL, the node is inserted as:

(*i*) next of save points to temp (save->next = temp)

(*ii*) next of temp points to ptr (temp->next = ptr)

If the ptr was equal to NULL, the node is added at the end of the priority queue as :

(*i*) next of rear points to temp ((*r)->next = temp)

(*ii*) rear becomes temp (*r = temp)

```
void enque (pque**f, pque**r, int d, int p)
{
pque*temp, *ptr, *save;
temp = alloc ( );
temp->data =d;
temp->prn =p;
temp->next NULL;
if(*r = = NULL)
{
*r= temp;
*f = *r;
}
else if ((*f)->prn   > =  p)
{
temp->next = *f;
*f = temp;
}
else
{
ptr = *f;
while (ptr ->prn    < =  p)
{
```

```
    save = ptr;
    ptr = ptr->next;
    if(ptr = = NULL)
            break;
    }
    save->next = temp;
    temp->next = ptr;
    if(ptr = = NULL)
    {
    (*r)->next = temp;
    *r = temp;
    }
    }
    }
```

2. Deletion in Priority Queue

The deletion operation in priority queue is similar to deletion in an ordinary queue so we will not discuss the same again.

```
    int deque(pque**f, pque**r)
    {
    int d;
    pque*temp;
    if(*f = = NULL)
    {
    printf ("Queue is empty\n");
    d = -1;
    }
    else
    {
    temp = *f;
    *f = (*f) ->next;
    if (*f = = NULL)
    *r = NULL;
    d = temp ->data;
    free (temp);
    }
    return d;
    }
```

9.4.2. Priority Queue Implementation Using 2-D Array

To implement priority queue using arrays we take one 2-D arrays and two 1-D arrays. The 2-D array is named pque and is of dimension 4 by 4. The first dimension denotes 4 priority queues and second dimension denote number of elements in each priority queue. The priority of first row (*i.e.* row 0) is 1, second row (*i.e.* row 1) is 2 and so on. For keeping track of rear and front of each priority queue we have two 1-D arrays FRONT and REAR, The element FRONT[i] and REAR[i] is front and rear for each priority queue respectively, where i is from 0 to 3. For example FRONT[0] and REAR[0] are front and rear for first priority queue *i.e.* pque[0] [i] (i is from 0 to 3).

We assume one function init that initializes rear and front of each priority queue to –1.

Array Representation of a Priority Queue

Another way to maintain a priority queue in memory is to use a separate queue for each level of priority (or for each priority number). Each such queue will appear in its own circular array and must have its own pair of pointers, FRONT and REAR. In fact, if each queue is allocated the same amount of space, a two-dimensional array QUEUE can be used instead of the linear arrays. Figure 9.6 indices this representation for the priority queue. Observe that FRONT[K] and REAR[K] contain, respectively, the front and rear elements of row K of QUEUE, the row that maintains the queue of elements with priority number K.

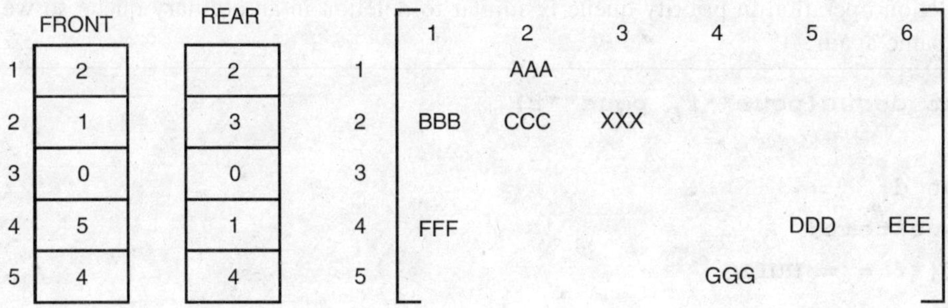

Fig. 9.9.

1. Insertion

For the insertion into any of the priority queue we have written the function insert. The function takes two integer arguments: one is the data to be insert (argument d) and second priority number of the queue. Note while inserting the elements into any of the priority queue we simply insert the elements into corresponding priority queue in the first come first serve basis as their priority is already fixed. Note how the elements are inserted:

```
REAR [prn - 1] ++;
if (REAR[prn-1] = = 0)
        FRONT [prn - 1] = 0;
pque[prn -1][REAR[prn -1]] = d;
```

The if condition turns out to be true in case when the first element is inserted. For insertion **REAR[prn - 1]** is incremented and element is inserted as :

```
pque[prn -1] [REAR[prn -1]] =d;
```

Consider following snapshot of the priority queue

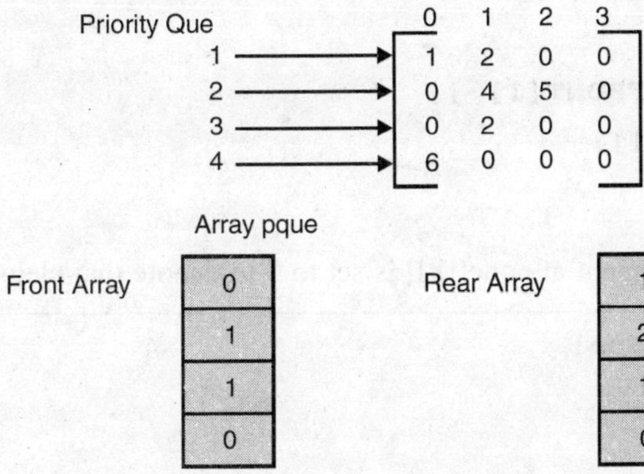

Fig. 9.10. Priority Queue representation.

Suppose prn is 1 and d is 4 so the above expression becomes:

```
pque[0][REAR[0] ] = 4;
pque[0] [2] = 4;
```

So the element 4 is inserted in first row, 3rd column.

```
void insert(int d, int prn)
{
if(REAR[prn -1] = = MAX-1)
{
printf ("Queue with priority %d is full\n", prn);
exit (0) ;
getch ( );
}
REAR[prn - 1] ++;
if(REAR[prn -1] = = 0)
FRONT[prn - 1] = 0;
pque[prn - 1] [REAR[prn -1] ] = d;
}
```

2. Deletion

For deletion we have written the function del. The function takes just one argument of integer type *i.e.* priority number of the queue from which deletion is to take place. We take out the element to be deleted in variable d and set that value to 0 as:

```
d = pque [prn -1] [FRONT [prn -1];
pque[prn -1] [FRONT[prn -1] ] = 0;
```

The front is then incremented to point to next element as:

```
FRONT [prn - 1]++;
```

To understand this, consider again the snapshot of the array pque as shown earlier. Assume prn is 2, so expression becomes:

```
d = pque[1] [FRONT[1] ];
d = pque[1] [1];
d = 4;
```

In the next statement element at pque[l][l] is set to 0 to denote that element has been deleted.

```
int del (int prn)
{
int d;
if(FRONT[prn -1] = = -1)
{
printf ("Queue with priority %d is empty\n", prn);
return -1;
}
d = pque[prn - 1] [FRONT[prn -1]];
pque[prn -1] [FRONT[prn -1]] = 0;
FRONT [prn -1] ++;
if(FRONT[prn - 1] = = -1)
        REAR[prn - 1] = -1;
return d;
}
```

9.5. DEQUE

A double ended queue (written as deque) is a queue where insertion and deletion can be performed at both end *i.e.* front pointer can be used for insertion (apart from its usual operation *i.e.* deletion) and rear pointer can be used for deletion (apart from its usual operation *i.e.* insertion). There are two variations of deque:

(*i*) **Input restricted queue :** In this type of deque as name suggests only deletion can be performed at both ends but insertion can only be performed through one end *i.e.* using rear only, That's why name input restricted.

(*ii*) **Output restricted queue :** In this type of deque as name suggests only insertion can be performed at both ends but deletion can only be performed through one end *i.e.* using front only. That's why name output restricted.

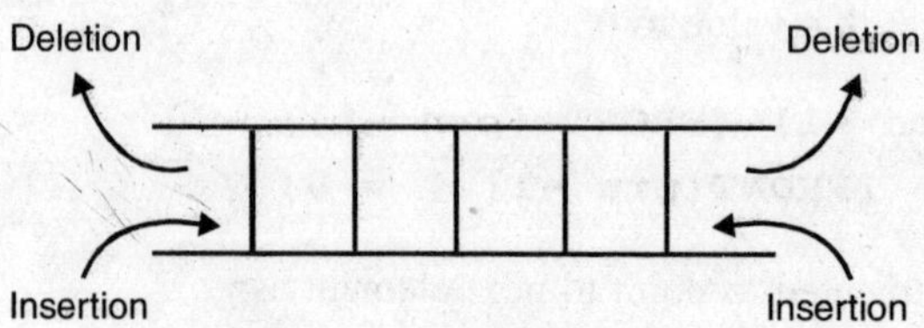

Fig. 9.11. A Deque.

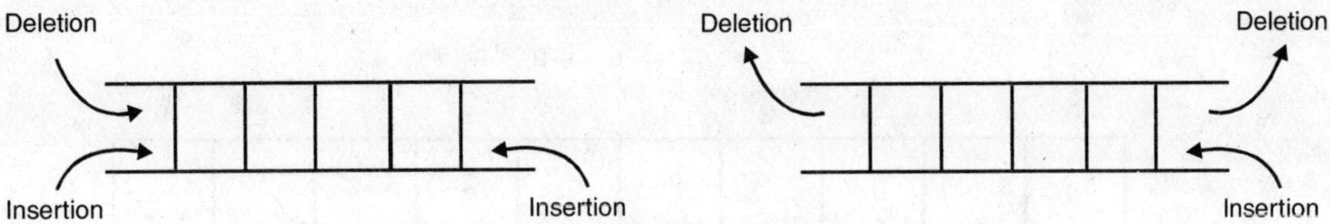

(*a*) IRD (*b*) ORD

Fig. 9.12.

9.5.1. Representation of Deque

A deque can be represented in computer's memory using array or linked list. Here we will discuss only the array representation. In the following we assume that we have an array arr that act as our deque of size 5. For performing insertion and deletion in the beginning and at the end, we have written four functions:

(*i*) **enque_beg** (for insertion in the beginning of deque)
(*ii*) **enque_end** (for insertion at the end of deque)
(*iii*) **deque_beg** (for deletion in the beginning of deque)
(*iv*) **deque_end** (for deletion at the end of deque)

9.6. SIMULATING TWO QUEUES'S USING ONE ARRAY

In the previous chapter we have seen how to simulate two stacks using one array. On the similar basis we can simulate two queues in one array. The structure for this implementation is shown below:

```
#define MAX 10
typedef struct
{
int arr [MAX];
int front[2];
 int rear[2];
} queue;
```

In the structure queue we have two arrays front and rear of size 2. For the first queue front and rear are denoted by front[0] and rear[0] and for the second queue front and rear are denoted by front [l] and rear[1] respectively. The arrays are initialized as:

```
queue Q = {{0}, {-1, MAX}, {-1, MAX}};
```

The size of the array is 10. The first queue will be having elements at position 0 to 4 and second queue will be having elements from 9 to 5. The front and rear of first queue will increment during deletion and insertion respectively and same will be decremented for second queue. This is shown in the Fig. 9.10.

The functions for insertion and deletion are shown below. They are self- explanatory as we have discussed the logic for insertion and deletion in queue in detail.

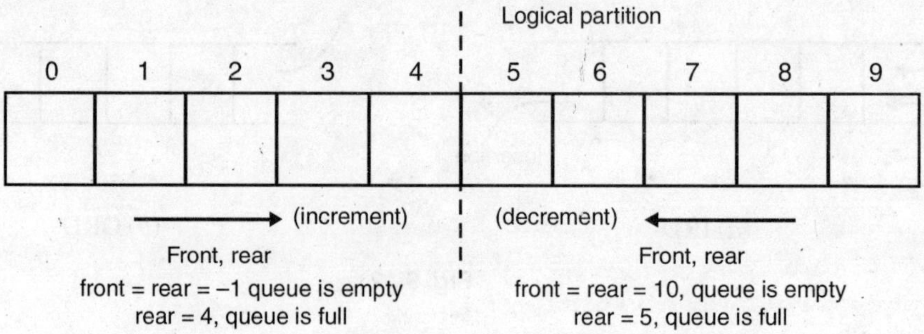

Fig. 9.13. Simulating two queues in one array.

```
void enque (queue *ptr, int d, jut SN)
{
switch (SN)
{
case 1:if(ptr->rear[SN–1] = = MAX/2-1)
{
printf("Queue is full, while inserting %d\n", d);
return;
}
ptr->rear [SN-1] ++;
ptr->arr[ptr->rear[SN-1] ]= d;
if(ptr->rear[SN-1] = = 0)
 ptr->front[SN-1] = 0;
break;
case 2: if(ptr->rear[SN-1] = = MAX/2)
{
printf("Queue is full, while inserting %d\n", d);
 return;
}
ptr->rear [SN-1] --;
 ptr->arr[ptr->rear[SN-1] = d;
 if(ptr->rear[SN-1] = = MAX-1)
 ptr->front[SN-1]  = MAX-1;
break;
}
}
int  deque(queue*ptr, int SN)
{
int d;
switch (SN)
```

```
{
case 1: if(ptr->front[SN-1] = = -1)
{
printf("Queue is empty\n");
return -1;
}
d = ptr->arr[ptr->front[SN-1] ];
ptr->arr[ptr->front[SN-1]]   = 0;
ptr->front  [SN-1] ++;
if  (ptr -> front  [SN -1] = = MAX/2)
{
ptr -> rear [SN -1 ] = -1;
ptr -> front  [SN -1] = -1;
}
break;
case 2: if(ptr->front[SN-1] = = MAX)
{
printf  ("Queue is empty\n");
return -1;
}
d = ptr->arr[ptr->front[SN- 1] ];
ptr->arr[ptr ->front[SN-1] ] = 0;
ptr->front  [SN-1] -;
if(ptr->front[SN - 1] = MAX/2-1)
{
ptr->rear[SN-1]  = MAX;
ptr->front[SN-1] = MAX;
}
break;
}
return d;
}
```

On the similar basis, we can simulate more than 2 queues in one single array.

9.7. APPLICATION OF QUEUES

The simple queue finds its application in operating system for the scheduling of various jobs/tasks / process for their execution by the CPU. Similarly the queues are there for various devices like printer, screen etc. For the scheduling of jobs as per the priority, priority queues are used by the operating system. For round robin scheduling circular queue can be used (get back to end of queue if execution is not over). In fact queue finds its application in any problem which characterizes the behavior of the queue.

SOLVED PROBLEMS

1. How to calculate number of elements in a circular queue given position of rear and front pointer. Considering circular queue implemented as array.

Ans. Number of elements in circular queue = Rear - Front+1

2. **What is the condition in circular to be full implemented as array ?**

Ans. Either Front = 1 and Rear = N, or Front = Rear+1

3. **Minimum number of queues needed to implement the priority queue?**

Ans. Two. One queue is used for actual storing of data and another for storing priorities.

4. **Consider the following QUEUE of chars, where QUEUE is a circular array which is allocated six memory cells :**

   FRONT = 2, REAR = 4

   QUEUE : _, A, C, D, _, _

   Describe the QUEUE as the following operation takes place in sequence :

   (*a*) E, F are added in sequence

   (*b*) 3 elements are deleted

   (*c*) G is added

   (*d*) Two elements are deleted

   (*e*) G is deleted

   (*f*) One element is deleted.

Ans. As the QUEUE is circular, therefore if the QUEUE's REAR POINTS to Nth location then the next element is to be inserted at the first location if it's free or empty and during insertion REAR increment by one and during deletion FRONT increments by one :

(*a*) E, F are added *i.e.*

QUEUE = _, A, C, D, E, F

FRONT = 2, REAR = 6

(*b*) 3 elements are deleted from the FRONT i.e.

QUEUE = _, _, _, _, E, F

FRONT = 5, REAR = 6

(*c*) G is added at the REAR, but REAR points to the last cell and FRONT of QUEUE, hence its inserted at the START *i.e.*

QUEUE = G, _, _, _, E, F

FRONT = 5, REAR = 1

(*d*) 2 elements are deleted from front *i.e.*

QUEUE = G, _, _, _, _, _

FRONT = 1, REAR = 1

(*e*) G is deleted *i.e.*

QUEUE = _, _, _, _, _, _

FRONT = 0, REAR = 0 i.e. QUEUE is empty.

(*f*) One element is deleted, and there is no element in the QUEUE and it returs **UNDERFLOW.**

5. **Consider a QUEUE, which is initially empty. Find the values of FRONT and REAR after each of the following operations :**

   (*i*) Insert A, B and C

   (*ii*) Delete 2 elements

   (*iii*) Insert D, E, F and G

   (*iv*) Insert H

   (*v*) Insert I, J

   Assume the QUEUE is implemented using arrays with MAXSIZE = 5.

Ans. Initially the QUEUE is empty

QUEUE = _, _, _, _, _

FRONT = 0, REAR = 0

(*i*) Insert A, B, C

```
QUEUE = A, B, C, _, _
FRONT = 1, REAR = 3
```

(*ii*) Delete 2 elements

```
QUEUE = _, _, C, _, _
FRONT = 3, REAR = 3
```

(*iii*) Insert D, E, F and G

```
QUEUE = F, G, C, D, E
FRONT = 3, REAR = 2
```

(*iv*) Insert H

```
OVERFLOW OCCURS AS QUEUE is FULL i.e.
if (FRONT = REAR + 1) then if
indicates that QUEUE is FULL.
```

6. What is Deques. How a priority QUEUE is stored in memory ?

Ans. If we want to insert or delete an element from the same side *i.e.* either from FRONT or REAR, then it is possible in Deques not in simple or circular queues.

The Deques used are of two types i.e.

(1) Input restricted Deques : In an input restricted deque, the insertion of an element is restricted from either one side while the deletion can take place from both side *i.e.*

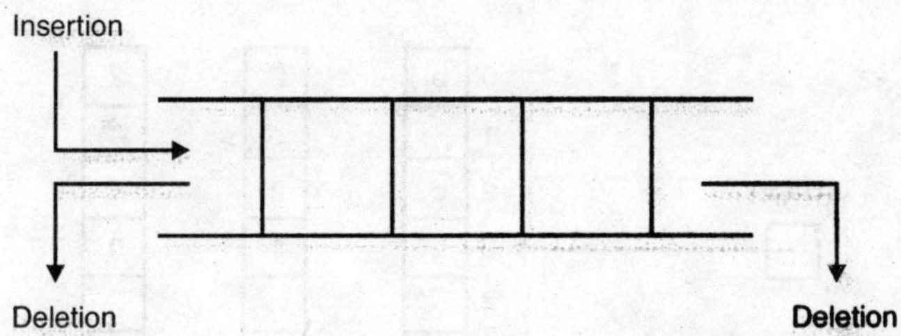

(2) Output restricted Deques : In an output restricted deques, the deletion of an element is restricted to be done from only one side while insertion can takes place from both the sides *i.e.*

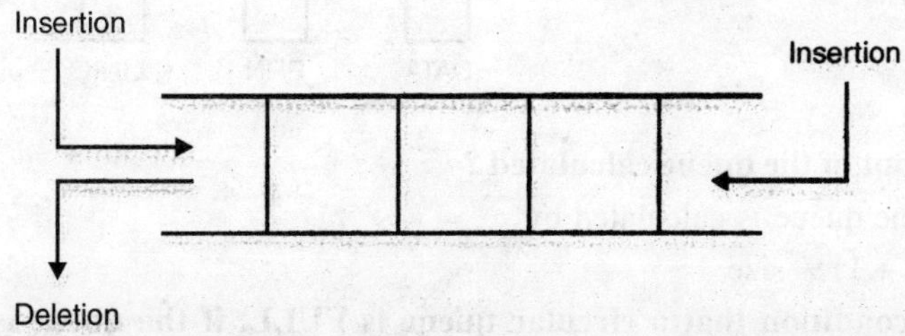

Priority Queues : If every data element of the given queue have assocaited with a priority number i.e. PRN, then the queue is implemented using priority queues using the following :

(1) Every element in the queue, assigns a PRN, lower the number highest is the priority.

(2) The element with highest priority is executed or deleted first.

(3) If two different elements are having the same PRN, then they are executed as they are inserted in the queue.

Priority Queues are stored in memory either using arrays or linked list.

Given Priority Queue :

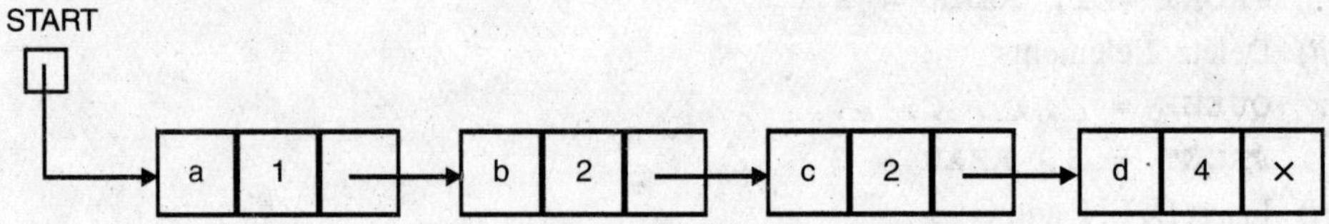

USING ARRAYS

Number of rows = Range of higher to lower PRN

Number of colums = Number of rows + 1

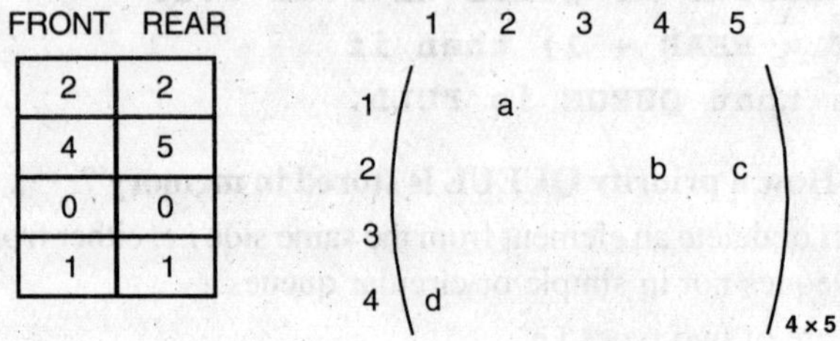

Using Linked List

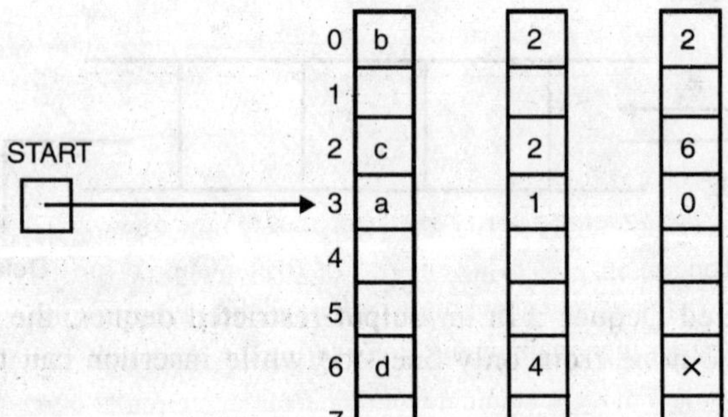

7. How is the front of the queue calculated ?

Ans. The front of the queue is calculated by

Front = (Front + 1) % size

8. What is the condition that a circular queue is FULL, if the queue is implemented using arrays ?

Ans. There are two conditions for queue full if queue is implemented using arrays.

(*i*) if FRONT = 1 and REAR = N

(*ii*) if FRONT = REAR + 1

SUMMARY

1. The queue is a linear data structure where operations of insertion and deletions and performed at separate ends that are known as front and rear.
2. Whenever a new item is added to the queue, rear pointer is used.
3. The front pointer is used when an item is deleted from queue.
4. A circular queue is similar to the ordinary queue with the difference that queue is circular; that is pointer rear can point to beginning of the queue when it reaches at the end of the queue. The advantages of this type of queue is that empty location left due to deletion of elements using front can again be filled using rear pointer.
5. A priority queue is a queue in each element is inserted or deleted on the basis of their priority. A higher priority element will be added first before any lower priority element. If priority of the two elements is same then they are added to the queue on first come first serve (FCFS) basis.
6. A double ended queue (written as deque) is a queue where insertion and deletion can be performed at both end *i.e.* front pointer can be used for insertion (apart from its usual operation *i.e.* deletion) and rear pointer can be used for deletion (apart from its usual operation *i.e,* insertion).
7. The simple queue finds its application in operating system for the scheduling of various jobs / tasks / process for their execution by the CPU.
8. Similarly the queues are there for various devices like printer, screen etc.

EXERCISES

1. What is queue? What are the various operations that can be performed over a queue?
2. Discuss all different types of queue in brief.
3. What are the various ways of representing queue in computer's memory?
4. Compare and contrast array and linked list represented of queue.
5. How to check end of circular queue in (*i*) array representation (*ii*) linked list representation.
6. What is deque? Discuss its various types?
7. How can we perform insertion and deletion in deque, represented as deque?
8. What is priority queue? How can we represent a priority queue?
9. Write a C program that implements priority queue as doubly linked list.
10. A circular queue is to be implemented using an array of 10 elements. Write the C code for the implementation of inserting an element in queue and checking whether queue is empty or not.
11. How circular queue is better then plain queue, considering the array representation?
12. Write a C routine that delete the Kth element from the front of a circular queue.
13. How can you implement
    (*i*) A stack of queue
    (*ii*) A Queue of stack
    (*iii*) A stack of stack
    (*iv*) A queue of queue
14. Write C code to simulate 3 queue's in one array.

❑❑❑

CHAPTER 10 TREES

INSIDE THIS CHAPTER

10.1. INTRODUCTION

In computer science, a tree is a widely-used non-linear data structure that emulates a tree structure with a set of linked nodes. The most popular type of tree is binary tree so we will discuss only binary tree and its various special types.

10.2. BINARY TREE

A binary tree is type of tree with finite number of elements and is divided into three main parts. The first part is called root of the tree and other two parts are itself binary tree which exists towards left and right of the tree. Each of the binary trees towards left and right of root are known as subtrees and subtrees may also have their root, left child and right child. Each of the elements in the binary tree is considered a node and a node will have three pieces of information: data, left pointer and right pointer. Figure shown some of the examples of binary tree:

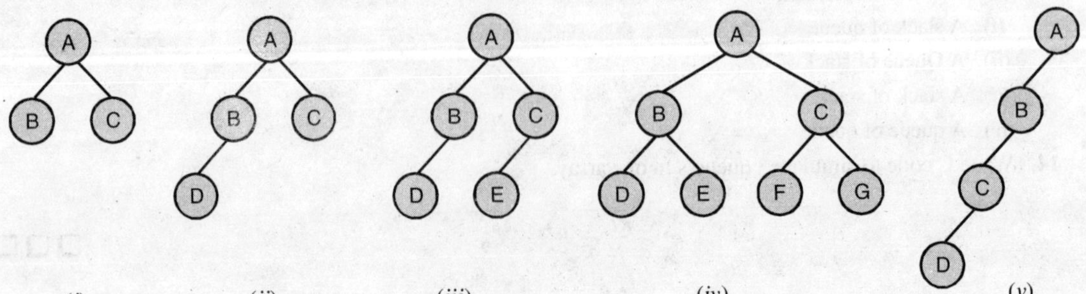

(i) (ii) (iii) (iv) (v)

Fig. 10.1. Binary Tree

10.2.1. Binary Tree Points and Terminology

(*i*) The binary starts at root of the tree and grows downward.

(*ii*) The topmost node in a tree is called the root node. Being the topmost node, the root node will not have parents. It is the node at which operations on the tree commonly begin. All other nodes can be reached from it by following edges or links.

(*iii*) The link between two nodes is termed as edge or arc.

(*iv*) If *X* is the root of the tree and *L* and *R* are its left child/son and right child/son respectively then *X* is the father/parent of *L* and *R*. *L* and *R* are termed as siblings or brothers. That is two nodes are said to be sibling if they have the same parent. This applies to every node with empty or non empty child.

(*v*) A node with no left or right child is termed as leaf node. It is also called as terminal node.

(*vi*) An internal node or inner node is any node of a tree that has child nodes and is thus not a leaf node.

(*vii*) A node *X* is termed as ancestor of node *Y* and *Y* is descendent of *X* if *X* is the parent of *Y* or parent of some ancestor of *Y*. Descendent is also termed as successor and ancestor is also termed as predecessor.

(*viii*) Following the point (*vii*) a node *Y* is a right descendent of node *X* if *Y* is either a right son of *X* or a descendent of a right son of *X*. Similar arguments applies to left descendent.

(*ix*) A path is sequence of edges from some node to other node with more that one edge.

(*x*) A path ending in a leaf in known as branch.

(*xi*) Going from root of the tree to any of the leaf node is termed as descending the tree.

(*xii*) Going from any leaf of the tree towards root of the tree is termed as climbing the tree.

(*xiii*) The root has level-0 and level of any other node is one more then the level of its parent.

(*xiv*) Each level in the binary tree corresponds to one generation and all nodes in that level said to belong to that generation.

(*xv*) The height of the binary tree is one more than the number of levels. Another definition of height is the maximum number of nodes in a branch of tree.

(*xvi*) The depth of the binary tree is the maximum level number.

(*xvii*) Number of sons for a node is considered as degree of that node

(*xviii*) Two binary trees are said to be copies when they have the identical structure and identical nodes at all levels of tree.

(*xix*) Two binary trees are said to be similar when they have the identical structures only

10.2.2. Strictly Binary Tree

It is a binary tree with non empty right and left subtrees. In other words it is binary tree with every node N has either 0 to 2 tree. The strictly binary tree is also termed as extended 2-Tree or simply 2-Tree. Sometimes nodes with 2 children's are known as internal nodes and nodes with 0 children are known as external nodes. Figure 10.2 shows some of the examples of strictly binary tree.

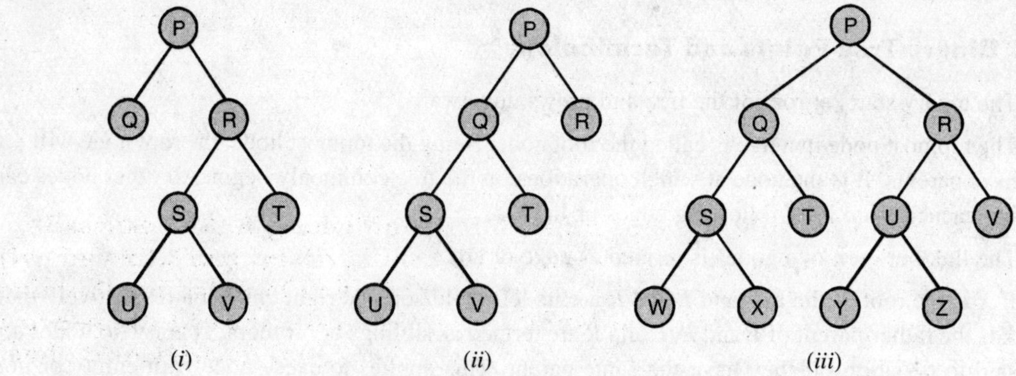

Fig. 10.2. Strictly Binary Tree.

Following figures show examples of binary trees that are not strict binary tress:

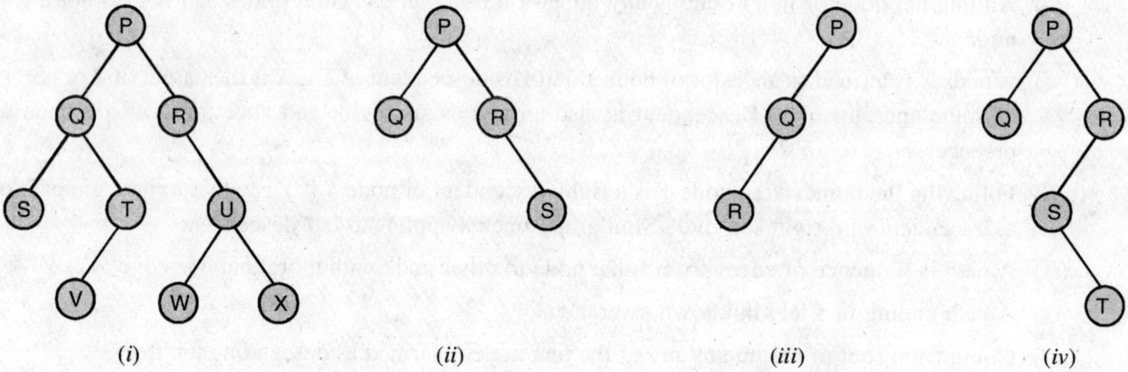

Fig. 10.3. Trees which are not Strictly Binary Trees.

Any binary tree that is not 2-tree can be converted into 2-tree by adding extra nodes to the parent so that the resulting tree will be a 2-tree. Due to extending the binary tree by adding new node the word "extended" is prefixed before 2-tree. Figure shows two examples of converting a non 2-tree into 2-tree:

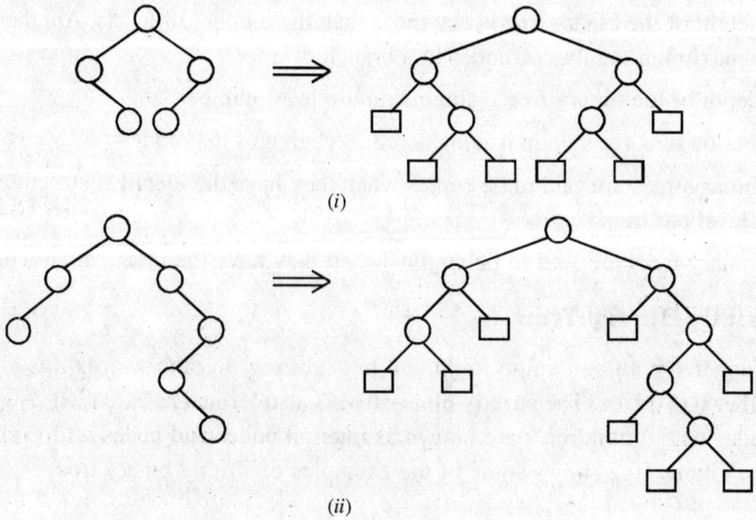

Fig. 10.4. Extended 2 Trees.

10.2.3. Complete Binary Tree

It is a special type of strictly binary tree where all the leaves of the tree reside at the same level. Using the depth we always say a complete binary tree of depth d where all leaves will be at level d. Figure shows some of the complete binary trees of various depths.

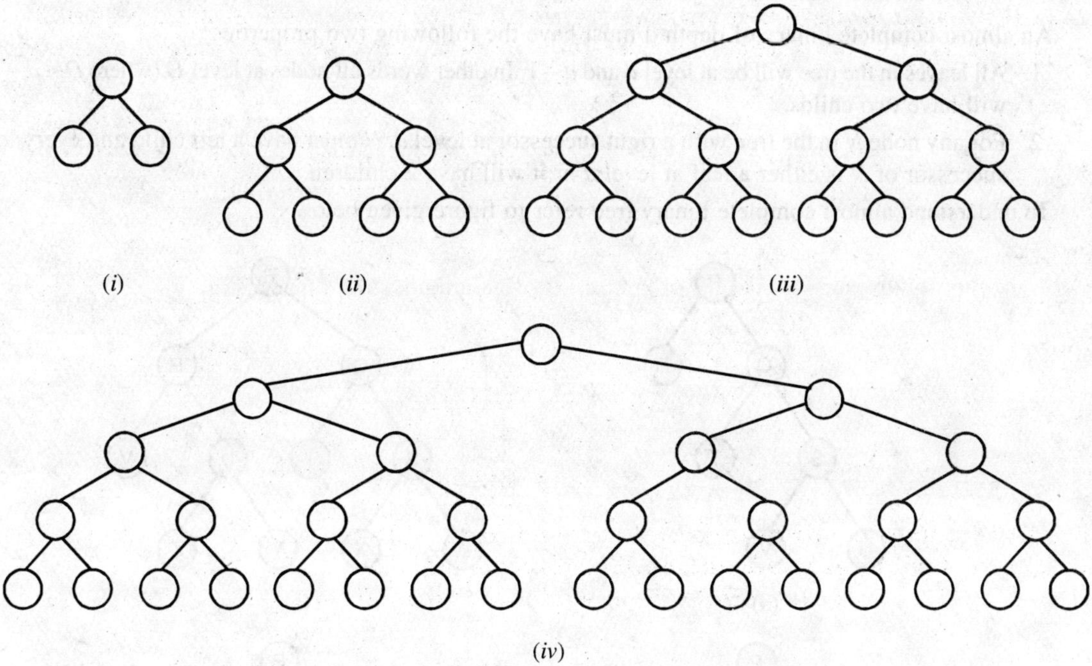

Fig. 10.5. Complete Binary Trees.

If a binary node has N nodes at a level L then it will have 2 N nodes at the next higher level *i.e.* at $L + 1$. In general at level L the complete binary tree will definitely have 2^L nodes. The total number of nodes T in a complete binary tree of depth d will be equal to sum of all nodes from level 0 to level d. Thus :

$$T = 2^0 + 2^1 + 2^2 + 2^3 \ldots 2^d$$

$$= 1 + 2\frac{(2^d - 1)}{2 - 1}$$

$$= 1 + 2^{d+1} - 2$$

$$T = 2^{d+1} - 1$$

$$2^{d+1} = T + 1$$

Taking log base 2 both sides

$$d + 1 = \log_2 (T + 1)$$

$$d = \log_2 (T + 1) - 1$$

From the above it is clear the number of nodes T in a complete binary tree of depth d is a function of $\log_2$ for example a complete binary tree of 31 nodes will have its depth 4 and with 15 nodes will have its depth 3.

The importance of complete binary tree comes from the fact that it is the only binary tree that has maximum number of nodes for a given depth *d*. Though depth will be very small for a large number of nodes; for example for a tree with 1000 nodes, depth will approximately be 9.

10.2.4. Almost Complete Binary Tree

An almost complete binary of depth d must have the following two properties :

1. All leaves in the tree will be at level *d* and $d - 1$. In other words all nodes at level *D* (where $D < d - 1$) will have two childs.
2. For any node *N* in the tree with a right successor at level *d*, *N* must have a left child and every left successor of *N* is either a leaf at level *d* or it will have 2 children.

To understand almost complete binary tree refer to figure given below :

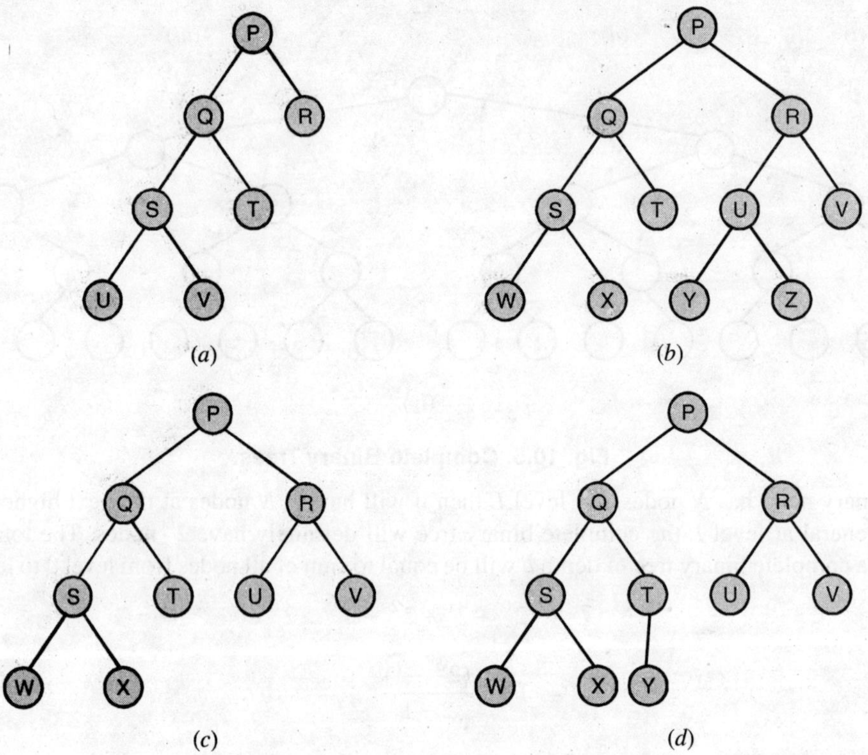

Fig. 10.6. Almost Complete Binary Trees.

In Fig. 10.6 (*a*), the depth of binary tree is 3, we have leaves at level 2 and 3 but also at level 1 (node *R* is a leaf node). Therefore violation of first property occurs. Thus Fig. 10.6(*a*) is not almost complete binary tree.

In Fig. 10.6(*b*), all leaves are at level 2 and 3 (depth of tree is 3) and no leaf at level 0 and 1. Thus first property is satisfied. To understand the second property, we consider say node *P*. Now its right successor at level 3 is say *Y*. Now every left successor of node *A* must be a leaf at level 3 or it must have two children. Node T is the left successor of node *A* and it is a leaf not at level 3. Thus violation of second property occurs, so Fig. 10.6(*b*) is not almost complete binary tree.

In Fig. 10.6 (*c*), first property is satisfied. To check for second property, consider any node whose right successor is at level 3 (depth of tree is 3); we have node *X*. Thus we consider a node *S* whose right successor *X* is at level 3. Now every left successor of node S must be a leaf at level 3 or it must have 2 childrens. We have left successor *W* of node S is at level 3 (which is a leaf). Thus second property is satisfied; therefore figure (c) is an almost complete binary tree.

Finally for Fig. 10.6 (*d*), first property is satisfied. To check for second property note that *S* has right successor *X* at level 3 and also its left successor *W* is a leaf at level 3. Thus figure (d) is an almost complete binary tree. But note one thing in Fig. 10.6 (*c*) and (*d*) that node Y is extra in figure (d). Note *Y* is left child of node *T* and node *T* has no right child. Thus Fig. 10.6 (*d*) is not a strictly binary tree whereas Fig. 10.6 (*c*) is. From this we can conclude that even if a binary tree is not strictly binary tree, it may be an almost complete binary tree.

10.2.5. Memory Representation of Binary Tree

For representing binary tree in memory, there are two popular ways. One is using sequential representation using arrays and second is using linked list. We will see both one by one.

1. Array Representation

The sequential representation of binary tree uses array for storing the data for each node. This is very simple. Assume we have an array by the name BTREE of size 20 and we want to store the following binary tree into the array.

Assuming array indexes starts at 1 then root of the tree is stored at index 1. Now if any parent node is stored at index I then its left child will be stored at 2*I and right child will be stored at 2*I + 1. As root of the tree is stored at 1 then its left child will be stored at 2 and right child at 3. Following in this way the data for every node is stored. This is shown in the array given below :

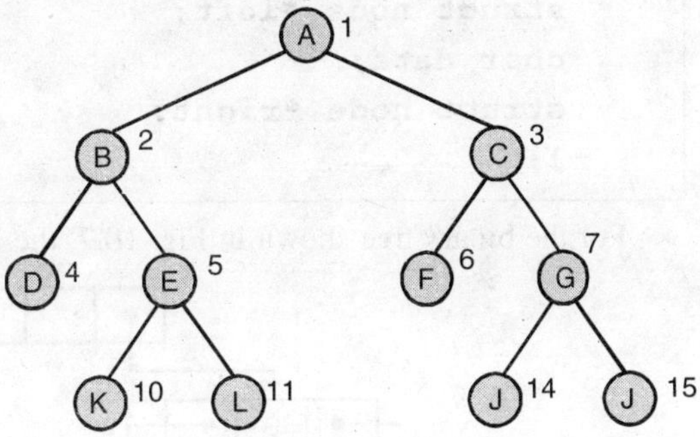

Fig. 10.7. A Binary Tree.

1	2	3	4	5	6	7	8	9	10	11	12	13	14	15	16	17	18	19	20
A	B	C	D	E	F	G			K	L			I	J					

As can be seen from the above array that most of the locations in array are empty; this causes wastage of memory space. That is, the array representation of binary tree is quite inefficient. There is one exception here, in case the array is complete or almost complete. This is because, in general for a binary tree with height H, the size of the array will approximately be 2^H. For example for the following binary tree

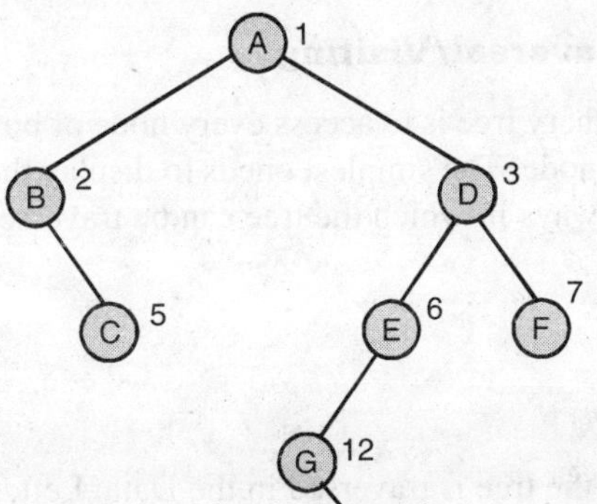

Fig. 10.8. A Binary Tree.

Number of nodes – 08 and height = 5. Then to store the following binary tree using sequential representation, we require an array of size $2^5 = 32$.

Note in the preceding discussion we assume array index starts at 1. When array index starts at 0, then if a parent node is stored at index I then left child will be stored at $2*I + 1$ and right child will be stored at index $2*I + 2$.

2. Linked Representation of Binary Tree

The linked representation is the most popular, efficient and frequently employed representation of binary tree. In the linked representation every node is represented by data, pointer to left child and a pointer to right child. In C for a node the structure can be declared as:

```
struct node
{
struct node *left;
char data;
struct node *right;
};
```

For the binary tree shown in Fig. 10.7, the linked representation will be :

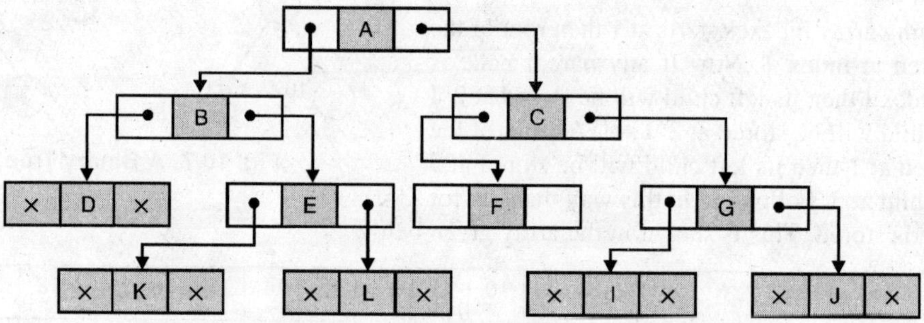

Fig. 10.9. Linked List Representation of Binary Tree.

As can be seen from the figure, the linked representation provides flexibility in terms of allocating memory for single node, insertion and deletion of a node. But some wastage of memory occurs due to null pointers in some of the node. This disadvantage can be removed using threaded binary tree that will be discussed later on.

10.2.6. Binary Tree Traversal/Visiting

A visit or a traversal of binary tree is to access every node of binary tree at most once and performing whatever processing over the node. The simplest one is to display the data of the node. For the binary tree traversal, there are 4 popular ways in which the tree can be traversed. They are :

1. Preorder
2. Inorder
3. Postorder
4. LevelOrder

In the Preorder traversal, the tree is traversed in the Data, Left, Right (DLR) fashion *i.e.*

(*i*) Process the data of the node say X

(*ii*) Go to left of the node X

(*iii*) Go to right of the node X

In the Inorder traversal, the tree is traversed in the Left, Data, Right (LDR) fashion *i.e.*

(*i*) Go to left of the node say X

(*ii*) Process the data of the node X

(*iii*) Go to right of the node X

In the Postorder traversal, the tree is traversed in Left, Right, Data (LRD) fashion *i.e.*

(*i*) Go to left of node say X

(*ii*) Go to right of node X

(*iii*) Process the data of the node X

Note in all three traversal methods you need to apply the method recursively on every node of the tree. Another way to use a stack for temporary holding the nodes.

In the Levelorder traversal all the nodes at one level say L are processed from left to right, then nodes at level $L + 1$ are processed from left to right and so on.

Example 10.1.

Consider the binary tree shown in figure 10.7, We find its preorder traversal. In preorder traversal, tree is traversed as DLR.

(*i*) Print the root node *i.e.* A

(*ii*) Go to left of A, follow DLR again, so print B

(*iii*) Go to left of B, follow DLR again, so print D

(*iv*) Go to right of B, follow DLR again, so print E

(*v*) Go to left of E, follow DLR again, so print K

(*vi*) Go to right of E, follow DLR again, so print L

(*vii*) Come back to the right of A, follow DLR again, print C

(*viii*) Go to left of C, follow DLR again, so print F

(*ix*) Go to right of C, follow DLR again, so print G

(*x*) Go to left of G, follow DLR again, so print I

(*xi*) Go to right of G, follow DLR again, so print J

Thus preorder traversal of the tree is : $A, B, D, E, K, L, C, F, G, I, J$

Example 10.2.

Consider the binary tree shown in figure 10.7. We find its inorder traversal, in inorder traversal, tree is traversed as LDR.

(*i*) Go to leftmost node by following A, B and D (L of LDR)

(*ii*) Print D (D of LDR, as left of D is NULL)

(*iii*) Comeback at B, print B (D of LDR)

(*iv*) Go to right of B (R of LDR)

(*v*) Go to leftmost node by following E and K

(*vi*) Print K (D of LDR, as left of K is NULL)

(*vii*) Comeback at E, print E (D of LDR)

(*viii*) Go to right of E (R of LDR)

(*ix*) Print *L* (no left or right child)

(*x*) Comeback to node *A*, print *A* (*D* of LDR)

(*xi*) Go to right of *A* (R of LDR)

(*xii*) Go to leftmost node by following *C*, and *F*

(*xiii*) Print *F* (*D* of LDR, as left of *F* is NULL)

(*xiv*) Comeback to node *C*, print *C* (*D* of LDR)

(*xv*) Go to right of *C* (*R* of LDR)

(*xvi*) Go to leftmost node by following *G*, and *I*

(*xvii*) Print *I* (*D* of LDR, as left of *I* is NULL)

(*xviii*) Comeback to node *G*, print *G* (*D* of LDR)

(*xix*) Go to right of *G* (*R* of LDR)

(*xx*) Print *L* (no left or right child)

Thus inorder traversal of the tree is : *D, B, K, E, L, A, F, C, I, G, J*

Example 10.3.

Consider the binary tree shown in figure 10.7. We find its postorder traversal. In postorder traversal, tree is traversed as LRD.

(*i*) Reach at leftmost node following *A*, *B* and *D*

(*ii*) Print *D* (D of LRD, as node *D* has no left and right child)

(*iii*) Go to right of *B i.e.* reach at *E*

(*iv*) Follow LRD at *E*, go to leftmost node

(*v*) Print *K* (*D* of LRD, as node *K* has no left and right child)

(*vi*) Go to right of *E i.e.* reach at *L*

(*vii*) Print *L*

(*viii*) Come back at *E*, print *E*

(*ix*) Come back at *B*, print *B*

(*x*) Comeback at *A*, go to right of *A*

(*xi*) Follow LRD at *C*, go to leftmost node

(*xii*) Print *F*

(*xiii*) Go to right of *C i.e.* reach at *G*

(*xiv*) Follow LRD at *G*, go to leftmost node

(*xv*) Print *I*

(*xvi*) Go to right of *G i.e.* reach at *J*

(*xvii*) Print *J*

(*xviii*) Come back at *G*, print *G*

(*xix*) Come back at *C*, print *C*

(*xx*) Come back at *A*, print *A*

Thus postorder traversal of the tree is : *D, K, L, E, B, F, I, J, G, C, A*

Example 10.4.

Consider the binary tree shown in Fig. 10.7. We find its levelorder traversal. In levelorder traversal, tree is traversed by following each level from left to right.

(*i*) Print *A*

(*ii*) Print *B* and *C* in order

(*iii*) Print *D*, *E*, *F*, *G* in order

(*iv*) Print *K*, *L*, *I*, *J* in order

Thus levelorder traversal is : *A, B, C, D, E, F, G, K, L, I, J*

In all the examples shown above, we have shown you traversal of binary tree using a step by step approach, though this was not necessary but for understating this for the first time we did choose this approach. You can directly find out the traversal manually once you know how to proceed. Figure shows one more example of traversal.

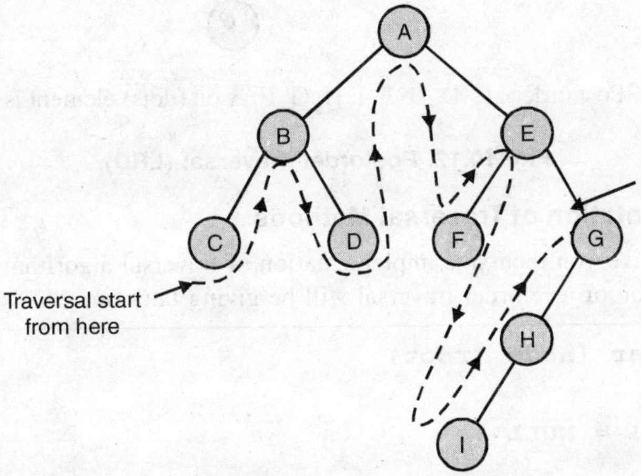

Inorder traversal : C B D A F E I H G on (dots) element is printed.

Fig. 10.10. Inorder Traversal (LDR).

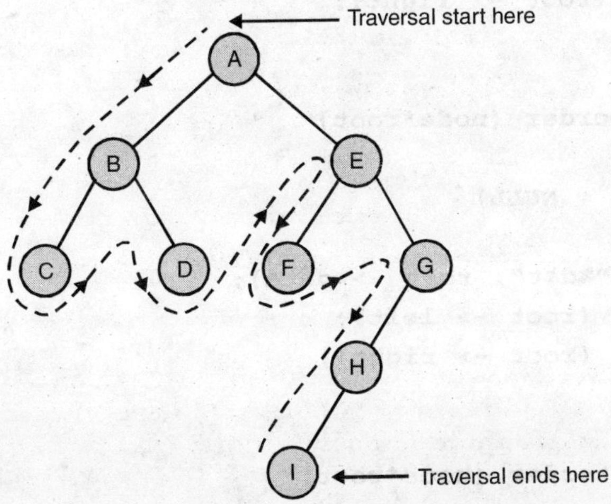

Pre order : A B C D E F G H I on (dots) element is pointed.

Fig. 10.11. Preorder Traversal (DLR)

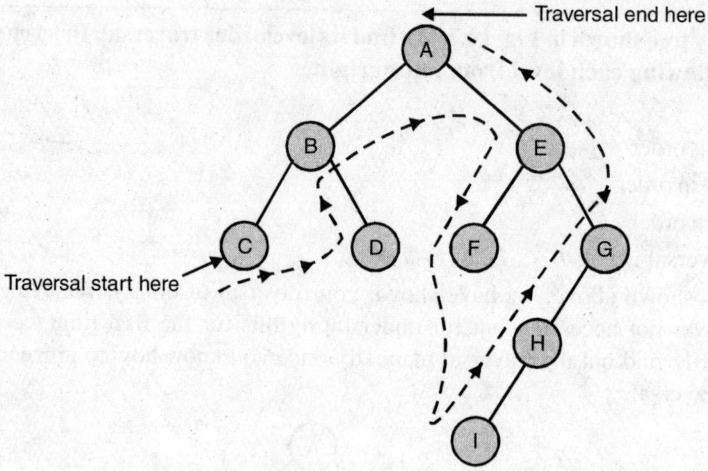

Post order : C D B F I H G E A on (dots) element is printed.

Fig. 10.12. Postorder Traversal (LRD).

10.2.6.1. C implementation of Traversal Methods

In this section we give you recursive implementation of traversal algorithms: preorder, inorder and postorder. Implementation of levelorder traversal will be giving later on.

```
void inorder (node *root)
{
if ( root ! = NULL )
{
inorder (root -> left);
printf ("%c", root -> data);
inorder (root -> right);
}
}
void preorder (node*root)
{
if(root! = NULL)
{
printf ("%d\t", root -> data);
preorder (root -> left);
preorder (root -> right);
}
}
void postorder (node*root)
{
if(root! = NULL)
{
```

```
postorder (root -> left);
postorder (root -> right);
printf ("%d\t", root -> data);
}
}
```

All three implementation look similar to each other, therefore we will simply show you working of only the inorder function. Consider the following binary tree:

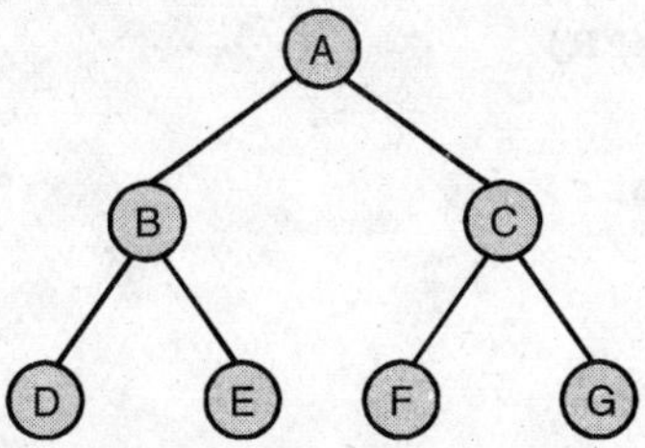

Fig. 10.13.

For the inorder traversal we proceed towards the left first (*L*), then process data (*D*) and then move to right (*R*) *i.e.* for the inorder traversal BST is traversed in LDR fashion. The pointer root is passed to the function inorder. As if condition is true, control enters into if block. The first statement inside the function is recursive call:

```
inorder (root->left)
```

The left of root is node *B*, so function inorder is called with root now pointing to *B*. As again if condition is true, recursive call takes place: inorder (root->left) with a pointer that points to left of node *B i.e.* node *D*. For the node *D*, when recursive call takes place

inorder(root->left), if condition fails and function returns. The function returns to the function call where root was pointing to node D, so next statement prints "D". Following the printf statement, again recursive call takes place and function is called as:

```
inorder (root->right);
```

As right of node D is NULL, the function returns to the function call where root was pointing to node B, printf statement executes and "B" is printed. Following the printf statement, again recursive call takes place and function is called as:

```
inorder (root->right);
```

As right of node *B* is E which is not NULL, if condition turns out to be true. Recursive call does not causes more recursion as left of *E* is NULL, so function returns and "*E*" is printed. As right of *E* is also NULL, function returns to function call where root was pointing to node *B*. Now function returns to the function call where root was pointing to node *A*. Node *A* is printed and the process is repeated for the right pointer of node *A*.

10.2.6.2. Stack Based Implementation

In the previous section we discussed recursive implementation of traversal algorithms. Here we discuss the non recursive versions. In non recursive version we make use of stack for storing the addresses of the node where the control will return back to start the traversal again. The first non recursive function we write for preorder traversal.

1. Preorder Traversal Using Stack

```
struct node
{
struct node *left;
int data;
struct node*right;
};
typedef struct node node;
void preorder (node*R)
{
node*stack [20], *ptr;
int top = 0;
stack[top] = NULL;
ptr = R;
while(ptr ! = NULL)
{
printf("%d", ptr ->data);
if(ptr ->right! = NULL)
{
top++;
stack[top] = ptr ->right;
}
if(ptr->left ! = NULL)
                ptr = ptr ->left;
else
{
ptr = stack[top];
top -;
}
}
}
```

The preorder traversal of BST using stack is very simple to understand. We take an array of pointers by the name stack (that also works like stack) for holding the pointers. Initially top is set to 0 and stack [top] is NULL. Pointer ptr points to root of the tree. The while loop runs till the pointer ptr is not NULL.

For the preorder traversal, the BST is traversed in the DLR fashion, so initially the data of the ptr is printed. If ptr is having any right child we push it onto the stack, so that we can get back to it. Next we move to left of the ptr till ptr does not become NULL. If ptr is NULL, we pop out the top element (actually address) from the stack and the while loop iterates.

For the following tree we show some of the iteration of the while loop.

For the first time control enters into while loop, ptr points to root node *i.e.* 11. The data 11 is printed. The root node has a right child (node 13); address of this node (ptr->right) is pushed onto the stack. As left of ptr is not NULL (node 6), ptr points to this node.

In the second iteration of while loop, data of ptr *i.e.* 6 is printed. Node 6 has a right child (node 9); it is pushed onto the stack; ptr points to its left (node 0).In the third iteration of while loop, data of ptr *i.e.* 0 is printed. Node 0 has a right child (node 5); it is pushed onto the stack; ptr points to its left (which is NULL) so else part executes. In the else part top of the stack is popped that was pointer to node 5.

Processing in the above manner continues and function prints the preorder traversal of the BST.

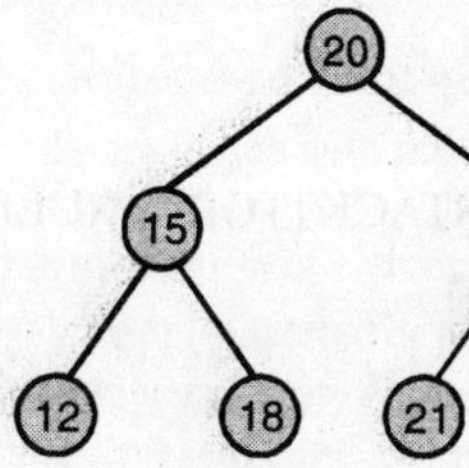

Fig. 10.14.

(*i*) 20 is printed, its right child 22 is pushed onto stack, ptr points to left of 20 *i.e.* 15.

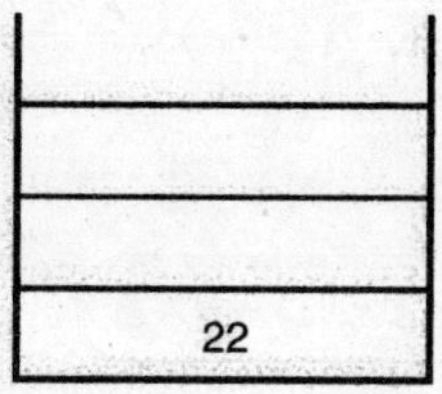

(*ii*) 15 is printed, its right child 18 is pushed onto stack, ptr points to left of 15 *i.e.* 12.

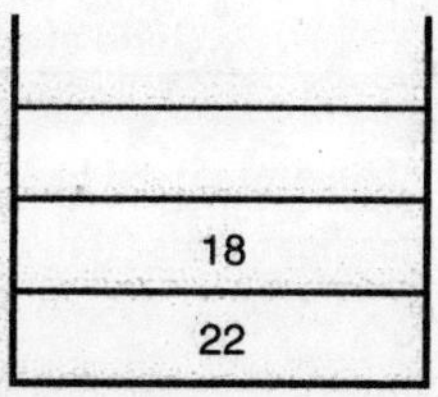

(*iii*) 12 is printed, its has not left or right child node is popped from stack ptr points to 18.

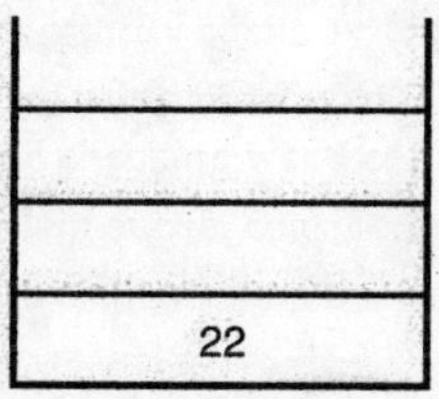

(*iv*) 18 is printed, it has no left or right child so node is popped from stack and points prints to 22.

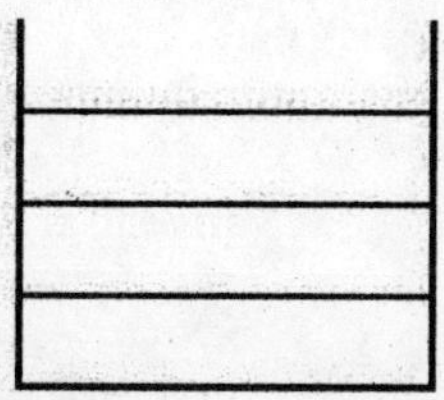

(*v*) 22 has a left child 21 ptr prints to it.

(*vi*) 21 is printed.

Preorder traversal : 20, 15, 12, 18, 22, 21

Fig. 10.15. Preorder Traversal Using Stack.

Algorithm for preorder traversal

Algorithm 10.1. PREORDER (ROOT, LEFT, INFO, RIGHT)....................................

ROOT: Starting node

LEFT: Points to Left Child

INFO: Information Part of a node

RIGHT: Points to Right Child

1. Set PTR = ROOT , TOP = 0, STACK [TOP] = NULL
2. Repeat while PTR != NULL
   (a) Process info [PTR]
   (b) If RIGHT [PTR] != NULL
   Set TOP = TOP+1 &
   STACK [TOP]=RIGHT [PTR]
   (c) Set PTR = LEFT [PTR]
3. IF (STACK[TOP] IS NOT NULL)
   Set PTR = STACK [TOP],
   TOP = TOP-1
4. Go to step 2
5. Exit

2. Inorder Traversal Using Stack

```
void inorder (node*R)
{
node*stack[20], *ptr;
int top = 0;
stack[top] = NULL;
ptr = R;
again:
while(ptr ! = NULL)
{
top++;
stack [top] = ptr;
        ptr = ptr->left;
}
ptr = stack[top];
top --;
while(ptr ! = NULL)
{
printf(“%d”, ptr->data);
if(ptr ->right! = NULL)
{
ptr = ptr->right;
```

```
goto again;
}
ptr = stack[top];
top —;
}
}
```

For the inorder traversal, the BST is traversed in the LDR fashion. For this the initial while loop causes pointer to move to its left repetitively and saving the pointers into stack till the ptr does not become NULL. For the tree shown below:

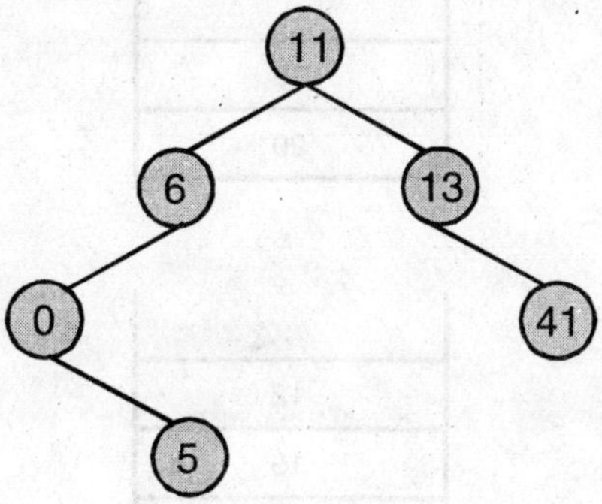

Fig. 10.16.

When ptr is NULL, three pointers for the node 11, 6, 0 have been pushed onto the stack, At this moment the top of the stack is popped up and ptr points to it. The data of the ptr is printed. If right of the pointer is not NULL then ptr points to its right and control is transferred back to the initial while loop. The sequence of LDR is repeated.

We show few iterations of the inorder traversal using stack:

Initially in the while loop, after pushing node 11, 6, and 0 onto the stack, ptr becomes NULL. Now top of the stack is popped up and ptr points to node 0. Inside the while loop the data for the node is printed *i.e.* 0 is printed. The right pointer of ptr is not NULL (it points to node 5); ptr now points to node 5 and control is transferred back to first while loop. Pointer is pushed onto the stack in the first iteration and ptr becomes NULL. Now the top of the stack is popped up and ptr points to node 5. As ptr is not NULL, control enters into the second while loop; data for the node 5 is printed. As right of the ptr is NULL, else part executes and top of the stack is popped up. Pointer ptr now points to node 6. As ptr is not NULL, data for the node is printed *i.e.* 6 is printed. Now as right of ptr (node 9) is not NULL, ptr points to it and control is transferred back to first while loop.

Processing in the above manner is repeated and inorder traversal of the BST is printed.

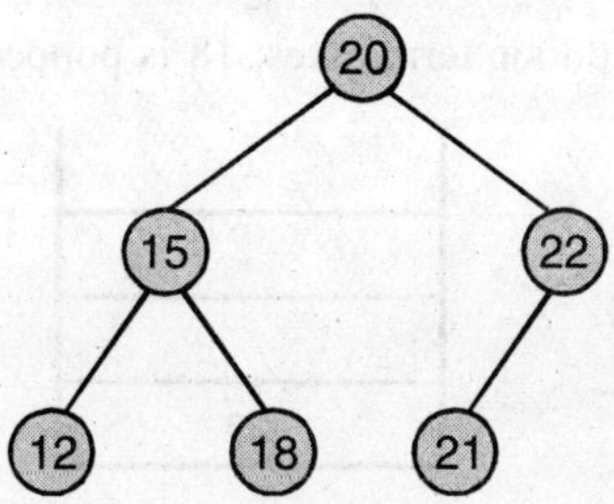

Fig. 10.17.

(*i*) 20 is pushed

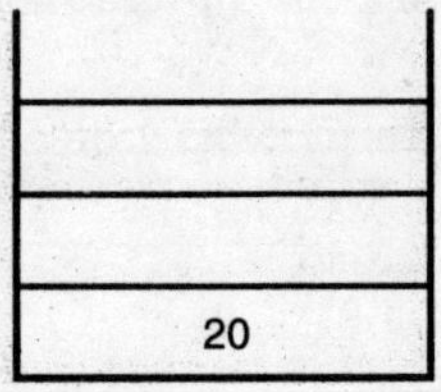

(*ii*) 15 is pushed

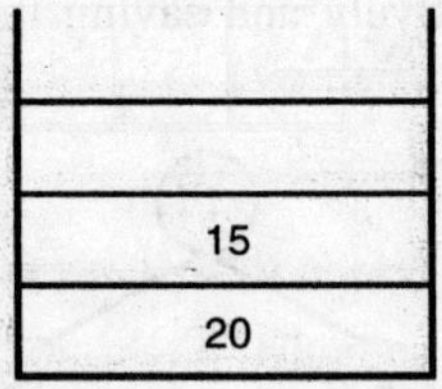

(*iii*) 12 is pushed

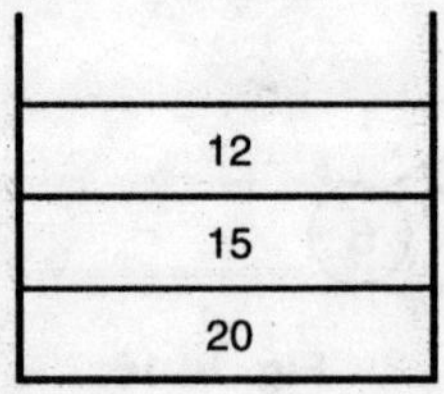

(*iv*) While loop terminates, pop from stack ptr points to 12, it is printed.

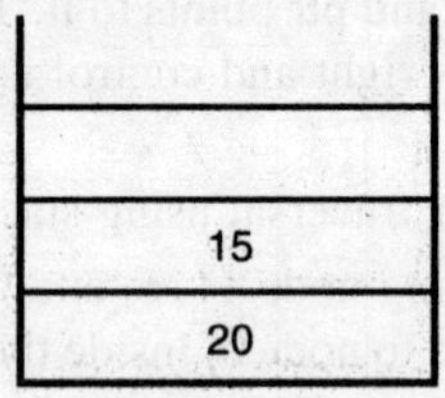

(*v*) 15 is popped and printed 15 has a right child 18, it is pushed onto stack.

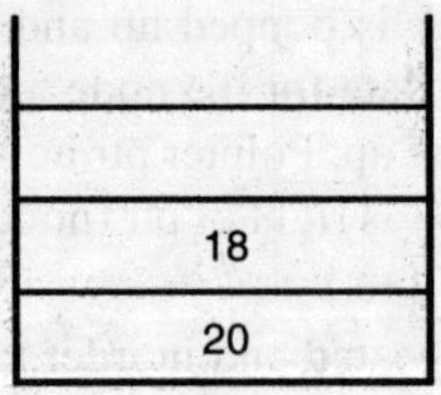

(*vi*) 18 has no left child, while loop terminates, 18 is popped and printed.

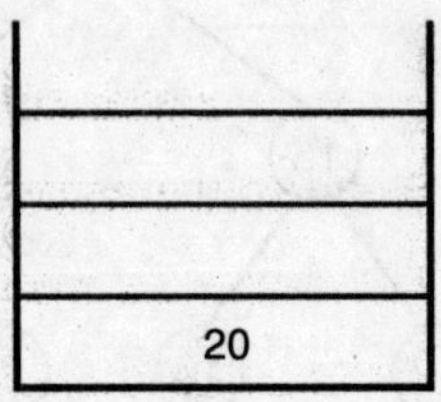

(*vii*) 20 is popped, ptr points to it, 20 is printed, 20 has a right child 22, ptr points.

(*viii*) 22 is pushed onto stack.

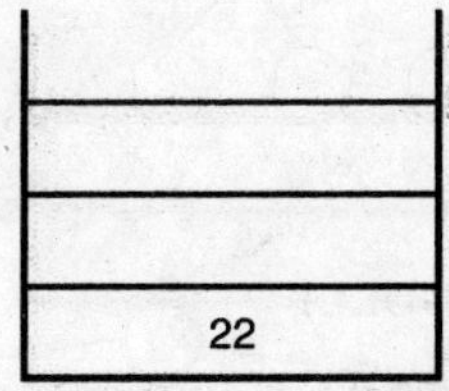

(*ix*) 22 has a left child 21, it is pushed onto stack, 21 has no left

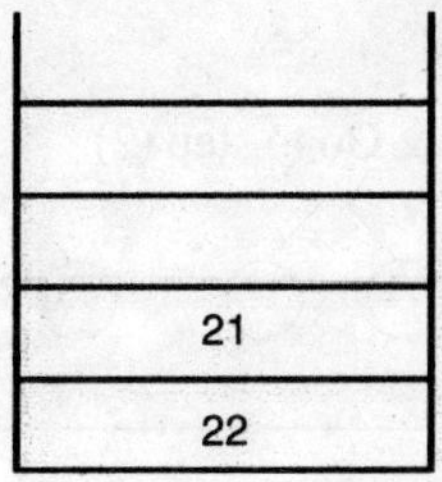

(*x*) 21 is pushed onto stack.

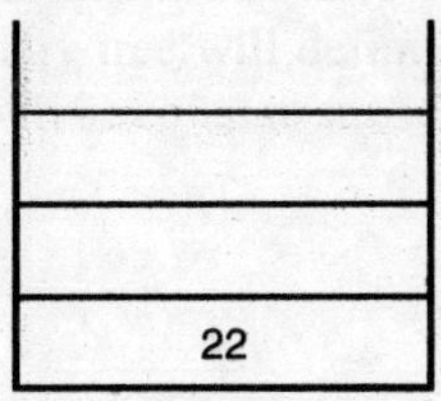

(*xi*) 22 is pushed onto stack.

Inorder traversal : 12, 15, 18, 20, 21, 22

Fig. 10.18. Inorder traversal using stack.

Algorithm for inorder traversal

Algorithm 10.2. INORDER (ROOT, LEFT, RIGHT, INFO) ....

ROOT: Starting node
LEFT: Points to Left Child
INFO: Information Part of a node
RIGHT: Points to Right Child

1. Set PTR = ROOT , TOP = 0, STACK[TOP] = NULL
2. Repeat while PTR != NULL
   (a) Set TOP = TOP + 1
   STACK [TOP]=PTR.
   (b) Set PTR = LEFT [PTR]
3. While (STACK[TOP] IS NOT NULL)
   (a) Set PTR = STACK [TOP]
   TOP=TOP-1
   (b) Process INFO [PTR]
   (c) If RIGHT [PTR] != NULL
   Set PTR = RIGHT [PTR] & Go to step (2)
4. Exit

3. Postorder Traversal Using Stack

```
void postorder(node*R)
{
typedef struct
{
node *pR;
char status;
}pnode;
pnode *stack[20];
int top = 0;
node *ptr = NULL;
stack[top]  = NULL;
pnode *temp;
pnode *myptr = NULL;
ptr = R;
again:
while(ptr ! = NULL)
{
top++;
temp = (pnode*) malloc (sizeof(pnode));
temp->pR = ptr;
```

```
temp->status = 'p';
stack[top] = temp;
if(ptr->right ! = NULL)
{
top++;
temp = (pnode*) malloc (sizeof(pnode));
temp->pR = ptr->right;
temp->status = 'N';
stack [top] = temp;
}
ptr = ptr->left;
}
myptr = stack[top];
ptr = myptr->pR;
top -;
while (myptr -> status = ='P')
{
printf("%d", ptr -> data);
myptr = stack [top] ;
ptr = myptr -> pR;
top -;
}
if (myptr->status = = 'N')
{
myptr -> status = 'p';
ptr = myptr -> pR;
goto again;
}
}
```

For the postorder traversal, the BST is traversed in LRD fashion. This traversal using stack is not easy as compared to other two traversals. In the function postorder we have created a temporary structure by the name pnode. The structure has just two data 'members: a pointer of type node and one char type field status. This structure is must as overall processing will depend on the field status and we did not want to alter the original structure node. The status field will have only two values, either 'N' (for negative) or 'P' (for positive).

We have an array of pointers of type pnode by the name stack. We also have two additional pointers temp and myptr of pnode type and one pointer ptr of node type. The logic is simple: Starting from root we traverse towards left, storing every node pointer inside stack with status set to 'P' and while doing this if any right child is found then this is also stored in the stack but with status field set to 'N'. This logic is implemented in the first while loop. Next when left becomes NULL, we take out the top pointer node from the stack, ptr points to pR field and top is decremented. We display all the nodes from the stack whose status is 'P'. As soon as some node (say myptr) with status is 'N' is found, its status is changed to 'P', ptr points to its pR field control transfer to initial while loop.

To better understand this considers the BST given below :

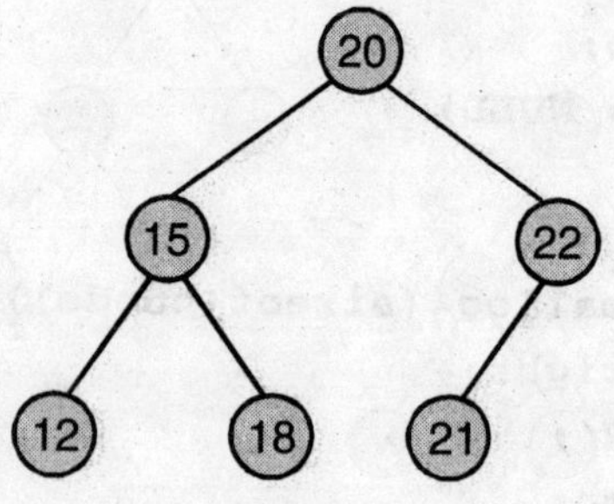

Fig. 10.19.

The steps for the postorder traversal for the above BST is shown below :

(*i*) 20 is pushed with status '*P*'

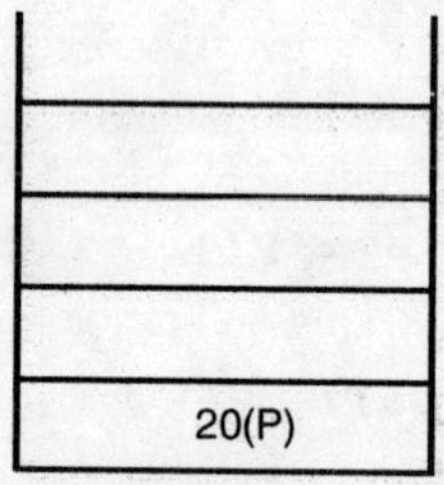

(*ii*) As 20 has right child 22 it is pushed with status '*N*'.

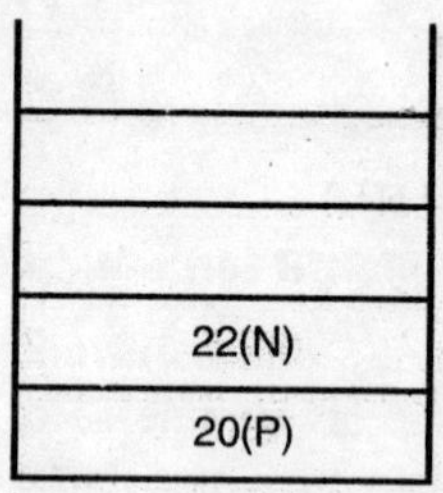

(*iii*) 15 is pushed with status '*P*'

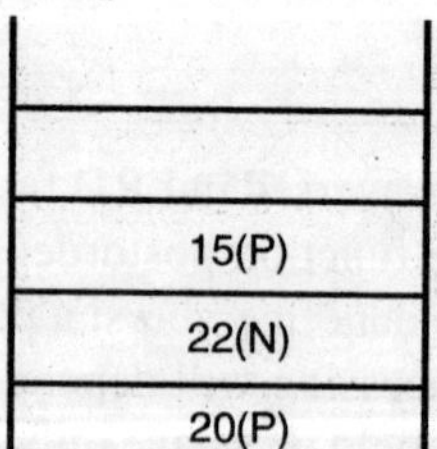

(*iv*) 15 has a right child 18 it is pushed with status '*N*'

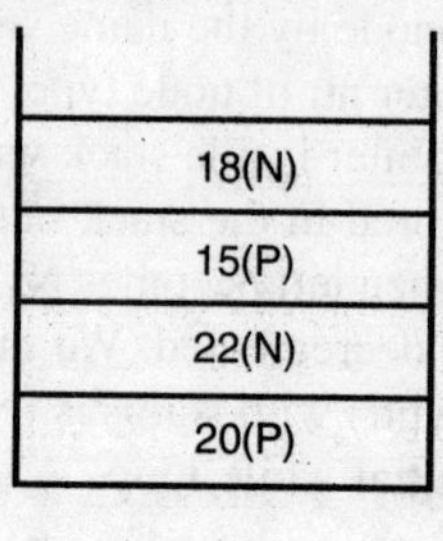

(*v*) 12 is pushed with status '*P*'

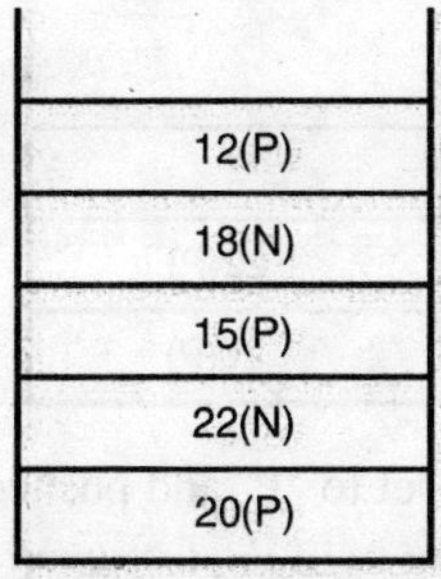

(*vi*) ptr is NULL, pop 12, display it as its status is '*P*' 12 is printed.

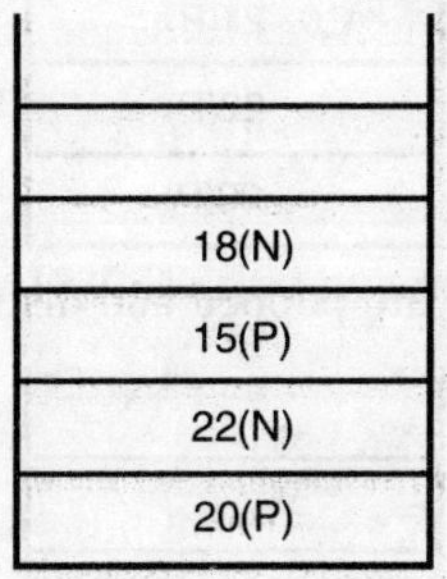

(*vii*) Next 18 is popped, as its status is '*N*', it is changed to '*P*' and pushed back to ack.

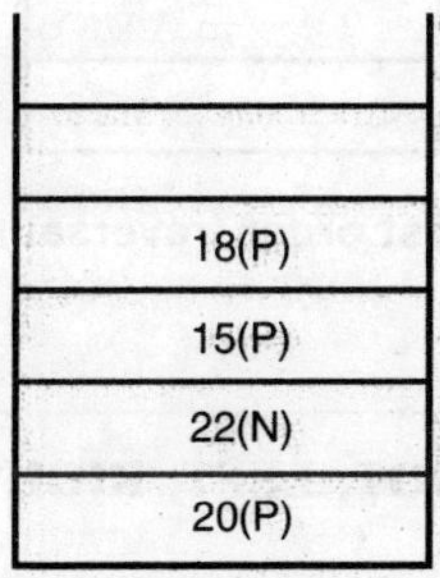

(*viii*) As 18 is popped and displayed as its status is '*P*'. 18 printed.

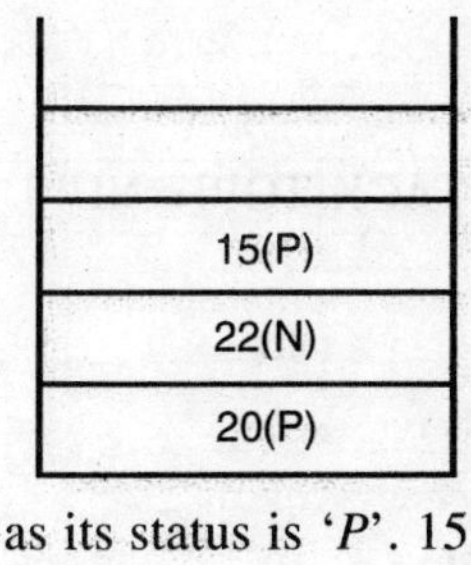

(*ix*) As 15 is popped and displayed as its status is '*P*'. 15 printed

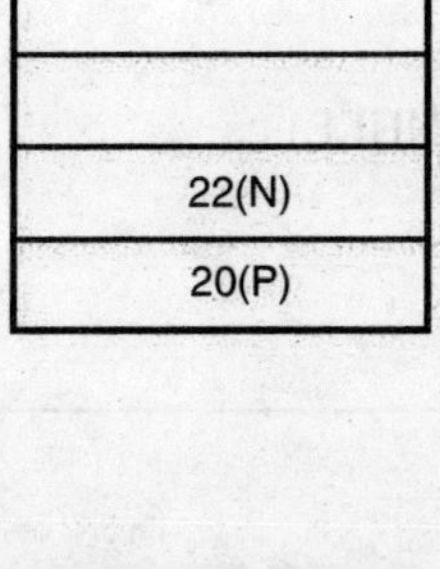

(*x*) 22 is popped, as the status is '*N*', its status is changed to '*P*', and pushed back.

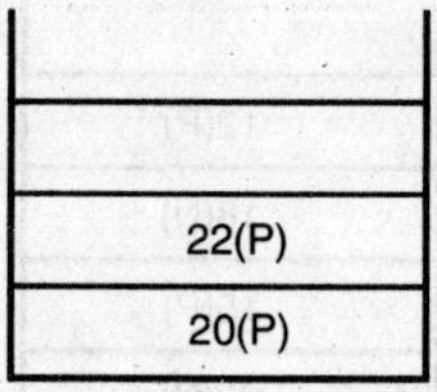

(*xi*) 22 has a left child 21, its status is set to '*P*' and pushed.

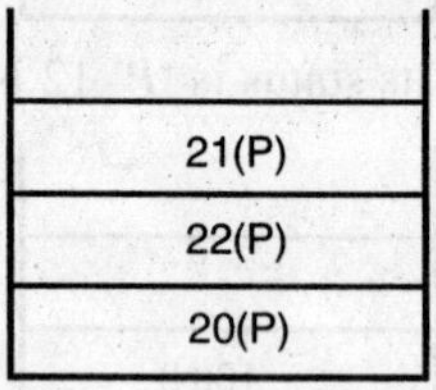

(*xii*) As all nodes have status '*P*', they are popped and displayed in order 21, 22, 20.

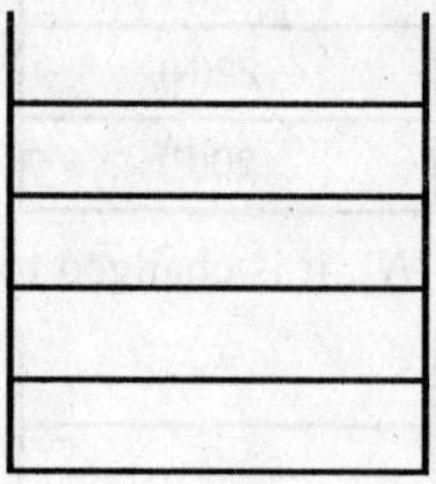

Fig. 10.20. Post order traversal using stack.

Algorithm for postorder traversal

Algorithm 10.3. POSTORDER (ROOT, LEFT, RIGHT, INFO) ....

ROOT: Starting node
LEFT: Points to Left Child
INFO: Information Part of a node
RIGHT: Points to Right Child

```
1.  Set PTR = ROOT , TOP = 0, STACK[TOP] = NULL
2.  Repeat while PTR != NULL
    (a) Set TOP = TOP + 1,
        Set STACK [TOP] = PTR
    (b) If RIGHT [PTR] != Null
                Set TOP = TOP+1
            Set STACK [TOP] = - RIGHT [PTR]
    (c) Set PTR = LEFT [PTR]
3.  While (STACK[TOP] IS NOT NULL)
    (a) Set PTR = STACK [TOP]
        TOP = TOP-1
    (b) If (PTR > 0)
```

(i) Process INFO [PTR]
(ii) Go to step (3)
(c) If (PTR < 0)
(i) Set PTR = - PTR
(ii) Go to step (2)
4. Exit

10.3. EXPRESSION TREES

Figure 10.21 shows an expression tree for $(a + b * c) + ((d * e + f) * g)$

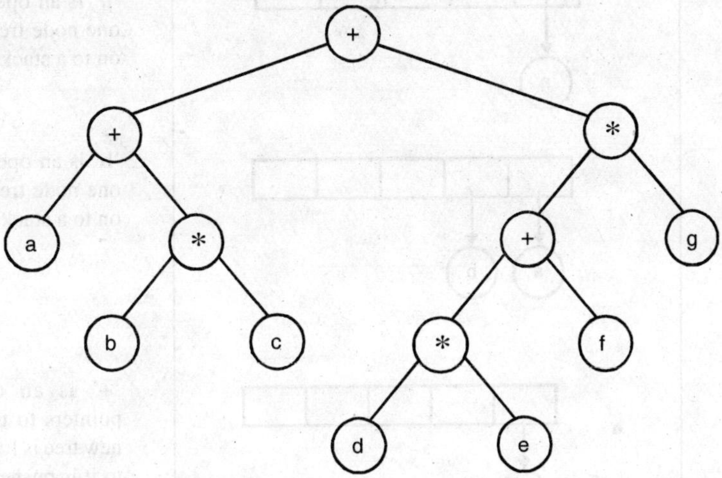

Fig. 10.21. Expression Tree for (a + b * c) + ((d * e + f) * g).

The leaves of an expression tree are operands, such as constants or variable names, and the other nodes contain operators. This particular tree happens to be binary, because all of the operations are binary, and although this is the simplest case, it is possible for nodes to have more than two children. It is also possible for a node to have only one child, as in the case with the unary minus operator. We can evaluate expression tree, T, by applying the operator at the root to the values obtained recursively by evaluating the left and right subtrees. In the example, the left subtree evaluates to (a + b * c) and the right subtree evaluates to ((d * c) + f) * g. The entire tree therefore represents :

(a + b * c) + ((d * e + f) * g)

inorder traversal (infix representation)

a + b * c + d * e + f * g

postorder traversal (postfix representation)

abc * + de* f + g * +

preorder traversal (prefix representation)

+ + a * b c * + * d e f g

Construction of an Expression Tree From a Postfix Expression

We will now discuss the conversion of postfix expression from into an expression tree. Since we already have an algorithm to convert infix to postfix, we can generate expression trees from the two common type of input. The method strongly resembles the postfix evaluation algorithm.

We read our expression one symbol at a time. If the symbol is an operand, we create a one node tree and push a pointer to it onto a stack. If the symbol is an operator, we pop pointers to two trees T_1 and T_2 from the stack (T_1 is popped first) and form a new tree whose root is the operator and whose left and right children point to T_2 and T_1 respectively. A pointer to this new tree is then pushed onto the stack.

Example 10.5.

Construct the tree from a given postfix expression

a b + c d e + **

Symbol scanned	Stack	Remark
a	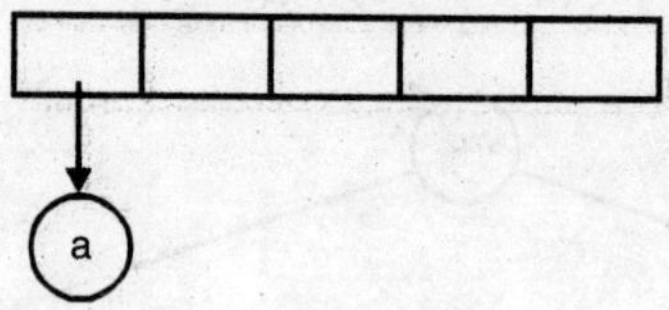	'*a*' is an operand, so we create one node tree and push pointer on to a stack.
b	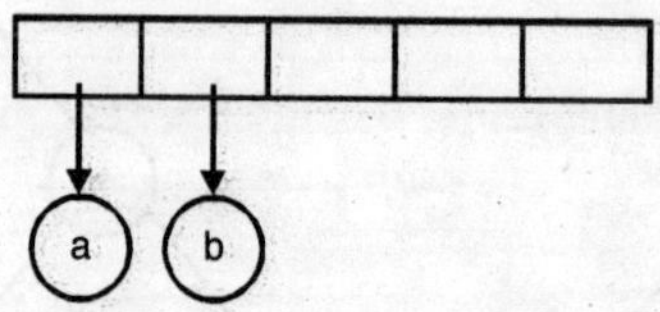	'b' is an operand, so we create one node tree and push pointer on to a stack
+	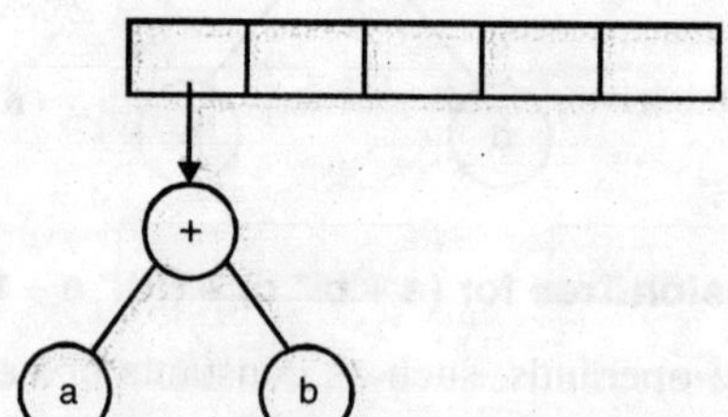	'+' is an operator, so two pointers to trees are popped, a new tree is formed, and a pointer to it is pushed on to the stack.
c	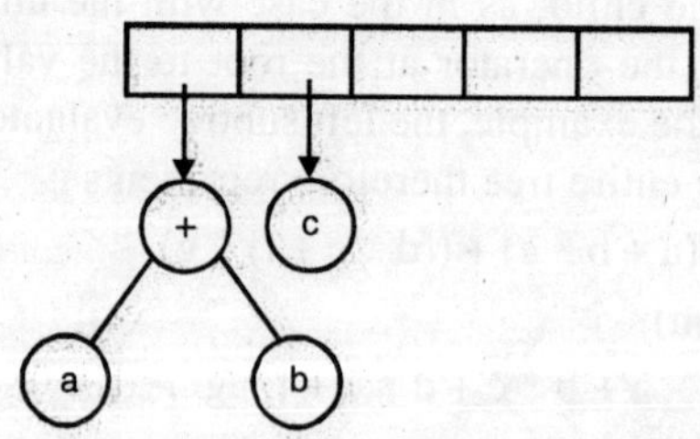	'c' is an operand, create a one node tree and a pointer to the corresponding tree is pushed on to the stack.
d	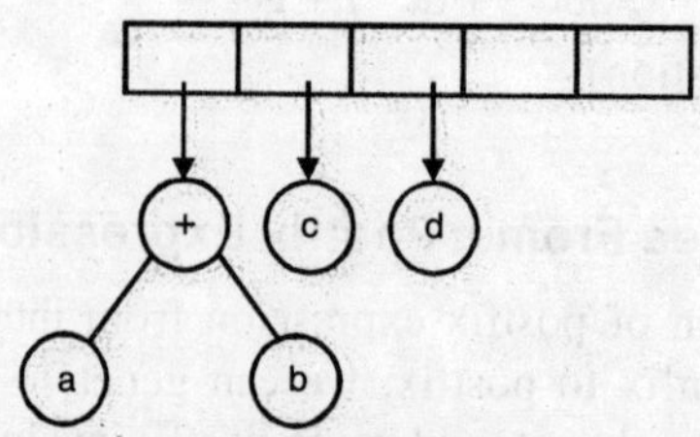	'd' is an operand, create a one node tree and a pointer to the tree is pushed on to the stack

Symbol scanned	Stack	Remark
e	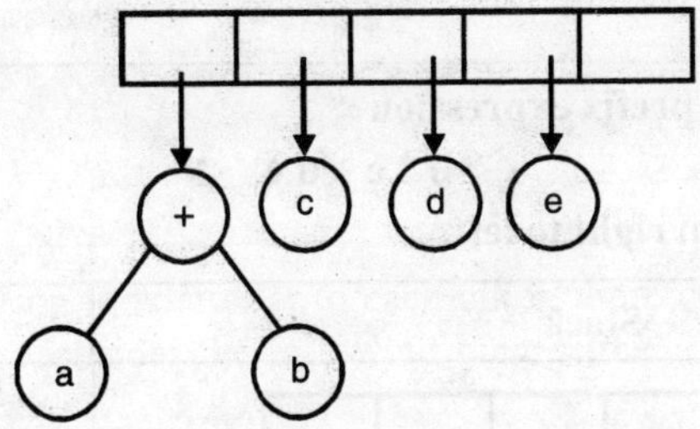	'e' is an operand, create a one node tree and a pointer to the tree is pushed on to the stack
+	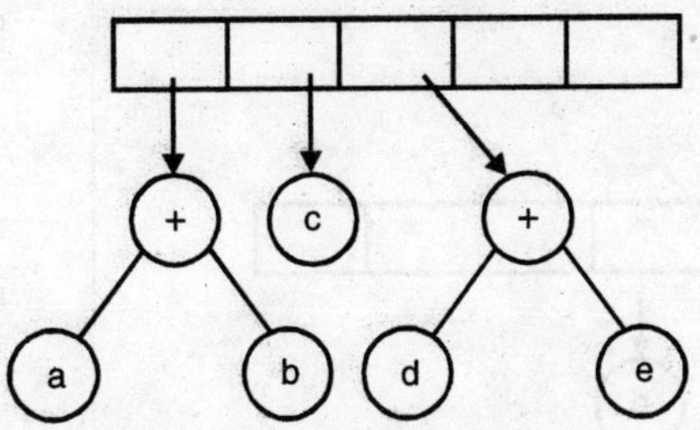	'+' is an operator, so two trees are merged.
*	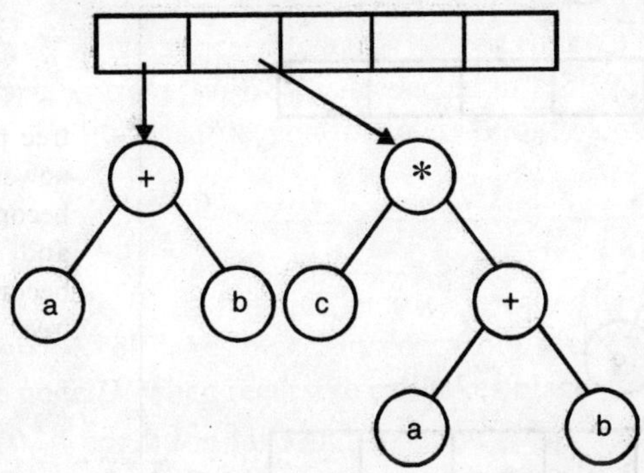	'*' is an operator, so we pop two tree pointers and form a new tree with a '*' as root node.
*	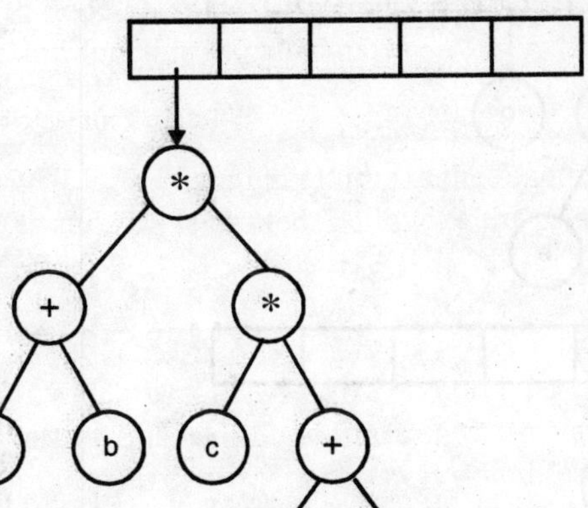	The last symbol '*' is an operator, two trees are merged, and a pointer to the final tree is left on the stack.

Construction of an Expression Tree from a Prefix Expression

In this we read our expression one symbol at a time. If the symbol is an operand, we create a one node tree and push a pointer to it onto a stack. If the symbol is an operator, we pop pointers to two trees T1 and

T2 from the stack (T1 is popped first) and form a new tree whose root is the operator and whose left and right children point to T1 and T2 respectively. A pointer to this new tree is then pushed onto the stack.

Example 10.6.

Construct the tree frm a given prefix expression

*** b * c + d e**

In this scan the expression from right to left.

Symbol scanned	Stack	Remark
e	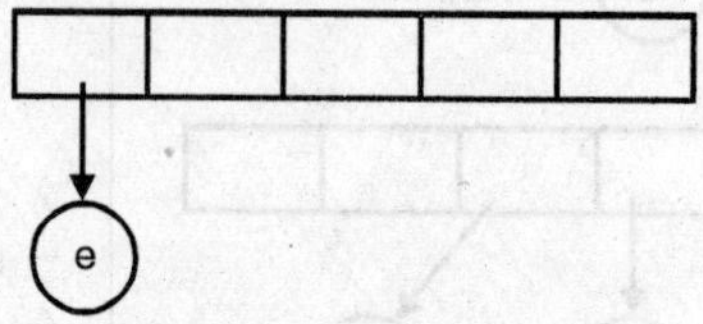	Since 'e' is an operand so create one node tree and push pointer on to stack
d	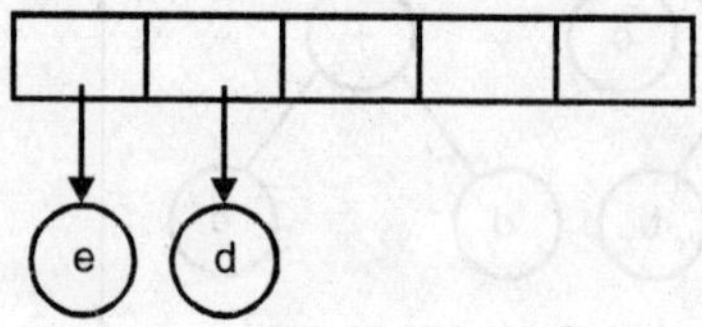	'd' is an operand, create one node tree and push pointer on to stack
+	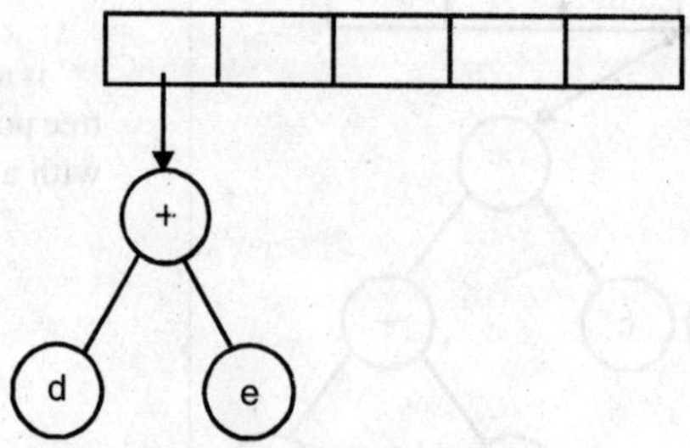	'+' is an operator, so we pop two tree pointers and from form a new tree. The 1st poped element becomes the left child of the tree and 2nd popped element becomes the right child of the tree.
c	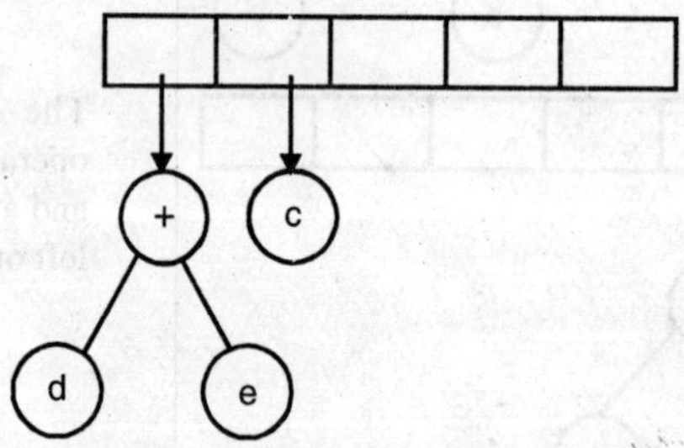	'c' is an operand, create one node tree and push pointer on to stack.
*	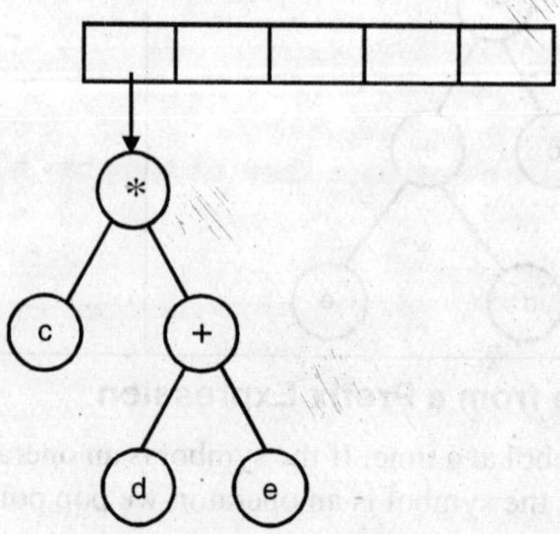	'*' is an operator, so we pop two tree pointers and form a new tree. The 1st popped element becomes the left child of the tree and 2nd popped element becomes the root of the right subtree of the node '*'

Symbol scanned	Stack	Remark
b	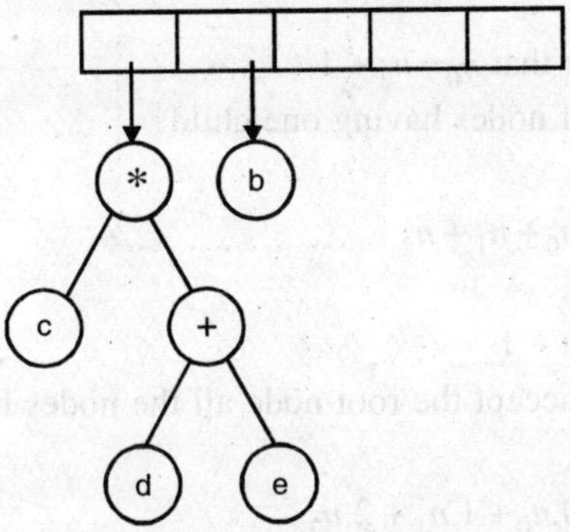	'b' is an operand, create one node tree and push pointer on to stack.
*	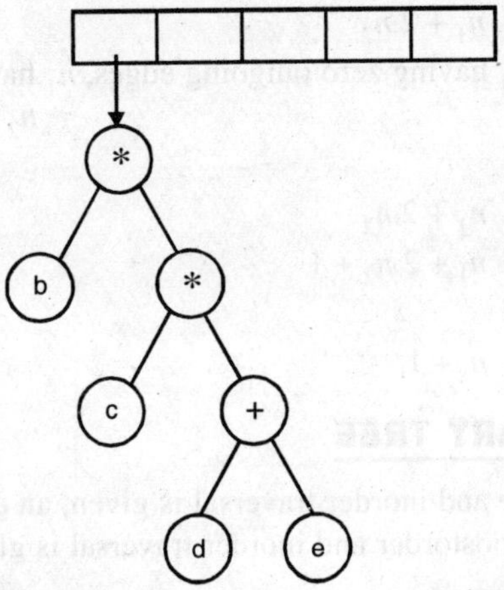	

10.4. PROPERTIES OF BST

10.4.1. The maximum number of nodes in a tree at level 'i' is 2^{i-1} where i >=1.

Step 1 : Put $i = 1$

At level 1, the maximum number of nodes = 2^{1-1} = 1 node (This is true)

Step 2 : Assume that the expression is also true for $i = i - 1$ as it is true for $i = 1$ shown in Step 1.

Hence at level (i-1), the maximum number of nodes=$2^{(i-1)-1} = 2^{i-2}$

Step 3 : The maximum number of nodes at level i = 2 * (maximum number of nodes at level i – 1)

$$i.e.\ \text{Max (level } i) = 2^* \text{ Max (level } i - 1)$$
$$= 2^*2^{i-2} \qquad \text{(using Step 2)}$$
$$= 2^{i-1}$$

10.4.2. The maximum number of nodes in a tree of height 'h' is equal to 2^h – 1 where $h \geq 1$

Maximum number of nodes in a tree of height 'h' =

Max (level 1) + Max (level 2) + Max (level 3) + ... + Max (level h)

$$= \sum_{i=1}^{h} \text{Max (level } i) = \sum_{i=1}^{h} 2^{i-1}$$

$$= 2^0 + 2^1 + ... + 2^{n-1} = 1.\ 2^h - 1 = 2^h - 1$$

10.4.3. If n_2 is the maximum number of nodes having two child and n_0 is the maximum number

Of nodes having zero child then prove that $n_0 = n_2 + 1$

Assume n_1 is the maximum number of nodes having one child

Total number of nodes (n) in a tree

$$n = n_0 + n_1 + n_2 \quad ...(1)$$

Total number of branches (B) in a tree,

$$B = n - 1 \quad ...(2)$$

(As in any tree accept the root node all the nodes having one incoming edge)

Total number of branches (B) in a tree,

$$B = 0.n_0 + 1.n_1 + 2.n_2$$

$$B = n_1 + 2.n_2 \quad ...(3)$$

(As n_0 having zero outgoing edges, n_1 having one outgoing edges and n_2 having two outgoing edges)

Using (2) and (3)

$$n - 1 = n_1 + 2.n_2$$

$$n = n_1 + 2.n_2 + 1 \quad ...(4)$$

Using (1) and (4)

$$n_0 = n_2 + 1$$

10.5. CONSTRUCTION OF BINARY TREE

A tree can be constructed if preorder and inorder traversal is given, an example of which is given in Fig. 10.22. Figure 10.23 shows a tree if postorder and inorder traversal is given.

Example 10.7.

Construction of the tree from its given preorder and inorder traversal.

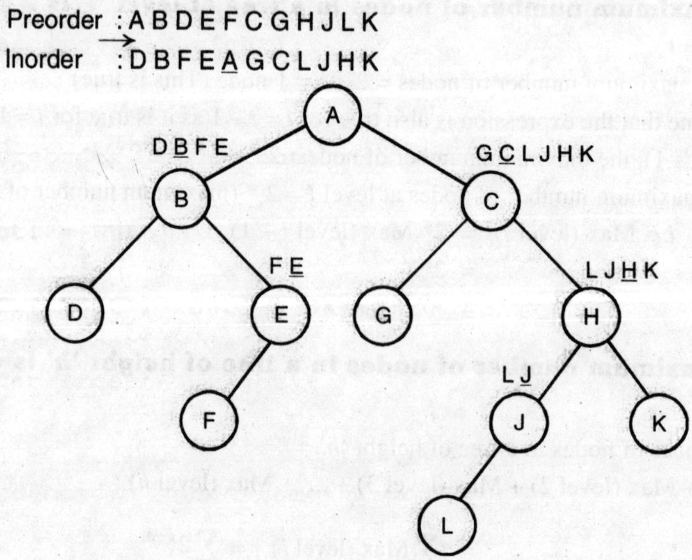

Fig. 10.22. Construction of a tree from its preorder and inorder traversal.

1. The first node in the preorder traversal is the root node of the tree (in the above figure it is A.)
2. Find the position of the root node (*i.e.* A) in inorder traversal are the nodes of the left subtree of the root node (*i.e.* DBFE) and the nodes succeed to the root node are the nodes in the right subtree of the root node (*i.e.* GCLJHK).
3. Now consider two sets of inorder and preorder traversals of the left and right subtrees of the root. The first set is the nodes appear in preorder traversal just after the root node (*i.e.* BDEF). The second set is the nodes appear after the root node in inorder traversal (*i.e.* GCLJHK) and the nodes in the preorder traversal except the root node and the nodes considered in the first set (*i.e.* CGHJLK).
4. Taking these two sets of preorder and inorder traversals for left and right subtrees, we will repeat steps 2 and 3 till the entire tree is constructed.

Example 10.8.

Constructionof the tree from its given postorder and inorder traversal.

Postorder : D F E B G L J K H C A←

Inorder : D B F E A G C L J H K

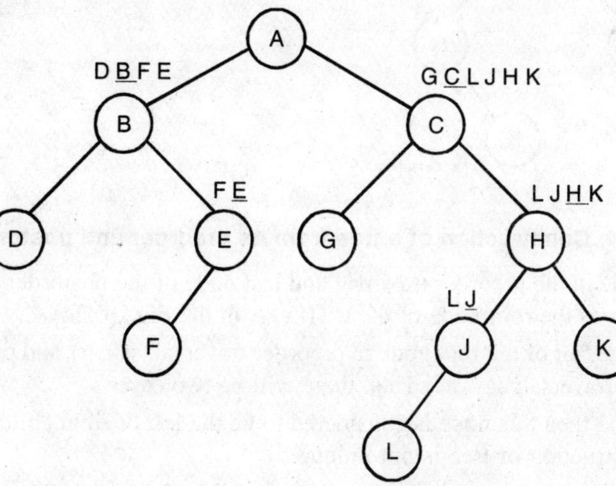

Fig. 10.23. Construction of a tree from its postorder and inorder traversal.

1. The last node in the postorder traversal is the root node in the tree (in the example it is *A*)
2. Find the position of the root node (*i.e.* A) in the above example in order traversal. The nodes precede to root node in the inorder traversal are the nodes in the left subtree of the root node (*i.e.* DBFE) and the nodes succeed to the root node are the nodes in the right subtree of the root node (*i.e.* GCLJHK).
3. Now consider two sets of inorder and postorder traversals of the left and right subtrees of the root. The first set is the nodes appear precede to the root node in inorder traversal (*i.e.* DBFE) and the sequence of same nodes that are in postorder traversal just before the root node (*i.e.* DFED). The second set is the nodes appear after the root node in inorder traversal (*i.e.* GCLJHK) and the sequence of nodes that are in the postorder traversal except the root node and the nodes considered in first set (*i.e.* CGHJLK).
4. Taking these two sets of postorder and inorder traversals for left and right subtrees, we will repeat steps 2 and 3 till the entire tree is constructed.

Example 10.9.

Construction of the tree from its given preorder and postorder traversal.

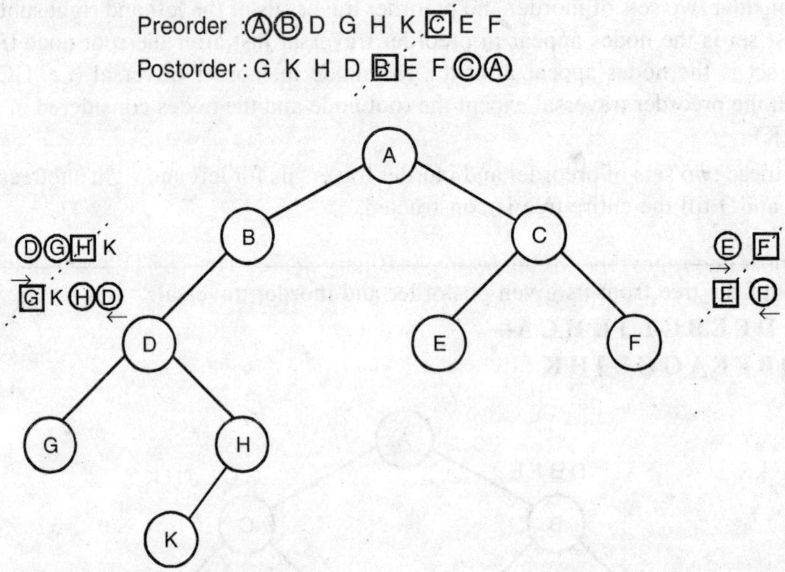

Fig. 10.24. Construction of a tree from its preorder and postorder traversal.

1. The first node in the preorder traversal and last node of the postorder traversal (which are same) is considered as the root node of the tre (*i.e.* A in the Fig. 10.24).
2. Find the successor of the root node in preorder traversal, say, n1 and predecessor of the root node in postorder traversal, say, n2. Then there will be two cases :
   (*i*) If n1 = n2 then this node is considered to be the left or right child of the root node, in which the construction of tree is not unique.
   (*ii*) If n1 ≠ n2 then n1 is considered as the left child (*i.e.* B) and n2 is considered as the right child (*i.e.* C) of the root node.
3. Find the positions of n2 (*i.e.* C) and n1 (*i.e.* B) in preorder and postorder traversals respectively. Now consider the two sets of preorder and postorder traversals of left and right subtrees of the root. The first set is the nodes appear after n1 and before n2 in preorder traversal (*i.e.* DGHK) and the nodes precede to node n1 is postorder traversal (*i.e.* GKHD). The second set is the nodes appear after n2 in preorder traversal (*i.e.* EF) and the nodes in between n1 and n2 in postorder traversal (*i.e.* EF).
4. Taking these two sets of preorder and postorder traversals for left and right subtrees we will repeat the steps 2 and 3 till the entire tree is constructed.

10.6. BINARY SEARCH TREE

Another application of the ADT binary tree is searching and sorting. By enforcing certain rules on the values of the elements stored in a binary tree, it could be used to search and sort. Consider the sample binary tree shown in Fig. 10.25.

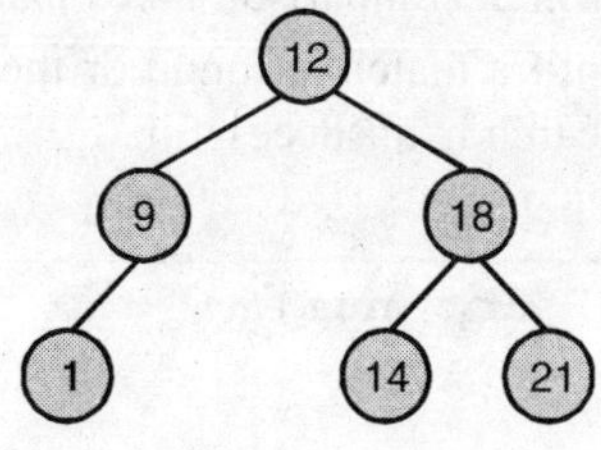

Fig. 10.25. A Binary Tree

The value of each node in the tree is greater than the value of the node in its left (if it exists) and it is less than the value in its right child (also, if it exists). In binary search tree, this property must always be satisfied.

One of the most important types of binary tree is Binary Search Tree (BST). As the name suggests they are used for searching purpose. For every node N in the BST, the following property will be true: The data at left node will be smaller then data at node N and data of right child will be greater then equal to data at node N.

10.6.1. Operations on Binary Search Trees

The most common operations on BSTs are insertion of a new key, deletion of an existing key, searching for a key and getting the data in sorted order. In this section, the first three are covered. The fourth operation could be done by traversing the BST in inorder.

In the three operations, it is an assumption that the binary search tree is not empty and it does not store multiple values. Also, the node structure BSTNode(left, info, right), as illustrated below, is used:

Fig. 10.26. BST Node.

The binary search tree used in this lesson has a list head as illustrated in the following figure:

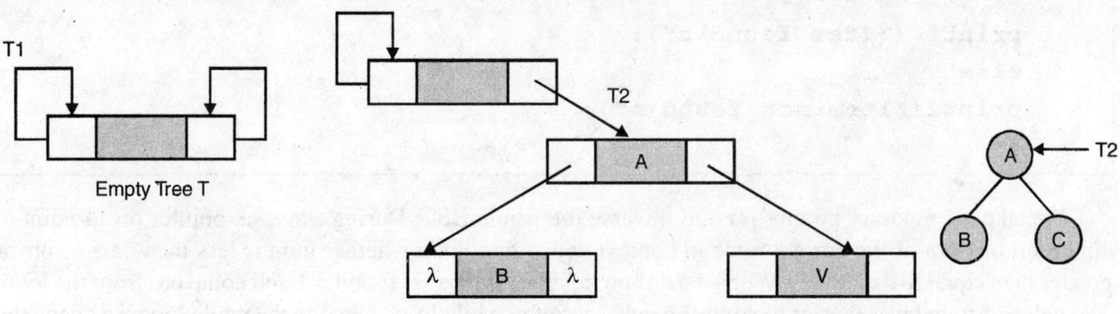

Fig. 10.27. Linked Representation of the Binary Search Tree T2.

If the binary search tree is empty, the left and right pointers of bstHead points to itself; otherwise, the right pointer points to the root of the tree.

10.6.2. Searching in Binary Search Tree

In searching for a value, say **k**, three conditions are possible :

■	if info[ptr] = K	value is at the node (successful search)
■	if info[ptr] > k	value is at the left subtree (search the left subtree)
■	if info[ptr] < k	value at the right subtree (search the right subtree)

Where info[ptr] is the value to which comparison takes place.

The same process is repeated until a match is found or the farthest leaf is reached but no match is found. In the case of the latter, the search is unsuccessful.

The C implementation is given below:

```
void search(node*R, int num)
{
node*ptr = R;
int found = 0;
if(R = = NULL)
{
printf("Tree is empty\n");
return;
}
while(ptr! = NULL)
{
if(ptr->data = = num)
{
found = 1;
break;
}
else if (ptr->data > num)
ptr = ptr->left;
else
ptr = ptr->right;
}
if (found = = 1)
printf ("Item found\n");
else
printf("Item not found\n");
}
```

We take a temporary pointer ptr and traverse the whole BST. During traversal pointer ptr to point to either left or right of the current node in context depending upon whether num is less than data of ptr or greater then equal to data of ptr. When data of ptr equals num we set found = 1 and come out from the loop. The value of found was 0 prior to control reaches into the while loop. Outside the while loop we check the value of found and display appropriate result.

The searching efficiency in BST comes from the fact that to search for a specific node the maximum number of comparisons will be limited by the height of the tree and we have seen that height of the BST is a function of $\log_2 N$, where N m number of nodes in binary tree or BST. Thus complexity of searching is $O(\log_2 N)$. In general if we want to perform searching over a linear array of size N then in the worst case we require N comparison and on an average we require N/2 comparisons. Thus complexity in this case is O(N). Comparing the two complexities it can easily be seen that for searching purpose BST outsmarts linear array by a larger fraction. Check for N = 10000.

10.6.3. Insertion in Binary Search Tree

In inserting a value, searching is performed in finding its proper location in the tree. However, if during the search process the key is found, the insertion will not be performed.

The following is the algorithm:

1. Start the search at the root node. Declare a node p and make it point to the root
2. Do the comparison:

```
if (k == info[p])  return false  // if key found, insertion
   not allowed else if (k < info[p])  p= left[p] // go left
else p = right[p]  // if (k > p.info) go right
```

3. Insert the node (p now points to the new parent of node to insert):

```
info[newNode]  = k
left[newNode]   = null
right[newNode]   = null
if (k < info[p])   left[p]  = newNode
else right[p]  = newNode
```

Creation of BST Using Linked List Representation

Creation of binary tree using linked list representation is very easy. To understand this we follow an example. Consider the following numbers given below :

Creation of binary tree using the linked representation is very easy. To understand this we follow an example. Consider the following numbers given :

12, 34, 5, 1, 15, 6, 7, 0, 13, 4, 53

The step by step development of BST is shown in the Fig. 10.28 given below

(*i*) 12, make root

(*ii*) 34 > 12, add as right of root (12)

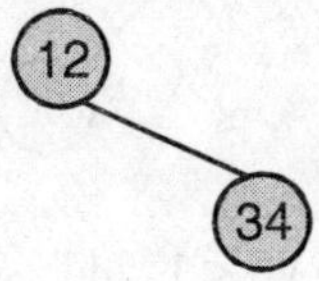

(*iii*) 5 < 12, add as left of root (12)

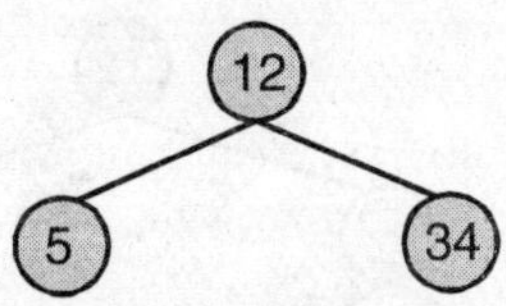

(*iv*) 1 < 12, reach at (5) 1 < 5 , add as left of node (5)

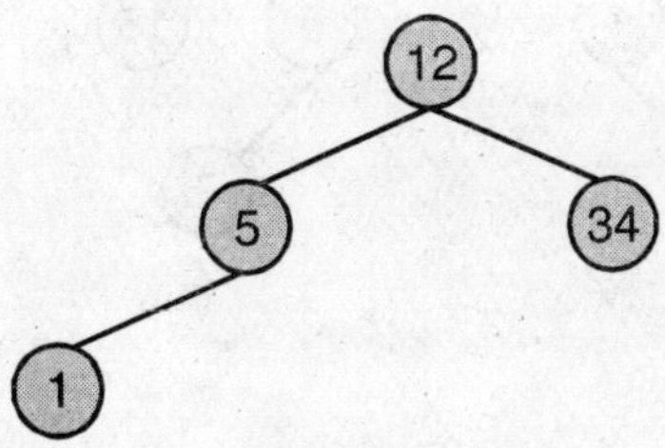

(*v*) 15 > 12, reach at (34) 15 < 34, add as left of node (34)

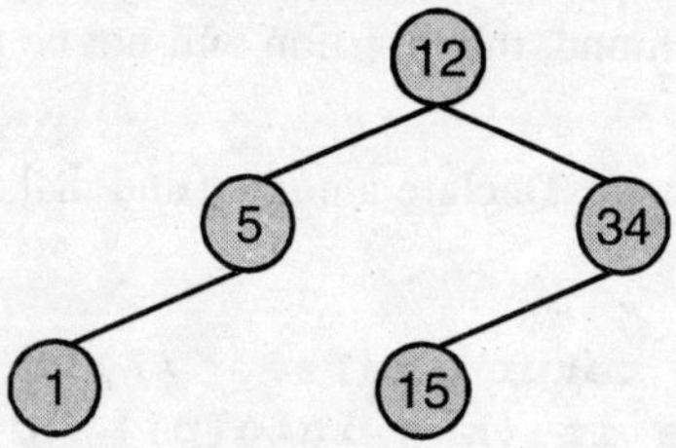

(*vi*) 6 < 12, reach at (5) 6 > 5, add as right of (5)

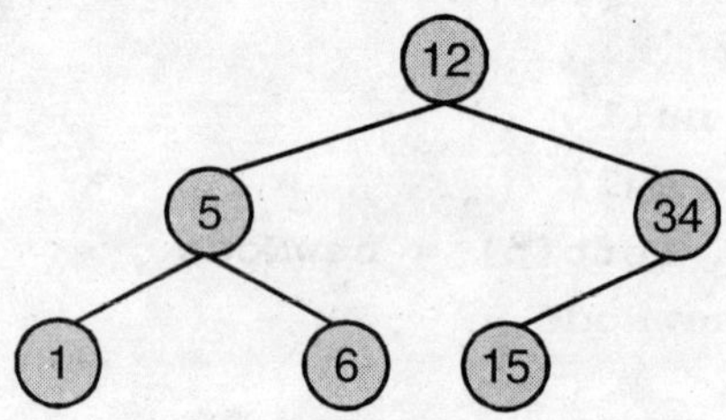

(*vii*) 7 < 12 reach at (5) 7 > 5, reach at (6) 7 > 6, add as right of node (6)

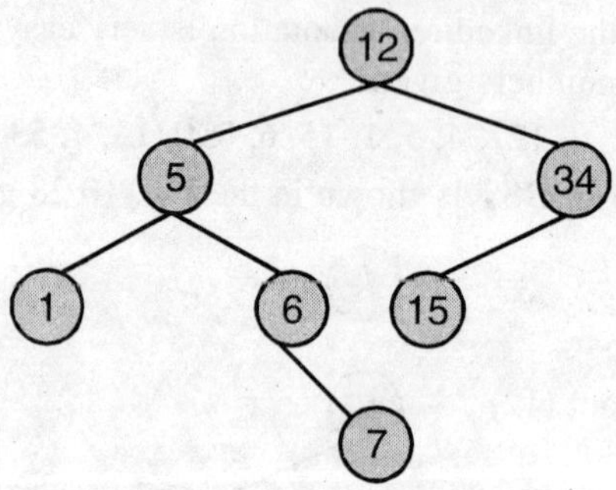

(*viii*) 0 < 12, reach at (5) 0 < 5 reach at (1) 0 < 1, add as left of 1

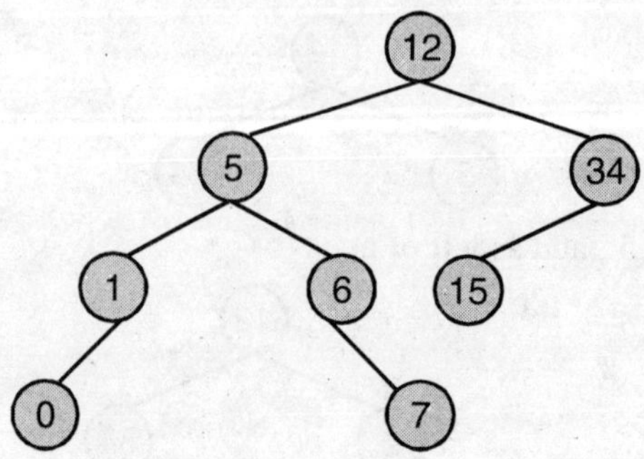

(*ix*) 13 > 12 reach at (34) 13 < 34 reach at (15) 13 < 15 add as left of (15)

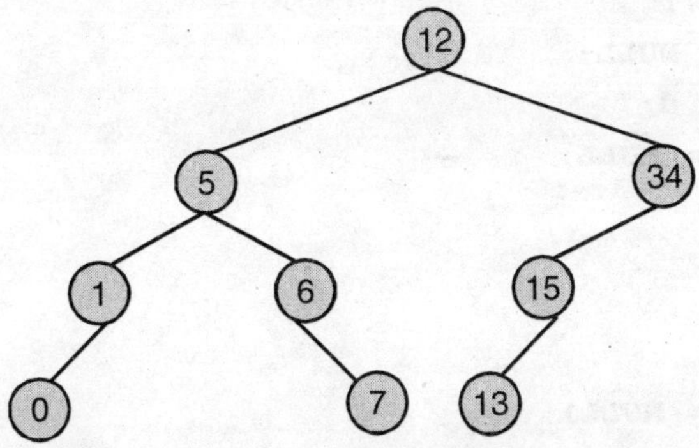

(*x*) 4 < 12, reach at (5) 4 < 5, reach at (1) 4 > 1 add as right of (1)

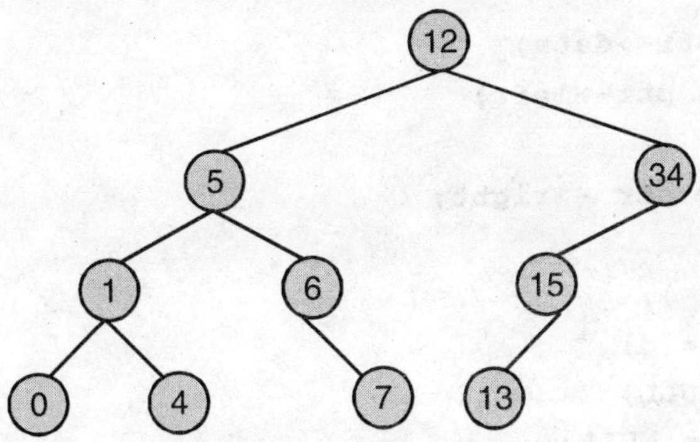

(*xi*) 53 > 12, reach at (34) 53 > 34, add as right of (34)

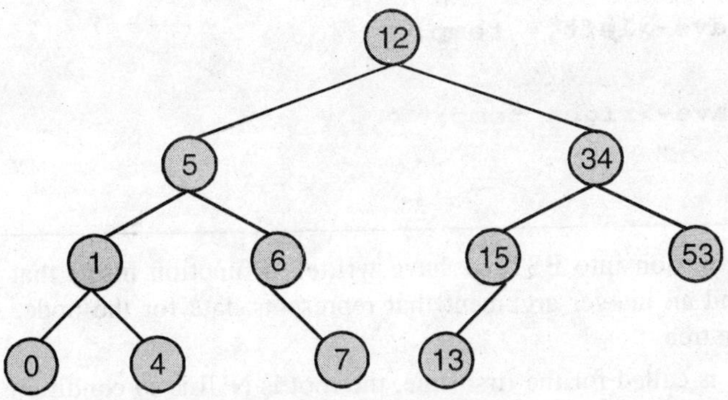

Fig. 10.28. Creation of BST.

Note every new data to be added is compared with root and then it follows its path towards left or right of the root and successive nodes till a suitable place for addition is not found.

The C implementation of creation of BST using linked list representation is given below:

```
void insert(node**R, int d)
{
node*ptr, *temp, *save;
if(*R = = NULL)
{
```

```
*R = alloc( );
(*R)->left = NULL;
(*R)->data = d;
(*R)->rjght = NULL;
}
else
{
ptr = *R;
while(ptr ! = NULL)
{
save = ptr;
if (d  <   ptr->data)
        ptr = ptr->left;
else
        ptr = ptr ->right;
}
temp alloc ( );
temp->data = d;
temp->left NULL;
temp->right = NULL;
if (d   <   save->data)
        save->left = temp;
else
        save->right temp;
}
}
```

For the creation/insertion into BST we have written a function insert that takes its argument of node** type and second an integer argument that represents data for the node. A special pointer root points to the root of the tree.

When the function is called for the first time, the root is NULL so condition *R = = NULL is true. Memory for the node is allocated and its various fields are set. When the function is called for the subsequent times, else part executes. We assume in the first call to insert function, data 11 was passed, so node with data 11 is root of our BST.

In the else part pointer ptr points to the root of the tree. We traverse the whole BST, saving ptr into the pointer save and advancing pointer ptr depending left or right upon whether data to be inserted is less than data of the ptr i.e. (d < ptr->data) or greater than equal to data of the ptr i.e. (d > = ptr->data). As in this context the variable d contains 6, if condition in the while loop is satisfied: save pointer points to the node 11 and ptr points to NULL; while loop terminates. Outside the while loop new node temp is created and all its fields are filled. Now depending upon whether d < save->data or d > = save->data, left of save points to temp or right of save points to temp. This is written as:

```
if (d < save->data)
save->left = temp;
else
save->right = temp;
```

Consider one more call to insert with d=13. This will be added as right of the node 11. In the next call to insert with d=0, when while loop terminates save will point to 6 and ptr will be NULL. Outside the loop left of save will point to node temp (with data 0). The complete insertion process is shown in the following Fig. 10.26.

(*i*) Insert 11

{11 is the root}

(*ii*) Insert 6

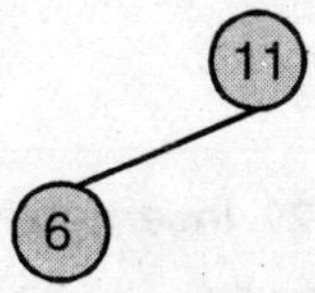

{6 is smaller than 11}

(*iii*) Insert 13

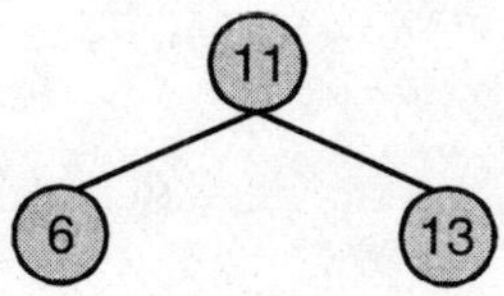

{13 is greater than 11}

(*iv*) Insert 0

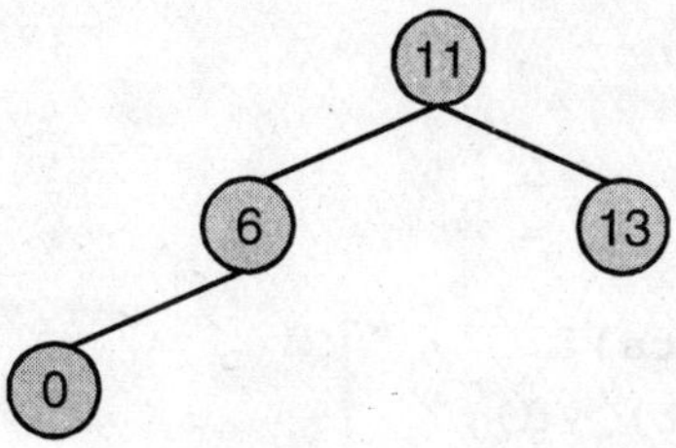

{0 < 11, go left of (11)
0 < 6, add as left of (6)

(*v*) Insert 5

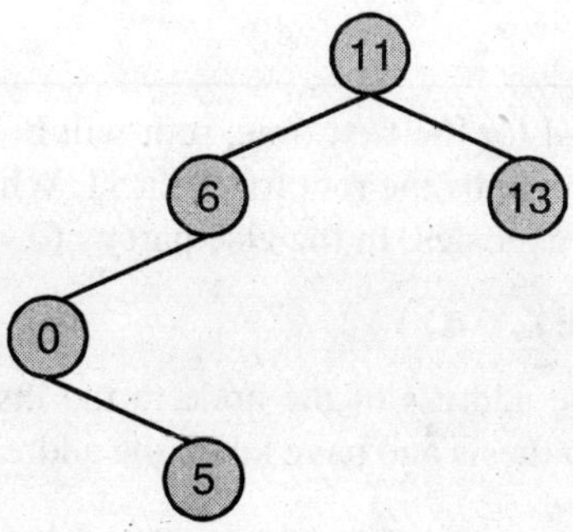

{5 < 11, go left of (11)
5 < 6 go left of (6)
5 > 0 add as right of (0)

(*vi*) Insert 41

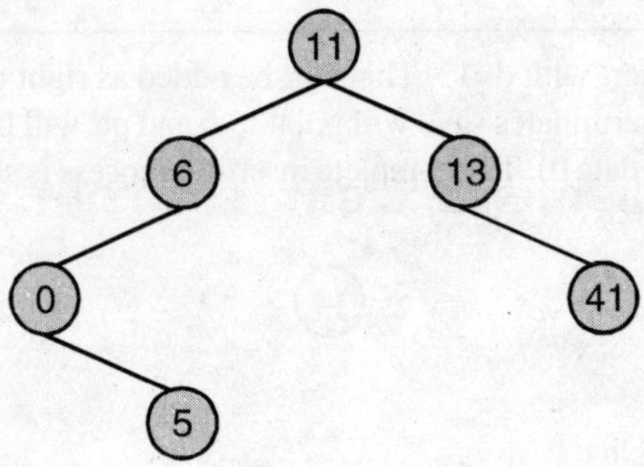

41>11, go right of (11)
41 > 13 add 41 as right of (13)

Fig. 10.29. Insertion in BST.

The Recursive version of insertion into BST is given below:

```
void insert(node**R, int d)
{
if (* R = = NULL)
{
*R = alloc ( );
(*R)->left = NULL;
(*R)->data = d;
(*R)->right = NULL;
}
else
{
if (d < (*R) -> data)
insert(&((*R)->left), d);
else
insert ( &(( * R) -> right ), d);
}
}
```

When the insert function is called for the first time, root will be NULL; if condition is satisfied and new node is created with data 2. This will be the root for the BST. When insert function is called with data 1, if condition will fail and else part executes. In the else part as (d < ptr->data), the statement

```
insert(&((*R)->left), d);
```

executes. As we need to pass the address of the node to the insert function we have put the whole expression (*R)->left inside the parenthesis and have taken the address of it. Now in the recursive call *R will be NULL as left of (*R) in the callee function was NULL. New node with data 1 is created and left of node 2 points to this node. Next time when insert is called with 6, statement

```
insert(& ((*R)->right), d)
```

executes and 6 is added to the right of the node 2.

All the elements are added in the similar fashion.

Algorithm for Insertion in BST

Algorithm 10.4. INSERT_BST (INFO, RIGHT.LEFT, ROOT, ITEM)..............................

ROOT: Starting node

LEFT: Points to Left Child

INFO: Information Part of a node

RIGHT: Points to Right Child

ITEM: ITEM to be inserted

(1) Set PTR = ROOT, PTR1 = ROOT

(2) Repeat while (PTR! = NULL)

  (a) IF INFO [PTR] > ITEM

    Set PTR1= PTR

    Set PTR = LEFT [PTR]

  (b) ELSE

    Set PTR1 = PTR

    Set PTR = RIGHT [PTR]

(3) (a) IF INFO[PTR1] > ITEM

    Set LEFT [PTR1] = ITEM

    Set LEFT [ITEM] = NULL

    Set RIGHT [ITEM] = NULL

  (b) ELSE

    Set RIGHT [PTR1] = ITEM

    Set LEFT [ITEM] = NULL

    Set RIGHT [ITEM] = NULL

(4) Exit

10.6.4. Deletion in Binary Search Tree

Deleting a key from the binary search tree is a bit more complex than the other two operations just discussed. The operation starts by finding the key to delete. If not found, the algorithm simply returns to tell that deletion fails. If the search returns a node, then there is a need to delete the node that contains the key that we searched for.

However, deletion is not as simple as removing the node found since it has a parent pointing to it. It is also a big possibility that it is the parent of some other nodes in the binary search tree. In this case, there is a need to have its children "adopted" by some other node, and the pointers pointing to it should also be adjusted. And in the process of reassigning pointers, the BST property of the order of the key values has to be maintained.

There are two general cases to consider in deleting a node **d** :

1. Node d is an external node (leaf) :

Action: update the child pointer of the parent **p**:
if **d** is a left child, set **left[p]** = null otherwise, set **right[p]** = null

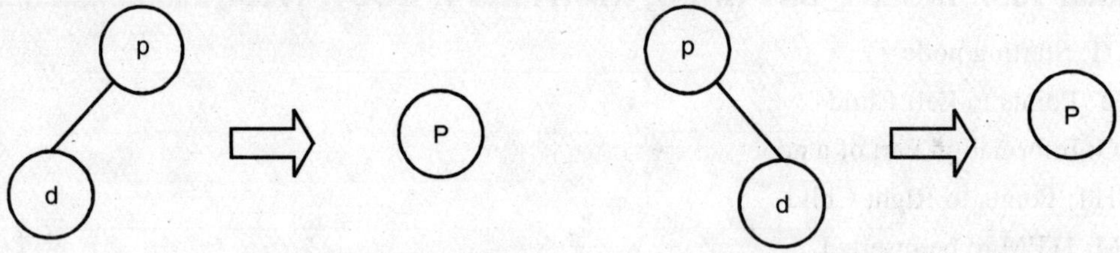

Fig. 10.30. Deletion of a Leaf Node.

Algorithm 10.5. DELETE_0CHILD (LEFT, RIGHT, INFO, ROOT, ITEM) ....

ROOT: Starting node

LEFT: Points to Left Child

INFO: Information Part of a node

RIGHT: Points to Right Child

ITEM: ITEM to be deleted having zero Child

The procedure deletes the node ITEM, where ITEM have zero children. The pointer PTR1 points to the parent of ITEM and pointer PTR points to the node ITEM to be deleted.

```
(1) Set PTR = ROOT, PTR1 = ROOT
(2) Repeat while (PTR! = NULL)
    (a) IF INFO [PTR] = ITEM
        Go to step (4)
    (b) ELSE IF INFO [PTR] > ITEM)
            Set PTR1= PTR
            Set PTR = LEFT [PTR]
    (c) ELSE
            Set PTR1 = PTR
            Set PTR = RIGHT [PTR]
(3) Element NOT FOUND and EXIT
(4) (a) IF INFO [PTR] > INFO [PTR 1]
            Set RIGHT [PTR 1] =NULL
```

```
        ELSE
              Set LEFT [PTR 1] =NULL.
    (b) FREE [PTR]
(5) Exit
```

2. Node is internal (there are two subcases) :

Case 1: If d has a left subtree,

1. Get the inorder predecessor, **ip**. Inorder predecessor is defined as the rightmost node in the left subtree of the current node, which in this case is **d**. It contains the preceding key if the BST is traversed in inorder.
2. Get the parent of **ip**, say **p_ip**.
3. Replace d with **ip**.
4. Remove **ip** from its old location by adjusting the pointers:
   (*a*) If **ip** is not a leaf node:
      1. Set the right child of **p_ip** to point to the left child of **ip**
      2. Set the left child of **ip** to point to the left child of **d**

   (*b*) Set the right child of **ip** to point to the right child of **d**

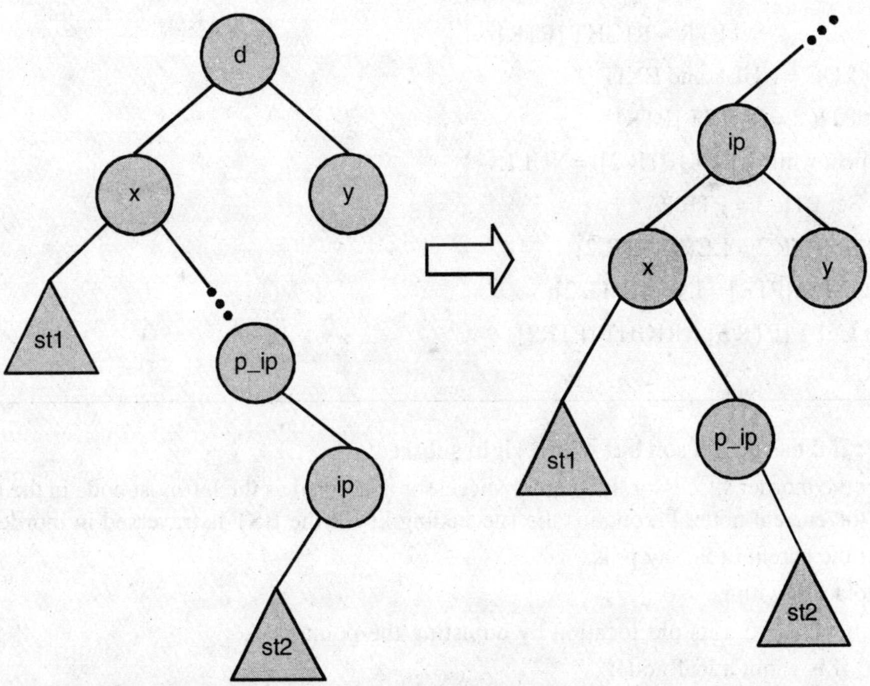

Fig. 10.31. Deletion of an Internal Node with a Left Subtree.

Algorithm 10.6. DELETE_TWOCHILD (LEFT, RIGHT, INFO, ROOT, ITEM) ....

ROOT: Starting node

LEFT: Points to Left Child

INFO: Information Part of a node

RIGHT: Points to Right Child

ITEM: ITEM to be deleted having two Childs

The procedure deletes the node ITEM, where ITEM have two children's. The pointer PTR1 points to the parent of ITEM and pointer PTR points to the node ITEM to be deleted.

(1) Set PTR = ROOT, PTR1 = ROOT

(2) Repeat while PTR! = NULL.

  (a) If INFO [PTR]= ITEM

    Go to step (4)

  (b) ELSE IF INFO [PTR] > ITEM

    Set PTR 1 = PTR

    Set PTR = LEFT [PTR]

  (c) ELSE

    Set PTR 1 = PTR

    Set PTR = RIGHT [PTR]

(3) Set LOC = NULL and EXIT

(4) Set PTR 2 = RIGHT [PTR]

(5) Repeat while LEFT [PTR 2]! =NULL

  (a) Set PTR 3 = PTR 2.

  (b) Set PTR 2 = LEFT [PTR 2]

(6) Set INFO [PTR] = INFO [PTR 2]

(7) Set LEFT [PTR 3] = RIGHT [PTR2]

(8) EXIT.

Case 2 : If d has no left son but with a right subtree

1. Get the inorder successor, **is**. Inorder successor is defined as the leftmost node in the right subtree of the current node. It contains the succeeding key if the BST is traversed in inorder.
2. Get the parent of **is**, say **p_is**.
3. Replace **d** with **is**.
4. Remove **is** from its old location by adjusting the pointers:

   (*a*) If **is** is not a leaf node:

   1. Set the left child of **p_is** to point to the right child of **is**
   2. Set the right child of **is** to point to the right child of **d**

   (*b*) Set the left child of **is** to point to the left child of **d**

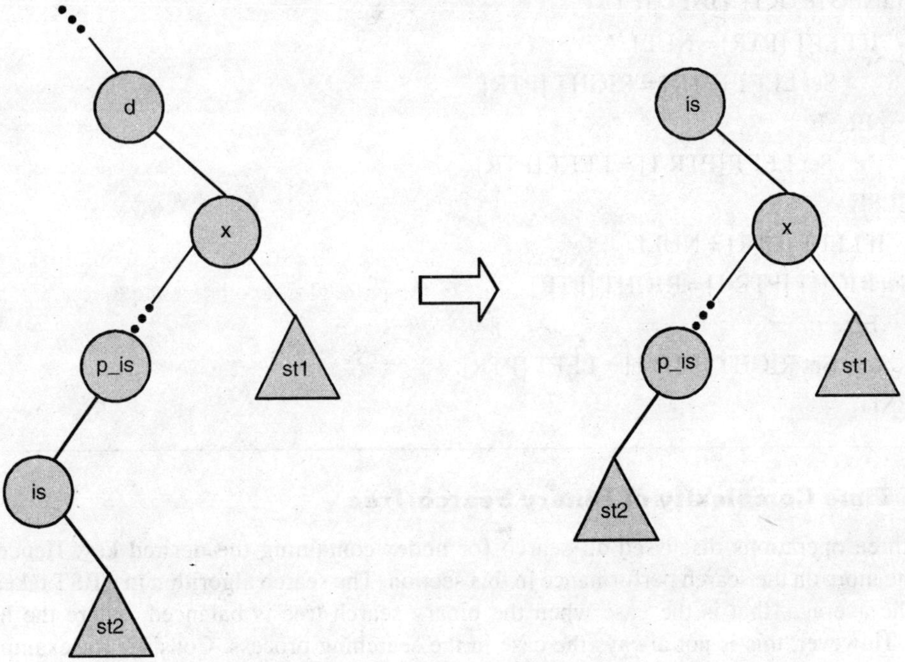

Fig. 10.32. Deletion of an Internal Node with No Left Subtree.

Algorithm 10.7. DELETE_ONECHILD (LEFT, RIGHT, INFO, ROOT, ITEM)..................

ROOT: Starting node

LEFT: Points to Left Child

INFO: Information Part of a node

RIGHT: Points to Right Child

ITEM: ITEM to be deleted having one child

The procedure deletes the node ITEM, where ITEM have one child. The pointer PTR1 points to the parent of ITEM and pointer PTR points to the node ITEM to be deleted.

(1) Set PTR = ROOT, PTR1 = ROOT

(2) Repeat while PTR! = NULL.

(a) If INFO [PTR]= ITEM

Go to step (4)

(b) ELSE IF INFO [PTR] > ITEM

Set PTR 1 = PTR

Set PTR = LEFT [PTR]

(c) ELSE

Set PTR 1 = PTR

Set PTR = RIGHT [PTR]

(3) Set LOC = NULL and EXIT

```
(4) If INFO [PTR 1] > INFO [PTR]
        If LEFT [PTR] = NULL
            Set LEFT [PTR] = RIGHT [PTR]
        ELSE
            Set LEFT [PTR 1] = LEFT [PTR]
    ELSE
        If LEFT [PTR] = NULL
    Set RIGHT [PTR 1] = RIGHT [PTR].
        Else
            Set RIGHT [PTR 1] = LEFT [PTR].
(5) EXIT.
```

10.6.5. Time Complexity of Binary Search Tree

The three operations discussed all search for nodes containing the desired key. Hence, we shall concentrate more on the search performance in this section. The search algorithm in a BST takes $O(\log_2 n)$ time on the average, that is the case when the binary search tree is balanced, where the height is in $O(\log_2 n)$. However, this is not always the case in the searching process. Consider for example the case wherein the elements inserted in a BST are in sorted (or reversed) order, then the resulting BST will have all of its node's left (or right) son set to NULL, resulting in a degenerated tree. In such a case, the time spent for searching deteriorates to $O(n)$, equivalent to that of sequential search, as in the case of the following trees :

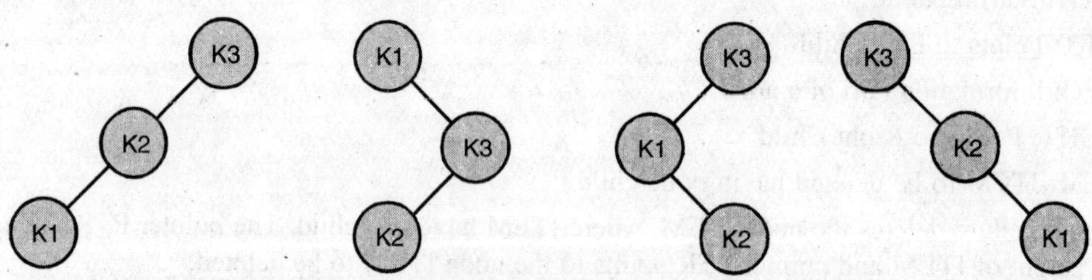

Fig. 10.33. Examples of Worst-Depth Trees.

10.7. HEADER NODES; THREADS

Consider a binary tree T. Variations of the linked representation of T are frequently used because certain operations on T are easier to implement by using the modifications. Some of these variations, which are analogous to header and circular linked lists, are discussed in this section.

Header Nodes

Suppose a binary tree T is maintained in memory by means of a linked representation. Sometimes an extra, special node, called a *header node,* is added to the beginning of T. When this extra node is used, the tree pointer variable, which we will call HEAD (instead of ROOT), will point to the header node, and the left pointer of the header node will point to the root of T. Figure 10.30 shows a schematic picture of the binary tree in Fig. 10.34 that uses a linked representation with a header node.

Suppose a binary tree T is empty. Then T will still contain a header node, but the left pointer of the header node will contain the null value. Thus the condition

```
LEFT[HEAD] = NULL
```

will indicate an empty tree.

Another variation of the above representation of a binary tree T is to use the header node as a sentinel. That is, if a node has an empty subtree, then the pointer field for the subtree will contain the address of the header node instead of the null value. Accordingly, no pointer will ever contain an invalid address, and the condition

```
LEFT[HEAD] = HEAD
```

will indicate an empty subtree.

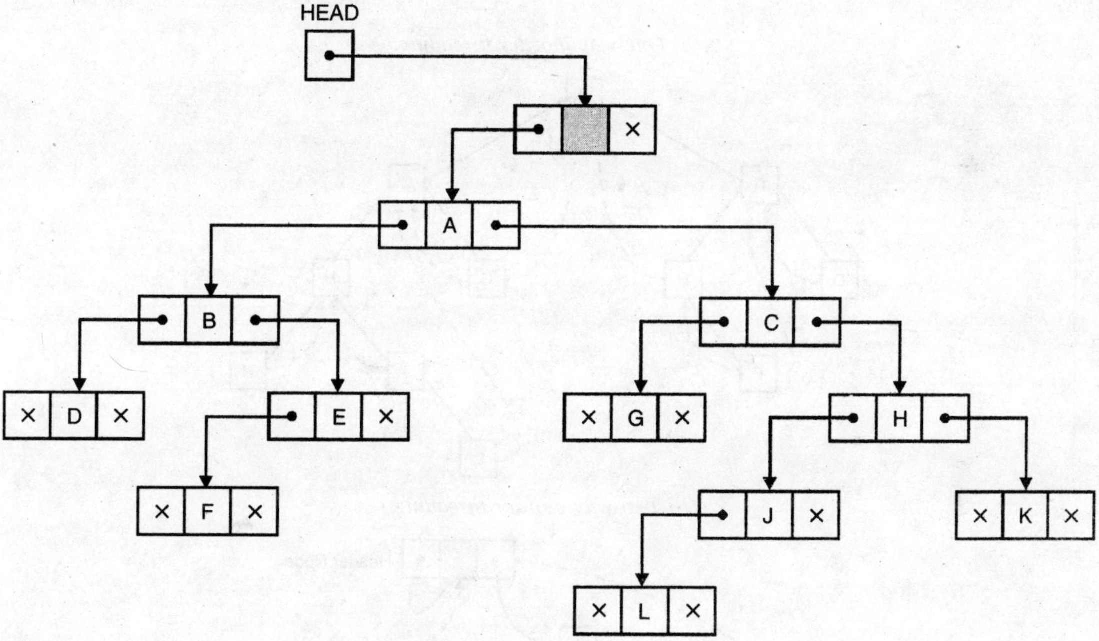

Fig. 10.34.

Threads; Inorder Threading

Consider again the linked representation of a binary tree T. Approximately half of the entries in the pointer fields LEFT and RIGHT will contain null elements. This space may be more efficiently used by replacing the null entries by some other type of information. Specifically, we will replace certain null entries by special pointers which point to nodes higher in the tree. These special pointers are called *threads,* and binary trees with such pointers are called *threaded trees.*

The threads in a threaded tree must be distinguished in some way from ordinary pointers. The threads in a diagram of a threaded tree are usually indicated by dotted lines. In computer memory, an extra 1-bit TAG field may be used to distinguish threads from ordinary pointers, or, alternatively, threads may be denoted by negative integers when ordinary pointers are denoted by positive integers.

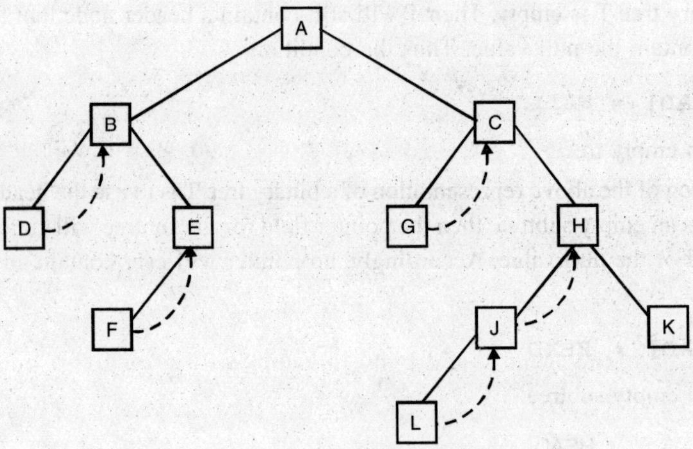

(a) One-way inorder threading.

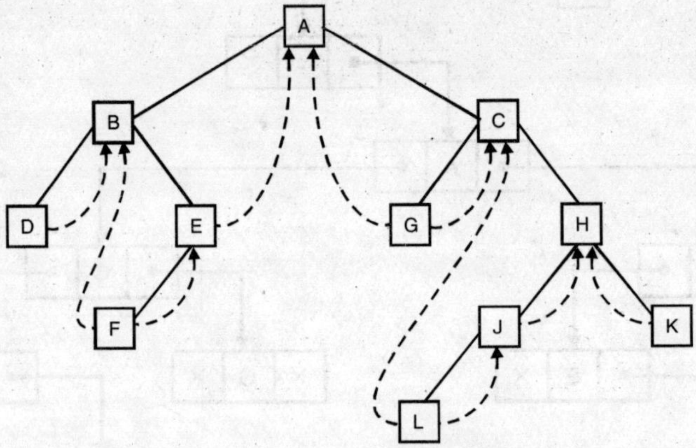

(b) Two-way inorder threading.

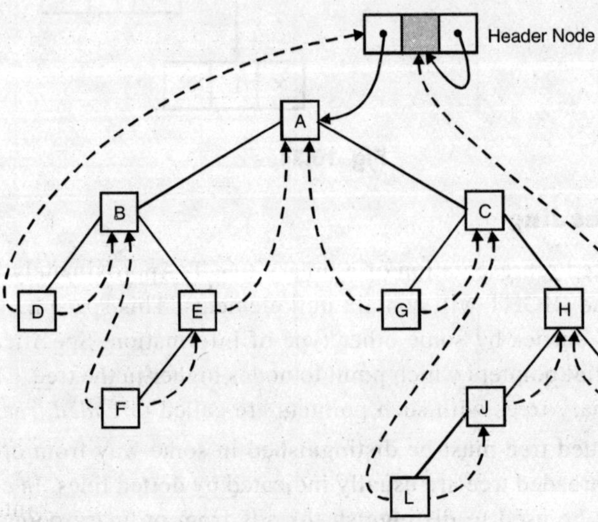

(c) Two-way threading with header node.

Fig. 10.35.

There are many ways to thread a binary tree T, but each threading will correspond to a particular traversal of T. Also, one may choose a one-way threading or a two-way threading. Unless otherwise stated, our threading will correspond to the inorder traversal of T. Accordingly, in the one-way threading of T, a thread will appear in the right field of a node and will point to the next node in the inorder traversal of T; and in the two-way threading of T, a thread will also appear in the LEFT field of a node and will point to the preceding node in the inorder traversal of T. Furthermore, the left pointer of the first node and the right pointer of the last node (in the inorder traversal of T) will contain the null value when T does not have a header node, but will point to the header node when T does have a header node.

There is an analogous one-way threading of a binary tree T which corresponds to the preorder traversal of T. On the other hand, there is no threading of T which corresponds to the postorder traversal of T.

Example 10.10.

Consider the binary tree T in Fig. 10.36.

HEAD: 20 (→ 20)

AVAIL: 8 (→ 8)

	INFO	LEFT	RIGHT
1	K	−17	−20
2	C	−3	6
3	G	−5	−2
4		14	
5	A	10	2
6	H	17	1
7	L	−2	−17
8		9	
9		4	
10	B	18	13
11		19	
12	F	−10	−13
13	E	12	−5
14		15	
15		16	
16		11	
17	J	7	−6
18	D	−20	−10
19		0	
20		5	20

Fig. 10.36.

(*a*) The one-way inorder threading of T appears in Fig. 10.35(*a*). There is a thread from node E to node A, since A is accessed after E in the inorder traversal of T. Observe that every null right pointer has been replaced by a thread except for the node K, which is the last node in the inorder traversal of T.

(*b*) The two-way inorder threading of T appears in Fig. 10.35(*b*). There is a left thread from node L to node (', since L is accessed after C in the inorder traversal of T. Observe that every null left pointer has been replaced by a thread except for node D, which is the first node in the inorder traversal of T. All the right threads are the same as in Fig. 10.35(*c*).

(*c*) The two-way inorder threading of T when T has a header node appears in Fig. 10.35(*c*). Here the left thread of D and the right thread of K point to the header node. Otherwise the picture is the same as that in Fig. 10.36.

10.8. BALANCED BINARY SEARCH TREES

It is the poor balance of a BST that makes it perform in $O(n)$ time. Therefore, to ensure that searching takes $O(\log_2 n)$, the balance has to be maintained. Balanced tree will result if half of the records inserted after any given record *r* with key *k* have keys smaller than *k* and similarly, half have keys greater than *k*. That is when no balance-handling is used during insertion. However, in real situation, keys inserted are in random order. Hence, there is a need to maintain the balance as keys are inserted or deleted.

Balance of a node is a very important factor in balance maintenance. It is defined as the height difference of the subtrees of a node, *i.e.*, the height of a node's left subtree minus the height of its right subtree.

10.8.1. AVL Tree

One of the most commonly used balanced BST is the AVL tree. It was created by G. Adel'son-Vel'skii and E. Landis, hence the name AVL. An AVL tree is a height-balanced tree wherein the the height difference between the subtrees of a node *i.e.* Balancing Factor (BF) = Height of Left Subtree (H_L) – Height of Right Subtree (H_R), for every node in the tree, is at most 1. Consider the binary search trees below where the nodes are labeled with the balance factors :

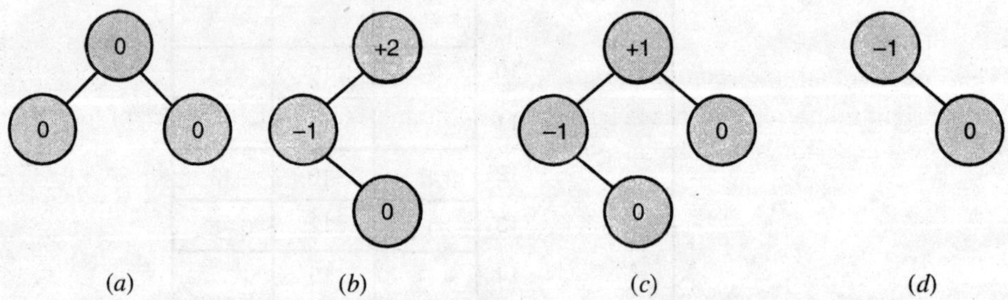

Fig. 10.37. BSTs with Balance Factors.

Binary search trees a, c and d all have nodes with balances in the range [–1, 1], *i.e.*, –1, 0 and +1. Therefore, they are all AVL trees. BST *b* has a node with a balance of +2, and since it is beyond the range, it is not an AVL tree.

In addition to having $O(\log_2 n)$ time complexity for searching, the following operations will also have the same time complexity if AVL tree is used :

- Finding the nth item, given *n*
- Inserting an item at a specified place
- Deleting a specified item

Tree Balancing

The following are examples of AVL trees :

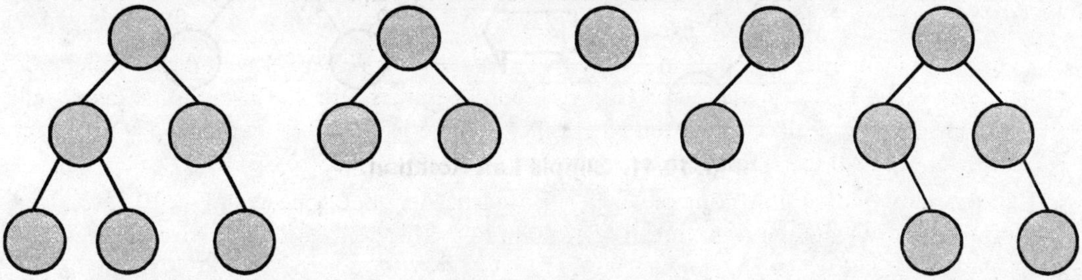

Fig. 10.38. AVL Trees.

The following are examples of non-AVL trees :

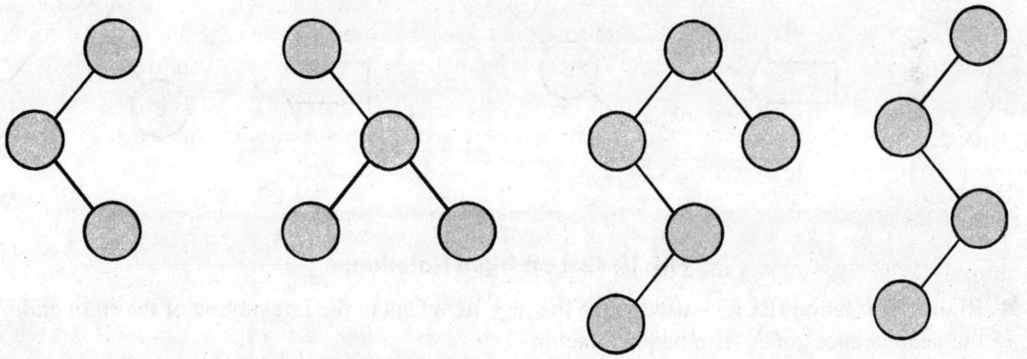

Fig. 10.39. Non-AVL Trees.

In maintaining the balance of an AVL tree, rotations have to be performed during insertion and deletion. The following are the rotations used :

- Simple right rotation (RR) – used when the new item C is in the left subtree of the left child B of the nearest ancestor A with balance factor +2

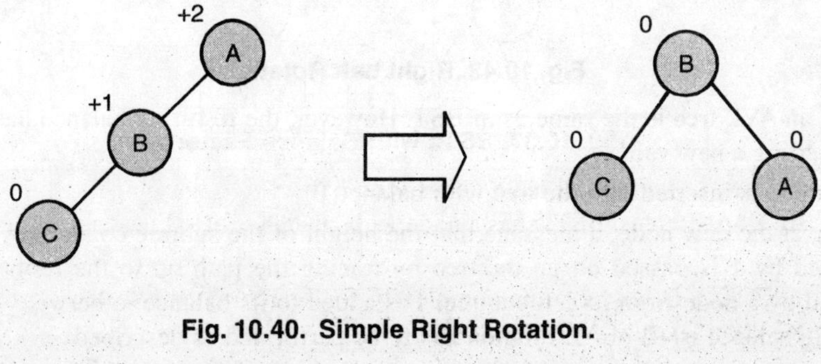

Fig. 10.40. Simple Right Rotation.

- Simple left rotation (LR) – used when the new item C is in the right subtree of the right child B of the nearest ancestor A with balance factor –2

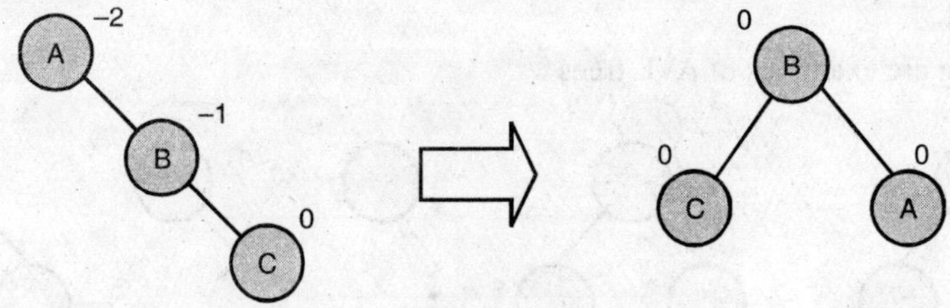

Fig. 10.41. Simple Left Rotation.

- Left-right rotation (LRR)– used when the new item C is in the right subtree of the left child B of the nearest ancestor A with balance factor +2

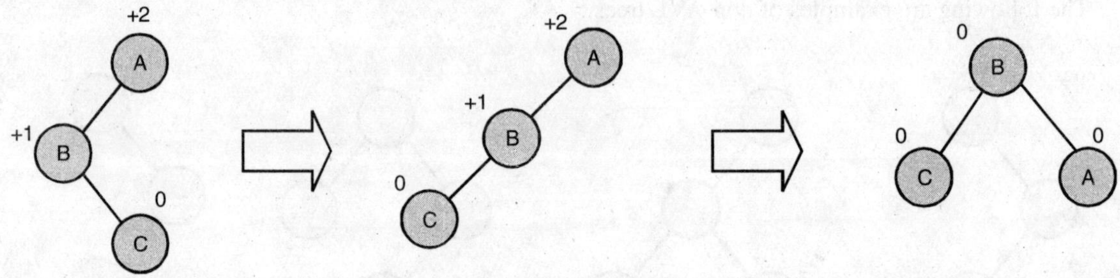

Fig. 10.42. Left Right Rotation.

- Right-left rotation (RLR) – used when the new item C is in the left subtree of the right child B of the nearest ancestor A with balance factor -2

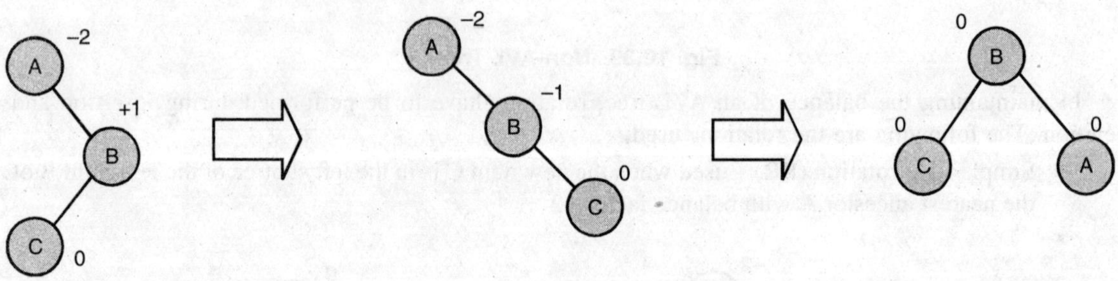

Fig. 10.43. Right Left Rotation.

Insertion in an AVL tree is the same as in BST. However, the resulting balance has to be checked in AVL tree. In inserting a new value,

1. A leaf node is inserted into the tree with balance 0
2. Starting at the new node, a message that the height of the subtree containing the new node has increased by 1 is passed on up the tree by tracing the path up to the root. If the message is received by a node from its left subtree, 1 is added to its balance, otherwise –1 is added. If the resulting balance is +2 or –2, rotation has to be performed as described.

For example, insert the following elements in an AVL tree:

4	10	9	3	11	8	2	1	5	7	6

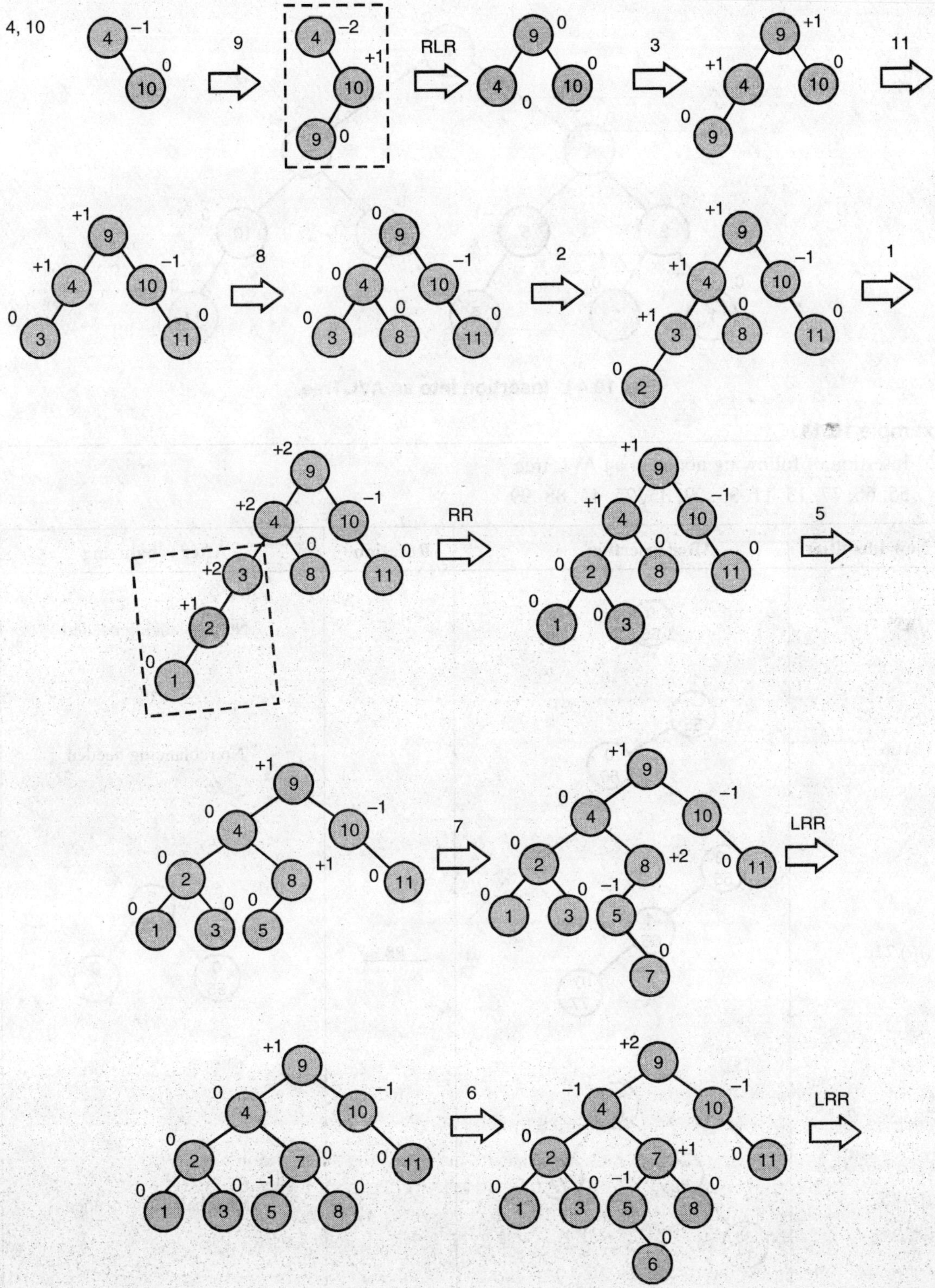
4, 10
9
RLR
3
11
8
2
1
RR
5
7
LRR
6
LRR

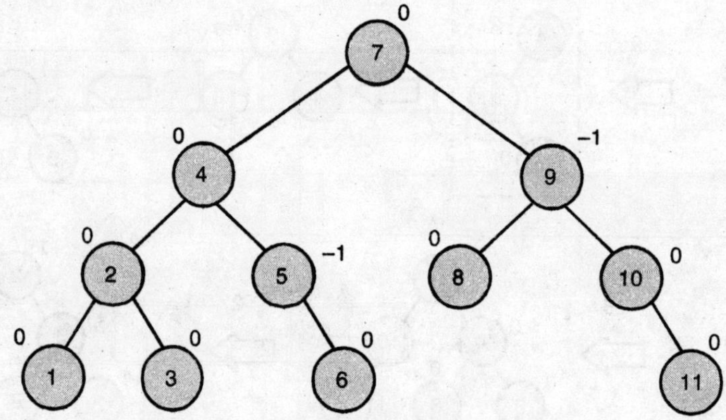

Fig. 10.44. Insertion into an AVL Tree.

Example 10.11.

Insertion of following nodes in an AVL tree
55, 66, 77, 15, 11, 33, 22, 35, 25, 44, 88, 99

New Identifier	After insertion	Rotation	After rebalncing
(*i*) 55			No rebalancing needed
(*ii*) 66	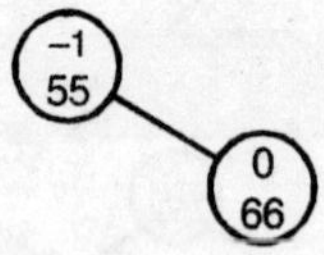		No reblancing needed
(*iii*) 77	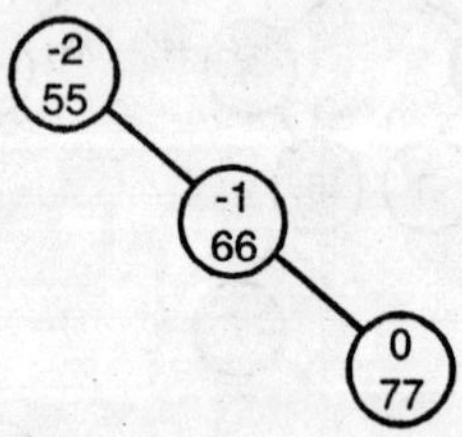	RR →	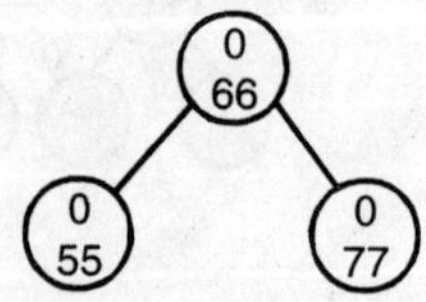
(*iv*) 15	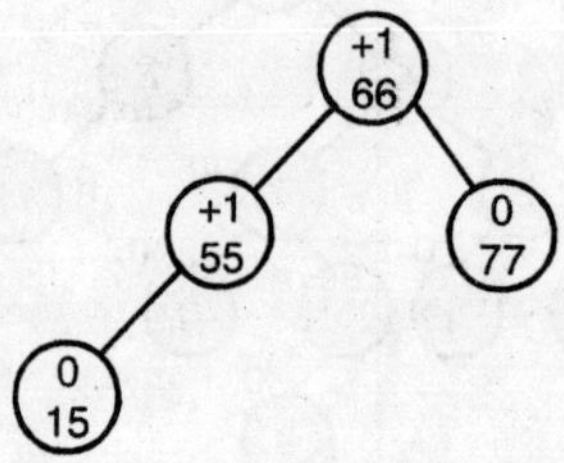		No rebalancing needed

New Identifier	After insertion	Rotation	After rebalncing
(*v*) 11	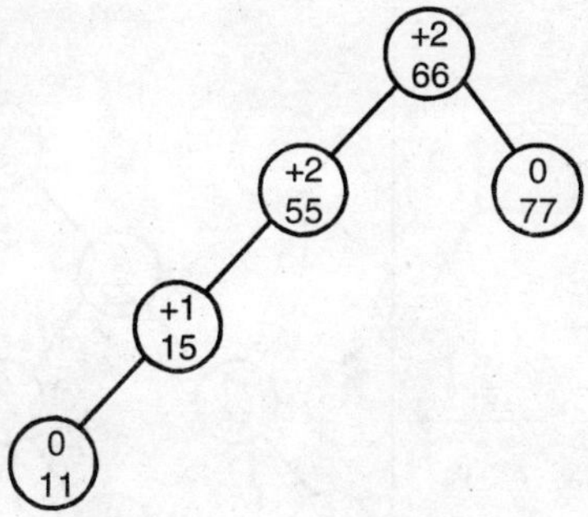	LL →	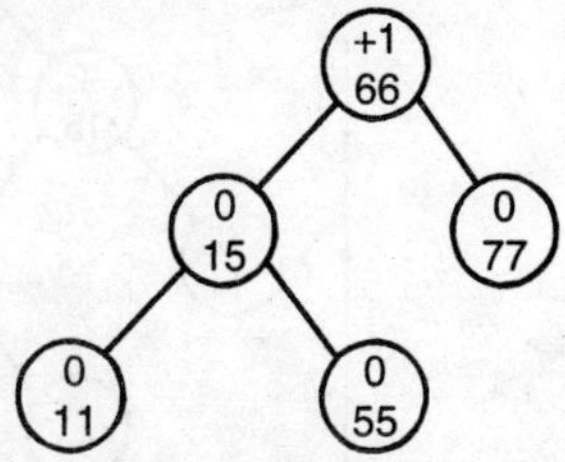
(*vi*) 33	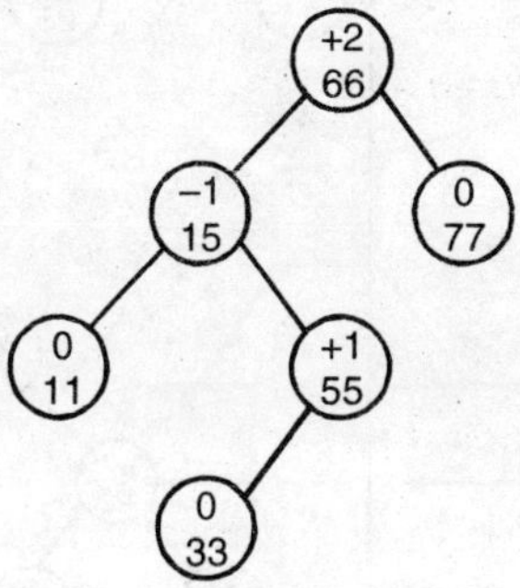	LR →	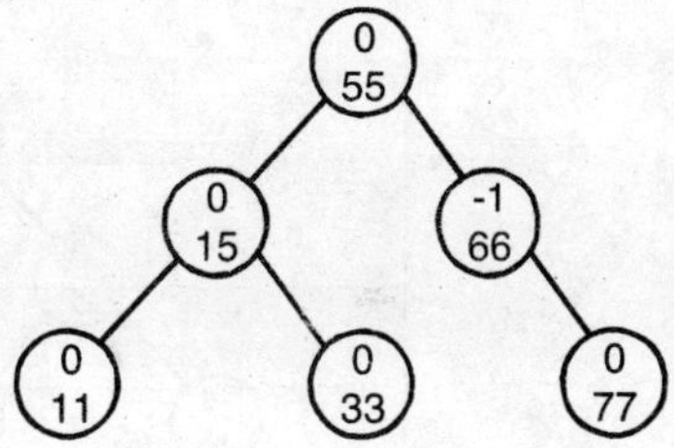
(*vii*) 22	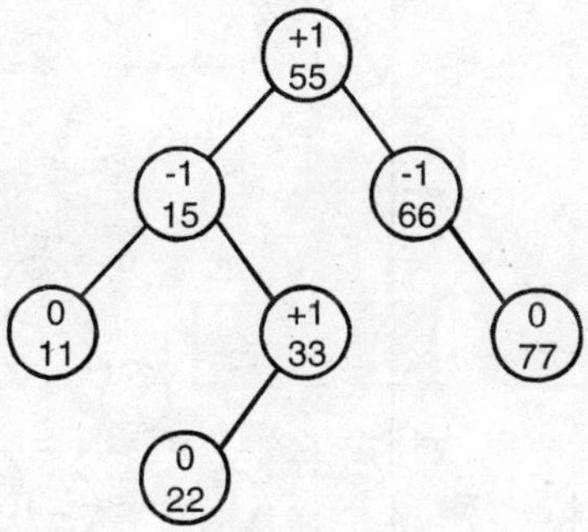		No rebalancing needed
(*viii*) 35	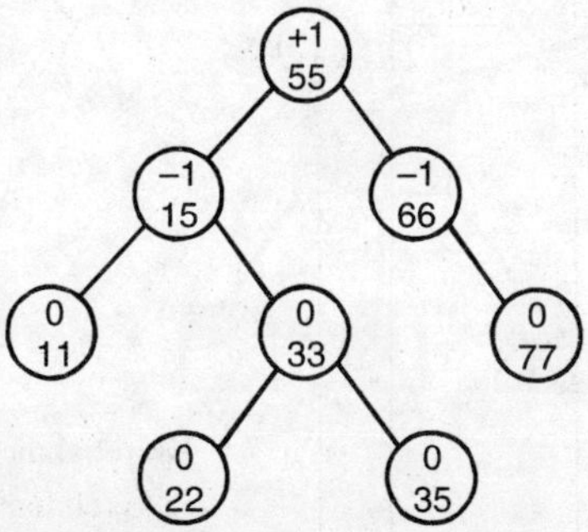		No rebalancing needed

New Identifier	After insertion	Rotation	After rebalncing
(*ix*) 25	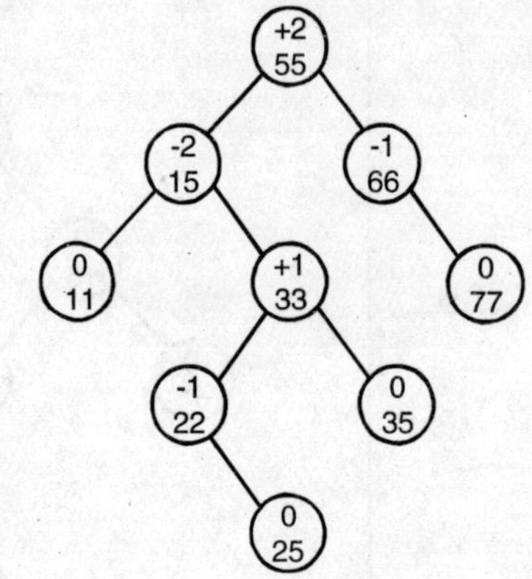	RL →	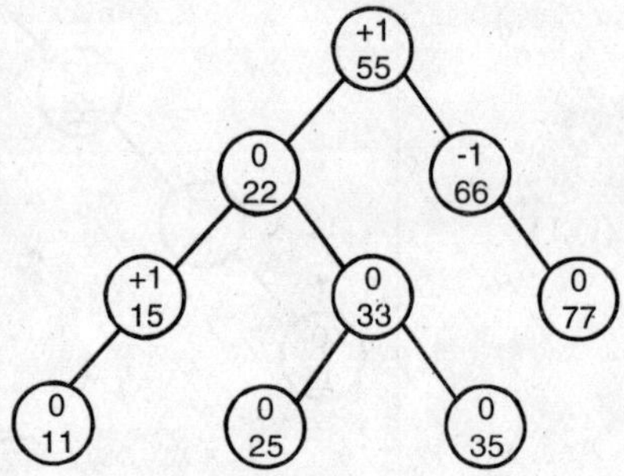
(*x*) 44	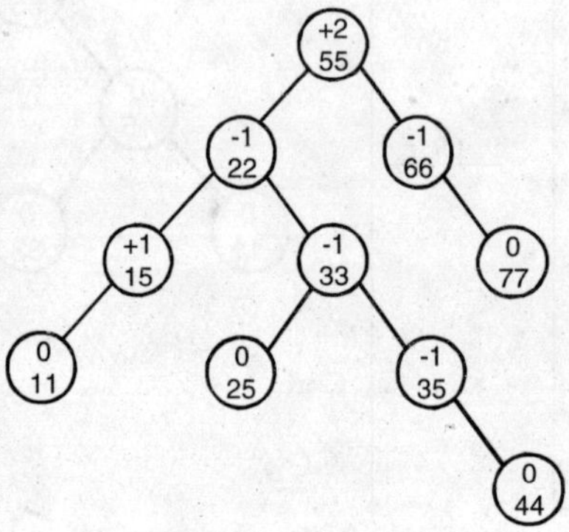	LR →	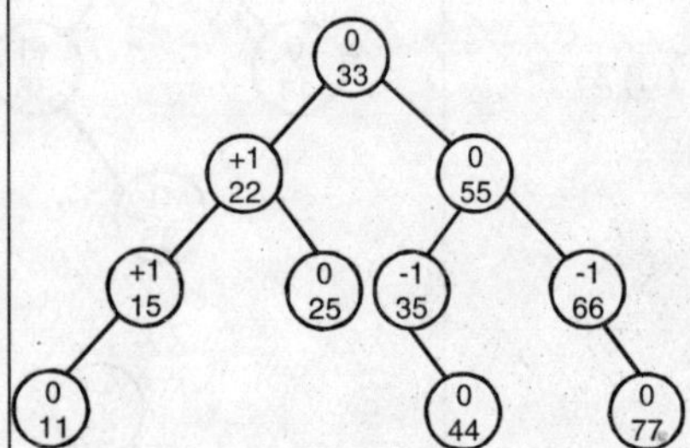
(*xi*) 88	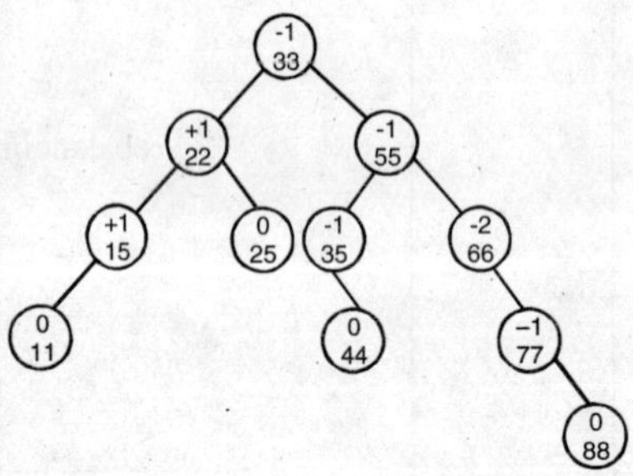		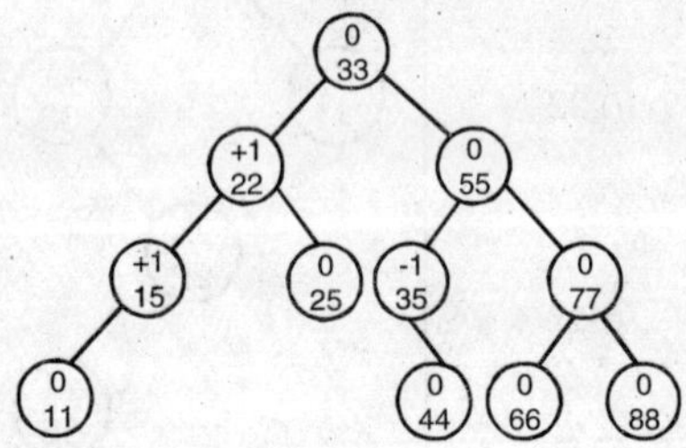
(*xii*) 99	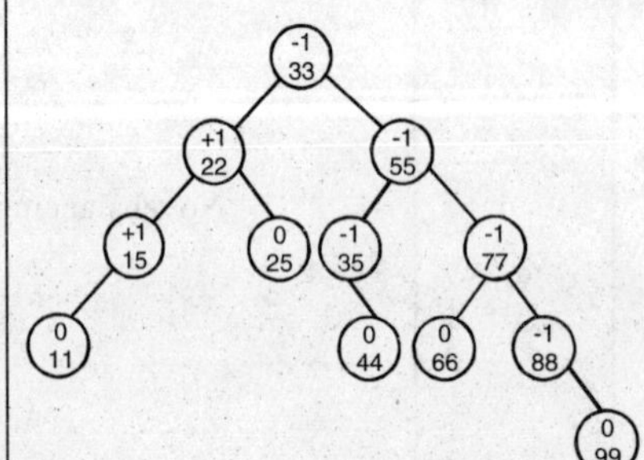		No rebalancing needed

Example 10.12.

Insertion of following nodes in an AVL tree

jan, feb, mar, apr, may, jun, jul, aug, sep, oct, nov, dec

New Identifier	After insertion	Rotation	After rebalncing
(*i*) jan			No rebalancing needed
(*ii*) feb	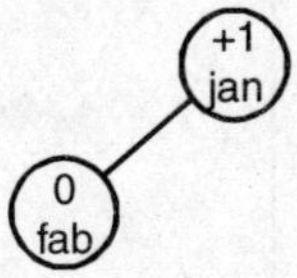		No rebalancing needed
(*iii*) mar	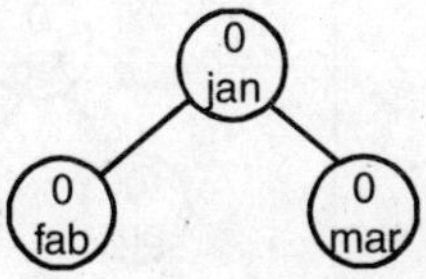		No rebalancing needed
(*iv*) apr	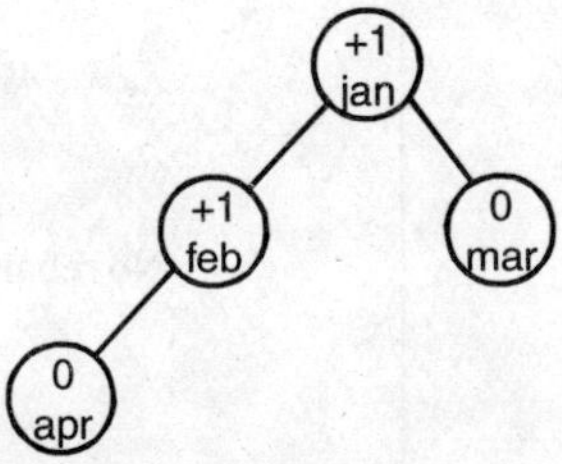		No rebalancing needed
(*v*) may	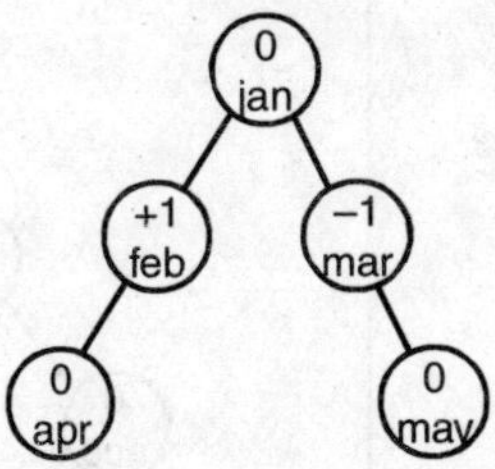		No rebalancing needed
(*vi*) jun	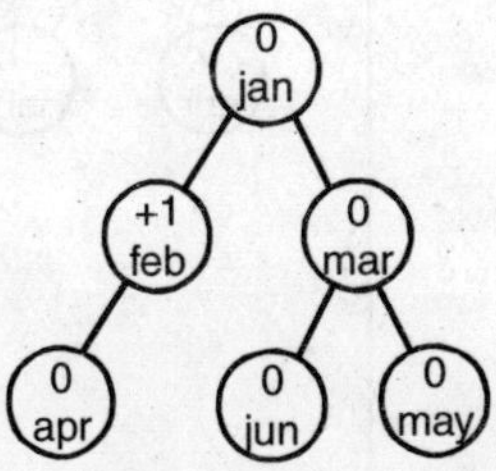		No rebalancing needed

<table>
<tr><th>New Identifier</th><th>After insertion</th><th>Rotation</th><th>After rebalncing</th></tr>
<tr><td>(vii) jul</td><td>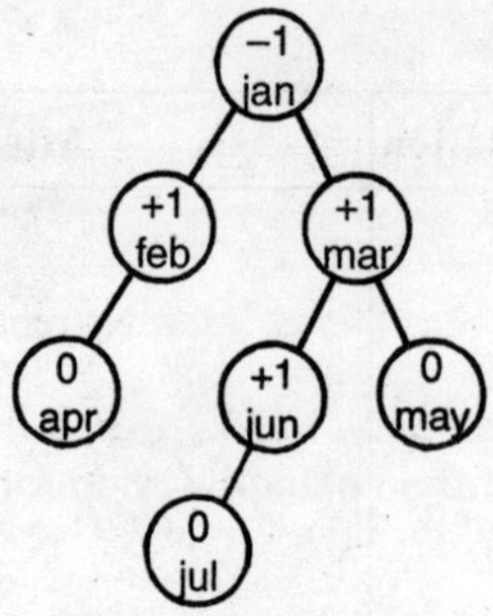
</td><td></td><td>No rebalancing needed</td></tr>
<tr><td>(viii) aug</td><td>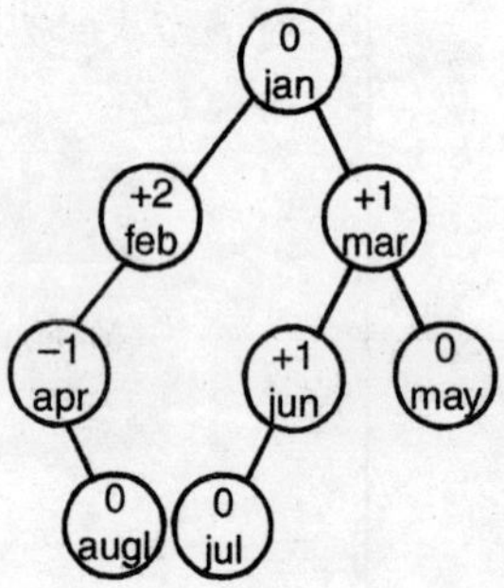
</td><td></td><td>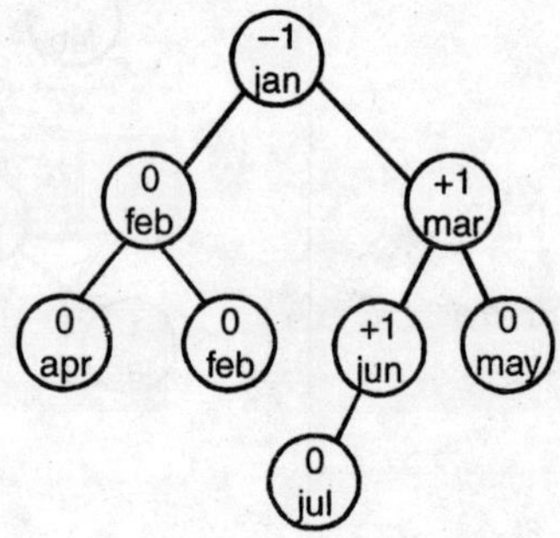
</td></tr>
<tr><td>(ix) sep</td><td>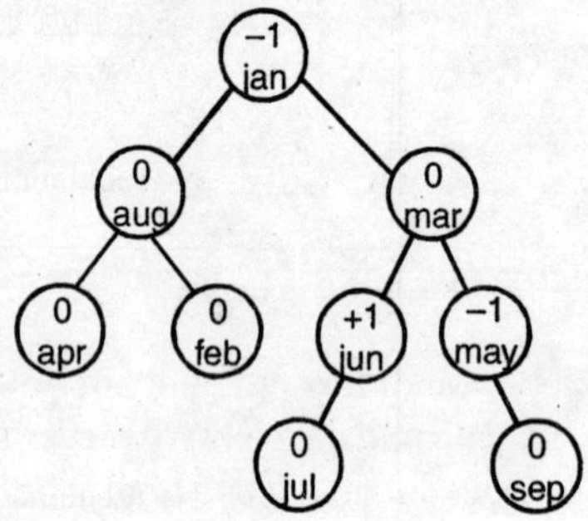
</td><td></td><td>No rebalancing needed</td></tr>
<tr><td>(x) oct</td><td>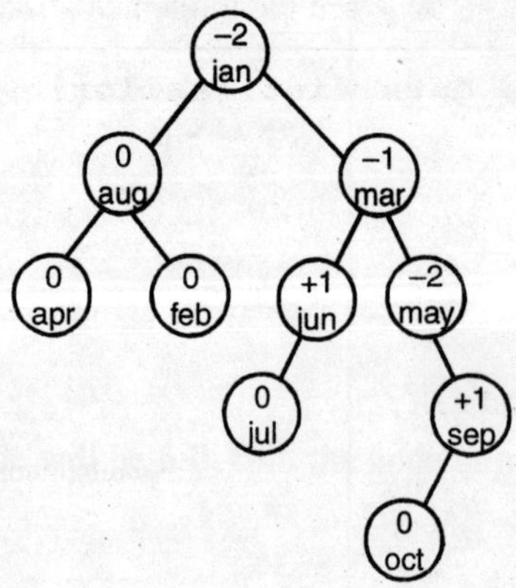
</td><td></td><td>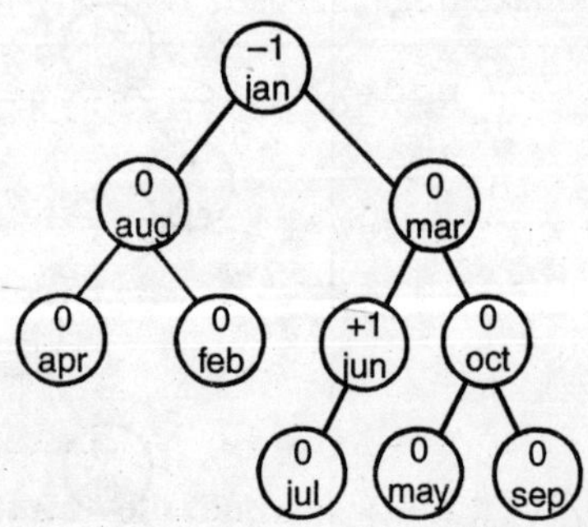
</td></tr>
</table>

New Identifier	After insertion	Rotation	After rebalncing
(*xi*) nov	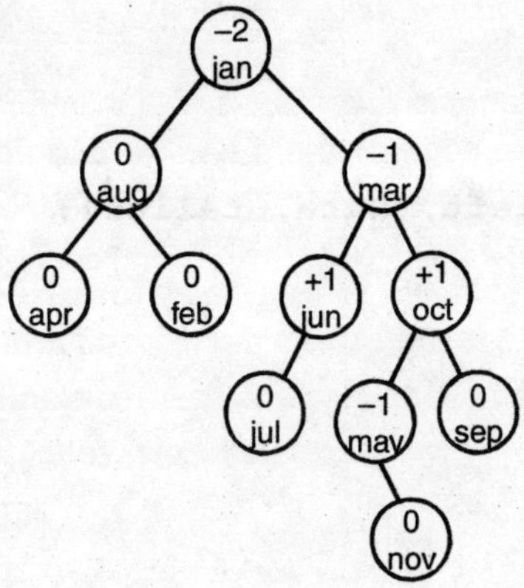		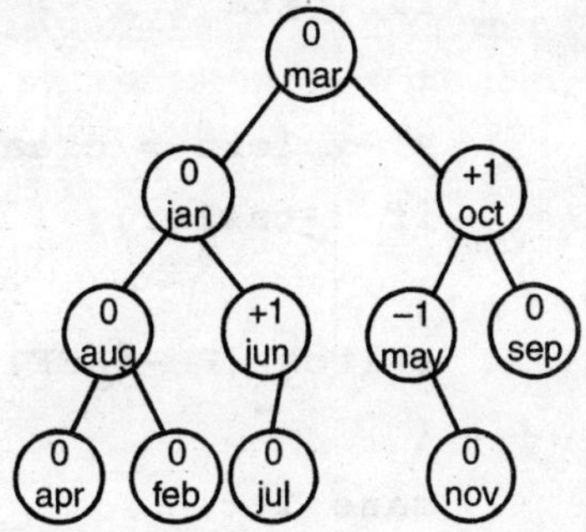
(*xii*) dec	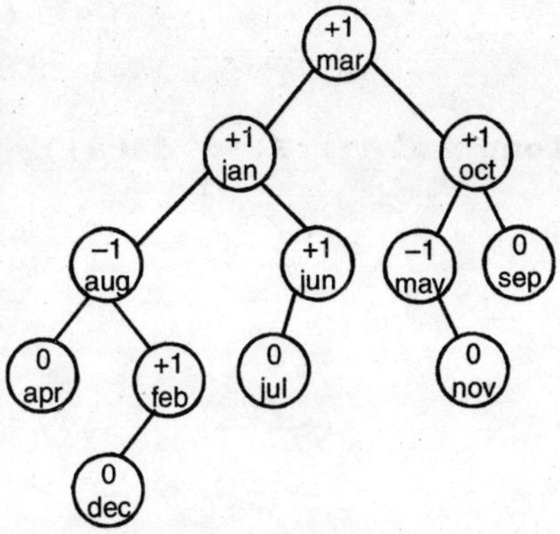		No rebalancing needed

10.8.2. Insertion in AVL Tree

The insertion procedure is similar to insertion in binary search tree but here after every insertion we have to check whether insertion has caused AVL tree to become unbalanced. Whenever the tree becomes unbalanced we have to check any of the 4 conditions as discussed above and apply the suitable rotation to make tree balanced. Manually we have seen this, but here we see the implementation in C.

```
node * createtree (node *R, int data, int *taller)
{
node *P1, *p2;
if (R = = NULL)
{
R = (node *) malloc (sizeof (node));
R -> data = data;
R -> left =  NULL;
R -> right = NULL;
R -> BF = 0;
*taller = YES;
```

```
return (R);
}
if (data < R -> data)
{
R —> left = createtree (R -> left, data, taller);
if (*taller);
{
switch (R -> BF)
{
case 1:
P1= R -> left;
if (P1 —> BF = = 1)
{
printf ("\nRight rotation along %d. ", R -> data);
R -> left = P1 -> right;
P1 -> right = R;
R->BF=0;
R=P1;
}
else
{
printf ("\n RL rotation, Left rotation along %d",P1 -> data);
P2 = P1 -> right;
P1 -> right = P2 -> left;
priritf ("\nthen Right rotation along %d. \n", R-> data);
P2 -> left = P1;
R -> left = P2 -> right;
P2 -> right = R
if (P2 -> BF = = 1)
R -> BF = -1;
else
R -> BF = 0;
if (P2 -> BF = = —1)
P1 -> BF = 1;
else
P1->BF= 0;
```

```
R = P2;
}
R->BF= 0;
*taller = NO;
break;
case 0:
R -> BF = 1;
break;
case -1:
R -> BF = 0;
*taller = NO;
}
}
}
if (data > R -> data)
{
R -> right = createtree (R -> right, data, taller);
if (*taller);
{
switch (R -> BF)
{
case 1:
R -> BF = 0;
*taller = NO;
break;
case 0:
R -> BF = -1;
break;
case -1:
P1 = R -> right;
if (P1 -> BF = -1)
{
printf ("\nLeft rotation along %d. ", R -> data);
R -> right = P1 -> left;
P1 -> left =R;
R -> BF = 0;
R = P1;
}
```

```
else
{
printf ("\nLR rotation, Right rotation along %d",P1 -> data);
P2 = P1 -> left;
P1 -> left = P2 -> right;
P2 -> right = P1;
printf ("\nthen Left rotation along %d. \n", R -> data);
R -> right = P2 -> left;
P2 ->left = R;
if (P2 -> BF = = -1)
R -> BF = 1;
else
R -> BF = 0;
if (P2 -> BF = = 1)
P1 -> BF = 1;
else
P1 -> BF = 0;
R = P2;
}
R -> BF = 0;
*taller = NO;
}
}
}
return (R);
}
void display (node *R)
{
if (R! = NULL)
{
printf ("%d\t", R -> data);
display (R -> left);
display (R -> right);
}
}
```

The function createtree that takes three arguments: a pointer of type node, an int pameter that represents data for the node and third parameter is an integer for checking the subtree has become taller after the insertion of data. The function returns the pointer to the root after insertion of data.

Initially when createtree is called for the first time with data 25, root wifi be NULL. A new node is created and root points to its, its various fields are filled; *taller becomes YES and BF is set to 0. When createtree function is called with data 20, following condition holds true:

```
if (data < R -> data)
```

Now function createtree is called with parameters R->left, data and taller. During the recursive call, first if condition will be satisfied (R->left is NULL), node is created and pointer R is returned back. When function returns to the callee function, recursion stops and R->left points to node with data 20. As *taller is 1, next if condition wifi be satisfied. This we are doing as after addition of node we have to set BF of root to some new value depending upon where the insertion took place. The BF of root was 0, insertion took place left of root, so BF of root wifi be changed to 1. This we have done in second case condition. Similarly elements 30, 20 and 15 will be added and their BF wifi be set. Now consider insertion of data 10. It will be added to the left of 15. R wifi be pointing to 25. BF of 25 is 1. The following code will execute:

```
P1 = R -> left;
if (P1 -> BF = = 1)
{
printf ("\nRight rotation along %d.", R -> data);
```

P1 points to node 20. As BF of P1 is 1, if condition will be true; left of R (left) will point to right of P1 (P1 -> right) and right of P1 will point to R(1 right = R). See the Fig. 10.45.

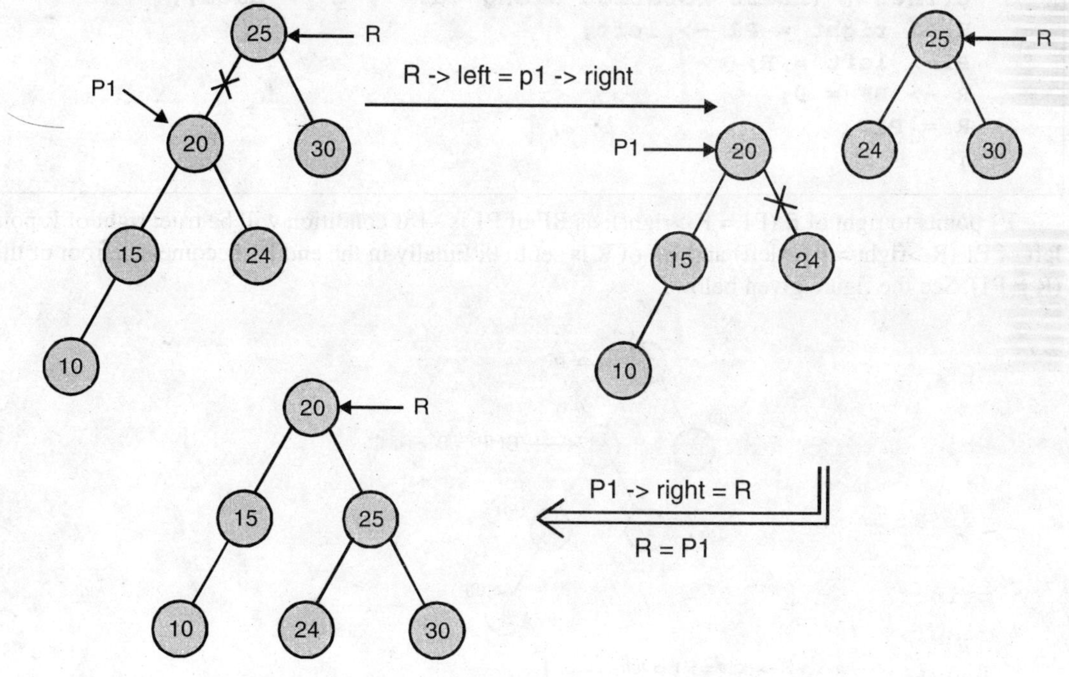

Fig. 10.45.

Note after the execution of the above code, after the else part, following code executes:

```
R -> BF = 0;
*taller = NO;
```

You can verify this from the figure given below, that it is indeed true.

To understand the Left rotation, consider the following figure :

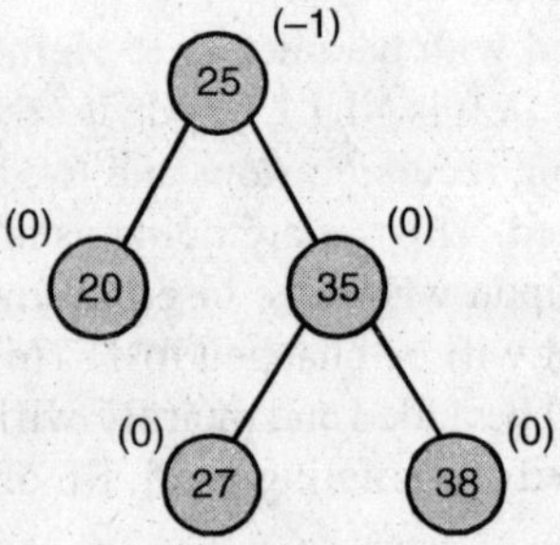

Fig. 10.46.

Suppose we want to insert nod2 with data 40. This will be added as right of node 38. Pointer R will be pointing to node 25; BF of R is -1; as shown below in a code

```
P1 = R -> right;
if (P1 -> BF = = -1)
{
printf ("\nLeft rotation along %d. ", R —> data);
R —> right = P1 —> left;
P1-> left = R;
R -> BF = 0;
R = P1;
}
```

P1 points to right of R (P1 = R->right); as BF of P1 is –1 if condition will be true; right of R points to left of P1 (R->right = P1->left) and BF of R is set to 0; Finally in the end P1 becomes the root of the tree (R = P1). See the figure given below:

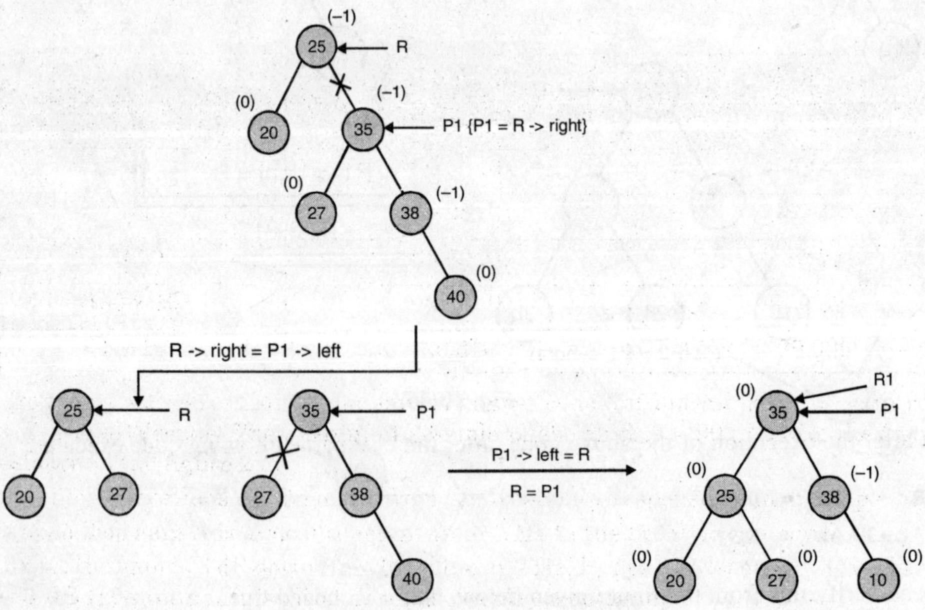

Fig. 10.47.

Note after the execution of the above code, after the end of switch-case block, following code executes :

```
R -> BF = 0;
*taller = NO;
```

Out of LR and RL rotation we discuss just RL rotation. After the discussion blows, it will be easy for you to figure out how code for LR works. Consider the following figure

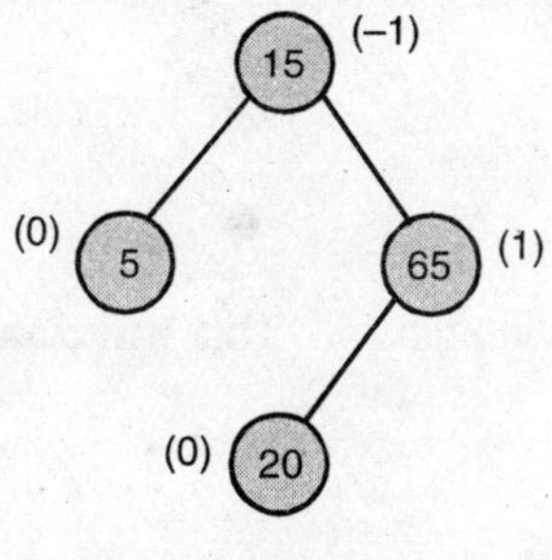

Fig. 10.48.

Suppose we want to insert node with data 25. It will be added as right of node 20. As node added to the right of node 20 that is to the left of node 65, we have RL rotation here shown below in a code.

```
P1 = P.-> left;
P2 = P1 -> right;
P1 -> right = P2 -> left;
P2 -> left = P1;
R -> left = P2 -> right;
P2 -> right  = R;
```

All these operations are shown below in the figure:

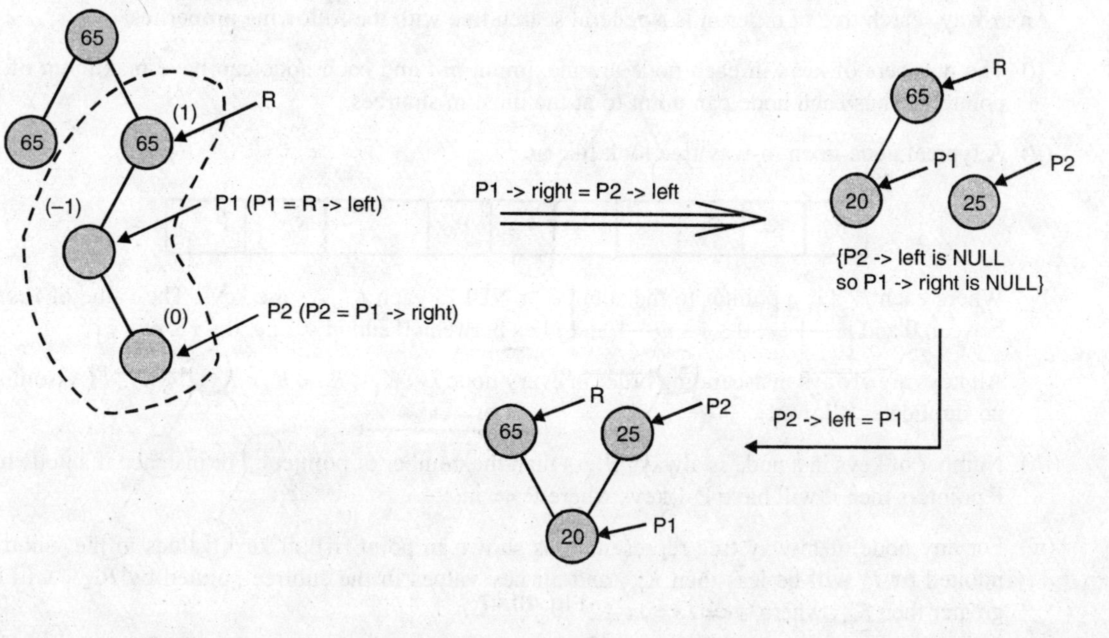

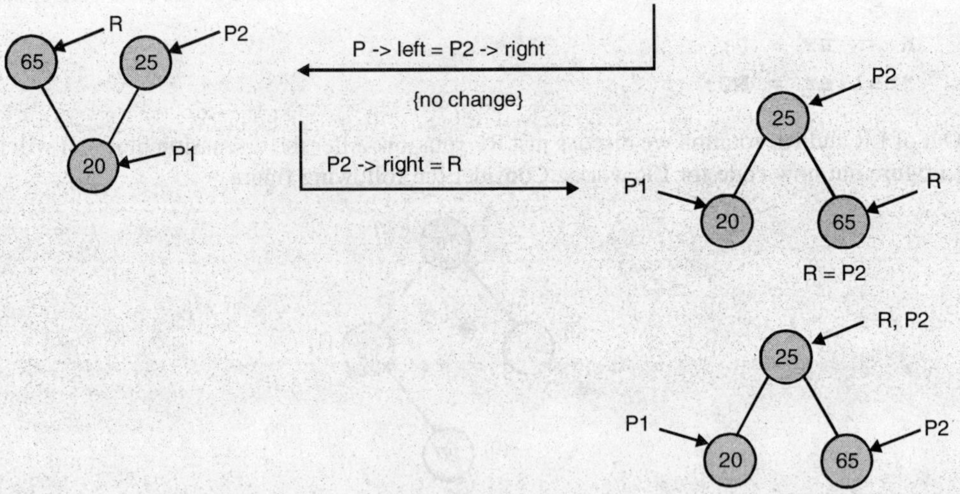

Fig. 10.49.

10.9. MULTIWAY (M-WAY) SEARCH TREE

M-way search tree are similar to binary search tree with the difference that instead of one key and two pointers (left and right) in each node an m-way tree can have m pointers and m-1 key in one node at the most. M-way trees are the general search trees where each node can accommodate more then one key and more than 2 pointers. This is governing by the order m of the tree. The advantage of this type of tree is that many keys can be searched in a single node instead of going down the tree. Further they are the trees which are used for external searching *i.e.* searching for the persistent data (residing onto the secondary device) instead of primary memory (internal searching). The basic definition of the m-way tree is as follows:

An m-way search tree of order m is a general search tree with the following properties:

(*i*) The numbers of keys in each node are maximum *m-i* and each node can have maximum of m pointers. Thus each node can point to at the most m subtrees.

(*ii*) A typical node in an m-way tree look like as :

P_0	K_1	P_1	K_2	P_2	K_3	P_3	......	K_{m-1}	P_{m-1}

Where each P_i's is a pointer to the subtree or NULL, each K_{j+1}'s are keys. The value of i is in beween 0 and $m - 1$ *i.e.* $0 \leq i \leq m - 1$ and j lies between 0 and $m - 2$ *i.e.* $0 \leq j \leq m - 2$

All keys are always in ascending order in every node *i.e.* $K_1 < K_2 < K_3 < K_4 ... < K_{m-1}$ (Assuming no duplicates allowed).

(*iii*) Number of keys in a node is always 1 less than the number of pointers. For instance if a node has P pointers then it will have P-1 keys where P <= m.

(*iv*) For any node of m-way tree represented as shown in point (ii) all keys values in the subtree pointed by P_i will be less then K_{i+1} and all key values in the subtree pointed by P_{i+1} will be greater then K_{i+1} where $0 <= i <= m - 2$

After studying properties of the m-way tree, let's check its properties through figures given below. In the figure there are 3 m-way trees.

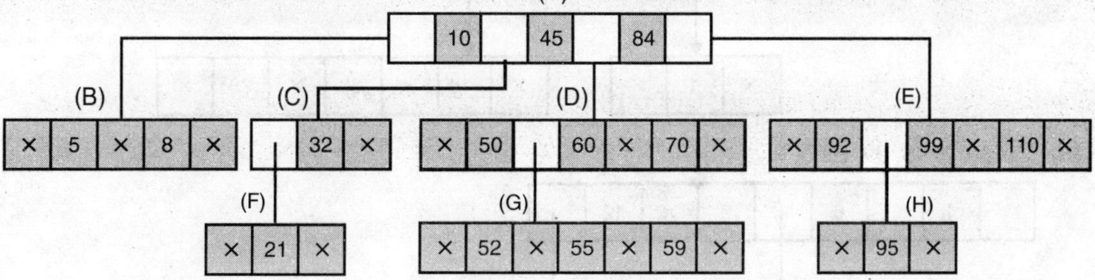

(a) M-way tree of order-4

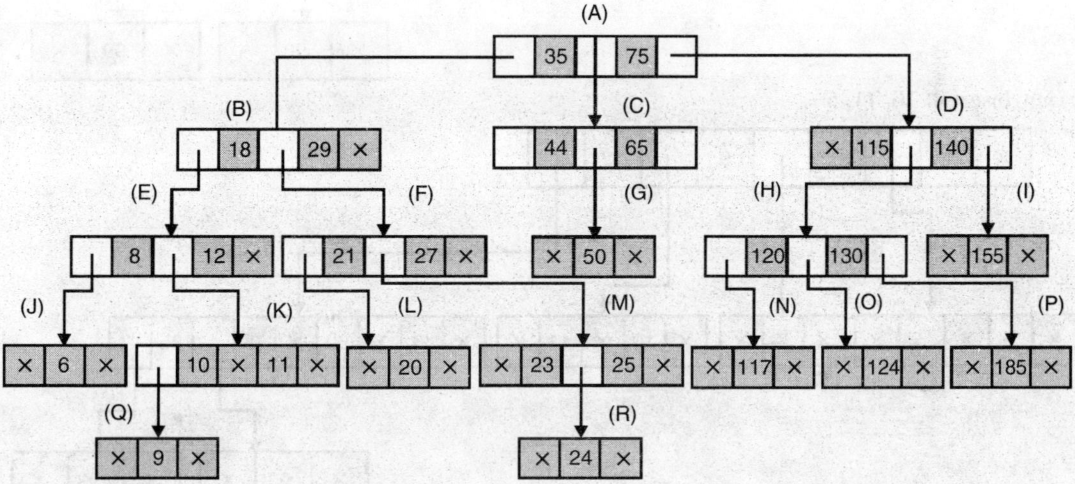

(b) Top down M-way tree of order-3

Fig. 10.50.

10.9.1. Creation/Insertion in an M-way Tree

Creation of an m-way tree is very simple as creation of binary search tree. To

understand this considers the creation of two m-way search tree of order 4 and 5. As long as a node has not got all the elements as specified by the order of the tree we keep on adding the element in that node. As soon as node becomes full, we create a new node appropriately pointed by any of the pointer of the full node. Thus during insertion the m-way tree grows downwards.

(*i*) Insert 12, 6, 9, 1

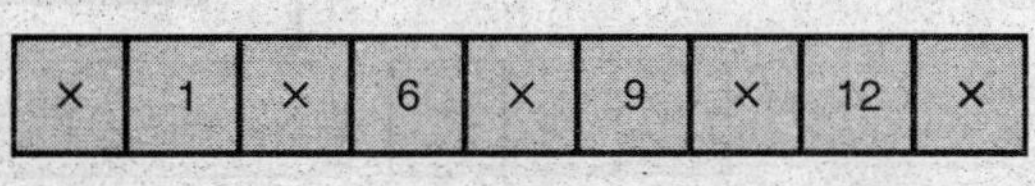

(*ii*) Inser 7, 14, 50, 78

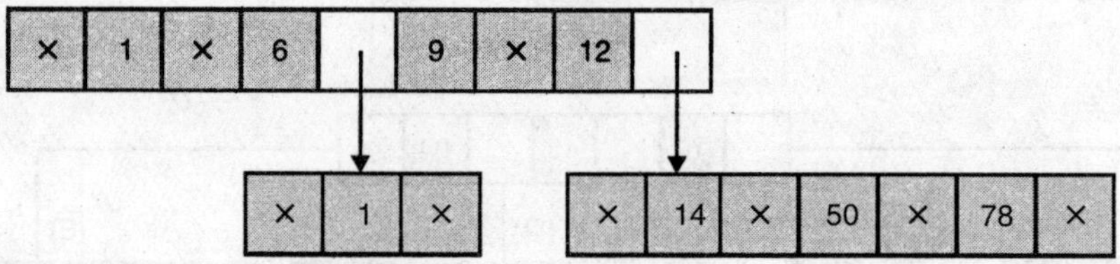

(*iii*) Insert 2, 95, 99, 93

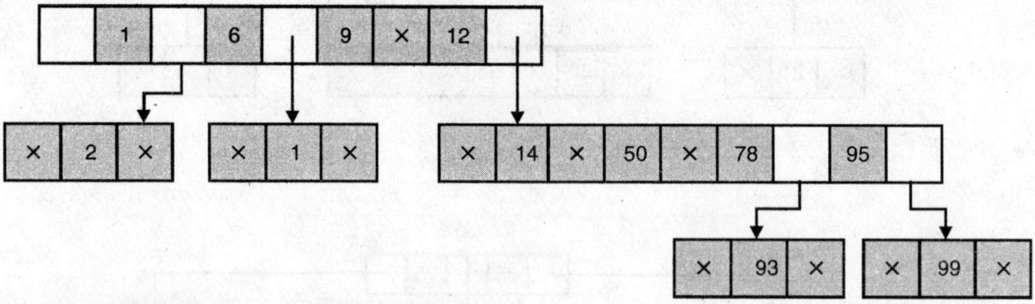

(*iv*) Insert 8, 10, 11, 5

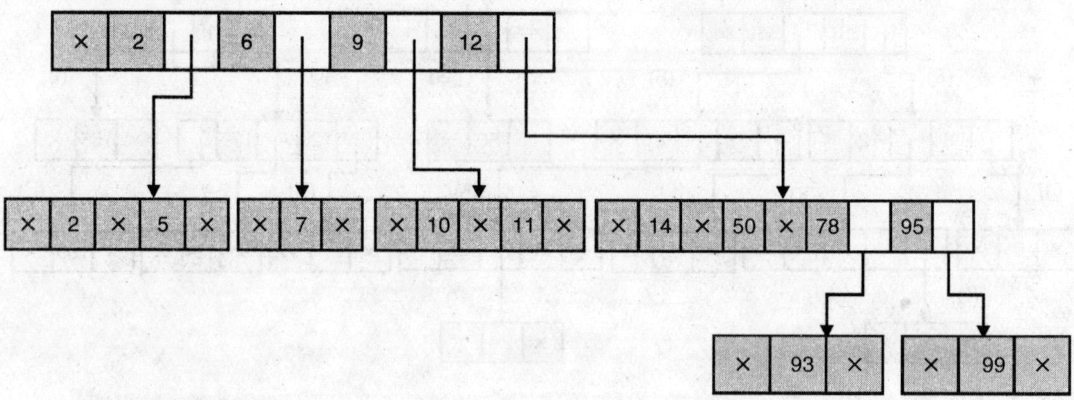

(*v*) Insert 13, 40, 43, 48

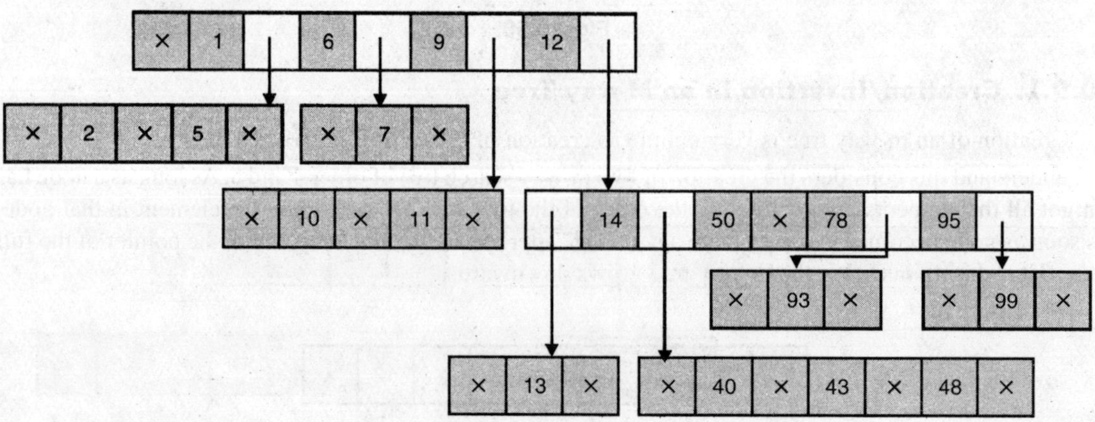

Fig. 10.51. Creation of M-way tree of order-5

1. Insert 12, 6

2. Insert 9

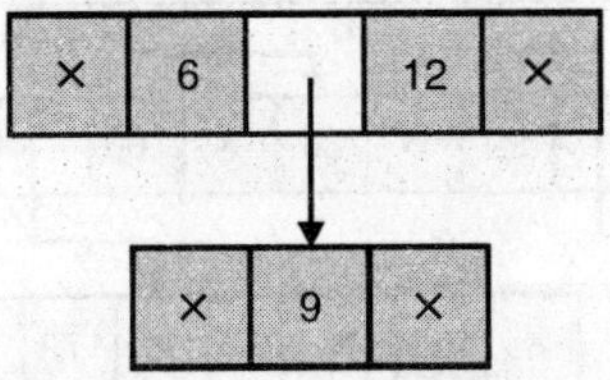

3. Insert 1, 7

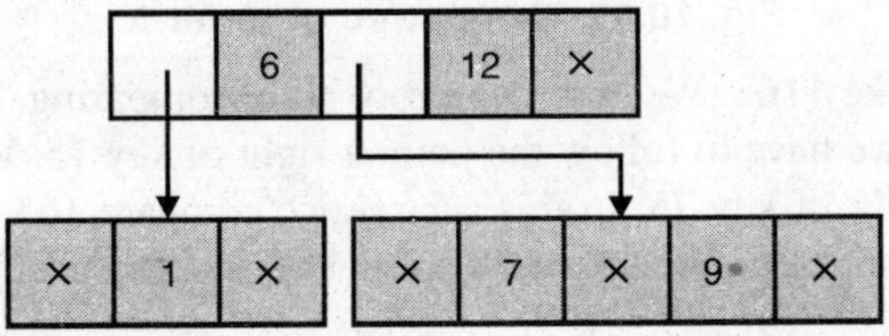

4. Insert 14, 50

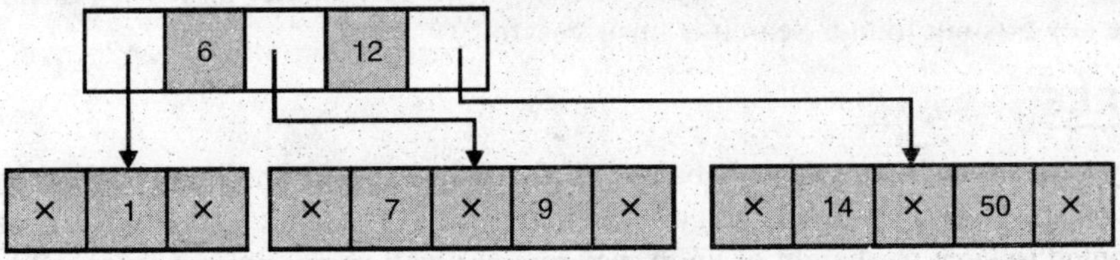

5. Insert 78, 2

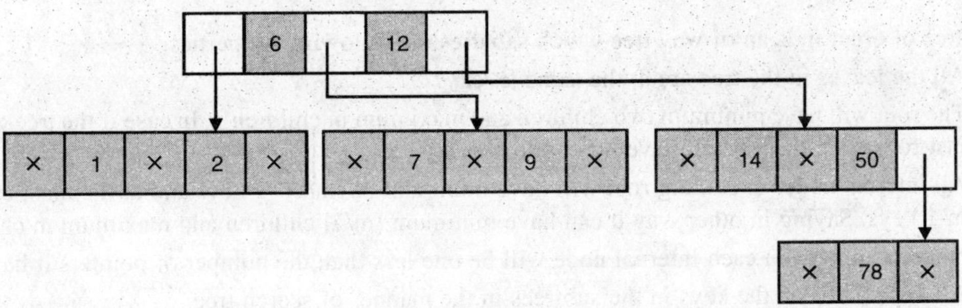

6. Insert 95, 8

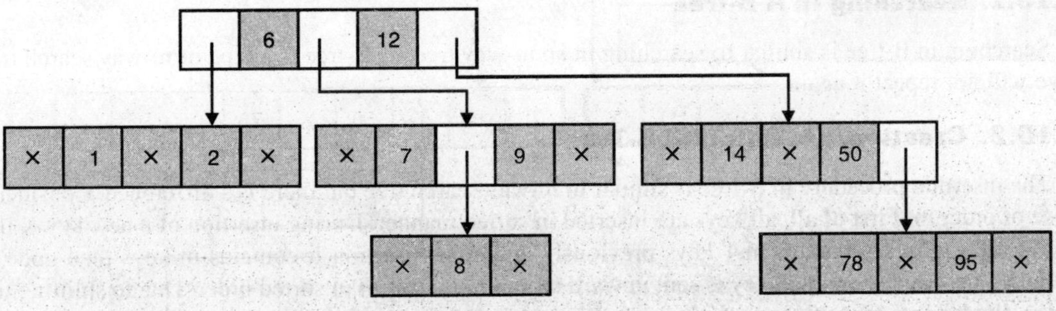

Fig. 10.52. Creation of M-way tree of order 3

Searching in an m-Way Tree

Searching in m-way tree is again similar to searching in binary search tree. To understand this consider the figure given below. It is an m-way tree of order 5.

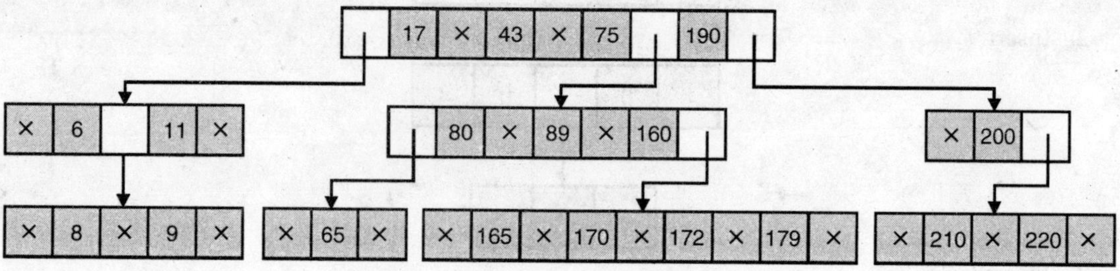

Fig. 10.53. M-way tree of order 5

Consider searching for a key 165. We start from root node comparing 165 to 17, 43 and 75. At this stage, we come to know that we have to follow the pointer right of key 75. We then search for the key in subtree pointed by pointer right of key 75. In this subtree we compare 165 with 70, 89 and finally 160. This causes us to follow the subtree pointed by right pointer of 160. In the subtree pointed by this pointer, key 165 is found.

Take another example and search for key 5. We follow the left pointer of key 17, then left pointer of key 6 and there key 5 is not found. Search is unsuccessful.

10.10. B TREE

B tree is type of m-way tree which is balanced (the B in B-tree). The tree is called balanced as all leaves are at the same level. Due to this height of B-tree is usua1 1y small. Actually B-tree is result of the fact that we wanted to keep the height of an m-way tree as small as possible so that searching would be faster.

A B tree of order m is an rn-way tree which satisfies the following properties:

(*i*) All the leaves in the tree are at the same level

(*ii*) The root will have minimum two children and maximum m children's. In case if the tree contains just root node then it will have no children at all.

(*iii*) All internal nodes excluding root will have minimum of [m/2] -1 keys and at the most can have m-1 keys. Saying in other way it can have minimum [m/2] children and maximum m childrens.

(*iv*) Number of keys in each internal node will be one less then the number of pointers it have.

(*v*) All keys partition the keys in the subtrees in the manner of search tree.

10.10.1. Searching In A B-tree

Searching in B-tree is similar to searching in an m-way tree, as B-tree is a type of m-way search tree so we will not repeat it again.

10.10.2. Creation/ Insertion of B-Tree

The insertion procedure in B-tree is similar to m-way search tree but there is a difference. Consider a B-tree of order m. First of all, all keys are inserted in sorted manner. During insertion of a new key K if a node is full *i.e.* node contains m-1 keys previously and after insertion it contains m keys then node is splitted at median key and that key is sent upward to join to its parent in sorted order. Due to splitting the node is divided into two sub nodes pointed by left and right pointer of median key. In parent node is also become full then the same procedure is repeated which in turn may causes splitting of root node and formation of new root.

Every time the splitting is done, tree grows upward towards root which is in contrast with the binary search tree and m-way search that grows downwards towards leaves.

To understand the insertion procedure in action, follow the examples given below. For the simplicity we have not shown the pointer fields.

B-Tree of Order 5 :

A. 1, 7, 6, 2

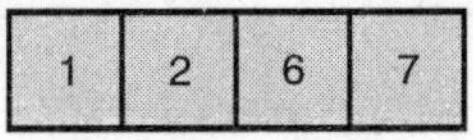

B. 11, node will be full, split the node at median 6

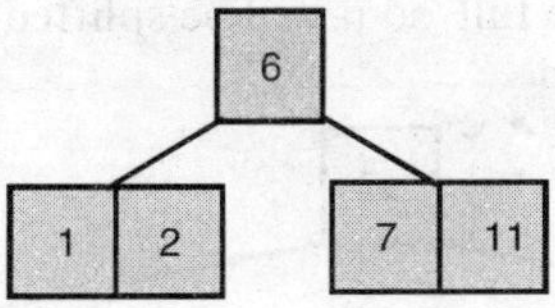

C. 4, 8, 13

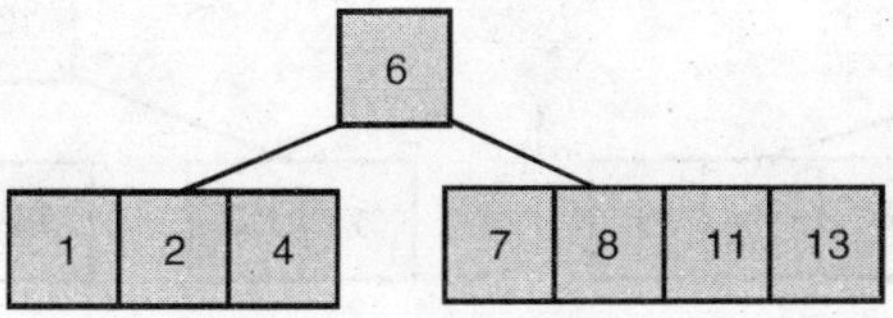

D. 10, node will be full, split the node at median 10. It will be joined with 6 in root node.

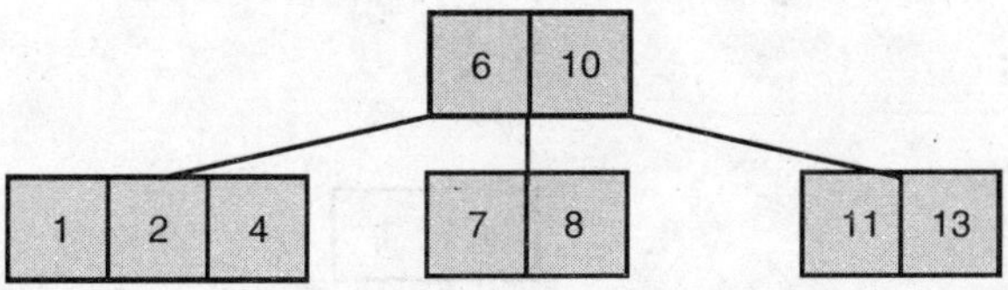

E. 5, 19, 9, 18

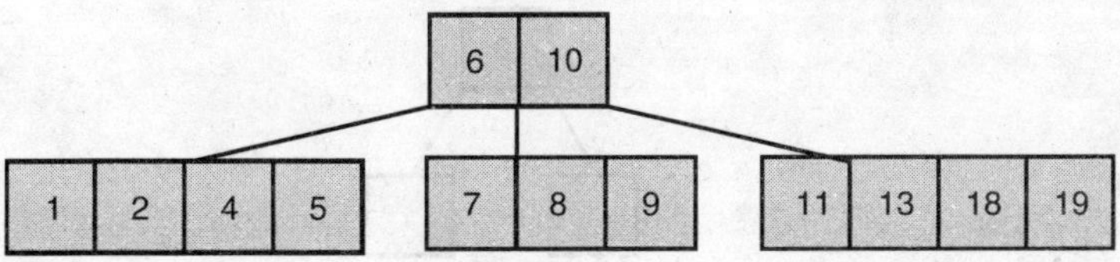

F. 24, node will be full, split the node at median 18, it will be joined with root node.

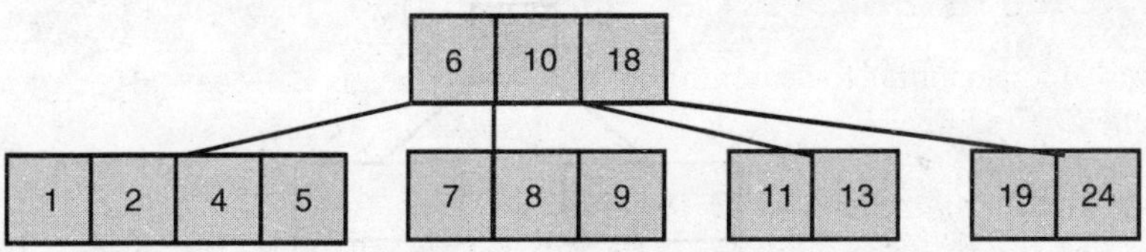

G. 3, node will be full split the node at median 3. It will be joined with root node.

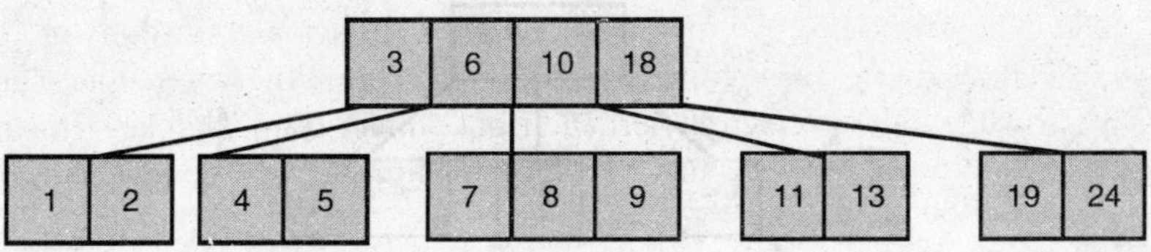

H. 12, 14, 20, 21

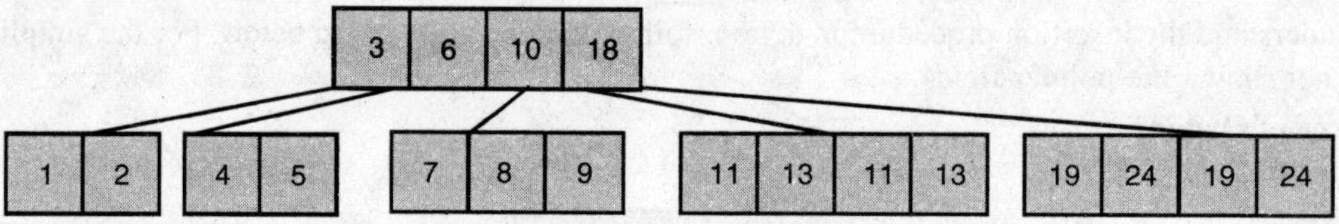

I. 16, node will full, split the node at median 13. It will be joined with root node. But on joining 13 to root node, the root node will also be full, so it will be spliffed up at median 10 and new root will be 10.

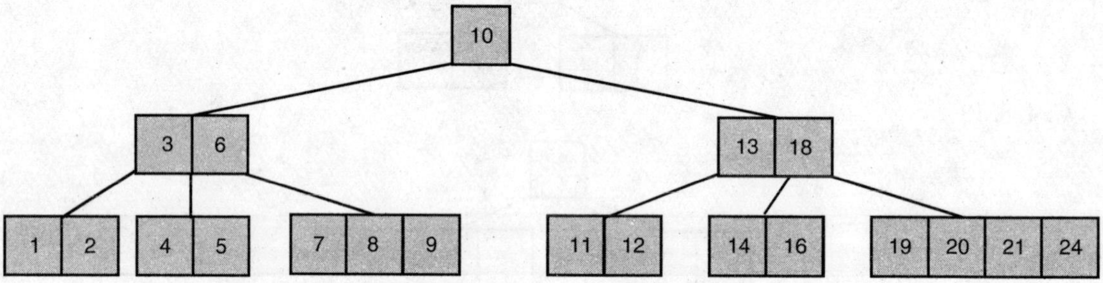

Fig. 10.54. B-Tree insertion order 5.

B-Tree of Order 3 :

A.

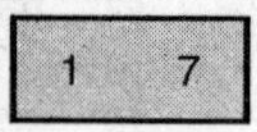

B. 6, node will be full, split the ode at median 6, root will be 6.

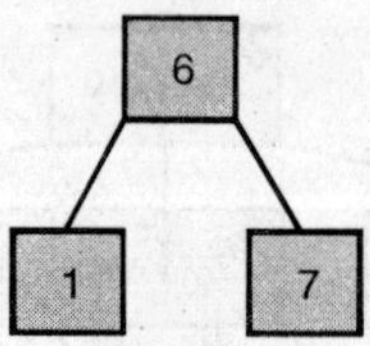

C. 2, 11

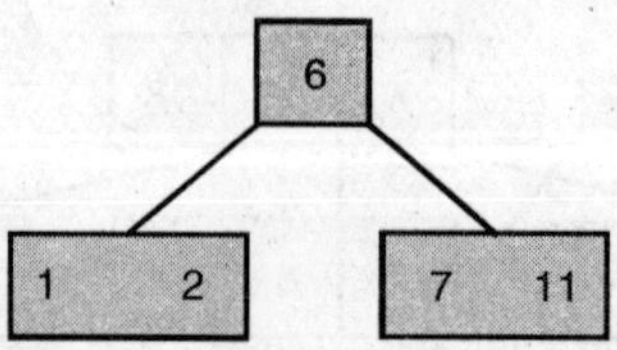

D. 4, node will be full, split the node at median 2, it will be joined with root

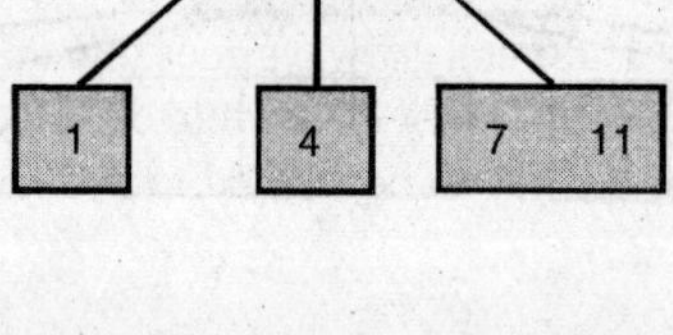

E. 8, double split, first at median 8 and second at median 6.

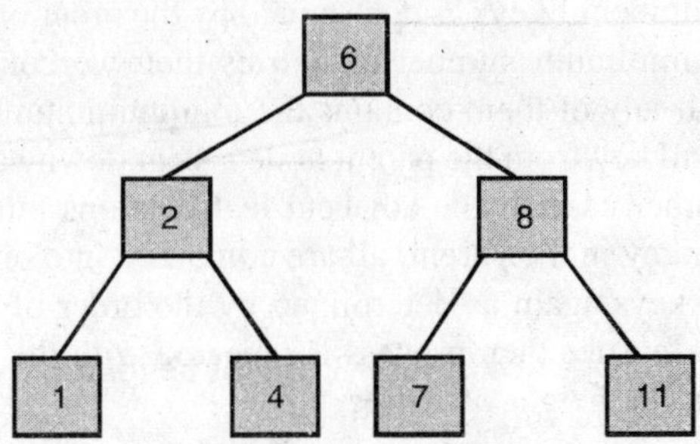

F. Insert 5, 15

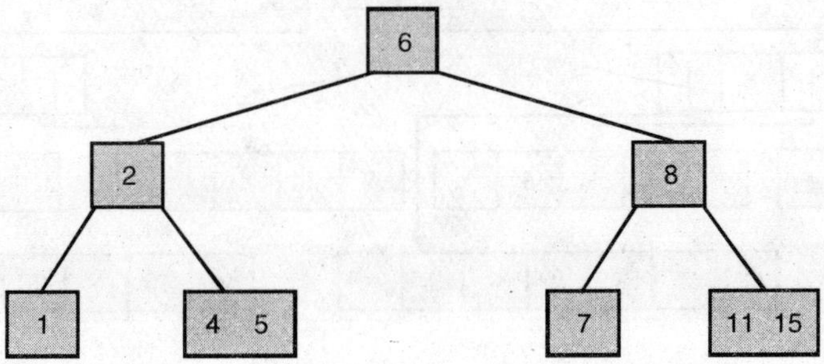

G. Insert 3 (split at median 4)

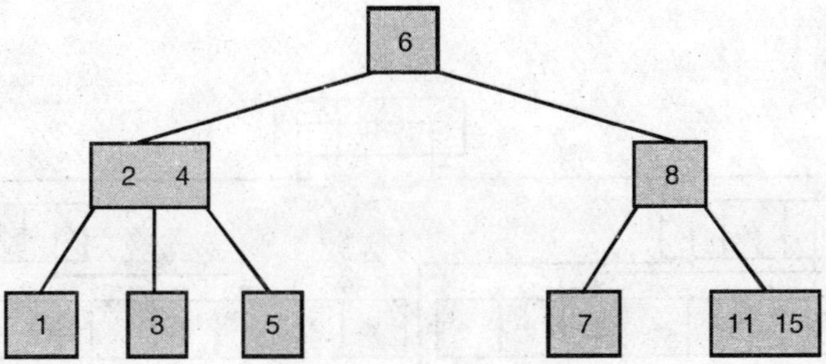

H. Insert 12, split at 12

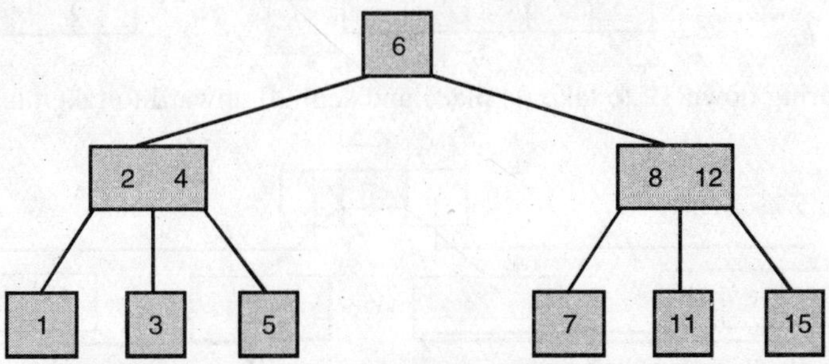

Fig. 10.55. B-tree insertion order-3

10.10.3. Deletion In B-tree

We have seen that during insertion a new entry always goes into the leaf node, similarly we wish to remove the key from the leaf node. But what if the entry to be deleted is not in the leaf node, then we can make immediate predecessor or successor of the key to be deleted to occupies the space of the deleted key.

Now during deletion we have to take into consideration that after deletion of specific key, the leaf node contains the minimum number of keys as determined by the order of the B-tree. In case after deletion the leaf node does not have minimum number of entries then we consider two immediately adjacent leaves with the same parent. If any of them contains the minimum number of keys then one can be sent upward into the parent node and key from the parent node is sent downwards into the node where deletion is taking place. Consider another case: If the adjacent leaf contains minimum number of keys then the both of the leaves and median key in the parent, all are combined into one new node. This new node will contain maximum number of keys again as determined by the order of the B-tree. If due to this action, parent node contains only few entries then process is repeated with the nodes in upward direction. This ultimately causes reduction in the height of B-tree.

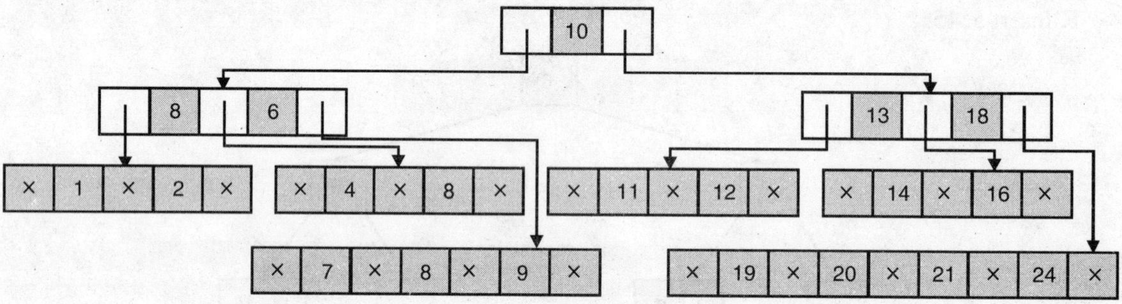

(*a*)

(*i*) Delete 8, simple deletion, no further change in the tree; delete 18, simply move 19 upward to parent node.

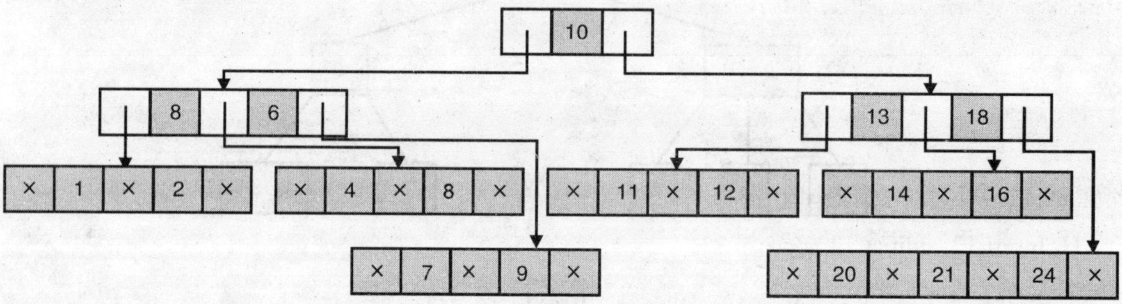

(*ii*) Delete 16, bring down 19 to take its place and sent 20 upward to take place of 19.

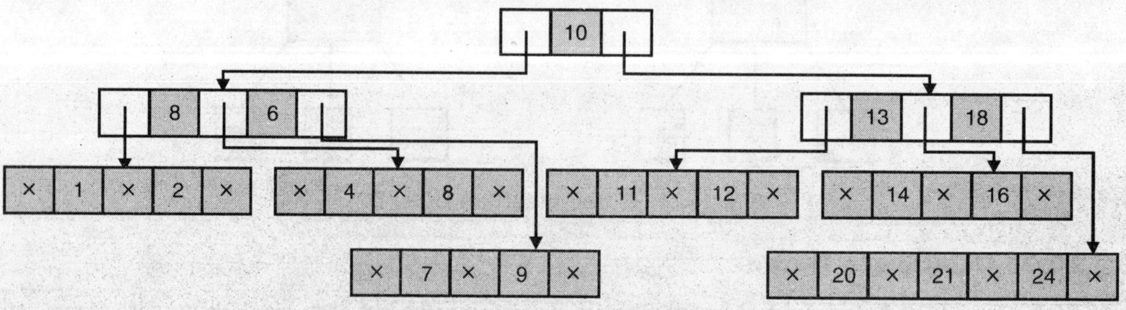

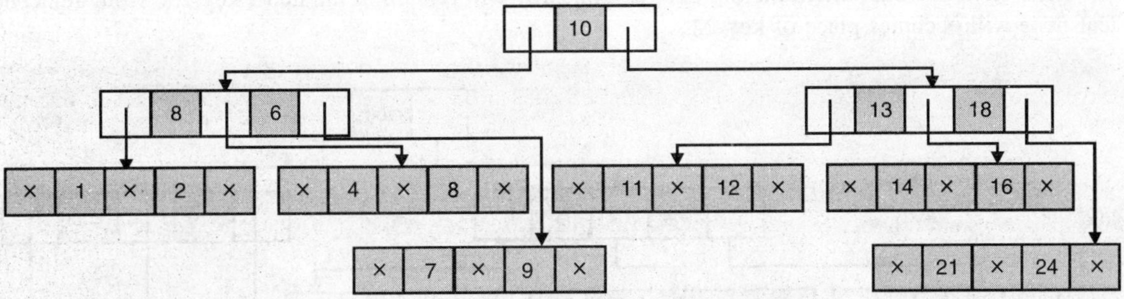

(*iii*) Delete 4, first combine at down side with keys (1, 2, 3, 5) then on the up side combine (6, 10, 13, 20)

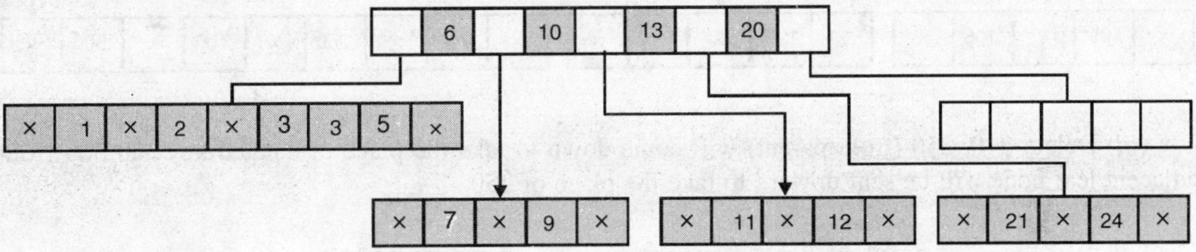

Fig. 10.56. B-tree insertion order-3.

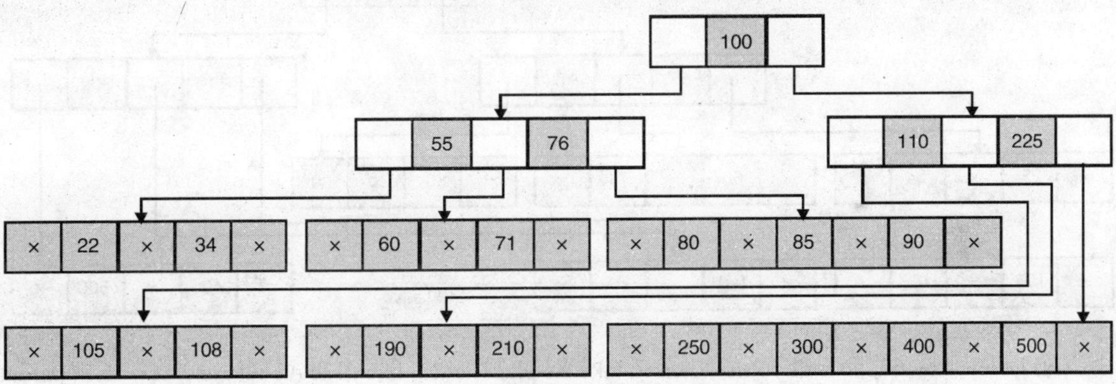

(*a*) B-Tree of Order 5

(*i*) Delete key 85, can be easily deleted as leaf node containing 85 has 3 keys (one more than minimum no. of keys)

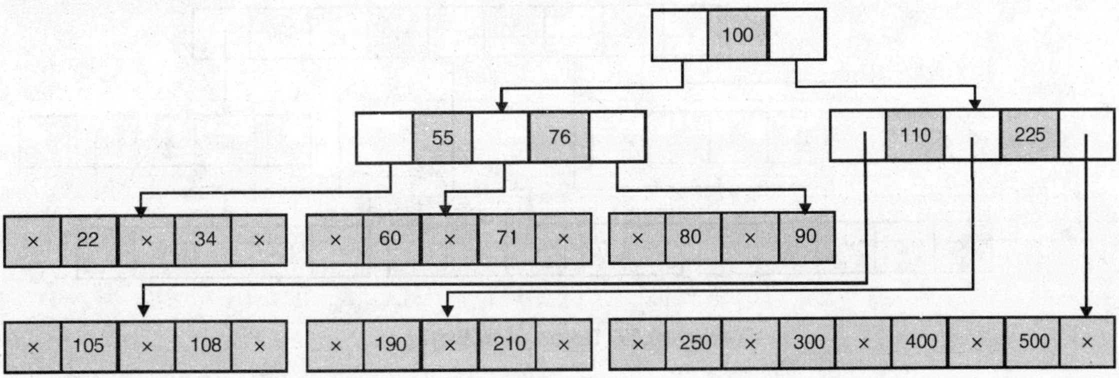

(*ii*) Delete key 225, after deleting 225, just one key will remain in the node, key 250 from adjacent leaf node will occupies place of key 225.

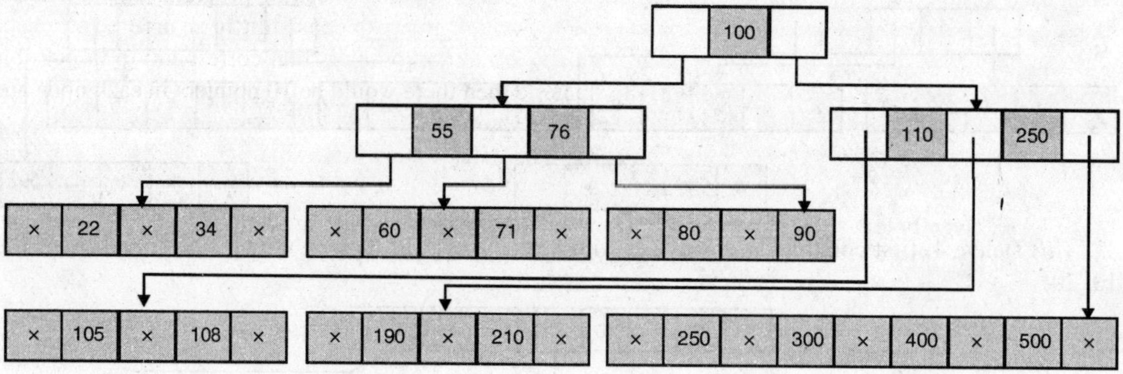

(*iii*) Delete 210, 250 (from parent) will come down to take the place of deleted key and 300 from adjacent leaf node will be sent upward to take the place of 250

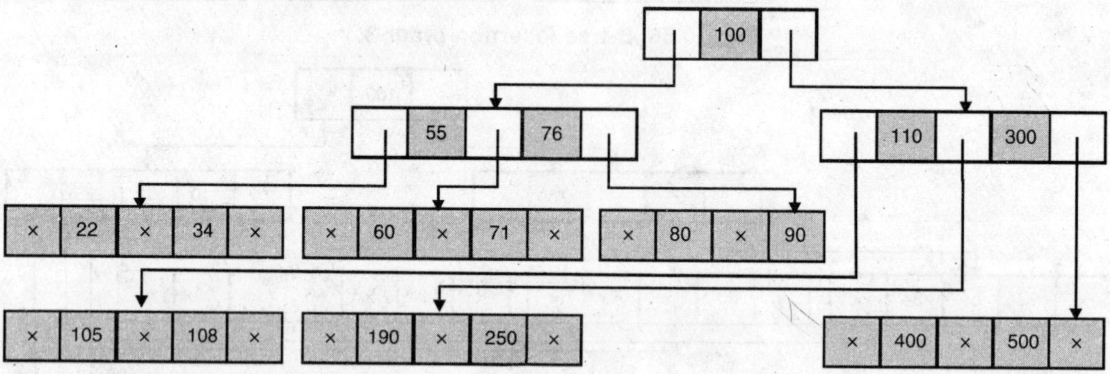

(*iv*) Delete key 60, 55 will come down to take its place, but this will leave just one key 76 in parent node, so a new node will be formed combining keys (76, 100, 110, 300). This new node will be new root. On the down side a new node consisting of keys (22, 34, 55, 71) will be formed.

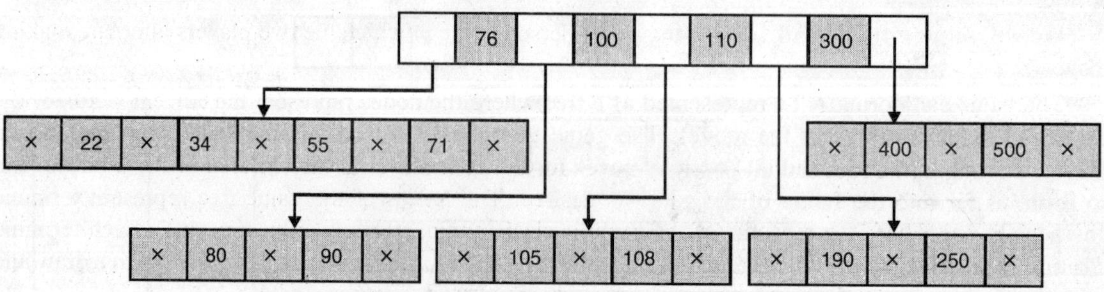

Fig. 10.57. Deletion in B-tree

10.11. TRIE

A trie (from retrieval), is a multi-way tree structure useful for storing string over an alphabet. It has been used to store large dictionaries of English (say) words in spelling-checking programs and in natural-language "understanding" programs. A trie of order m is either empty or consisting of an ordered sequence of exactly m tries of order m. In a trie each node can contain m pointers that correspond to m possible symbols in each of the node. If the symbols are numeric then there would be 10 pointers in each node and in case of alphabets there would be 26 pointers. Each pointer in a node is associated with a particular symbol value on the basis of its position in the node from lowest symbol value to highest symbol value. Tries are also known as lexicographic search trees.

Figure given below shows an example of trie created with the following strings:

an, ant, all, allot, alloy, aloe, are, ate, be

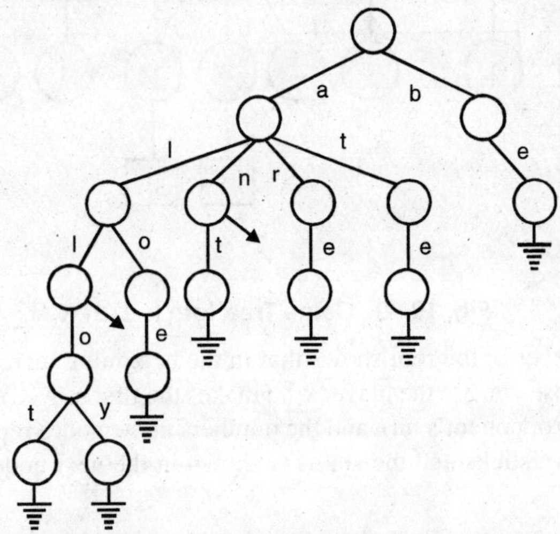

Fig. 10.58. An example of trie.

The idea is that all strings sharing a common stem or prefix hang off a common node. When the strings are words over {a. z}, a node has at most 7 children - one for each letter plus a terminator.

The elements in a string can be recovered in a scan from the root to the leaf that ends a string. All strings in the trie can be recovered by a depth-first scan of the tree. The height of a trie is the length of the longest key in the trie.

10.12. GAME TREE

We will set up a framework for playing a two-person game in which the two players alternate making moves.

The game can normally be represented as a tree where the nodes represent the current status of the game and the arcs represent the moves. The game tree consists of all possible moves for the current players starting at the root and all possible moves for the next player as the children of these nodes, and so forth, as far into the future of the game as desired. The leaves of the game tree represent terminal positions as one where the outcome of the game is clear (a win, a loss, a draw, a payoff). Each terminal position has a score. High scores are good. For example, we may associate 1 with a win, 0 with a draw and –1 with a loss.

Example : Game of NIM

In this game, several piles of sticks are given. We represent the configuration of the piles by a monotone sequence of integers, such as (1, 3, 5, 7) or (2, 2, 3, 9, 110). A player may remove, in one turn, any number of sticks from one pile of his/her choice. Thus, (1, 3, 5, 7) would become (1, 1, 3, 5) if the player were to remove 6 sticks from the last pile. The player who takes the last stick loses. The NIM game (1, 2, 2) can be presented by Fig. 10.59.

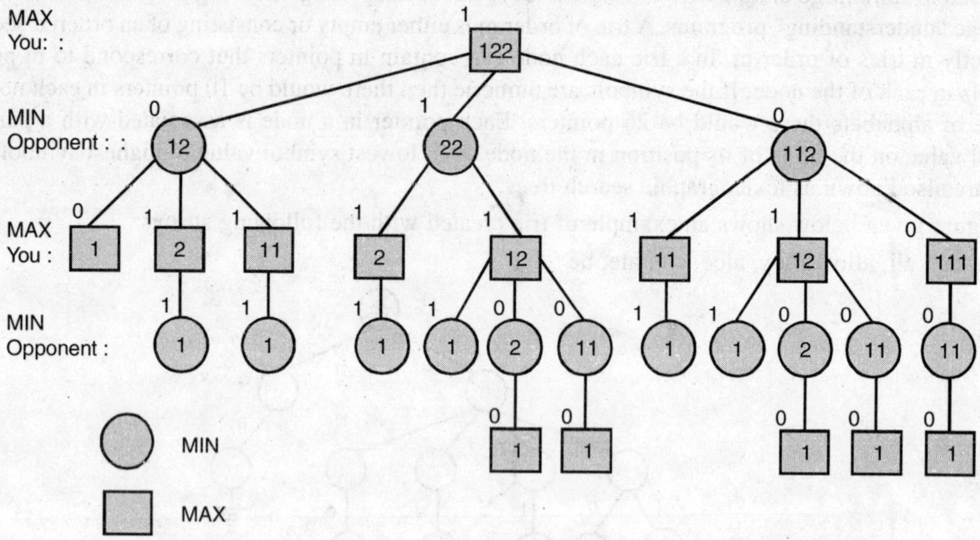

Fig. 10.59. Game Tree for (1, 2, 2) NIM.

In Fig. 10.59, the number in the root shows that in the beginning there are five sticks which consists of three sets, 1, 2, 2. Suppose you are the player who makes the first move. You may take one or two sticks. After your move, it is your opponent's turn and the numbers in the nodes represent the sticks left. Then the opponent moves one or two sticks and the status is shown in the next nodes and so on until there is one stick left.

Now, we can use this game tree to analyze the best possible move. For each player, the best move is to make the opponent lose and make himself/herself win. So, one should make the move to get the **MAX** score and force their opponent to get the **MIN** score. A loss is presented by "0" and a win is presented by "1". The **MAX** nodes represent the position of the current player and the **MIN** nodes represent the position of the opponent. Since the goal of this game is that the player who removes the last stick loses, the scores are assigned to "0" if the leaves are at **MAX** nodes and the scores are assigned to "1" if the leaves are **MIN**. Then we back up the scores to assign the internal nodes from the bottom nodes. At **MAX** nodes, choose the **MAXIMUM** score of the children; at **MIN** nodes, choose the **MINIMUM** score of the children. In this manner, we may compute the scores of the non_leaf nodes from the bottom up. In the example of figure 1, the root node is "1", and thus corresponds to a win for the first player. The first player should pick a child position that corresponds to a "1".

MiniMax Game Tree

In a MiniMax tree, one can view in its entire form, the score values at each of the levels of the tree at any given point during the game. By viewing this tree, a player may be able to forsee which moves are more advantageous and beneficial for themselves. The root of the tree represents the position of the current player, thus, depending on the number of levels that is to be searched, all odd levels represent the first player while the even levels represent the second player.

In a two player game, the first player moves with **MAX** score and the second player moves with **MIN** score. A minimax search is used to determine all possible continuations of the game up to a desired level. A score is originally assigned to the leaf, then by evaluating each possible set of moves, a score is assigned to the upper level by the minimax algorithm. The minimax algorithm performs a preorder traversal and computes the scores on the fly. The same would be obtained by a simple recursive algorithm. The rule is as follows:

```
minmax(u)
  {    //u is the node you want to score
     if u is a leaf return score of u;
           else if  u in a min node
                 for all childern of u: v1, .. vn ;
                        return min{ minmaxv1),..., minmax(vn)}
           else for all childern of u: v1, .. vn ;
                        return max{ minmax(v1),..., minmax(vn)}
     }
```

Note : If we know the scores of all the nodes, it would be futile and boring to play the game because we would know the outcomes already (against a clever opponent). The scores are the same as the outcome of a game that would result in two infinitely clever players would play each other starting from that position.

The complexity of the algorithm is equal to the number of nodes in the tree.

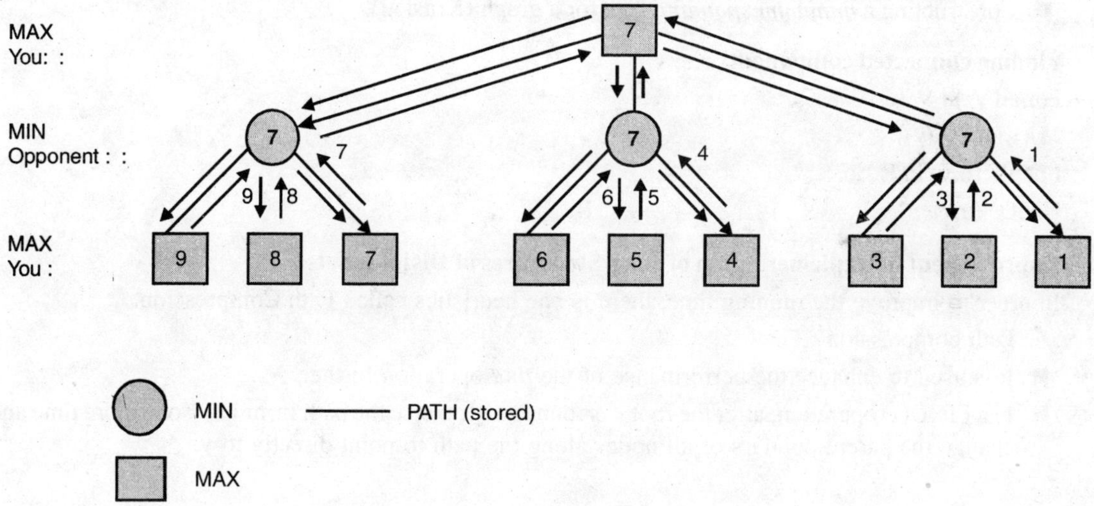

Fig. 10.60. Minimax Search.

Figure 10.60 shows how the minimax algorithm works. There is a game tree which consists of all possible moves. The root represents the current player and the children of the root represents the opponent. And the leaves contains the scores. We want to find the best move for the current player. So at the position of the current player, we choose the MAX score of its children; and at the position of opponent, we choose the **MIN** score of its children. For **MIN** Nodes, the score computed starts with **+infinity** and decreases with time. For MAX nodes, scores starts with **-infinity** and increases with time.

10.13. DISJOINT SET

The Disjoint Set ADT

Objects: A collection of nonempty *disjoint* sets $S = S_1, S_2, ..., S_k$, *i.e.*, each S_i is a nonempty set that has *no* element in common with any other S_j. Each set is identified by a *unique element* called its *representative*.

Operations :

- MAKE-SET(x): *Given an element x that does not already belong to one of the sets,*
- FIND-SET(x): *Given an element x,*
- UNION(x,y):
  - *Given two distinct elements x and y*
  - *let S_x be the set that contains x and S_y be the set that contains y.*
  - *UNION(x, y) is an operation that results in a new set consisting of*

Note : If both *x* and y belong to the same set already (*i.e.*, $S_x = S_y$), nothing is done by this operation.

Applications :

- Maintaining the set of *connected components* of a graph.
- Maintain lists of *duplicate copies of webpages.*
- Constructing a *minimum spanning tree* for a graph (Kruskal).

Finding connected components :

For all v in V do
MAKE-SET(v)
For all (u,v) in E do
UNION(u,v)

Improvement in Implementation of Data Structures of Disjoint Sets :

In order to improve the running time, there is one heuristics called Path Compression.

1. Path compression :

- It is used to enhance the performance of the *find* operation further.
- In a FIND(*x*) operation, after the root *y* is found, we traverse the path from *x* to *y* one more time and change the parent pointers of all nodes along the path to point directly to *y*.

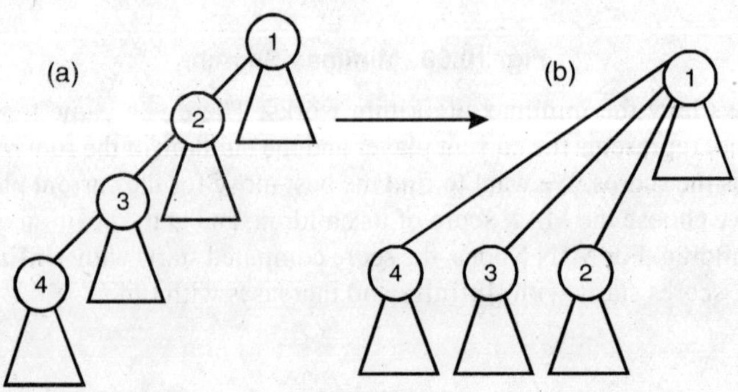

Fig. 10.61.

FIND(4) with path compression

- The name of the set (root) is found to be the *fig a at the right side*
- Therefore, the parent pointer of each node on the path from 4 to 1 is reset so that it points to node 1 as shown in Fig. 10.61(*b*)

10.14. HUFFMAN ALGORITHM

Input : A set of symbols $S = \{s_1, s_2, \dots s_m\}$ and a set of corresponding weights w_i, where w_i is the weight associated to the symbol s_i.

Output : A binary tree representing a prefix code for the set of symbols S whose codewords have minimum average weighted length.

Initialize F to be a forest of isolated vertices, labeled $s_1, \dots s_m$ with respective weights $w_1, \dots, w_m$.
For $i = 1$ to $m - 1$ do

1. Choose from forest F two trees, T and T', of smallest weights (ties settled by some arbitrary ordering of the trees in F).
2. Create a new binary tree whose root has T and T' as its left and right subtrees.
3. Label the edge to T with a 0 and the edge to T' with a 1.
4. Assign the new tree the weight equal to the sum of the weights of T and T'.
5. Replace trees T and T' in forest F by the new tree.

Return F.

The binary tree produced by this algorithm is called a ***Huffman tree***, and the corresponding prefix code is called a ***Huffman code***.

Lemma 10.1. If the leaves of a binary tree are assigned weights, and if each internal vertex is assigned a weight equal to the sum of its children's weights, then the tree's average weighted depth equals the sum of the weights of its internal vertices.

Theorem 10.1. For a given list of weights $w_1, w_2, \dots, w_1$, a Huffman tree has the smallest possible average weighted depth among all binary trees whose leaves are assigned those weights.

SOLVED PROBLEMS

1. What is left and right skewed trees ?

Ans. See the figure given below:

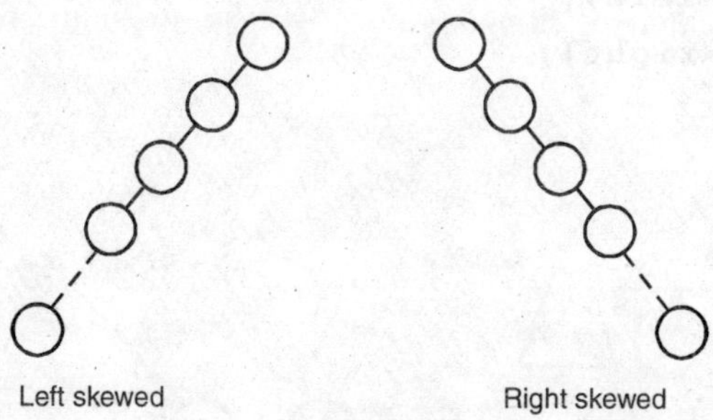

Fig. 10.62.

2. How can you find minimum and maximum element of a BST non- recursively?

Ans. Simply by traversing the BST is LDR manner *i.e.* in inorder and storing first and last element in two different variables.

3. How can you find minimum and maximum element of a BST without traversing it in inorder?

Ans. The C function for both is given below:

```
int getmin (node*R)
{
node*ptr=R;
while (ptr->left ! =NULL)
                ptr=ptr->left;
return ptr->data;
}
int getmax (node*R)
{
node*ptr=R;
while (ptr->right  ! = NULL)
                ptr=ptr->right;
return ptr->data;
}
```

4. Write C routine to find height of the binary tree?

Ans.

```
int height (node*R)
{
int LH,RH;
if(R= =NULL)
return 0;
else
{

LH=height (R->left);
RH=height (R->right);
if (LH>RH)
return ++LH;
else
return ++RH;
}
}
```

5. Write C routine to count number of nodes in binary tree recursively?

Ans.

```
int countR (node*R)
{
```

```
if (R= =NULL)
return 0;
else
return (countR(R->left) + countR(R->right)+1);
}
```

6. Write C routine to find number of nodes in binary tree non-recursively?

Ans. Take any traversal method using stack and count the nodes.

```
int countNR (node*R)
{
node*stack[201 ,*ptr;
int top=0,N=0;
stack[top] = NULL;
ptr=R;
while (ptr !=NULL)
{
N++;
if (ptr->right! =NULL)
{
top++;
stack[top] = ptr->right;
}
if (ptr ->left ! =NULL)
        ptr = ptr ->left;
else
{
ptr = stack [top];
top -;
}
}
return N;
}
```

7. What is formula for finding number of possible binary trees given N number of nodes ?

Ans. The formula is : $\frac{2n!}{n! \times (n+1)!}$

8. Draw all possible binary trees of 3 nodes:

Ans. There will be 5 such trees (using the above formula)

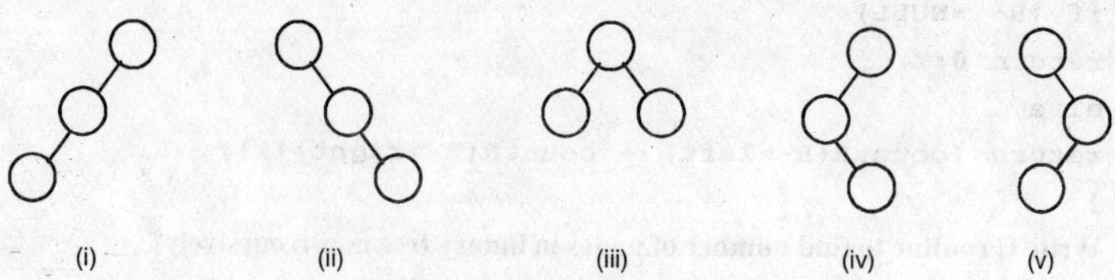

Fig. 10.63.

9. Consider the following characters:

'B','I','N','A','R','Y','S','E','A','R','C','H','T','R','E','E'

Construct the BST from above:

Ans. The final tree is show below:

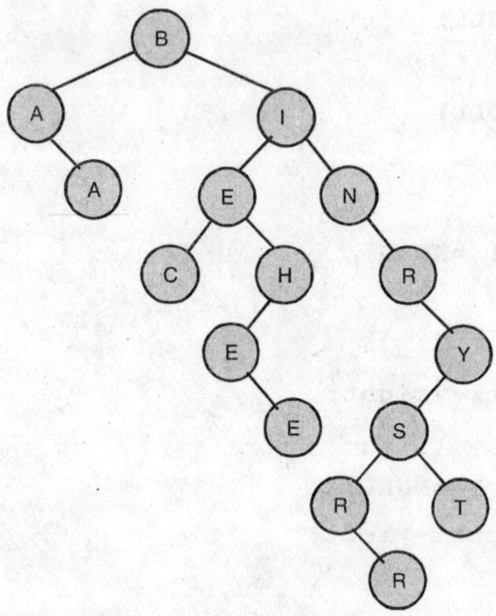

Fig. 10.64.

10. Consider the inorder and post order of a binary tree given below:

Inorder : F G B A D C E

Postorder: G F B D E C A

Ans. inorder traversal is done as LDR and postorder traversal is done as LRD, so A will be the root of the binary tree.

(*i*)

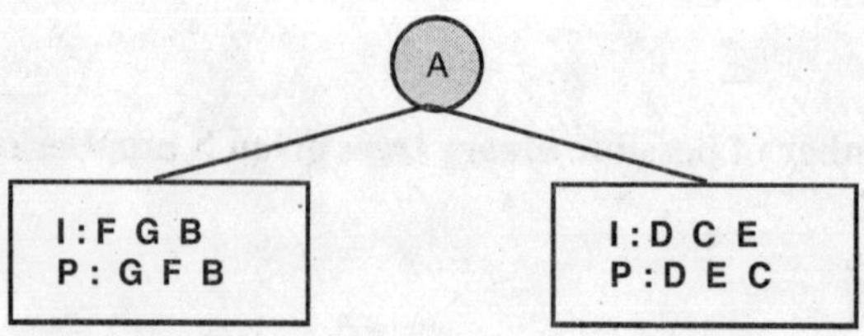

(ii)

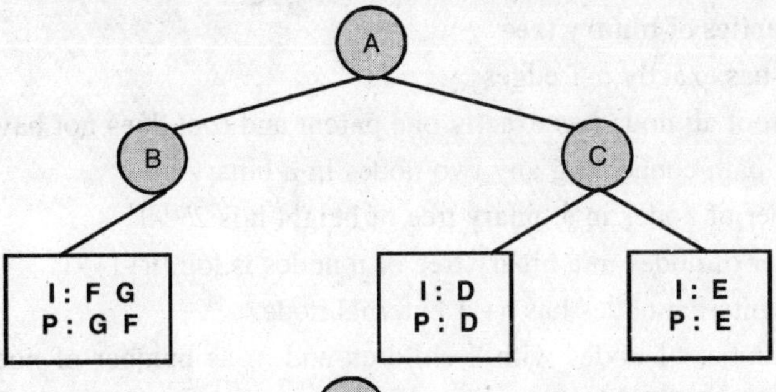

(iii)

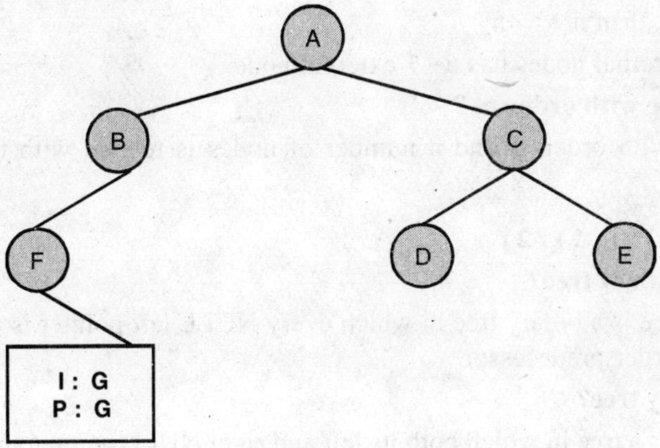

(iv)

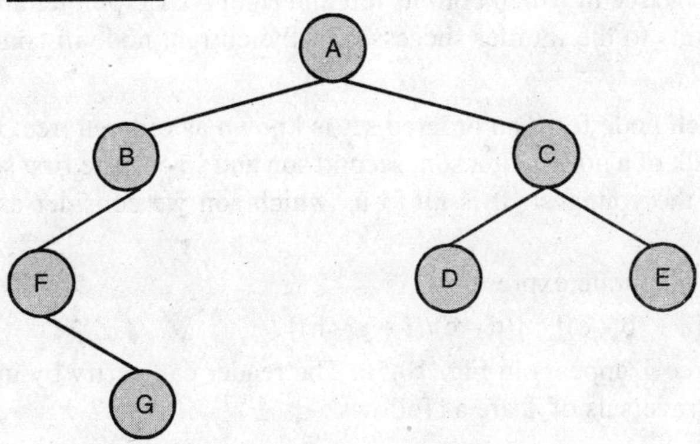

11. Write recursive C routine for searching in a BST.

Ans.

```
int search(node *R, int num)
{
if (R = = NULL)
return 0;
 if (R -> data = = num)
return 1;
if (num  <  R->data)
return search(R ->left, num);
else
return search(R ->right, num);
}
```

12. Write some of the properties of binary tree

Ans. 1. A tree with n nodes has exactly n-1 edges.

2. In a tree except the root all node has exactly one parent and root does not have any parent.
3. There is exactly one path connecting any two nodes in a binary tree.
4. The maximum number of nodes in a binary tree of height h is 2^{h+1}-1
5. The minimum number of nodes in a binary tree of n nodes is $\log_2(n+1)$ -1
6. A binary tree with n internal nodes has n+ 1 external node.
7. Considering n_2 as number of nodes with 2 children and n_0 as number of nodes with zero children *i.e.* leaf nodes, then $n_2+1 = n_0$
8. A binary tree with n internal nodes has n+ 1 external node.

13. What is height h of a B tree with order m ?

Ans. The height h of a B tree with order m and n number of nodes is related with the following inequality:

```
h = log em / 2 u ((n+ 1)/2)
```

14. What is left in-threaded binary tree?

Ans. A left in-threaded binary tree is a binary tree in which every NULL left pointer is replaced by a thread that points to its inorder predecessor.

15. What is in-threaded binary tree?

Ans. An in-threaded binary tree is a tree in which both its left and right NULL pointer are replaced by thread. The right thread points to the inorder successor of the current node in context.

16. What is an ordered tree?

Ans. A tree where subtrees of each node form an ordered set is known as ordered tree. It is a type of general tree where we can talk of a node's first son, second son and so on. The first son will be the oldest and last son will be the youngest. It is up to us which son we consider as youngest or oldest.

17. Let E denote the following algebraic expression :

$$[a + (b - c)] * [(d - e)/(f + g - h)]$$

Ans. The corresponding binary tree T appears in Fig. 10.56. The reader can verify by inspection that the preorder and postorder traversals of T are as follows :

(Preorder) * + a1 – b c / – d e – + f g h

(Postorder) a b c – + d e – f g + h – / *

The reader can also verify that these orders correspond precisely to the prefix and postfix Polish notation of E as discussed in Sec. 6.4. We emphasize that this is true for any algebraic expression E.

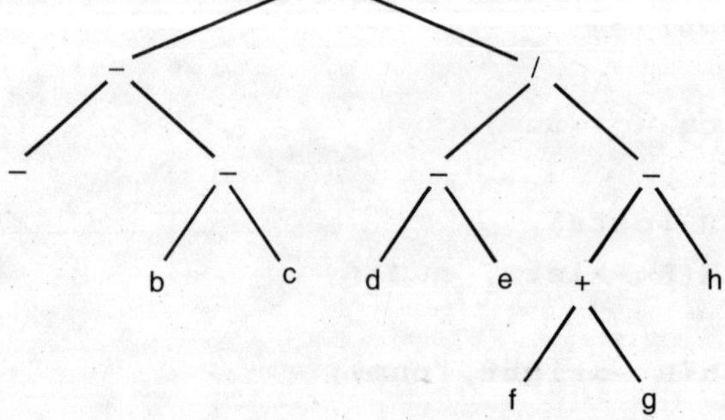

Fig. 10.65.

18. Algebraic Expressions

Consider any algebraic expression E involving only binary operations, such as

$$E = (a - b)/((c * d) + e)$$

E can be represented by means of the binary tree T pictured in Fig. 6.63. That is, each variable or constant in E appears as an "internal" node in T whose left and right subtrees correspond to the operands of the operation. For example :

(*a*) In the expression E, the operands of + are c * d and e.

(*b*) In the tree T, the subtrees of the node + correspond to the subexpression c * d and e.

Clearly every algebraic expression will correspond to a unique tree, and vice versa.

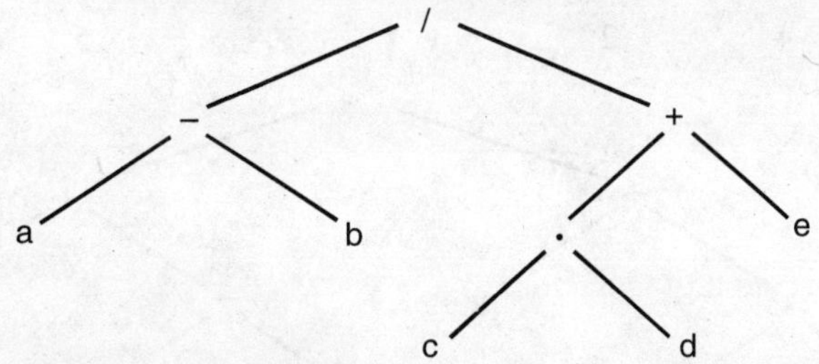

Fig. 10.66. E = (a – b)/((c * d) + e).

19. A binary tree T has 9 nodes. The inorder and preorder traversals of T yield the following sequences of nodes :

Inorder	**:**	**E**	**A**	**C**	**K**	**F**	**H**	**D**	**B**	**G**
Preorder	**:**	**F**	**A**	**E**	**K**	**C**	**D**	**H**	**G**	**B**

Draw the tree T.

Ans. The tree T is drawn from its root downward as follows.

(*a*) The root of T is obtained by choosing the first node in its preorder. Thus F is the root of T.

(*b*) The left child of the node F is obtained as follows. First use the inorder of T to find the nodes in the left subtree T1 of F. Thus T1 consists of the nodes E, A, C and K. Then the left child of F is obtained by choosing the first node in the preorder of T1 (which appears in the preorder of T). Thus A is the left son of F.

(*c*) Similarly, the right subtree T2 of F consists of the nodes H, D, B and G, and D is the root of T2, that is D is the right child of F.

Repeating the above process with each new node, we finally obtain the required tree in Fig. 10.67.

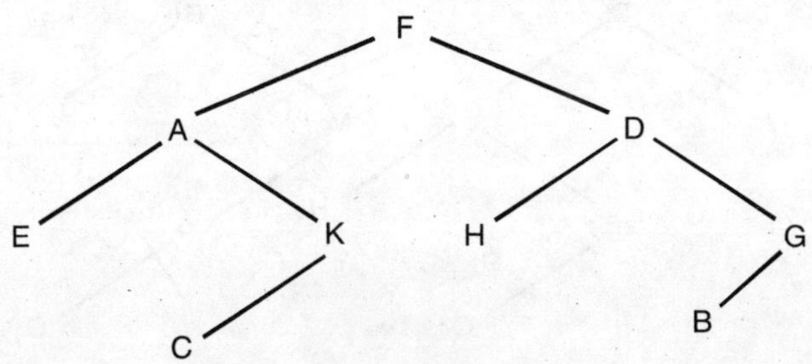

Fig. 10.67.

20. Consider the algebraic expression E = (2x + y) (5*a* – *b*)^3.

(*a*) Draw the tree T which corresponds to the expression E.

(*b*) Find the scope of the exponential operator; *i.e.* find the subtree rooted at the exponential operator.

(*c*) Find the prefix Polish expression P which is equivalent to E, and find the preorder of T.

Ans. (*a*) Use an arrow (–) for expression and an asterisk (*) for multiplication to obtain the tree shown in Fig. 10.68.

(*b*) The scope of - is the tree shaded in Fig. 10.68. It corresponds to the subexpression (5a – b)^3.

(*c*) There is no difference between the prefix Polish expression P and the preorder of T. Scan the tree T from the left, as in Fig. 10.68, to obtain :

* + * 2 x y - – * 5 a b 3

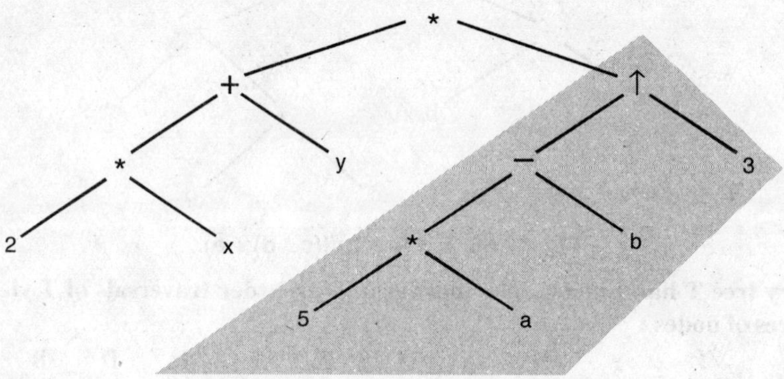

Fig. 10.68.

21. Suppose the following list of letters is inserted in order into an empty binary search tree :

J, R, D, G, T, E, M, H, PO, A, F, Q

(*a*) Find the final tree T and (b) find the inorder traversal of T.

Ans. (*a*) Insert the nodes one after the other to obtain the tree in Fig. 10.69.

(*b*) The inorder traversal of T follows :

A, D, E, F, G, H, J, M, P, Q, R, T

Observe that this is the alphabetical listing of the letters.

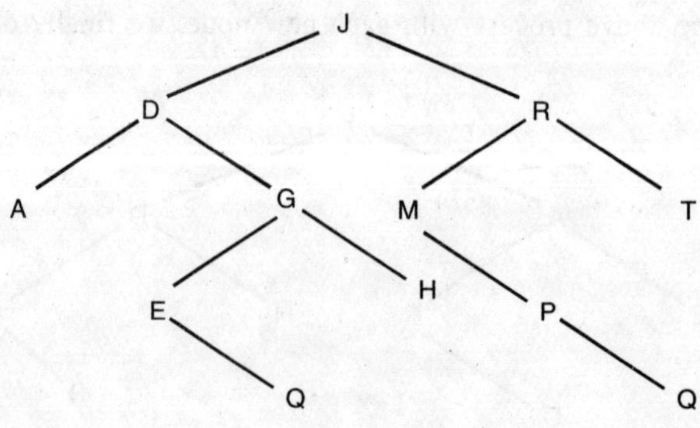

Fig. 10.69.

22. Define the degree of a node as the number of its non-empty children. Prove by induction that the number of degree 2 nodes in any binary tree is one less than the number of leaves.

Ans. Use as the base case the tree of one leaf node. The number of degree-2 nodes is 0, and the number of leaves is 1. Thus, the theorem holds.

For the induction hypothesis, assume the theorem is true for any tree with $n - 1$ nodes.

For the induction step, consider a tree T with n nodes. Remove from the tree any leaf node, and call the resulting tree T. By the induction hypothesis, T has one more leaf node than it has nodes of degree 2.

Now, restore the leaf node that was removed to form T. There are two possible cases.

(1) If this leaf node is the only child of its parent in T, then the number of nodes of degree 2 has not changed, nor has the number of leaf nodes. Thus, the theorem holds.

(2) If this leaf node is the child of a node in T with degree 2, then that node has degree 1 in T. Thus, by restoring the leaf node we are adding one new leaf node and one new node of degree 2. Thus, the theorem holds.

By mathematical induction, the theorem is correct.

23. Why is the BST Property defined so that nodes with values equal to the value of the root appear only in the right subtree, rather than allow equal-valued nodes to appear in either subtree ?

Ans. If equal valued nodes were allowed to appear in either subtree, then during a search for all nodes of a given value, whenever we encounter a node of that value the search would be required to search in both directions.

24. What are the minimum and maximum number of elements in a heap of height *h* ?

Ans. The minimum number of elements is contained in the heap with a single node at depth $h - 1$, for a total of 2^{h-1} nodes.

The maximum number of elements is contained in the heap that has completely filled up level $h - 1$, for a total of $2^h - 1$ nodes.

25. Prove that the number of leaf nodes in a 2-3 tree with *k* levels is between 2^{k-1} and 3^{k-1}.

Ans. The 2-3 tree of k levels will have the fewest nodes if no parent has three children. We know from Chapter 5 that the complete binary tree with k levels has at least 2^{k-1} leaves.

The 2-3 tree of k levels will have the most nodes if every parent has three children. A 3-ary tree with k levels can have as many as 3^{k-1} nodes.

26. Difference between B-Tree and B^+-Tree?

Ans.

B-Tree	B^+-Tree
In a B-tree, pointers to data records exist at all levels of the tree.	In a B+-tree, all pointers to data records exists at the leaf-level nodes.
B-Tree have more levels than B+-Tree.	A B+-tree can have less levels (or higher capacity of search values) than the corresponding B-tree.
These data structures are variations of search trees that allow efficient insertion and deletion of new search values.	These data structures are variations of search trees that allow efficient insertion and deletion of new search values.
Not Possible in B-Tree.	The leaf nodes are chained together to provide fast sequential access to the data records.

27. What are the characteristics of multi-way search trees?

Ans. To make the processing of m-way trees easier some type of order will be imposed on the keys within each node, resulting in a **multi way search tree of order m** (or an **m-way search tree**). By definition an m-way search tree is an m-way tree in which:

- Each node has m children and *m*–1 key fields
- The keys in each node are in ascending order.
- The keys in the first i children are smaller than the ith key
- The keys in the last *m-i* children are larger than the ith key

4-way search tree

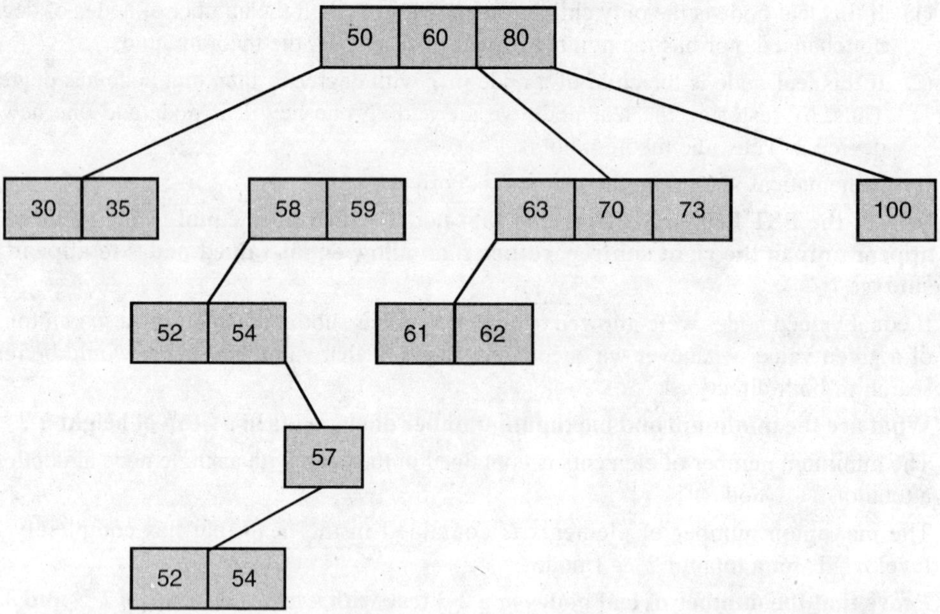

M-way search trees give the same advantages to m-way trees that binary search trees gave to binary trees - they provide fast information retrieval and update. However, they also have the same problems that binary search trees had - they can become unbalanced, which means that the construction of the tree becomes of vital importance.

28. What are B-Trees?

Ans. An extension of a multi way search tree of order *m* is a **B-tree of order m**. This type of tree will be used when the data to be accessed/stored is located on secondary storage devices because they allow for large amounts of data to be stored in a node.

A B-tree of order *m* is a multi way search tree in which:

1. The root has at least two subtrees unless it is the only node in the tree.
2. Each nonroot and each nonleaf node have at most m nonempty children and at least m/2 nonempty children.
3. The number of keys in each nonroot and each nonleaf node is one less than the number of its nonempty children.
4. All leaves are on the same level.

These restrictions make B-trees always at least half full, have few levels, and remain perfectly balanced.

The nodes in a B-tree are usually implemented as a class that contains an array of m-l cells for keys, an array of m pointers to other nodes, and whatever other information is required in order to facilitate tree maintenance.

Operation on B-Trees :

1. Searching
2. Insertion
3. Deletion

29. What is the difference between strictly and complete binary tree?

Ans. Complete Binary tree :

- All leaf nodes are found at the tree depth level

- All nodes (non-leaf) have two children

Strictly Binary tree :

- Nodes can have 0 or 2 children

30. List the properties of Binary trees?

Ans. Properties of binary trees :

- The number of nodes *n* in a perfect binary tree can be found using this formula: $n = 2^{h+1} - 1$ where *h* is the height of the tree.
- The number of nodes *n* in a complete binary tree is minimum: $n = 2^h$ and maximum: $n = 2^{h+1} - 1$ where *h* is the height of the tree.
- The number of nodes *n* in a perfect binary tree can also be found using this formula: $n = 2L - 1$ where *L* is the number of leaf nodes in the tree.
- The number of leaf nodes *n* in a perfect binary tree can be found using this formula: $n = 2^h$ where *h* is the height of the tree.
- The number of NULL links in a Complete Binary Tree of n-node is (n+1).
- The number of leaf node in a Complete Binary Tree of n-node is *UpperBound*(*n* / 2).
- For any non-empty binary tree with n_0 leaf nodes and n_2 nodes of degree 2, $n_0 = n_2 + 1$

31. List the types of binary trees?

Ans. Types of binary trees

- A **rooted binary tree** is a rooted tree in which every node has at most two children.
- A **full binary tree** (sometimes proper binary tree or 2-tree or strictly binary tree) is a tree in which every node other than the leaves has two children.
- A **perfect binary tree** is a full binary tree in which all leaves are at the same depth or same level. (This is ambiguously also called a complete binary tree.)
- A **complete binary tree** is a binary tree in which every level, except possibly the last, is completely filled, and all nodes are as far left as possible.
- A **balanced binary tree** is where the depth of all the leaves differs by at most 1. Balanced trees have a predictable depth (how many nodes are traversed from the root to a leaf, root counting as node 0 and subsequent as 1, 2... depth). This depth is equal to the integer part of $\log_2(n)$ where n is the number of nodes on the balanced tree. Example 1: balanced tree with 1 node, $\log_2(1) = 0$ (depth = 0). Example 2: balanced tree with 3 nodes, $\log_2(3) = 1.59$ (depth=1). Example 3: balanced tree with 5 nodes, $\log_2(5) = 2.32$ (depth of tree is 2 nodes).
- A **rooted complete binary tree** can be identified with a free magma.
- A **degenerate tree** is a tree where for each parent node; there is only one associated child

node. This means that in a performance measurement, the tree will behave like a linked list data structure.

32. What is Depth, Height, Siblings, ancestor, descendant, size, in-degree & out-degree with respect to trees?

Ans.
- The **depth** of a node n is the length of the path from the root to the node. The set of all nodes at a given depth is sometimes called a level of the tree. The root node is at depth zero.
- The **height** of a tree is the length of the path from the root to the deepest node in the tree. A (rooted) tree with only a node (the root) has a height of zero.
- **Siblings** are nodes that share the same parent node.
- If a path exists from node p to node q, where node p is closer to the root node than q, then p is an **ancestor** of q and q is a **descendant** of p.
- The **size** of a node is the number of descendants it has including itself.
- **In-degree** of a node is the number of edges arriving at that node.
- **Out-degree** of a node is the number of edges leaving that node.

33. If there are 1000000 numbers of nodes in a complete tree then find its depth?

Ans. Depth of the tree = $\log_2 (1000000) = 19$

34. What are the advantages of Height Balancing Binary tree over others?

Ans.
- Height-balanced binary trees are "balanced"
- Operations that run in time proportional to the height of the tree are **O(log n)**, n the number of nodes with limited performance variance
- Variance is a very important concern in real time applications, e.g. connecting calls in a telephone network

35. A binary tree with 20 nodes has null branches?

Ans. 21

A binary trees with n nodes have exactly n+ 1 null node.

36. How many different trees are possible with 10 nodes?

Ans. 1014

For example, consider a tree with 3 nodes (n=3), it will have the maximum combination of 5 different (*i.e.*, 23 – 3 = 5) trees.

In general:

If there are n nodes, there exist 2n-n different trees.

37. List out few of the Application of tree data-structure?

Ans.
1. The manipulation of Arithmetic expression,
2. Symbol Table construction,
3. Syntax analysis.

38. There are 8, 15, 13, 14 nodes were there in 4 different trees. Which of them could have formed a full binary tree?

Ans. 15.

In general:

There are $2n - 1$ nodes in a full binary tree.

By the method of elimination : Full binary trees contain odd number of nodes. So there cannot be full binary trees with 8 or 14 nodes, so rejected. With 13 nodes you can form a complete binary tree but not a full binary tree. So the correct answer is 15.

Note : Full and Complete binary trees are different. All full binary trees are complete binary trees but not vice versa.

39. In RDBMS, what is the efficient data structure used in the internal storage representation?

Ans. B+ tree. Because in B+ tree, all the data is stored only in leaf nodes, that makes searching easier. This corresponds to the records that shall be stored in leaf nodes.

40. A binary tree T has 9 nodes, the in order and preorder traversal of tree T yields the following sequence of nodes :

INORDER :	**E**	**A**	**C**	**K**	**F**	**H**	**D**	**B**	**G**
PREORDER	**E**	**A**	**F**	**K**	**C**	**D**	**H**	**G**	**B**

Draw the tree T. Show all the steps ?

Ans. To Draw a binary tree T using *inorder* and *preorder.*

(*i*) In PREORDER, the first node is a root node *i.e.* 'E' is a root node of the binary tree T, search the node E in the in order and all nodes at the left of 'E' are at left subtree while all the nodes at the right side of 'E' are at right subtree.

(*ii*) Now can the sequence of PREORDER from left to right and search the node encounted in INORDER and check whether the encounted node is either left or right of root node (E) as shown below :

(*a*) First node in preorder is 'E' 

(*b*) Next node in pre order is A and the node A is at right side of 'E' in order, so insert it at the right of E as right of E is null.

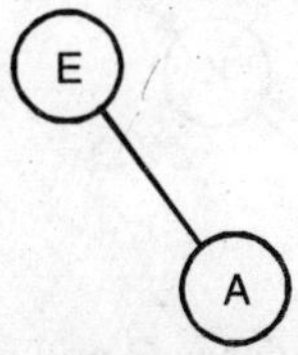

(*c*) Next node in pre order is F and the node F is at right of root node 'E' in inorder and in Binary tree T, right of 'E' is not null, therefore check the node A and node F is at right of A, hence F is inserted as the right child of A as it is null.

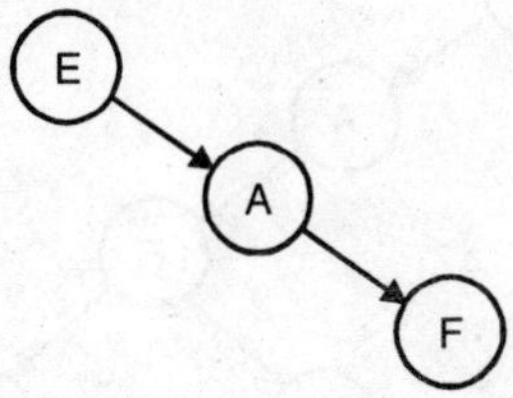

(*d*) Generally, next node in preorder is K and it's inserted left of F as K is at the right of E and A and left of F in inorder.

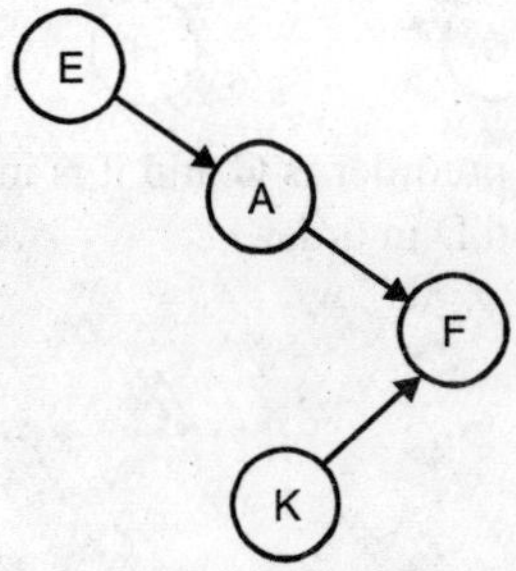

(*e*) Generally, next node in preorder is 'C' and its inserted at the left child of K (*a*) 'C' is at right of E & C and at left of F and K, in inorder.

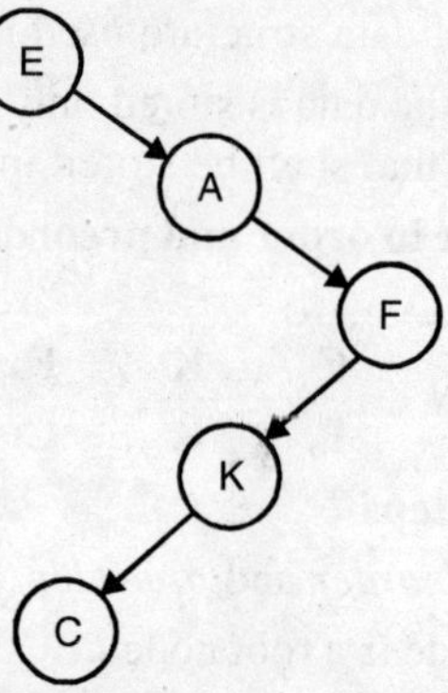

(*f*) Generally, next node in preorder is D and its inserted at the right of 'F' as node D is at right of E, A and F in inorder.

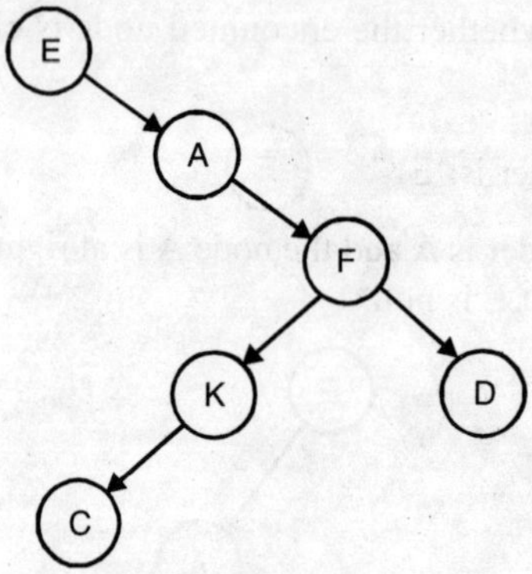

(*g*) Generally, next node in preorder is II and it's inserted at the left child of D as it is at right of E, A and F and at left of D in inorder.

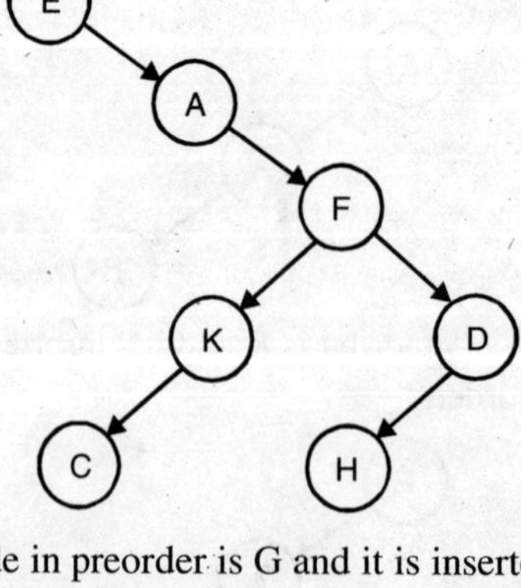

(*h*) Generally, next node in preorder is G and it is inserted at the right child of D as node G is at the right of E, A, F and D in order.

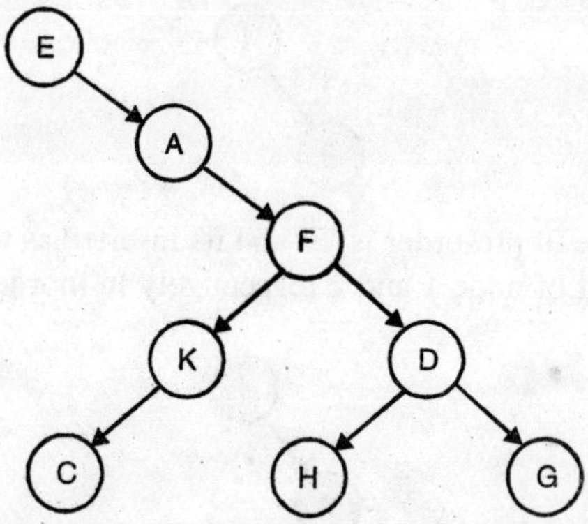

(*i*) Generally, the last node in preorder 'B' is inserted at the left child of node G as shown below :

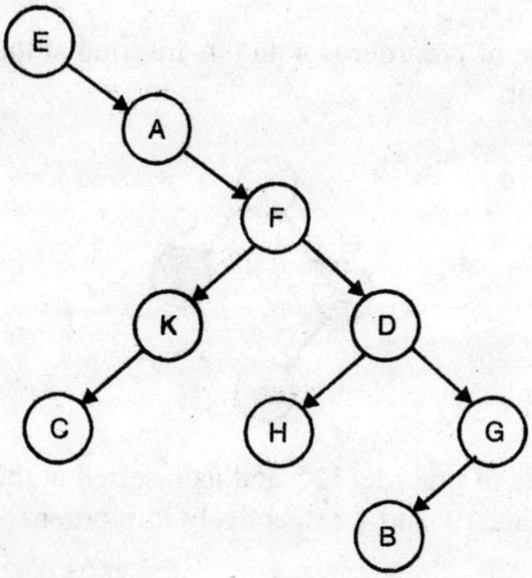

the required **Binary Tree 'T'**.

41. Give various traversing techniques of binary tree. Suppose following traversed sequence of a binary tree are given :

Pre-order : 1, 2, 3, 4, 5, 6, 7, 8, 9

In-order : 2, 3, 1, 5, 4, 7, 8, 6, 9

Construct the binary tree ?

Ans. The techniques used to traverse a binary tree are :

(*a*) PRE-ORDER [ROOT-LEFT-RIGHT]

(*b*) IN-ORDER [LEFT-ROOT-RIGHT]

(*c*) POST-ORDER [LEFT-RIGHT-ROOT]

(*i*) The first node of preorder *i.e.* '1' is the root node of bianry tree T as

(*ii*) Next node of preorder is '2' and it is inserted at the left of '1' as node 2 is at left of '1' in inorder *i.e.*

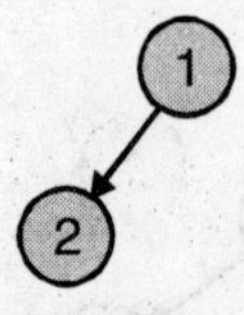

(*iii*) Generally next node of pre-order is '3' and its inserted as the right child of node '2' as node 3 is at left and right of node 1 and 2 respectively in inorder.

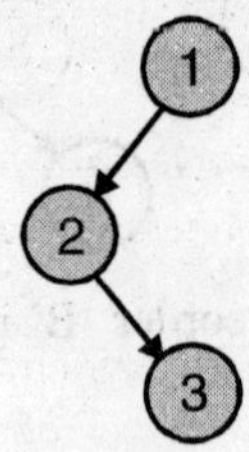

(*iv*) Generally, next node of preorder is 4 and its inserted at the right of node '1' as node 4 is at right of '1' in inorder.

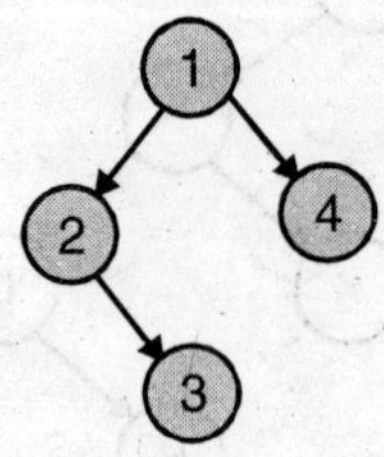

(*v*) Generally, next node of preorder is 5 and its inserted at the left child of node 4 as node 5 is at right and left of node 1 and 4 respectively in inorder.

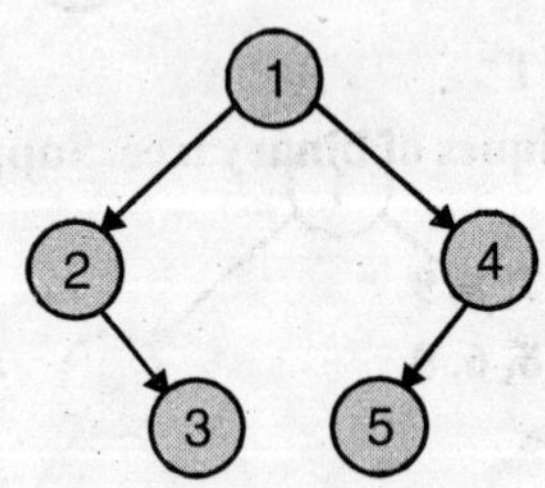

(*vi*) Generally, next node in preorder is 6 and its inserted at the right of node 4 as node 6 is at right of node 1 and 4 in inorder :

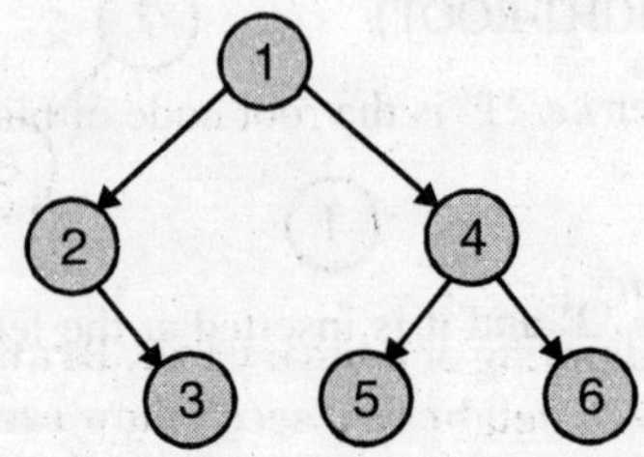

(*vii*) Generally, next node is preorder is 7 and its inserted at the left of node 6 as node 7 is at right of 1 and 4 and left of node 6 in inorder :

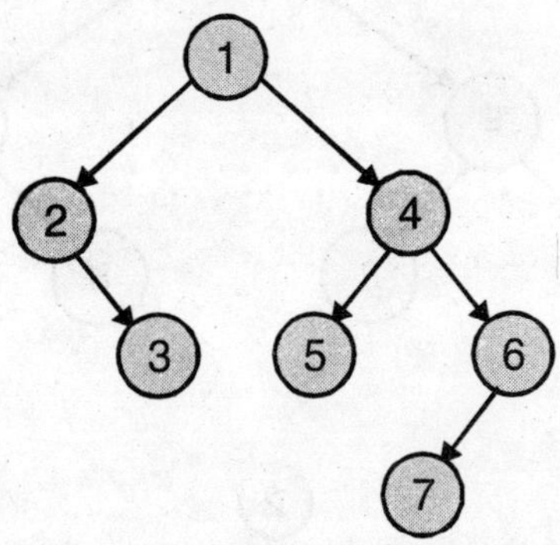

(*viii*) Generally, next node in pre-order is 8 and it's inserted at the right of node 7.

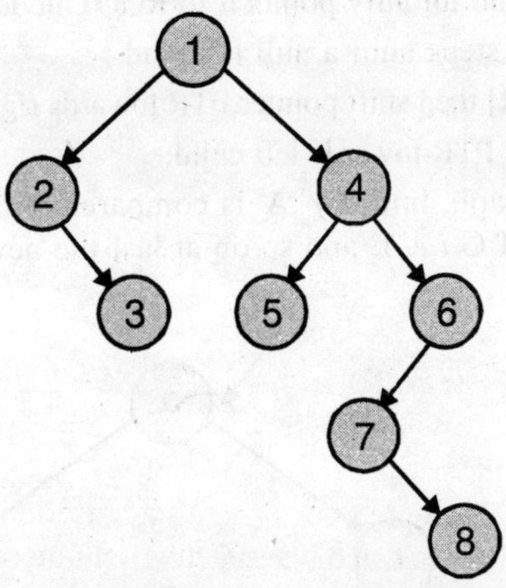

(*ix*) Generally, next node of preorder is '9' and it's inserted at the right of node 6.

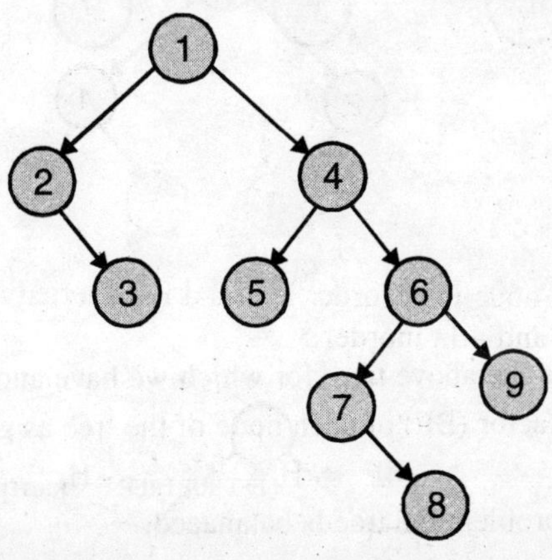

this is *required binary tree 'T'*.

42. Consider the following drawing of a binary tree. Draw the tree after the data item 'A' has been inserted and tree has been height balanced (show each step) ?

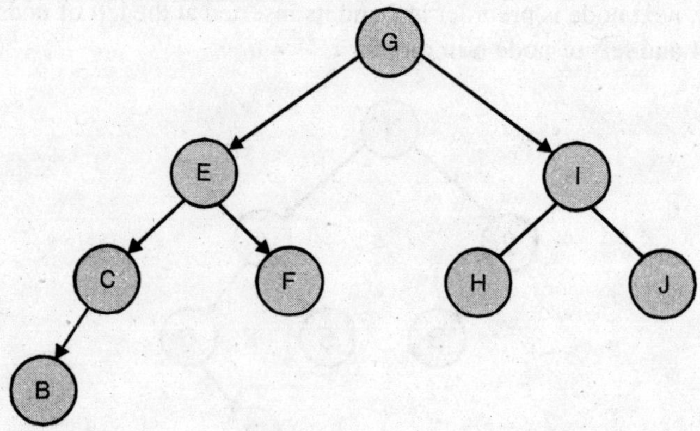

Ans. To insert an data item 'A' in the above binary tree, use the following procedure starts from root node.

(1) Take a pointer PTR and initially points it to ROOT node.

(2) Repeat the following steps until a null is found :

(*a*) if A > INFO [PTR] then shift pointer PTR towards right child.

(*b*) else shifts pointer PTR towards left child.

As in the above given graph, initially 'A' is compared with G as G is greater then compare again A with left child of G *i.e.* E and so on at last the new node 'A' is inserted left of B as shown below :

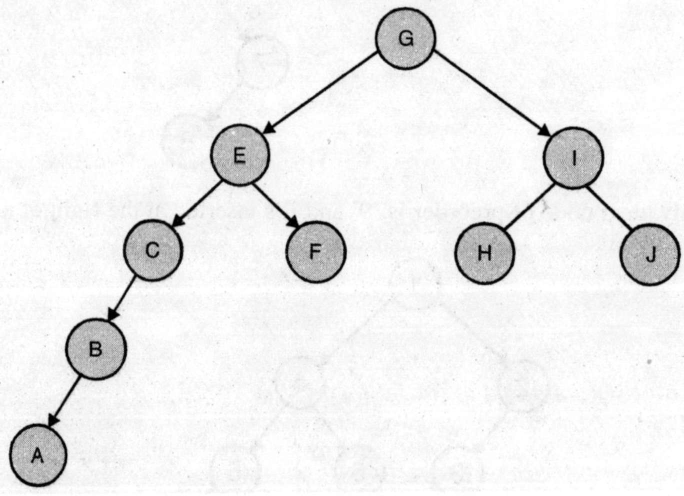

Now, we have to balance the above tree, for which we have another algorithm i.e.

(1) Find the balancing factor (BF) of each node of the tree as given below :

$$BF = H_{LEFT\ SUBTREE} - H_{RIGHT\ SUBTREE}$$

(2) If BF = –1, 0, 1, no problem the tree is balanaced.

(3) else if BF of any node is –2, –3, ..., –∞, then rotate the tree anticlockwise using various rotations.

(4) elese if BF of any node is +2, +3, ..., +∞, then also rotate the tree clockwise using various rotations.

As shown below the above tree along with their balancing factors of each node.

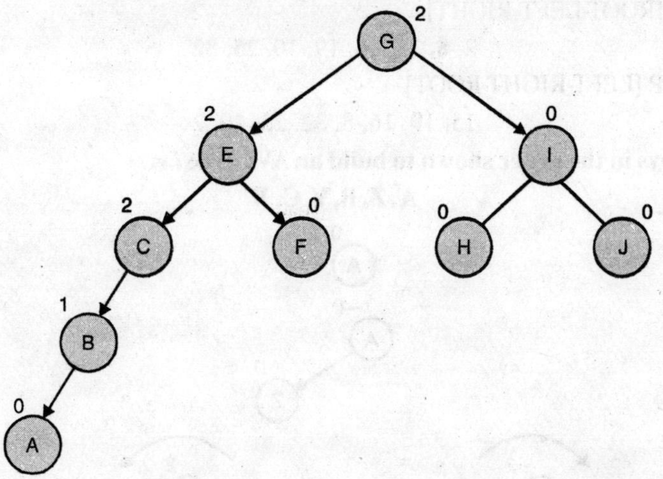

In the tree, node C, E and G are unbalanced, so rotation will perform at the lowest level node *i.e.* node C *i.e.*

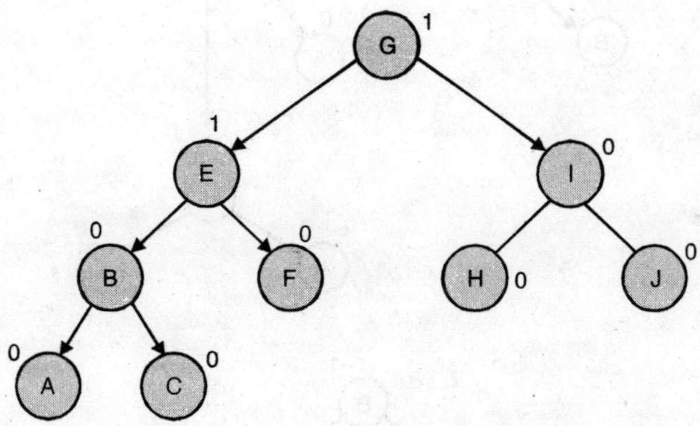

Now all the nodes of the tree are having their BF's either 1, 0 or –1 hence the tree is balanced.

43. What is the In-order traversal of the following tree :

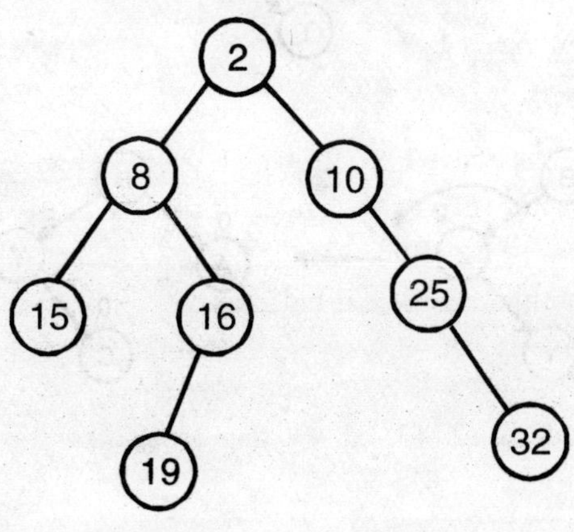

Ans. IN-ORDER [LEFT-ROOT-RIGHT]

15, 8, 19, 16, 2, 10, 25, 32

PREORDER [ROOT-LEFT-RIGHT]

2, 8, 15, 16, 19, 10, 25, 32

POST-ORDER [LEFT-RIGHT-ROOT]

15, 19, 16, 8, 32, 25, 10, 2

44. Insert the keys in the order shown to build an AVL tree *i.e.*

A, Z, B, Y, C, X

Ans. (I) Insert A

(II) Insert Z

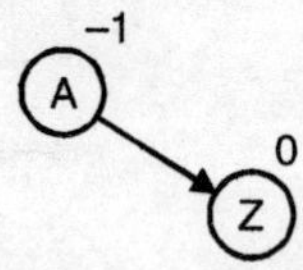

(III) Insert B

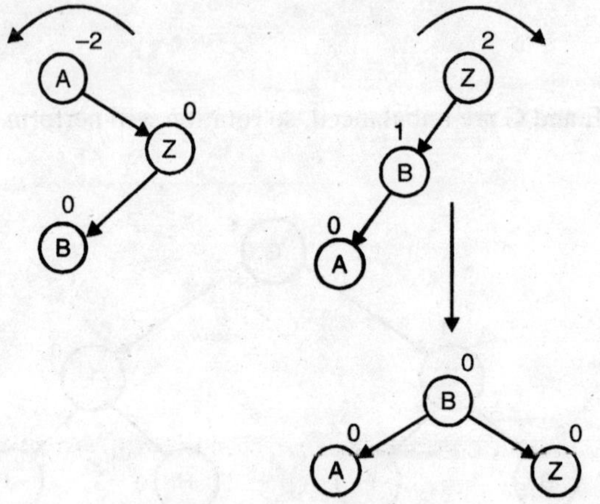

(IV) Insert Y

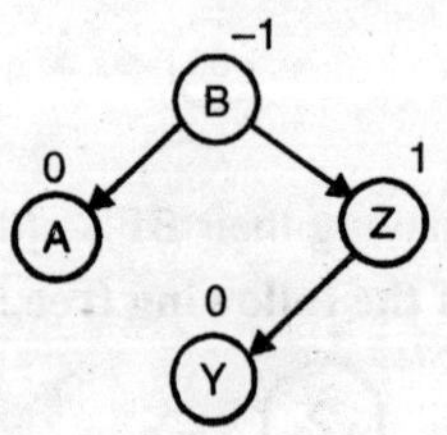

(V) Insert C

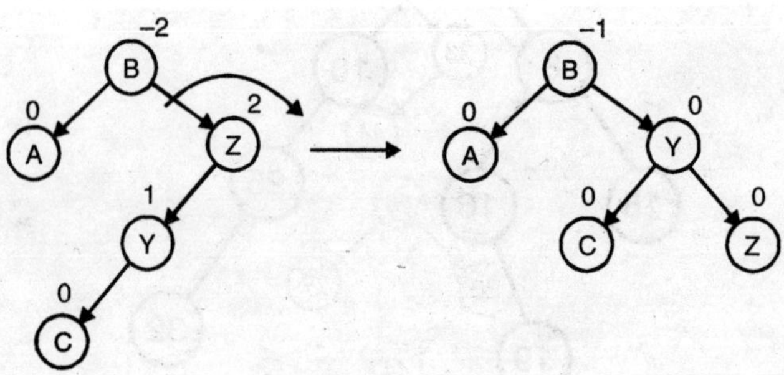

(VI) Insert X

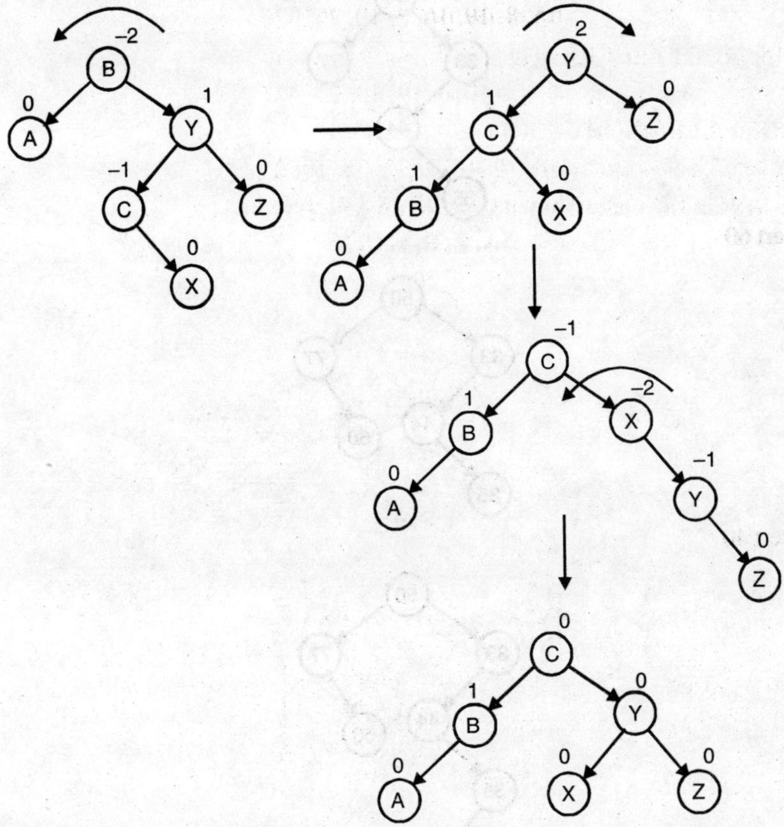

45. Suppose the following seven numbers are inserted in order into an empty binary search tree T, then draw the binary search tree :

50, 33, 44, 77, 35, 60, 40

Ans. (I) Insert 50

(II) Insert 33

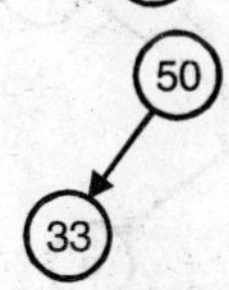

(III) Insert 44

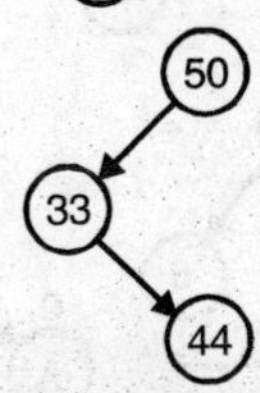

(IV) Insert 77

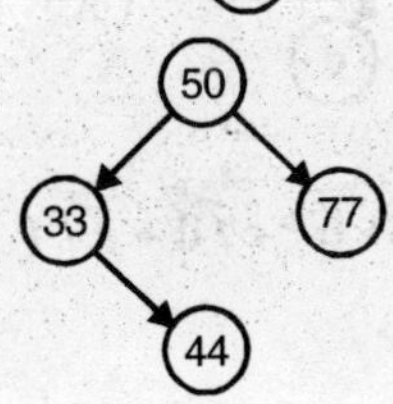

(V) Insert 35

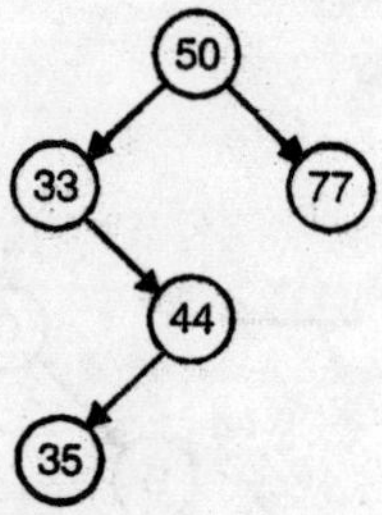

(VI) Insert 60

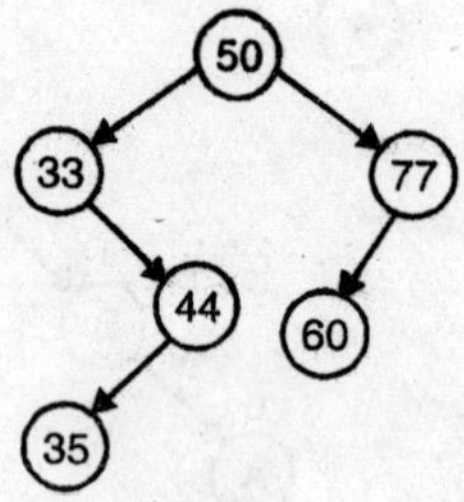

(VII) Insert 40

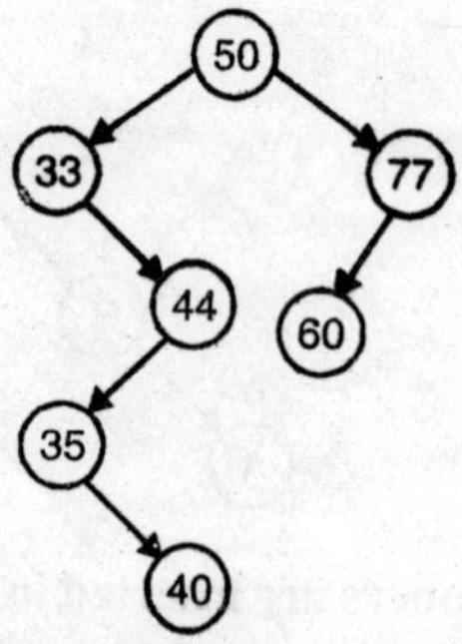

Application of BST, it will sort the given list.

46. Show the pre-order, in-order and post-order notation of the following expression tree ?

Ans.

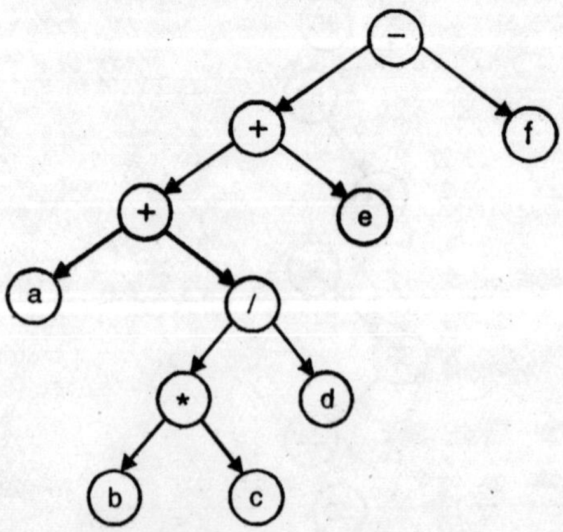

IN-ORDER [LEFT-ROOT-RIGHT]

a + b * c / d + e – f

PRE-ORDER (ROOT-LEFT-RIGHT]

− + + a / * b c d e f

POST-ORDER (LEFT-RIGHT-ROOT]

a b c * d / + e + f −

47. Insert the keys, in the order shown to build an AVL trees :

55, 60, 65, 25, 20, 40, 30, 45, 35

Ans. (I) Insert 55

(II) Insert 60

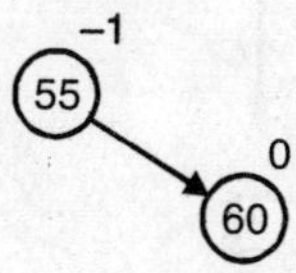

(III) Insert 65

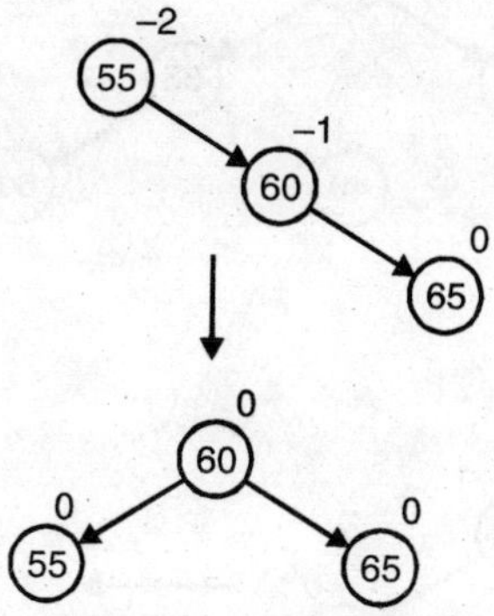

(IV) Insert 25

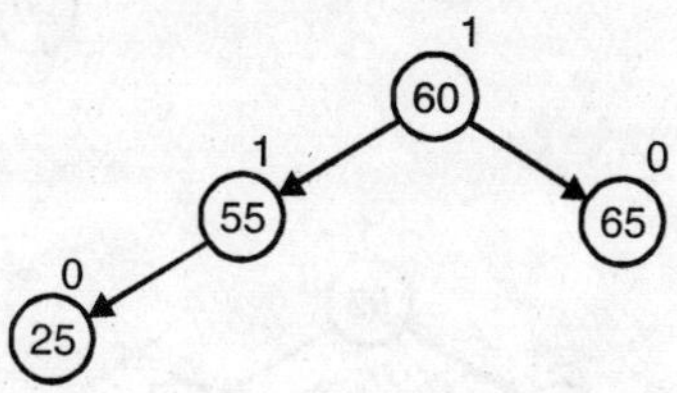

(V) Insert 20

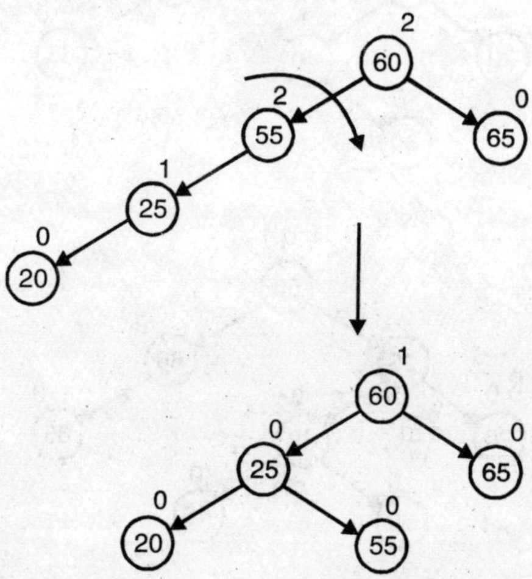

(VI) Insert 40

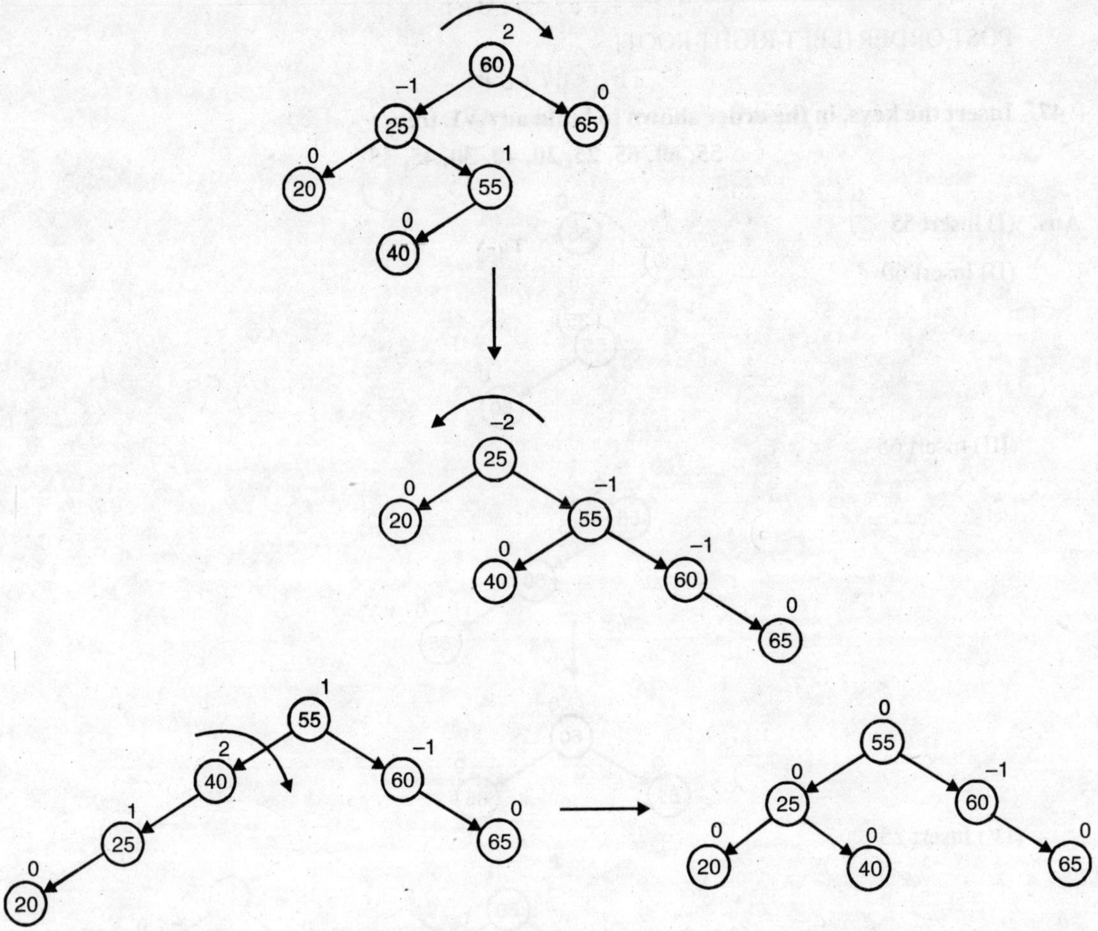

(VII) Insert 30

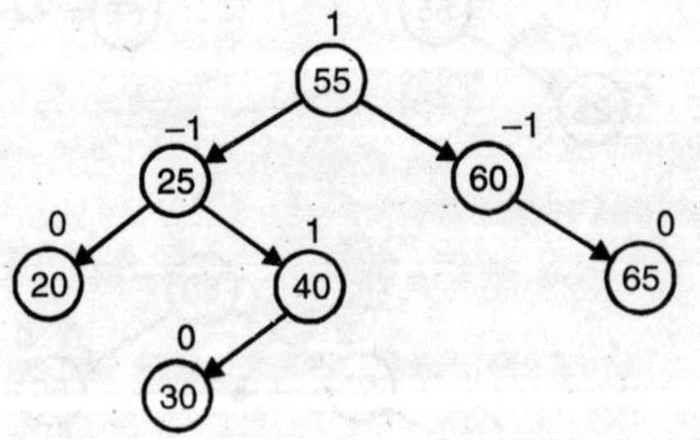

(VIII) Insert 45

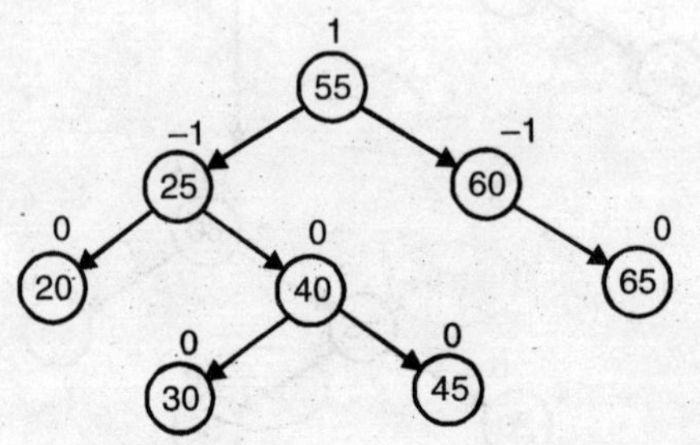

(IX) Insert 35

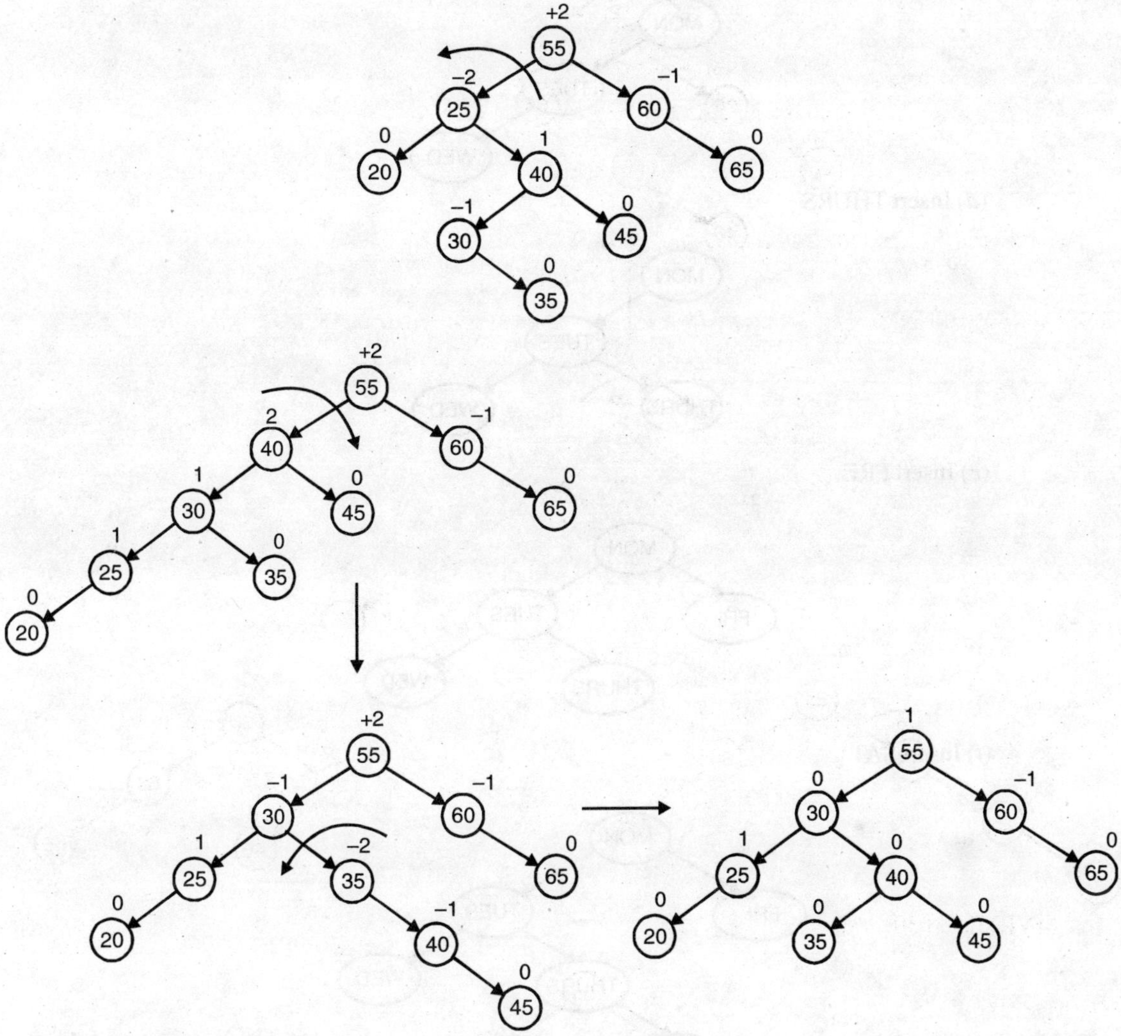

48. Implement the following for the sequence :

MON, TUES, WED, THURS, FRI, SAT, SUN

(*i*) Binary Search Tree

(*ii*) AVL tree.

Ans. (*i*) **Binary Search Tree**

(*a*) Insert MON

(*b*) Insert TUES

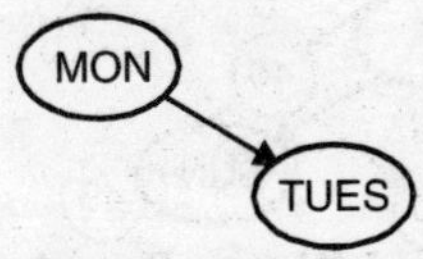

(*c*) Insert WED

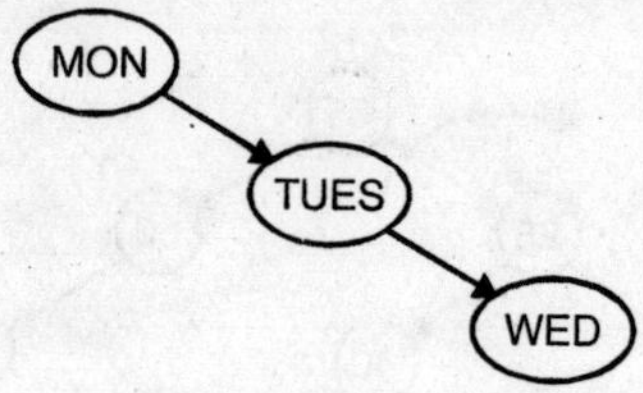

(*d*) Insert THURS

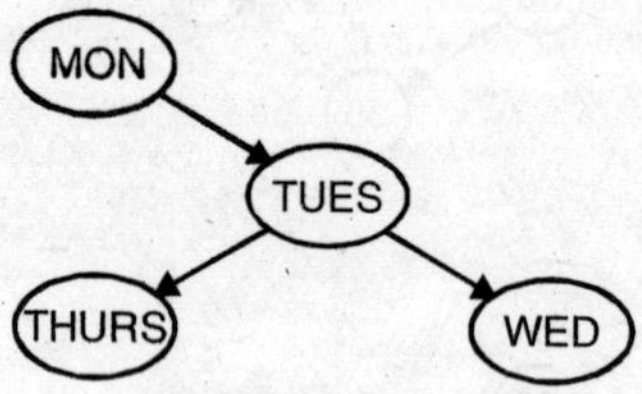

(*e*) Insert FRE

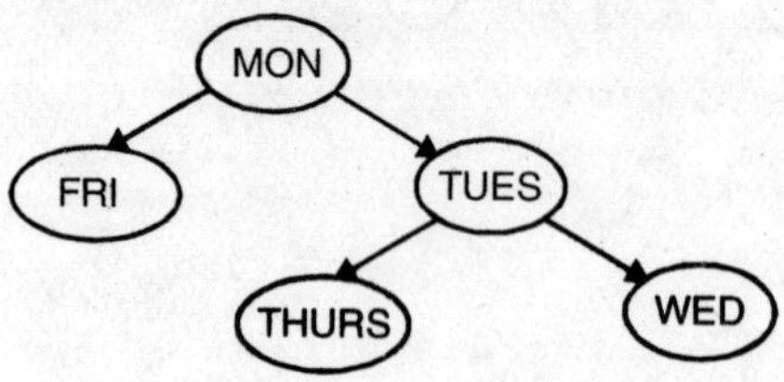

(*f*) Insert SAT

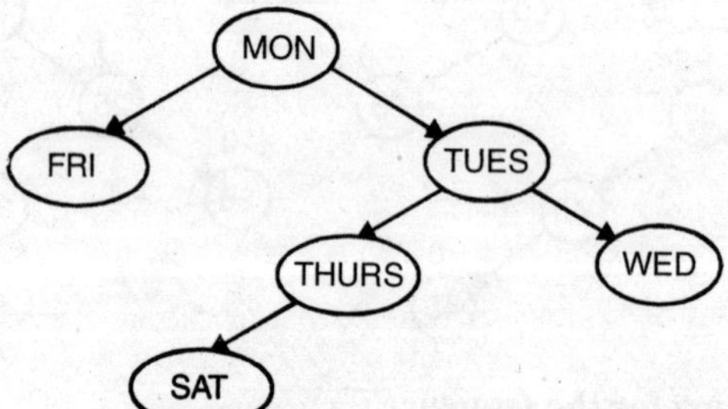

(*g*) Insert SUN

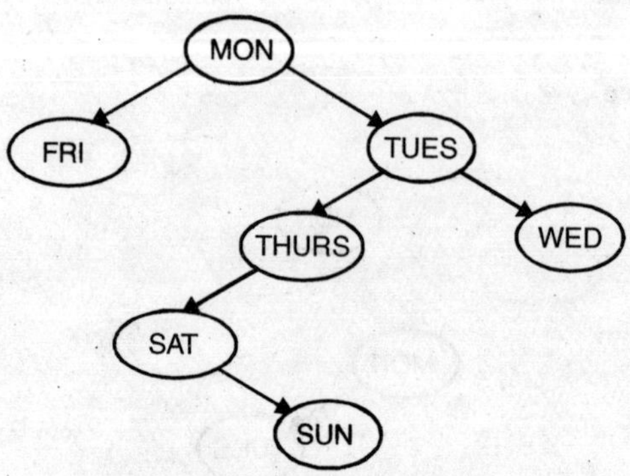

Required Binary Search Tree

(*ii*) AVL Tree

(*a*) Insert MON

(*b*) Insert TUES

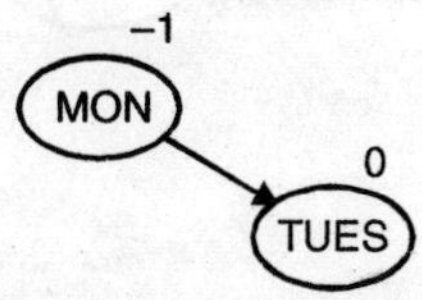

(*c*) Insert WED

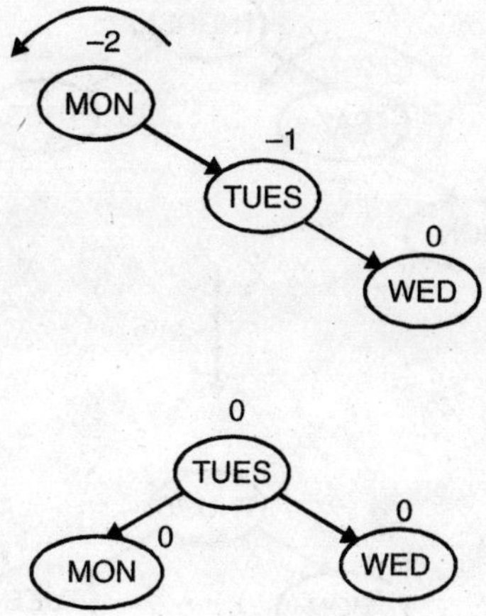

(*d*) Insert THURS

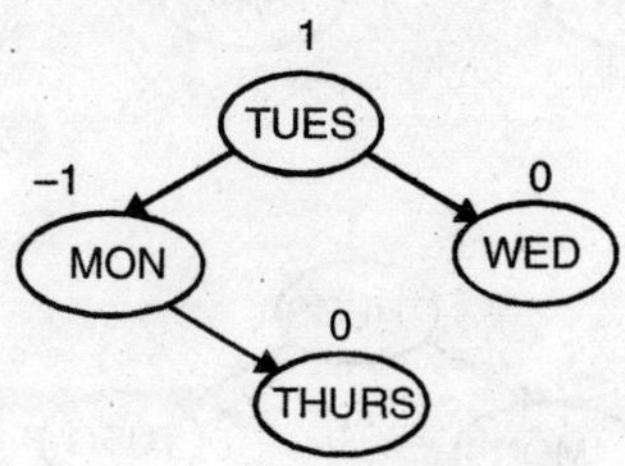

(*e*) Insert FRI

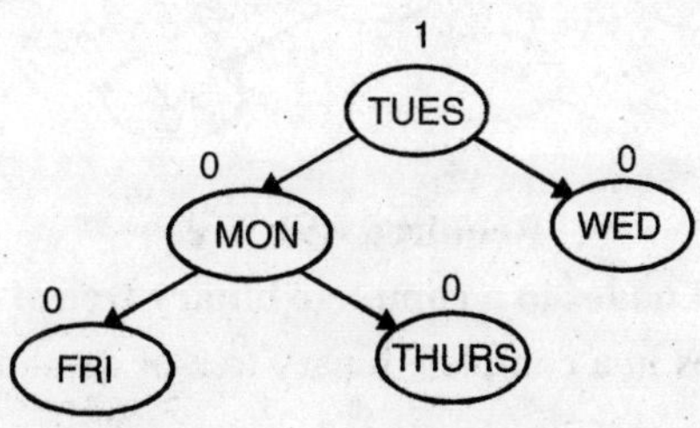

(*f*) Insert SAT

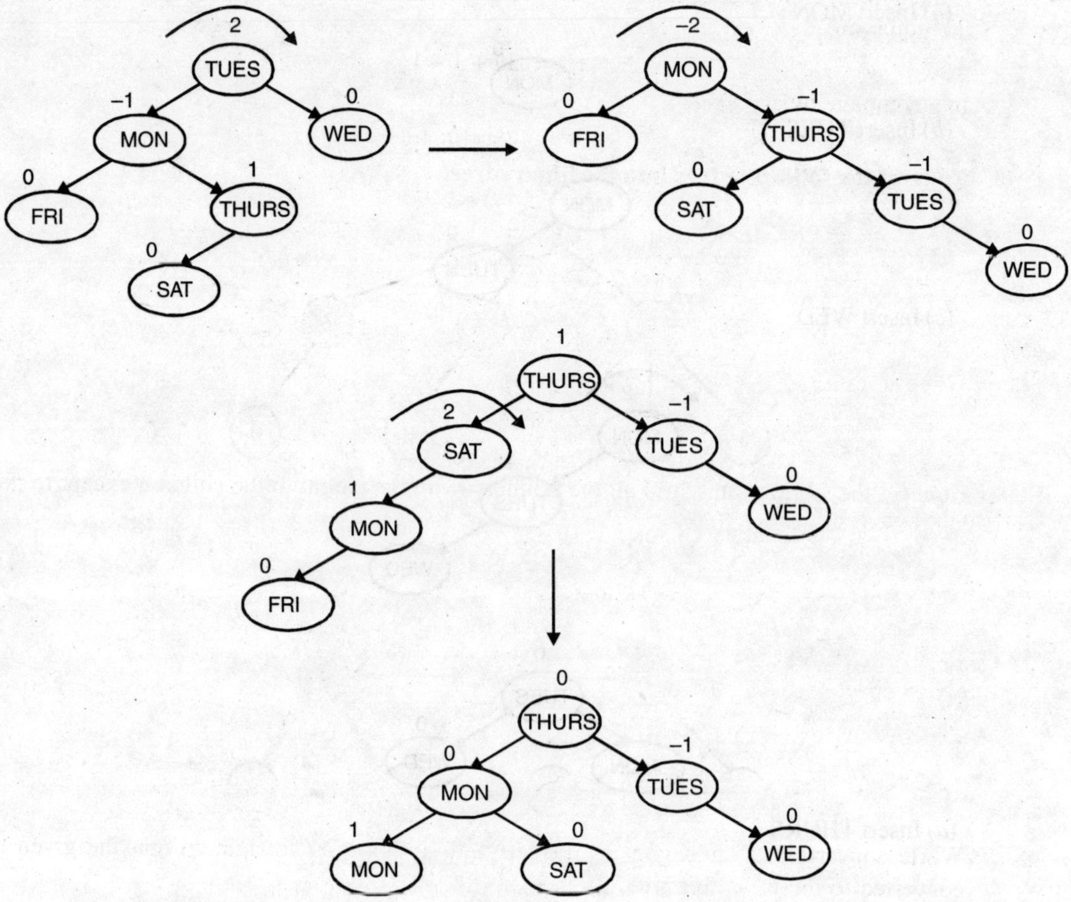

(*g*) Insert SUN

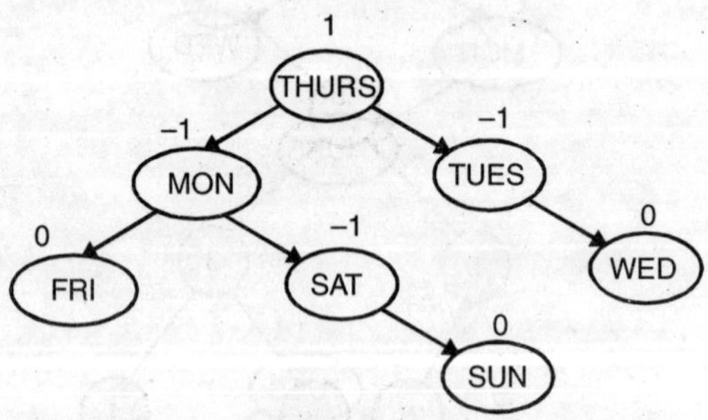

Required AVL Tree

49. Find the total number of nodes in a complete binary tree of depth d ?

Ans. The total number of nodes in a complete binary tree of depth *d*.,

Thus $$t_n = 2^0 + 2^1 + ... + 2^d$$

i.e. sum of number of nodes at each level between 0 and *d*

Since all leaves in such a tree are at level '*d*', the tree contains 2*d* leaves and, therefore 2*d* – 1 non-leaf nodes.

If the number of nodes, t_n, in a complete binary tree is known, we can compute its depth d, from the equation,

$$t_n = 2d + 1 - 1$$

In a complete binary tree,

$$d = \log_2 (tn + 1) - 1$$

50. Convert the following tree into the binary tree.

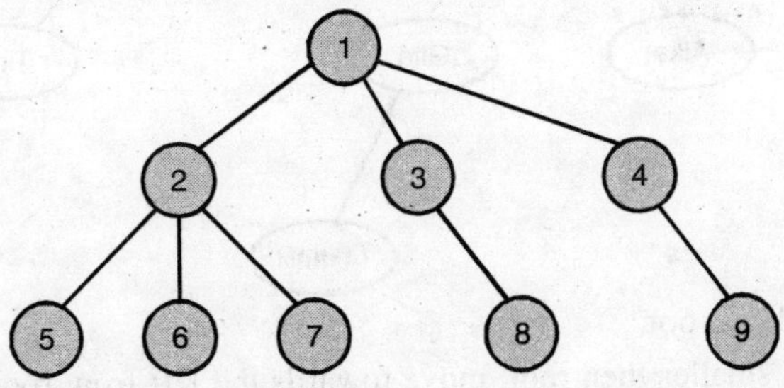

Ans. Connect the siblings and drop all the pointers from the parent to the children except to the first child as in following figure.

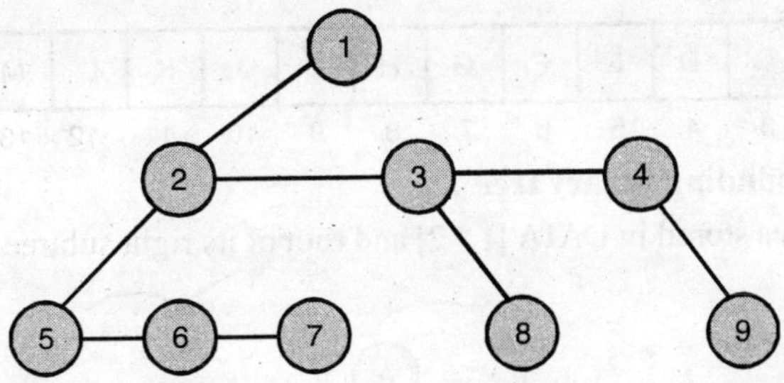

While converting it child goes to the left and sibling goes to right so that the given tree is converted to the following tree

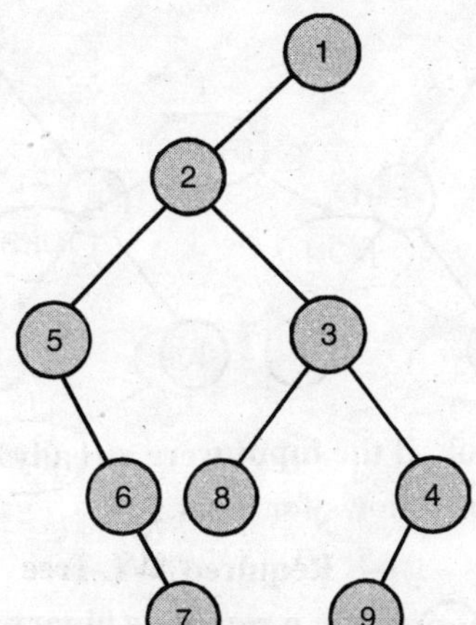

51. Create a binary search tree for the following sequence :

Pinki, Dolly, Gita, Sita, Hari, Alka, Tina

Ans.

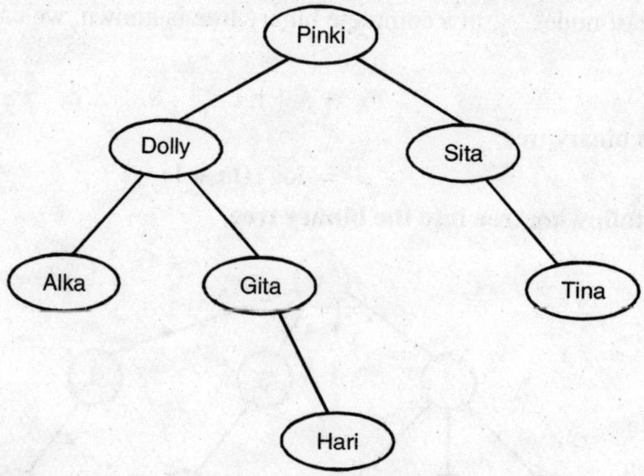

Place the first node as root.

If coming node is smaller then root, move towards the left from root, otherwise move towards right. Place the node where you find null pointer along the path.

52. An array DATA representing a binary tree is given as follows :

A	B	C	D	E	F	G	H	I	J	K	L	M	N	O
1	2	3	4	5	6	7	8	9	10	11	12	13	14	15

Draw the corresponding binary tree ?

Ans. For any given nodes stored in DATA [I * 2] and root of its right subtree will be in DATA [I * 2 + 1] as shown below :

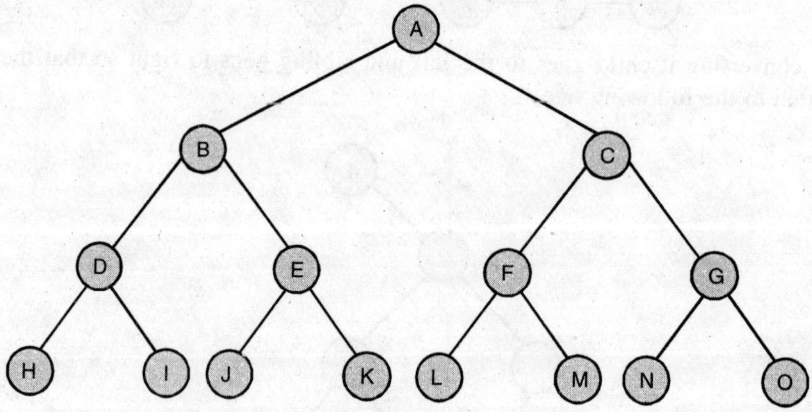

53. How would a binary tree look, if the input were already ordered in ascending order ?

Ans. The tree will then look like as follows for input :

5, 10, 15, 20, 25

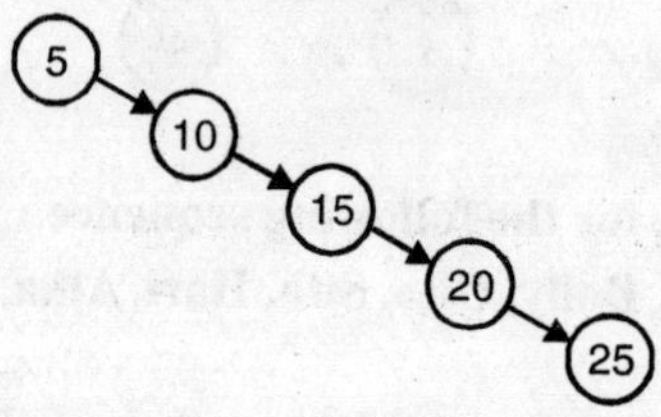

54. Consider the following tree :

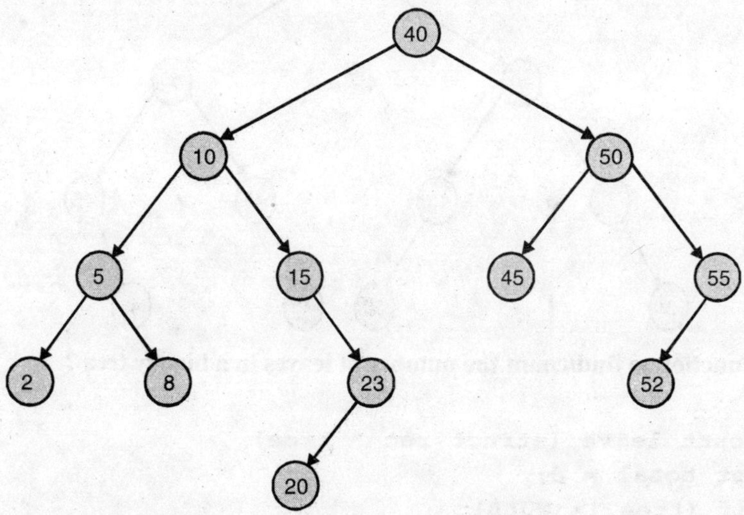

(*a*) Show how the tree will look after we insert node 4, 54 and 60.

(*b*) Show how the tree will look after we delete 8, 23 and 50. Start with the original tree ?

Ans. (*a*)

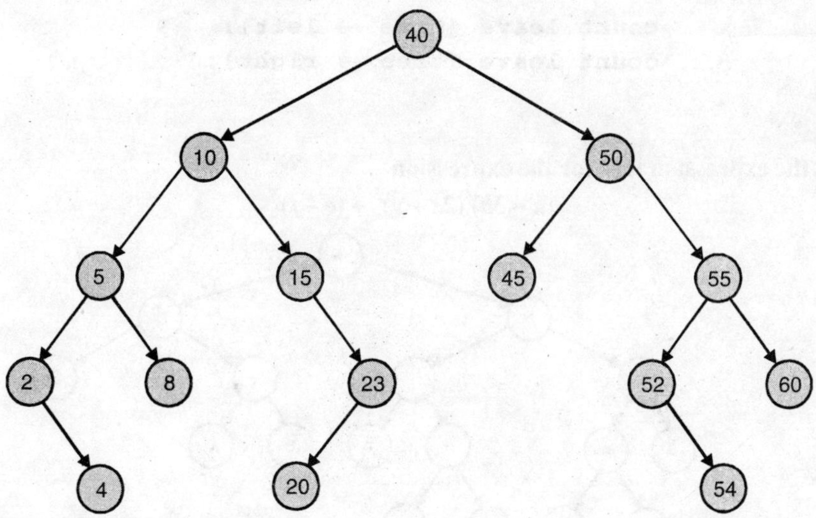

(*b*) Node 8 is a leaf node, so its deletion, doesn't cause any change in the structure of the tree.

When we delete 23, its left child is connected to the parent node of 23. Since 20 is greater than 15, it's connected as its right child.

When we delete 50, then the leftmost node *i.e.* 52 of its right subtree is replaced by 50 and right-child of 52 replaces node 52.

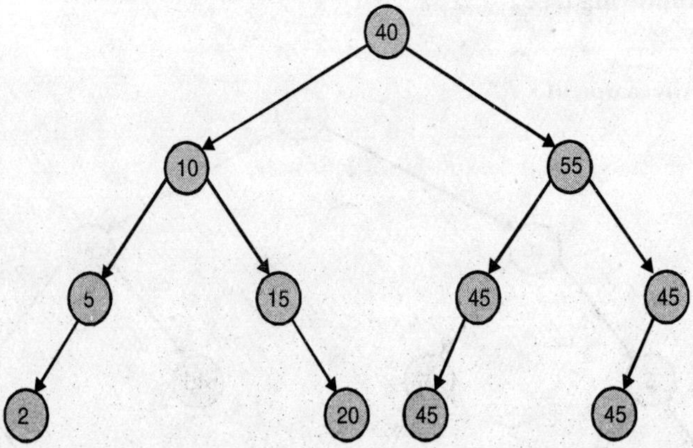

55. Write a function to find/count the number of leaves in a bianry tree ?

Ans.

```
void count leave (struct rec * tree)
  { int total = 0;
     if (tree != NULL)
     {
     if ((tree → left = = NULL) &&
        (tree → right = = NULL))
            total ++;
        else
            count leave (tree → left);
            count leave (tree → right);
     }
  }
```

56. Draw the expression tree for the expression

$$(a - 3b)\,(2x - y)^5 + (e - f)^3$$

Ans.

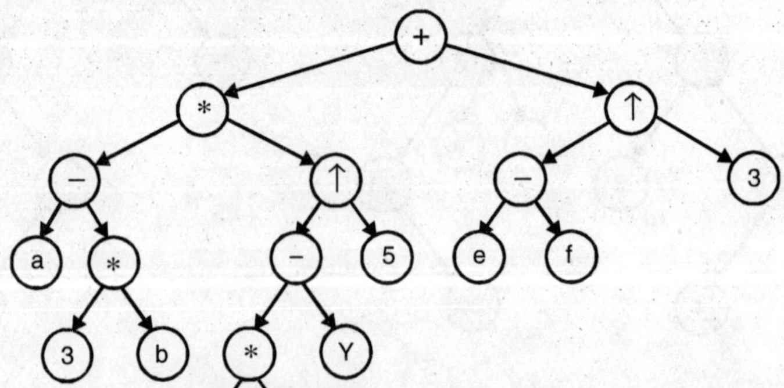

Required Expression Tree

Pre-order :

$+ * - a * 3\, b \uparrow - * 2\, X Y 5 \uparrow - e f 3$

Post-order :

$a\, 3\, b * - 2\, X * Y - 5 \uparrow * e f - 3 \uparrow +$

57. Form a binary tree for the expression :

$$Y : a + b - c * (D - E) + F$$

and also find the scope of - ?

Ans. The binary tree are formed such that all the operators are at internal nodes of the tree, while all the operands are placed as the external or leaf node of the binary tree.

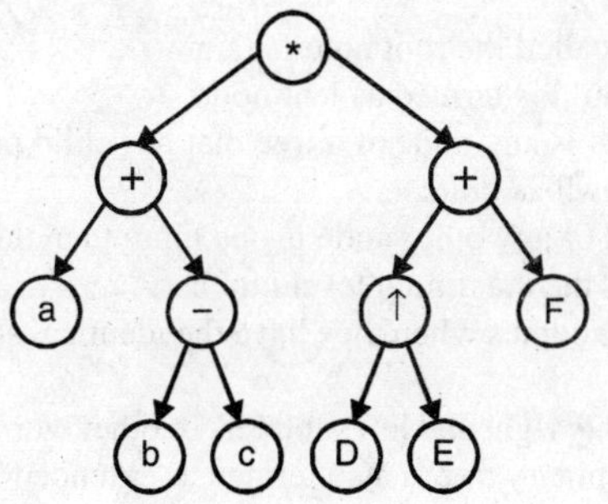

Scope of – are :

D and E

i.e. its left and right subtrees.

58. Form a binary tree for the expression :

(*a*) $Y : (2a + b)(C + D)^4$

(*b*) $Y : (3a + b - c)(d + e)^3$

Ans. (*a*) $Y : (2a + b)(C + D)^4$

$(2a + b) * (C + D) \uparrow 4$

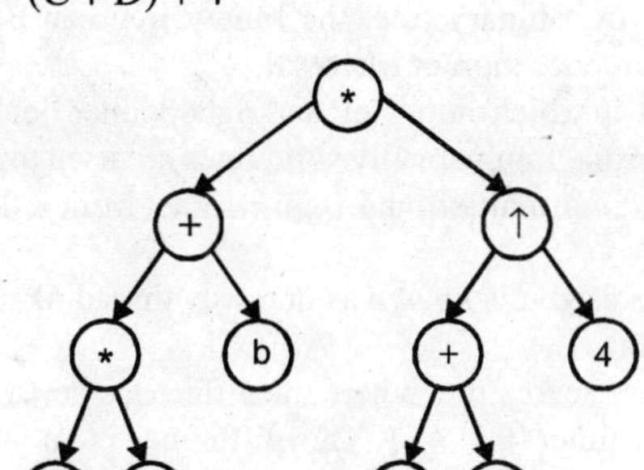

(*b*) $Y : (3a + b - c)(d + e)^3$

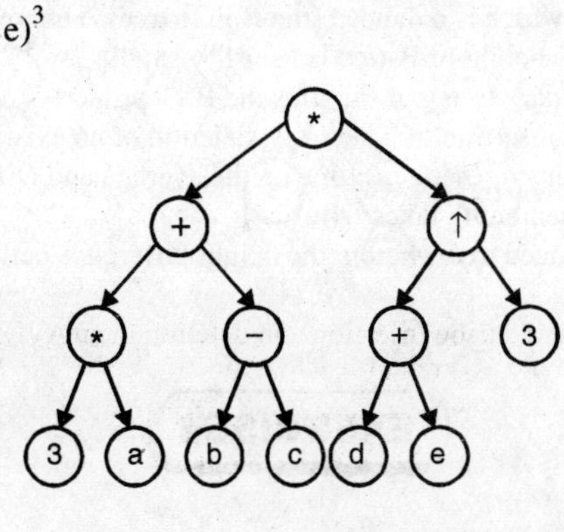

SUMMARY

1. A binary tree is type of tree with finite number of elements and is divided into three main parts. The first part is called root of the tree and other two parts are itself binary tree which exists towards left and right of the tree.
2. Each of the binary trees towards left and right of root are known as subtrees and subtrees may also have their root, left child and right child.
3. The topmost node in a tree is called the root node.
4. A node with no left or right child is termed as leaf node.
5. An internal node or inner node is any node of a tree that has child nodes and is thus not a leaf node.
6. A path ending in a leaf in known as branch.
7. The root has level-0 and level of any other node is one more then the level of its parent.
8. The depth of the binary tree is the maximum level number.
9. Two binary trees are said to be copies when they have the identical structure and identical nodes at all levels of tree.
10. It is a binary tree with non empty right and left subtrees. In other words it is binary tree with every node N has either 0 to 2 tree. The strictly binary tree is also termed as extended 2-Tree or simply 2-Tree.
11. It is a special type of strictly binary tree where all the leaves of the tree reside at the same level. Using the depth we always say a complete binary tree of depth d where all leaves will be at level d.
12. Expression tree are the simple binary tree that represent an arithmetic expression. Here we consider only the unary and binary operator but if an expression contains a function or operator with more than 2 operands then the corresponding expression tree will be general tree.
13. The maximum number of nodes in a tree at level 'i' is 2^{i-1} where i >=1.
14. The maximum number of nodes in a tree of height 'h' is equal to 2^h-1 where $h \leq 1$.
15. If n2 is the maximum number of nodes having two child and n0 is the maximum number of nodes having zero child then prove that $n_0 = n_2 + 1$.
16. 16. For every node N in the BST, the following property will be true: The data at left node will be smaller then data at node N and data of right child will be greater then equal to data at node N.
17. Given a preorder & inorder of a binary tree, the binary tree can be reconstructed the tree can also be reconstructed given its post order & inorder traversal.
18. A threaded binary tree is a tree in which empty left and right pointer field are utilized to points to higher nodes. In general to show binary tree diagrammatically solid lines are used to represent left and right node from the parent node. To differentiate a normal left and right pointer from a thread, the threads are represented by dashed lines.
19. When just one pointer field is used it is known as one-way threading and when both the pointers are used it is known as two-way threading.
20. An AVL tree is a type of binary search tree where the difference between the height of left subtree (H_L) and height of right subtree(H_R) is either 0, 1 or -1. This difference is known as balance factor (BF).
21. M-way search tree are similar to binary search tree with the difference that instead of one key and two pointers (left and right) in each node an m-way tree can have m pointers and m-1 key in one node at the most.
22. B tree is type of m-way tree which is balanced (the B in B-tree). The tree is called balanced as all leaves are at the same level. Due to this height of B-tree is usual l y small.
23. A Binary Search Tree is a binary tree that satisfies the BST property.
24. Operations on BSTs include insertion of a new key, deletion of an existing key, and searching for a key.
25. The BST search algorithm runs in $O(\log_2 n)$ time on the average and $O(n)$ time on the worst case.
26. Balanced BSTs ensure that searching takes $O(\log_2 n)$
27. An AVL tree is a height-balanced tree wherein the height difference between the subtrees of a node for every node in the tree is at most 1.
28. Rotations have to be performed during insertion and deletion in an AVL tree to maintain its balance

EXERCISES

1. What is digital search tree?
2. What is top down multiway search tree? Write algo / C code for insertion in B-tree.

3. What are the differences among strictly binary trees complete binary tree and almost complete binary tree?
4. Insert the following numbers to create an AVL Tree: 28 73 89 75 74 13 10 5
5. Describe storage schemes for threaded binary tree. Which scheme is better and why? Give an example to justify your answer?
6. Take 10 consecutive words from your favorite dictionary and construct an AVL tree from them.
7. Given inorder: 5 1 3 11 6 8 4 2 7 and Preorder: 6 1 5 11 3 4 8 7 2, construct Binary tree.
8. Consider the algebraic expression $E = (4A + B)(3X-Y)^3$
   (*i*) Draw the tree that corresponds to expression E
   (*ii*) Find the prefix polish expression P which is equivalent to E, & find the preorder of T.
9. Explain threaded binary. How to distinguish between a thread and a pointer in such a tree?
10. Write an algorithm for deleting a nude from a binary search tree. Take all possible cases.
11. What is multi-way tree? How does it differ and resemble with a binary search tree. Discuss the process of inserting a new key in a multi-way tree?
12. Prove that a strictly binary tree with n leaves contains 2n-1 nodes.
13. Two binary trees are similar if they are both empty or if they are both nonempty and the left and right subtrees are similar. Write an algorithm if two binary trees are similar.
14, Distinguish between full binary tree and complete binary tree using an example.
15. Draw all possible non similar Trees T-where:
   (*i*) T is binary tree with 3 nodes
   (*ii*) T is a 2-Tree with 4 external nodes
16. What is almost complete binary tree? Explain with suitable example.
17. Insert the following list of integers in a binary search tree:
   7, 39, -2, 0, 3, 42, 20, 5, 40
   Make inorder traversal of the tree and print the values. State the algorithm also.
18. Differentiate between trie and dictionary.
19. Insert the following list of elements in an AVL tree:
   3, 5, 11, 8, 4, 1, 12, 7, 2, 6, 10
20. What is threaded binary tree? How it can be represented in computer memory?
21. From the following integer elements: 2 5 7 90 34 23 1 34 25 10 3 7 11, create a B tree of order 3 and order 5.
22. Discuss theoretically the problem of deletion in B tree.
23. Show all different types of rotation in an AVL tree using a practical example.
24. Given the following data:
   Jan, Feb, Mar, Apr, May, Jun,, Jul. Aug, Sep, Oct, Nov, Dec
   Create AVL tree, Binary search tree, B tree of order 3
25. What are the disadvantages of using array for representation of binary tree?
26. Compare and contrast binary tree with binary search tree.
27. Insertion of a new node in a BST will be a leaf node or any internal node?
28. How can we make a new node to be inserted in a search tree to appear as internal node?
29. Name the various operations that alter the binary tree.
30. How the performance of binary search tree is affected by its performance?
31. Name any operation(s) that can only be performed over BST and not on binary tree.
32. Consider the preorder and inorder given for a binary tree
   Preorder : 1, 2, 12, 13, 11, 14, 16, 17
   Inorder : 12, 2, 13, 1, 14, 11, 17, 16
   Find the binary, tree
33. Consider the Inorder and postorder for a binary tree
   (*i*) Inorder-: 4, 3, 11, 5, 1, 8- 2, 17 -10, 9
   (*ii*) Postorder : 4, 11, 5, 3, 81V 10, 9. 2, 1
   Find the binary tree
34. Consider the Preorder and Inordër given for a binary tree:
   (*i*) Preorder: ABCEIFJDGHKL
   (*ii*) Inorder: EICFJBGDKLA
   Find (*i*) height of binary tree (*ii*) Number of leaf and internal nodes

❑❑❑

CHAPTER 11

GRAPHS

11.1. INTRODUCTION

A graph G is composed of a finite set of vertices and a finite set of edges. A graph is denoted as G = (V, E). The elements of V are the vertices and those of E are the edges. The vertex set of V are denoted by V_G and edge set is denoted by E_G. Thus $G = (V_G, E_G)$

In general graphs are known as simple graphs. An edge e is denoted by e = (u, v) where u and v are the nodes, endpoints, links or vertices. The two vertices u and v are said to be adjacent or neighbors to each other. An edge is usually associated with either 1 or 2 vertices. An edge where u and v are same is known as loop *i.e.* associated with just one vertex. The number of vertices in a graph is called its order and number of edges is called its size.

A graph can be represented by drawing it as plane figure with nodes and edges. The edges are drawn between end points. Edges sometimes can be given name or are denoted by the endpoints they contain and vertices are encircled. For example for the graph shown below:

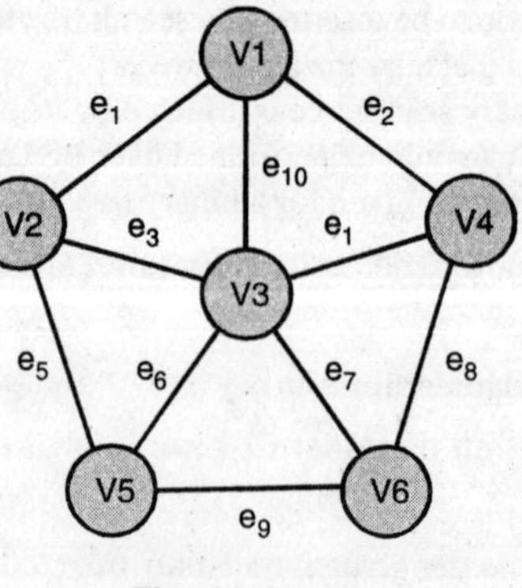

Fig. 11.1.

we have $V_G = \{V_1. V_2, V_3, V_4, V_5, V_6\}$

and $E_G = \{e_1, e_2, e_3, e_4, e_5, e_6, e_7, e_8, e_9, e_{10}\}$

11.2. GRAPH TERMINOLOGY

(*i*) An edge associated with one vertex is called a loop.

(*ii*) Two distinct edges with the same endpoints are called parallel edges.

(*iii*) Two vertices that are connected by an edge are called adjacent.

(*iv*) An edge is called incident on each of its endpoints. Two edges incident on the same endpoints are called adjacent.

(*v*) A vertex with no edges incident on it is isolated.

(*vi*) A graph with no vertices is empty graph or null graph.

(*vii*) Number of edges incident on a vertex v is known as degree of the vertex and denoted as deg [v]. A loop is counted a 2. For an isolated vertex degree is 0.

(*viii*) The degree of a graph is the sum of all the degrees of the vertices in the graph.

(*ix*) The total degree of a graph G is always an even number.

(*x*) A sequence of edges that begins at an initial vertex and ends at a terminal vertex is known as path. A path of length N starting from a vertex u to vertex v can be defined as: P (v_0, v_1, v_2, v_3......... v_N). Here u = v_0 and v = v_N, Number of edges in the path is known as length of the path. In the path v_0 will be adjacent to v1, v2 will be v3 and finally v_{n-1} to V_N. In general v_{i-1} will be adjacent to v_i for all i = 1, 2, 3 .... N.

(*xi*) A path is said to be closed when first and last vertex is same *i.e.* $V_0 = V_N$.

(*xii*) A cycle is a closed simple path with length > = 3.

(*xiii*) A path is simple when all vertices are distinct.

(*xiv*) Reach ability in a graph describes a relationship between two vertices in which one vertex is reachable from the other via a path.

(*xv*) A connected graph is a graph in which for any given vertex in the graph, all the other vertices are reachable from it that is there is a path between any two vertices.

(*xvi*) A graph G is known as strongly connected for any two pair of vertices u and v if there is a path from u to v and a path from v to u.

(*xvii*) Complete graph is simple graph in which every pair of distinct vertices are adjacent. A complete graph will always be connected.

(*xviii*) A complete graph with N vertices will have N (N -1) /2 edges in it.

(*xix*) A graph is said to be labeled when edges are assigned some label. In general the label will be weight or length of the edge (u, v). Such a graph is known as weighted graph. By default when no weight is specified then weight of every edge is assumed to be 1.

(*xx*) Two edges el and e2 are said to be multiple/parallel edges when el = (u, v) and e2 = (u, v) *i.e.* their end points are same.

(*xxi*) A graph with multiple edges and loop is known as multigraph.

(*xxii*) A simple graph G is a graph that doesn't have loops or parallel edges.

(*xxiii*) A directed graph or digraph G (written as DAG)is a finite set of vertices and a finite set of directed edges, where each edge is associated with an ordered pair of vertices called its endpoints. In directed graph for an edge e = (u, v), u is the source, origin or starting point and v is destination, terminal point or ending point; v is adjacent to u.

(*xxiv*) In a directed graph indegree (written as indeg (v)) of a vertex is the number of edges coming to it and outdegree (written as outdeg (v)) of a vertex is number of edges going out from it.

(*xxv*) A node in DAG is known as source when its indegree is 0 and outdegree >0

(*xxvi*) A node in DAG is known as sink when its outdegree is 0 and indegree >0

(*xxvii*) A graph H is a subgraph of G if every vertex in H is also in G, every edge in H is also an edge is G, and every edge in H has the same endpoints as in G

(*xxviii*) A walk from v to w is a finite alternating sequence of adjacent vertices and edges of a graph. Less formally a walk is any route through a graph from vertex to vertex along edges. A walk can end on the same vertex on which it began or on a different vertex. A walk can travel over any edge and any vertex any number of times.

(*xxix*) A path is a walk that does not include any vertex twice, except that its first vertex might be the same as its last.

(*xxx*) A trail is a walk that does not pass over the same edge twice. A trail might visit the same vertex twice, but only if it comes and goes from a different edge each time.

(*xxxi*) A closed walk is a walk that begins and ends at the same vertex.

(*xxxii*) A circuit is a trail that begins and ends on the same vertex. Another definition of circuit is that a circuit is a closed walk that does not contain a repeated edge.

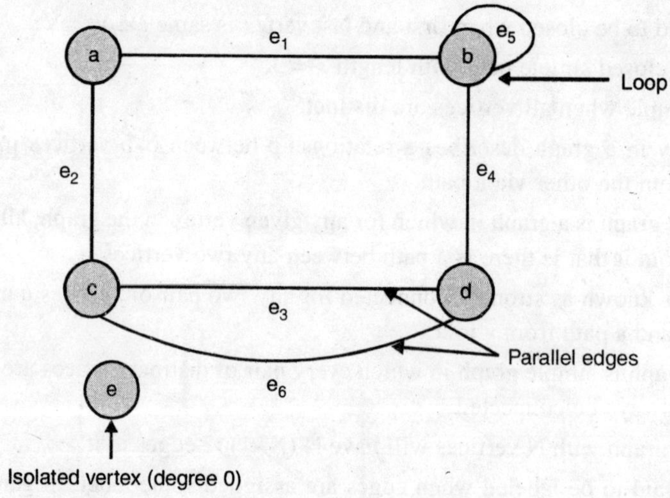

(A)

e1 and e2 are adjacent
e2 and e3 are adjacent
e1 and e4 are adjacent
e3 and e4 are adjacent
degree of (b) is 2 + 2 = 4
degree of (a), is 2
degre of (c) and (d) is 3

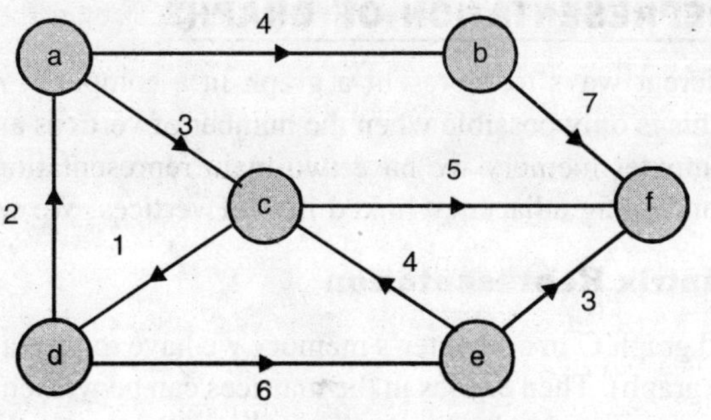

(B)

In-deg [c] = 2, out-deg [c] = 2
in-deg [f] = 3 out-deg [f] = 0 (sink)
a -> c -> d -> e -> f is a path
a -> c -> d -> a is a cycle
Directed graph, weighted graph

Complete graph of 4 vertices — A subgraph of above graph

(C)

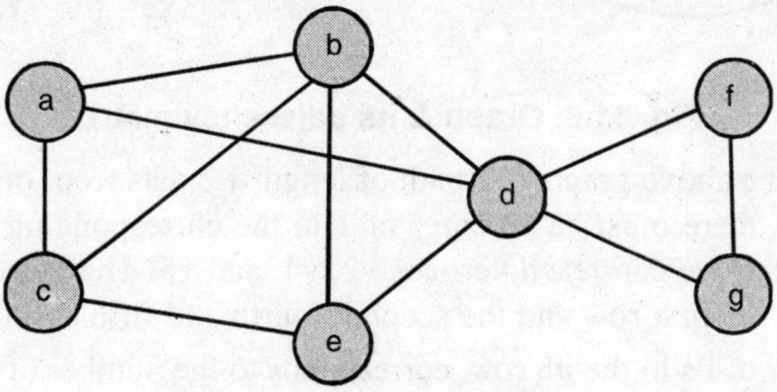

(D)

a, b, d, f, g, d is a walk
d, f, g, d, e, d, f, is a walk
a, b, c, e, b, d, g is a triail
a, b, c, e, b, d, a is a circuit

Fig. 11.2. Graph Terminology

11.3. COMPUTER REPRESENTATION OF GRAPHS

There are several different ways to represent a graph in a computer. Although graphs are usually shown diagrammatically, this is only possible when the number of vertices and edges is reasonably small. To represent graphs in computer memory we have two main representations. The first one is by using adjacency matrix and second using adjacency linked lists of vertices. We will take both one by one.

11.3.1. Adjacency Matrix Representation

To represent a directed graph G in computer's memory we have to take a matrix of dimension N by N (N is number of vertices in graph). Then entries in the matrices can be written on the basis of the following assuming A is our adjacency matrix for the graph.

A [i, j] = 1 if there is a directed edge from vertex v_i to vertex v_j

A [i, j] = 0 if there is no directed edge from vertex v_i to vertex v_j

Here it is assumed that vertices are ordered from 1 to N. The matrix thus obtained contains only entries 0's and l's and is known as adjacency matrix, Boolean matrix or bit matrix. The adjacency matrix obtained depends upon the how we do the numbering of the vertices. A different ordering of vertices can result into different adjacency matrix. Adjacency matrix representation is also known as sequential representation.

Example 11.1.

Consider the graph given below and its adjacency matrix representation on right side :

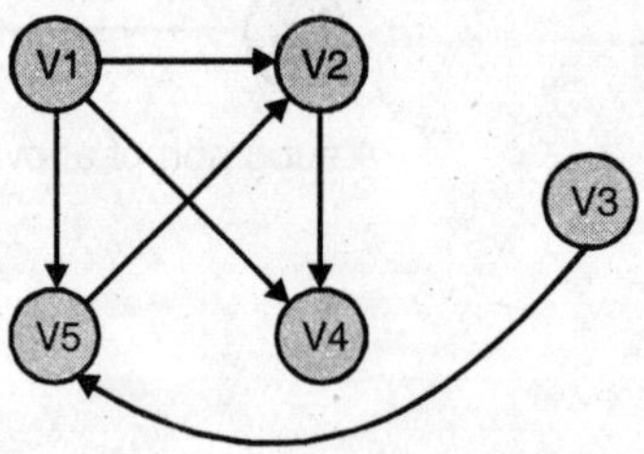

	V_1	V_2	V_3	V_4	V_5
V1	0	1	0	1	1
V2	0	0	0	1	0
V3	0	0	0	0	1
V4	0	0	0	0	0
V5	0	1	0	0	0

Fig. 11.3. Graph & its adjacency matrix.

As you can see from the above graph, if a path of length 1 exists from one vertex to another (*i.e.* the two vertices are adjacent), there must be an entry of 1 in the corresponding position in the matrix. For example, from the vertex v1, we can reach vertices v2, v4, and v5. Therefore, we have a corresponding entry of 1 in the matrix in the first row and the second, fourth and fifth columns.

In general, the number of l's in the ith row, corresponds to the number of edges leaving the vertex v_i, and the number of l's in the jth column, corresponds to the number of edges entering the vertex v_j.

The major advantage of matrix representation is that the calculation of paths and cycles can easily be performed using well known operations of matrices. However, the disadvantage is that this form of representation takes away from the visual aspect of graphs. It would be difficult to illustrate in a matrix, properties that are easily illustrated graphically. Another disadvantage of this form of representation is that for large number of graphs, lot of entries in the matrix will be 0's so matrix will be sparse. This will result in wastage of lot of memory space. Further addition and deletion in graph will be time consuming operations and won't be easy.

11.3.2. Calculating A Path Between Vertices

As shown in the previous example, the existence of an edge between two vertices v_i and v_j is shown by an entry of 1 in the ith row and j^{th} column of the adjacency matrix. This entry represents a path of length 1 from v_i to v_j We denote this matrix by A^1. To compute a path of length 2, the matrix of length 1 must be multiplied by itself, and the product matrix is the matrix representation of path of length 2 and the matrix denoted is A^2 Similarly when this matrix A^2 is multiplied by A^1 we can get matrix A^3. An entry A^3 [i,j] in the matrix A^3 denotes number of paths of length 3. Similarly an entry A^2 [i, j] denote number of paths of length 2 in matrix A^2.

In general, to generate the matrix of path of length N, take the matrix of path of length N-1, and multiply it with the matrix of path of length 1.

Example 11.2.

Consider the graph shown below. We compute all matrices A^1, A^2, A^3 and A^4 for this graph.

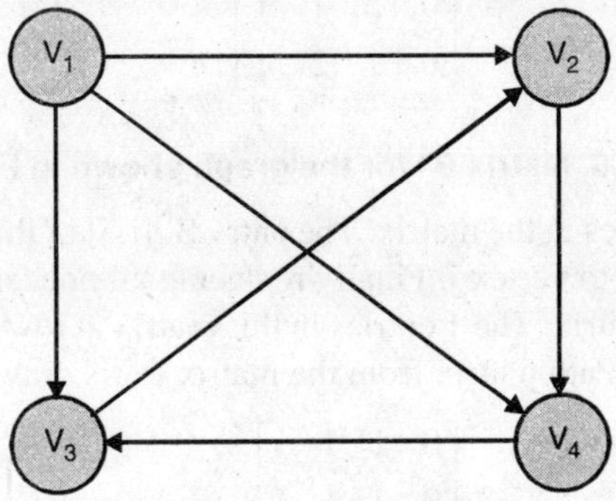

$$A^1 = \begin{array}{c} \\ V_1 \\ V_2 \\ V_3 \\ V_4 \end{array} \begin{array}{c} \begin{array}{cccc} V_1 & V_2 & V_3 & V_4 \end{array} \\ \begin{bmatrix} 0 & 1 & 1 & 1 \\ 0 & 0 & 0 & 1 \\ 0 & 1 & 0 & 0 \\ 0 & 0 & 1 & 0 \end{bmatrix} \end{array}$$

$$A^2 = \begin{bmatrix} 0 & 1 & 1 & 1 \\ 0 & 0 & 1 & 0 \\ 0 & 0 & 0 & 1 \\ 0 & 1 & 0 & 0 \end{bmatrix}$$

$$A^3 = \begin{bmatrix} 0 & 1 & 1 & 1 \\ 0 & 1 & 0 & 0 \\ 0 & 0 & 1 & 0 \\ 0 & 0 & 0 & 1 \end{bmatrix}$$

$$A^4 = \begin{bmatrix} 0 & 1 & 1 & 1 \\ 0 & 0 & 0 & 1 \\ 0 & 1 & 0 & 0 \\ 0 & 0 & 1 & 0 \end{bmatrix}$$

Fig. 11.4. Graph & its A^i matrices (i = 1 + 4)

To understand more about the above matrices A^i's, consider entry $A^2[2,3]$. This entry is 1, it means there is one path of length 2 from vertex v_2 to v_3. This path is v_2 -> v_3v_4. Similarly consider an entry $A^3[1,4]$. This entry is 1, it means there is one path of length 3 from v_1 to v_4. This path is v_1->v_3->v_2 ->v_4.Finally consider an entry $A^4[3,2]$.This entry is 1, it means there is one path of length 4 from vertex v_3 to v_2. This path is v_3->v_2->v_4->v_3->v_2.

Now finally we add all the matrices A^i's where i=1,2,3,4 then we get a matrix B as:

$$B^4 = A^1 + A^2 + A^3 + A^4$$

In general matrix B is defined as :

$$B^N = A^1 + A^2 + \ldots + A^N$$

$$B^4 = \begin{bmatrix} 0 & 4 & 4 & 4 \\ 0 & 1 & 1 & 2 \\ 0 & 2 & 1 & 1 \\ 0 & 1 & 2 & 1 \end{bmatrix}$$

Fig. 11.5. Matrix B^4 for the graph shown in Fig. 11.3.

Where N is number of vertices in the matrix. The entry $B^N[i, j]$ of this matrix gives us number of paths of length N or less from vertex i to vertex j. Finally replacing all nonzero entries in matrix B^N by 1 gives a matrix that is known as path matrix. The 1 entries in this matrix at row i and column j simply tells us that node j is reachable from node i. Path matrix from the matrix B^N is drawn below:

$$P = \begin{bmatrix} 0 & 1 & 1 & 1 \\ 0 & 1 & 1 & 1 \\ 0 & 1 & 1 & 1 \\ 0 & 1 & 1 & 1 \end{bmatrix}$$

Fig. 11.6. Path matrix for the above graph.

Examining the matrix P, we see that the node v2 is not reaches from any of the other nodes.

Recall that a directed graph G is said to be strongly connected if, for any pair of nodes u and v in G, there are both a path from u and v and a path from v to u. Accordingly, G is strongly connected if and only if the path matrix P and G has no zero entities. Thus the graph G in Fig. 8.3. is not strongly connected.

The **transistive closure** of a graph G is defined to be the graph G′ such that G′ has the same nodes as G and there is an edge (vi, vj) in G¢ whenever there is a path from vi and vj in G. Accordingly, the path matrix P of the graph G is precisely the adjacency matrix of its transitive closure G¢. Furthermore,k a graph G is strongly connected if and only if its transitive closure is a complete graph.

Remark : The adjacency matrix A and the path matrix P of a graph G may be viewed as logical (Boolean) matrices, where 0 represents "false" and 1 represents "true." Thus, the logical operations of Ù (AND) and Ú (OR) may be applied to the entries of A and P. The value of Ù and Ú appear in Fig. 11.7. These operations will be used in the next section.

∧	0	1
0	0	0
1	0	1

(*a*) AND

∨	0	1
0	0	1
1	1	1

(*b*) OR

Fig. 11.7.

11.3.3. Linked Representation of Graph

Let G be a directed graph with m nodes. Figure 11.7 (a) shows the graph and Fig. 11.7(b) shows each node in G followed by its adjacency list. Fig. 11.7(c) shows a schematic diagram of linked representation of G in memory. The linked representation will contains two lists,

(*i*) a node list Node and

(*ii*) an edge list Edge.

Node list

Each element in the list node will correspond to a node in G, and it will be a record of the form :

node	next	adj	

Hence *node* will be the name or key value of the node, *next* will be a pointer to the next node in the list *node* and *adj* will be a pointer to the first element in the adjacency list of the node, which is mtainined in the list *Edge.* The shaded area indicates that there may be other information in the record, such as the indegree indeg of node, the outdegree *outdeg* of the node, the *status* of the node during the execution of an algorithm, and so on. (*Node* is an array of records containing fields such as *name, indeg, outdeg, status, ...)*

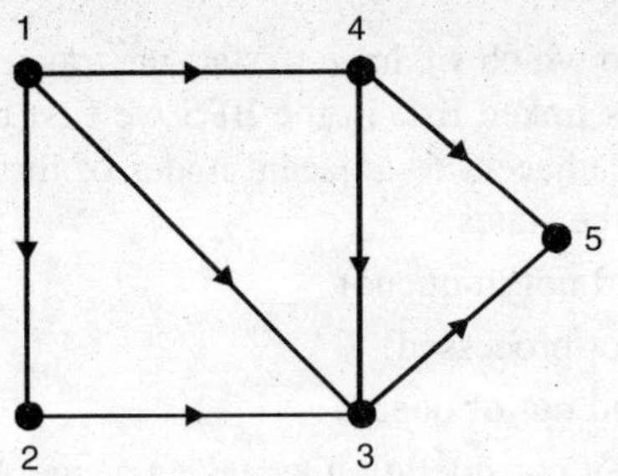

(*a*) *Graph G*

Node	Adjacent List
1.	2, 3, 4
3.	3
3.	
4.	3, 5
5.	3

(*b*) *Adjacent list of G*

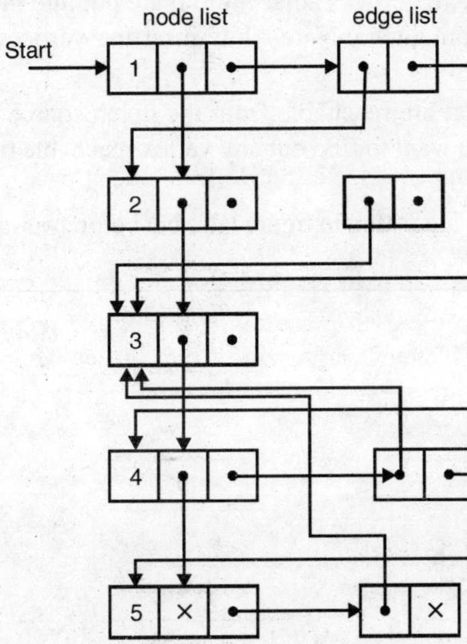

(*c*) *Linked representation of G*

Fig. 11.8.

Edge list

Each element in the list *Edge* will correspond to an edge of G and will be a record of the form :

dest	link	

The field *dest* will point to the location in the list *Node* of the destination or terminal nodes in the same adjacency list. The field *link* will link together the edges with the same initial node. *i.e.* the nodes in the same adjacency list. The shaded area indicates that there may be other information in the record corresponding to the edge, such as a field edge containg the labelled data of the edge when G is a labelled graph, a field *weight* containing the weight of the edge when G is a weighted graph, and so on.

11.4. GRAPH TRAVERSAL

Traversing a graph means accessing every vertex of graph and edge and performing some processing if desired (for example to display it) For that we have two popular graph traversal algorithms. The first one is known as Breadth First Search (BFS) and another is Depth First Search (DFS). We will examine both one by one.

11.4.1. Breadth First Search

In BFS we take a source vertex or root vertex from which we have to start the traversal. For keeping track of the vertices, we take a queue implemented as linked list. In the BFS we first traverse the root vertex, then all its adjacent nodes, then all the nodes adjacent to adjacent nodes of the root and so on. During the traversal the vertices may be one of the three states:

(*i*) Status 0, means ready state (not processed and not in queue)

(*ii*) Status 1, means waiting state (in queue but not processed)

(*iii*) Status 2, means processed state (processed and out of queue)

For holding the vertices adjacent to the vertex under consideration we take a queue. Initially the root vertex is considered; its status is changed from 0 to 1 and inserted into queue. The root vertex is processed, deleted from queue (status changed to 2) and all vertices adjacent to it are put into the queue. The process is repeated for every front vertex deleted from queue. Note while inserting vertices into the queue only those vertices whose status is 0 are inserted.

The BFS traversal shows all vertices that are reachable from the given source vertex. It does show only for the vertex passed as argument. If you want to find out any vertex reachable from any other vertex, the algorithm needs to be modified.

We show the step by step execution of the algorithm in figure later but before we see the implementation of BFS in C.

```
#define MAX 10
struct graphnode
{
int vname;
struct graphnode*next;
};
struct Queue
{
int data;
struct Queue*next;
```

```
};
typedef graphnode Gnode;
typedef Queue Que;
Gnode*alloc ( )
{
return (Gnode*)malloc(sizeof(Gnode));
}
Que*front, *rear;
void initque ( )
{
front = NULL;
rear = NULL;
}
void enque (Que**f, Que**r, int d)
{
Que*temp;
temp = (Que*)malloc (sizeof(Que));
temp->data = d;
temp->next = NULL;
if(*r = = NULL)
{
*r = temp;
*f = *r;
}
else
{
(*r)->next = temp;
*r = temp;
}
}
int deque(Que**f, Que**r)
{
int d;
Que*temp;
if(*f = = NULL)
        d = -1;
else
{
temp = *f;
*f = (*f) ->next;
if (*f = = NULL)
    *r = NULL;
```

```
d = temp->data;
free (temp);
}
return d;
}
void BFS(Gnode*G[], int src)
{
Gnode *ptr;
int i, u, v, status[MAX];
initque ( );
printf("\nThe BFS of graph is\n");
for(i =0;i<MAX; i++)
        status[i] = 0;
status[src] = 1;
enque (&front, &rear, src);
u = deque(&front, &rear);
while (u ! = -1)
{
}
ptr = G[u-1];
while (ptr ! = NULL)
{
v = ptr->vname;
if (status[v-1] = = 0)
{
status[v-1] = 1;
enque(&front, &rear, v);
}
ptr = ptr-> next;
}
printf("%d" , u);
status[u-1] = 2;
u = deque(&front, &rear);
}
}
```

During the traversal, the vertices may be in one of the three states: 0, 1 or 2. For that we have taken an array status of size MAX. The value of status [i] will only be 0, 1, or 2. If status[i] for a vertex i is 0, it means ready state, 1 status denotes waiting state and status 2 denotes processed state. Initially for all vertices, status is 0; status for the source vertex is set to 1.

For applying BFS we consider the graph shown below:

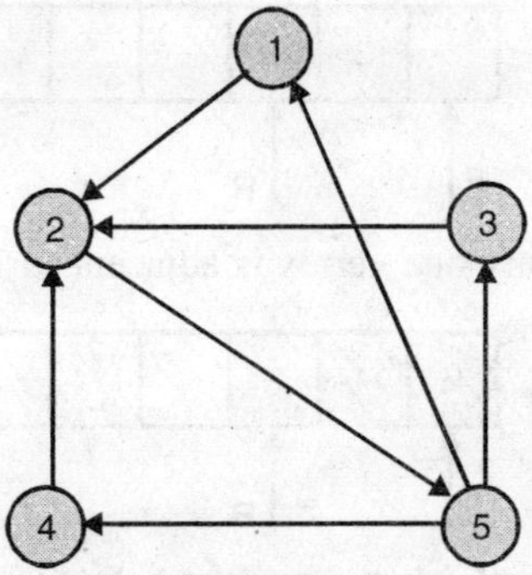

Fig. 11.9.

The source vertex is inserted into queue. The same is deleted from queue and is stored in u. Before deletion the queue will have just one element 5 :

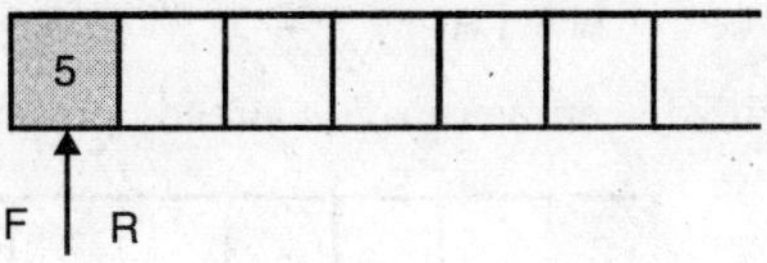

Queue with just one vertext the source vertex

Now while loop runs till queue does not become empty (denote by condition when u contains-1). The pointer ptr points to start of adjacency linked list of vertex u. Now every node of this linked list is examined, If status of any of the vertex is 0 then its status is changed to 1 and is inserted into the queue. As vertex 5 has three vertices adjacent to it, all three will be inserted into the queue with their status 1. We can simply understand the thing as:

(*i*) Initially all nodes have status 0

(*ii*) Nodes have status 1 when they are in queue

(*iii*) Any node deleted from queue has been processed with status changed from 1 to 2.

At this instant the queue will look like as

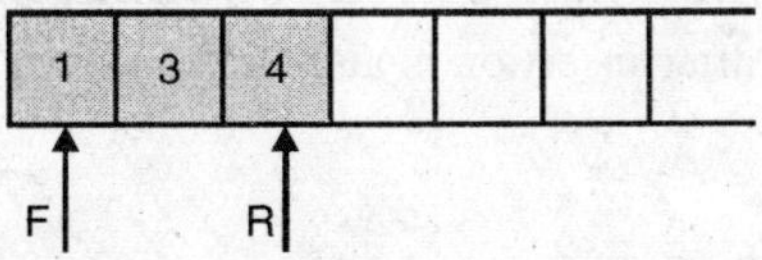

Outside the inner **while** loop, **us** is printed (5 is printed); the status is changed to 2. Front node is deleted and u has the value 1. The process is repeated. The entire processing is shown in the figure given below

(*i*) 5 is inserted

(*ii*) 5 is taken out and printed, all its adjacent vertices are inserted into queue.

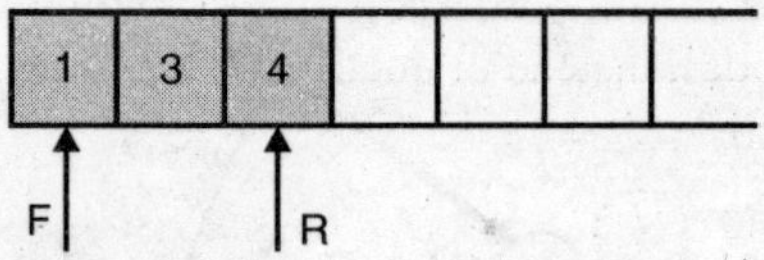

(*iii*) 1 taken out and printed, only one vertex is adjacent to 1 *i.e.* 2, it is inserted into queue.

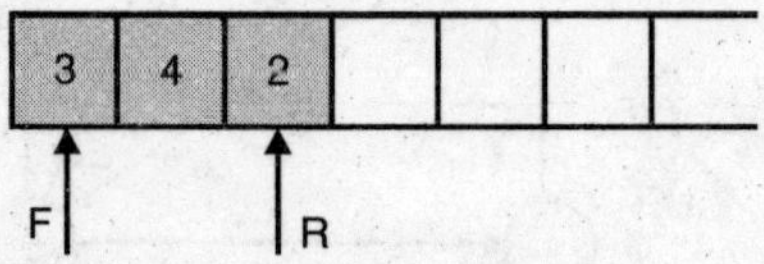

(*iv*) 3 is taken out and printed, its adjacent vertex 2 is already in queue, so it is not inserted again.

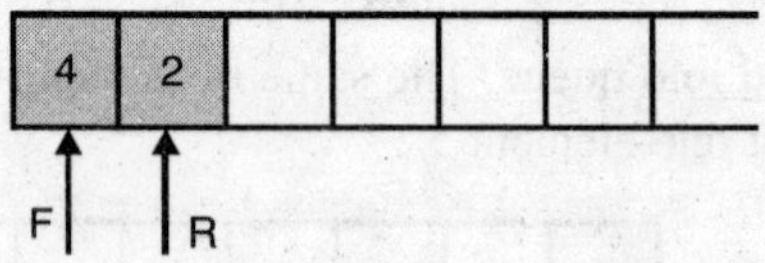

(*v*) 4 is taken out & printed, it adjacent vertex 2 is already present in queue, so it is not inserted again

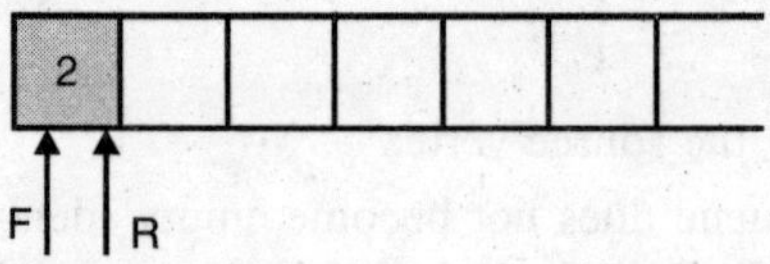

(vi) Finally 2 is taken out & printed; its adjacent vertex 5 is already processed.
BFS stating from source vertex is 51342

Fig. 11.10. BFS in action.

Traversing of a Graph

STATUS 0: Initial states of a vertex 'v' (if the vertex is neither insert in either STACK or QUEUE nor deleted from either STACK or QUEUE)

STATUS 1: a node 'v', if insert in either STACK or QUEUE, then its STATUS changes to 1

STATUS 2: a node 'v', if after insertion is also deleted from either STACK or QUEUE, then its STATUS changes to 2.

Breadth First Search :

1. Initially set the STATUS of all the vertices equal to 0
2. Select a starting vertex 'v' & Insert it in QUEUE Q and change its STATUS to 1.
3. Repeat steps 4 & 5 while QUEUE Q is not empty
4. Delete a vertex 'v' from the FRONT of a QUEUE Q. Process vertex 'v' and change its STATUS to 2.
5. Insert all the neighbor of the vertex 'v' at the REAR of a QUEUE Q, which have its STATUS 0 and changes their STATUS to 1
6. Exit

11.4.2. Depth First Search

In the DFS we start from source vertex and reach to one of its adjacent node; we then move to node adjacent to adjacent to root and so on *i.e.* in DFS we do not process all adjacent nodes first instead we go deeper inside the graph by following one adjacent node to another. That's why it can be used to find out nodes that are reachable from a given node. The implementation of DFS is similar to the BFS with the difference that for storing the nodes, instead of queue we have stack data structure. To understand the DFS we consider the following graph:

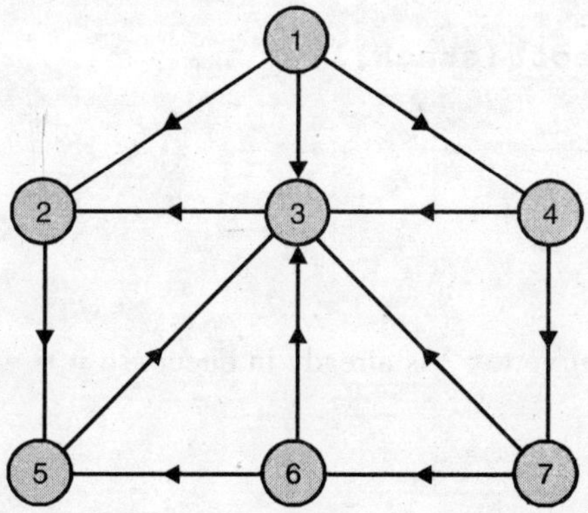

Directed Graph

Adjacency string representation.

vertex 1 →234
vertex 2 →5
vertex 3 →2
vertex 4 →37
Vertex 5 →3
Vertex 6 →53
Vertex 7 →63

Fig. 11.11. Graph for DFS.

The implementation in C given below:

```
# define Max 10
struct graphnode
{
int vname;
struct graphnode*next;
};
struct stack
{
int data;
struct stack*next;
};
typedef struct graphnode Gnode;
typedef struct stack stack;
Gnode*alloc ( )
{
return (Gnode*)malloc(sizeof(Gnode));
}
stack*top;
void push(stack**T, int d)
{
stack *temp;
if(*T = NULL)
{
(*T) = (stack*)malloc(sizeof (stack));
(*T)->data = d;
```

```
(*T)->next = NULL;
}
else
{
temp = (stack*)malloc(sizeof (stack));
temp->data = d;
temp->next = (*T);
(*T)  = temp;
}
}
int pop(stack **T)
{
stack *temp;
int d;
if(*T = = NULL)
            return -1;
d = (*T)->data;
temp = *T;
(*T) = (*T)->next;
free(temp);
return d;
}
void DFS(Gnode*G[ ], int src)
{
Gnode *ptr;
int i, u, v, status[MAX];
top = NULL;
printf("\nThe DFS of graph is\n");
for(i = 1;i< =MAX;i++)
            status [i] = 0;
status[src] = 1;
push(&top, src);
u = pop (&top);
while (u ! = -1)
{
printf("%d", u);
status[u] = 2;
ptr = G[u];
while(ptr ! = NULL)
{
v = ptr -> vname-48;
```

```
if(status{v] = = 0)
{
status[v] = 1
push(&top, v);
ptr = ptr->next;
}
else
break;
}
u = pop(&top);
}
}
```

In the graph we take the source vertex as 4. Initially this source vertex is pushed onto the stack and then popped out. The same is displayed inside the while loop; its status is changed to 2. A pointer ptr points to its adjacency linked list. All its adjacent nodes are traversed, their status is changed to 1 (if earlier status was 0) and they are pushed onto the stack. As node 4 has node 3 and 7 adjacent to it, they are pushed onto the stack. The contents of stack will be as:

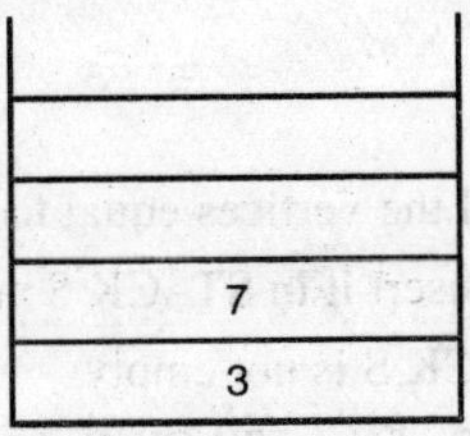

Next 7 is propped out & printed from stack, its status is changed to 2 and all its adjacent node who were in ready state.

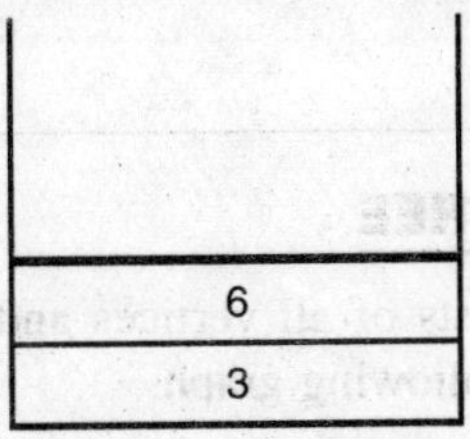

Only 6 will be pushed

Repeating the step again 6 will be popped out & printed. Its adjacent nodes (ready state) will be pushed.

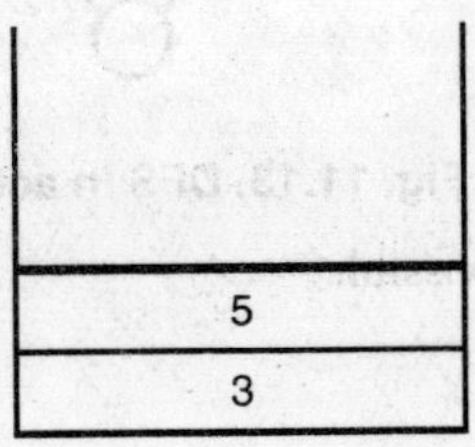

Repeating the steps again, 5 will be popped out & printed.Its adjacent node 3 is already in stack. 3 will be popped out 2 printed. Its adjacent nodes (in ready state) will be pushed onto stack.

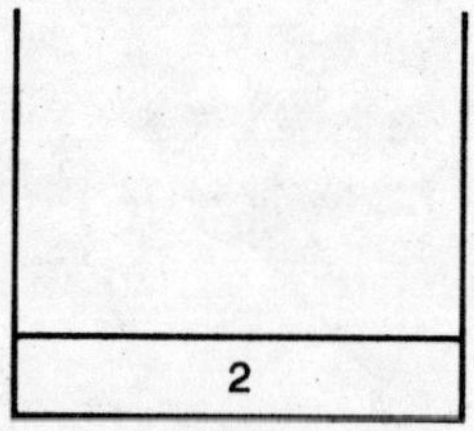

Finally 2 is popped out & printed. Thus DFS of graph with source vertex 4 is 476532 *i.e.* above are the set of vertices reachable from 4.

Fig. 11.12. DFS in action.

Traversing of a Graph

STATUS 0: Initial states of a vertex 'v' (if the vertex is neither insert in either STACK or QUEUE nor deleted from either STACK or QUEUE)

STATUS 1: a node 'v', if insert in either STACK or QUEUE, then its STATUS changes to 1

STATUS 2: a node 'v', if after insertion is also deleted from either STACK or QUEUE, then its STATUS changes to 2

Depth First Search

1. Initially set the STATUS of all the vertices equal to 0
2. Select a starting vertex 'v' & Insert it in STACK S and change its STATUS to 1.
3. Repeat steps 4 & 5 while STACK S is not empty
4. POP a vertex 'v' from the TOP of the STACK S. Process or traverse vertex 'v' and change its STATUS to 2.
5. PUSH all the neighbor of the vertex 'v' at the TOP of STACK S, which have its STATUS 0 and changes their STATUS to 1
6. Exit

11.5. MINIMUM SPANNING TREE

A Spanning tree of a graph consists of all vertices and some of the edges so that graph does not contain a cycle. For example for the following graph.

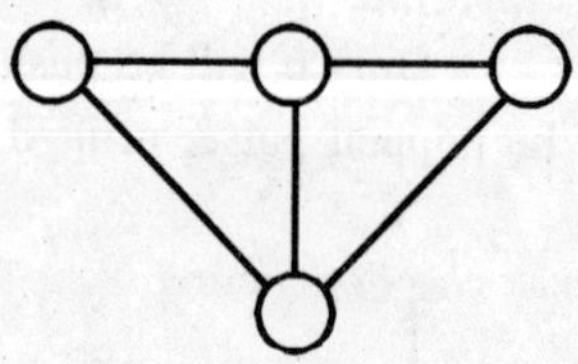

Fig. 11.13. DFS in action.

Following spanning tree may be possible.

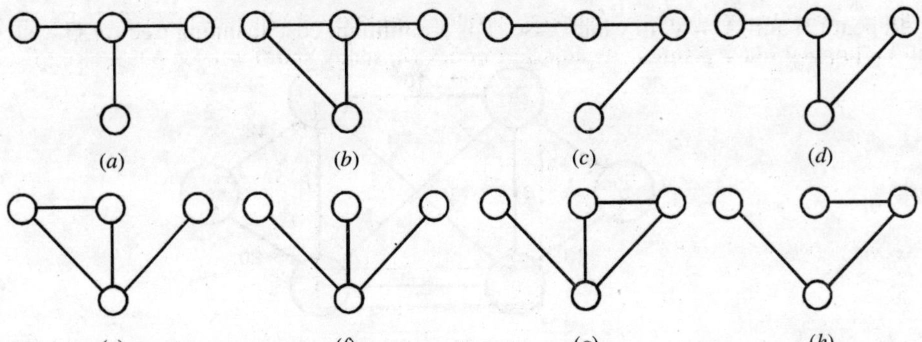

Fig. 11.14.

A minimum spanning tree (MST) is defined for a weighted graph. It is the spanning tree for a weighted graph whose sum of weight of edges in the spanning tree is minimum. For example for the following graph

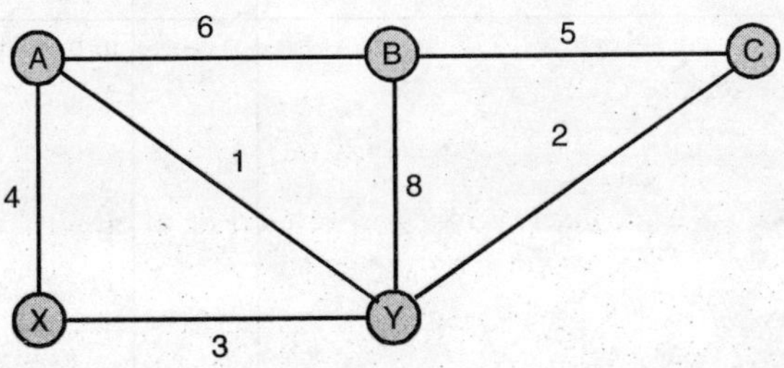

Fig. 11.15. A weighted graph.

The MST will be :

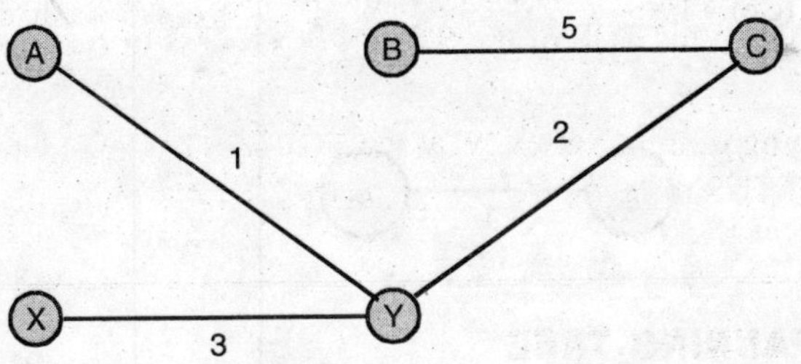

Fig. 11.16. MST of graph shown in previous figure.

For finding a MST for a graph, there are two most popular algorithms given by Prims and Kruskal's. In this section we will understand Prim's algorithm and in the next section we will see algorithm of Kruskals.

11.5.1. Prim's Algorithm

This algorithm finds the edge of least cost connecting some vertex u in U to some vertex v in (V – U) at each step of the algorithm :

Let G = (V, E) be a connected, weighted, undirected graph. Let U denote the set of vertices chosen and T denote the set of edges already taken in at any instance of the algorithm.

1. Choose initial vertex from V and place it in U.
2. From among the vertices in V – U choose that vertex, say v, which is connected to some vertex, and say u, in U by an edge of least cost. Add vertex v to U and edge (u, v) to T.

3. Repeat (2) until U = V, in which case, T is a minimum cost spanning tree for G.

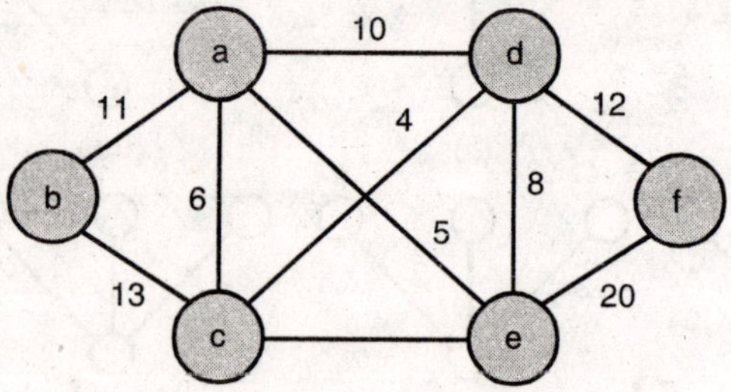

Fig. 11.17.

The following shows the execution of Kruskal's algorithm in solving the MST problem of the graph above :

Edges	MST	U	V – U	Remarks
(c, e) – 1 (c, d) – 4 (a, e) – 5 (a, c) – 6 (d, e) – 8 (a, d) – 10 (a, b) – 11 (d, f) – 12 (b, c) – 13 (e, f) – 20		—	a, b, c, d, e, f	List of edges in non-decreasing order of weight
(c, e) – 1 **accept** (c, d) – 4 (a, e) – 5 (a, c) – 6 (d, e) – 8 (a, d) – 10 (a, b) – 11 (d, f) – 12 (b, c) – 13 (e, f) – 20	(c,e) – 1 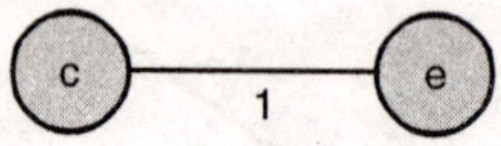	c, e	a, b, d, f	c and e not in U
(c, d) – 4 **accept** (a, e) – 5 (a, c) – 6 (d, e) – 8 (a, d) – 10 (a, b) – 11 (d, f) – 12 (b, c) – 13 (e, f) – 20	(c, e) – 1 (c,d) – 4 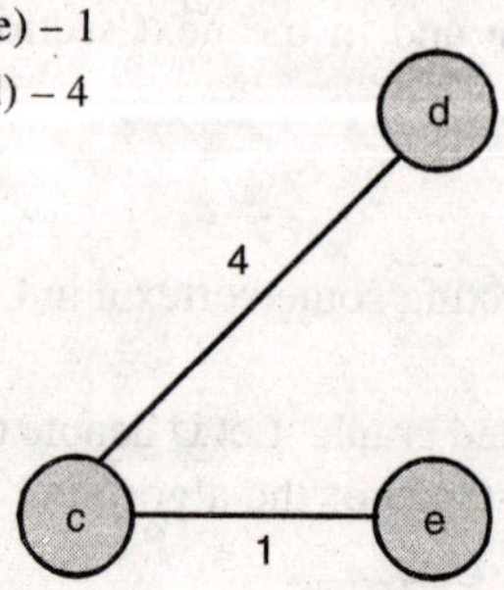	c, d, e	a, b, f	**d** not in U

Edges	MST		U	V – U	Remarks
(a, e) – 5 **accept** (a, c) – 6 (d, e) – 8 (a, d) – 10 (a, b) – 11 (d, f) – 12 (b, c) – 13 (e, f) – 20	(c, e) – 1 (c, d) – 4 (a, e) – 5	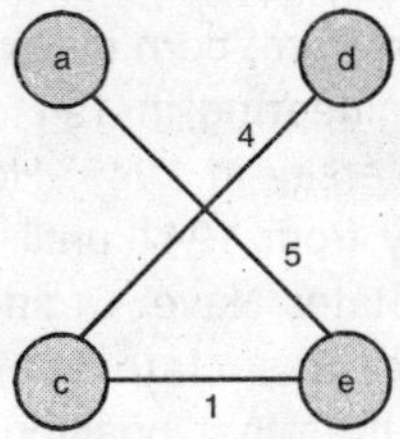	a, c, d, e	b, f	a not in U
(a, c) – 6 **reject** (d, e) – 8 (a, d) – 10 (a, b) – 11 (d, f) – 12 (b, c) – 13 (e, f) – 20	(c, e) – 1 (c, d) – 4 (a, e) – 5	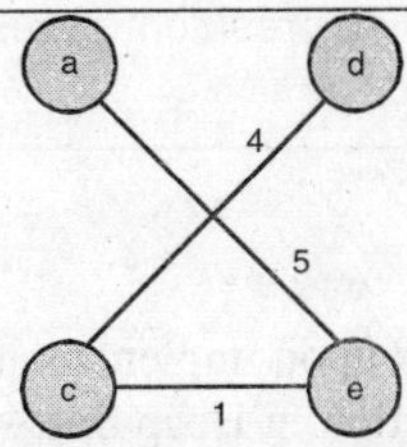	a, c, d, e	b, f	a and c are in U
(d, e) – 8 **reject** (a, d) – 10 (a, b) – 11 (d, f) – 12 (b, c) – 13 (e, f) – 20	(c, e) – 1 (c, d) – 4 (a, e) – 5	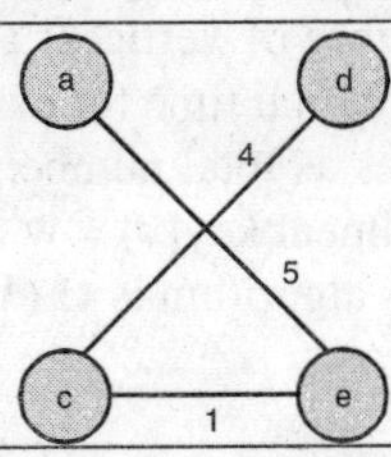	a, c, d, e	b, f	d and e are in U
(a, d) – 10 **reject** (a, b) – 11 (d, f) – 12 (b, c) – 13 (e, f) – 20	(c, e) – 1 (c, d) – 4 (a, e) – 5	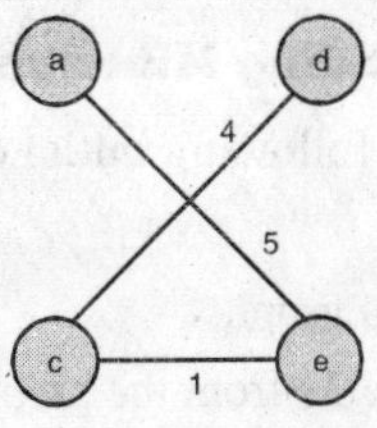	a, c, d, e	b, f	a and d are in U
(a, b) – 11 **accept** (d, f) – 12 (b, c) – 13 (e, f) – 20	(c, e) – 1 (c, d) – 4 (a, e) – 5 (a, b) – 11	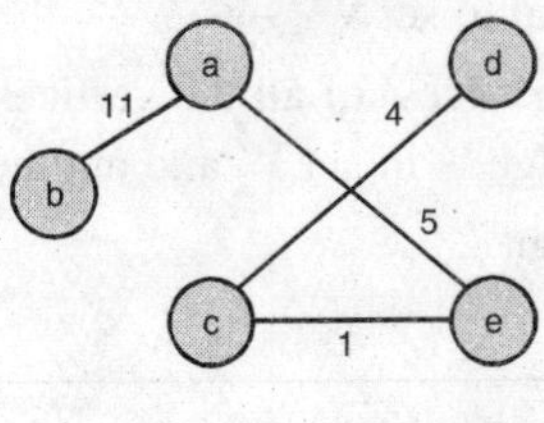	a, b, c, d, e	f	b not in U
(d, f) – 12 **accept** (b, c) – 13 (e, f) – 20 (b, c) – 13 (e, f) – 20	(c, e) – 1 (c, d) – 4 (a, e) – 5 (a, b) – 11 (d,f) - 12	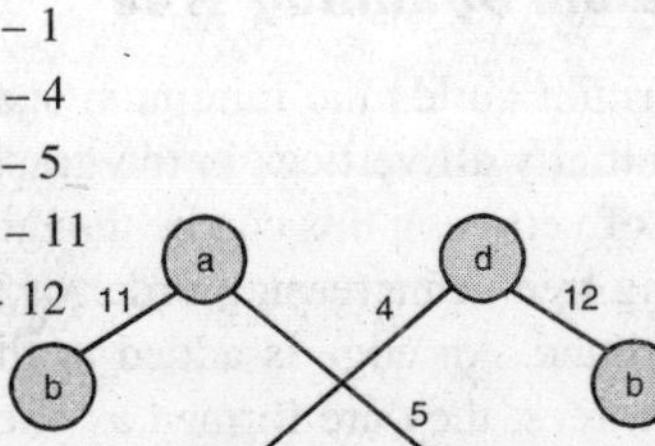	a, b, c, d, e, f		f not in U All vertices are now in U COST = 33

Result of Applying Prim's Algorithm on above Graph

DO YOU KNOW

ROBERT CLAY PRIM(BORN 1921) Robert Prim, born in Sweetwater, Texas, received his B.S. in electrical engineering in 1941, and his Ph.D in mathematics from Princeton University in 1949. He was an engineer at the general Electric company from 1941 until 1944, an engineer and mathematics at the United States Navel Ordinance Lab from 1944 until 1949, and a researcher associate at Princeton University from 1948 until 1949. Among the other positions he has held are director of mathematics and mechanics research at Bell Telephone Lab from 1958 until 1961 and vice president of research at Sandia Corporation. He is currently retired.

Performance of Prim's Algorithm

The performances of prim's algorithm depend upon implementation of priority queue Q. Assuming it is implemented using heaps (the Build Heap function in Heap Sort sorting, refer chapter "Sorting"). The main loop (line no. 6) executes N times (N number of vertices) and each operation for taking out the vertex with minimum key takes $O(\log_2 V)$ time. So total time for extracting the vertex with minimum key is $O(V \log_2 V)$. The for loop executes O (E) times as total number of vertices adjacent to all vertices is 2 * | E |. The last operation in for loop (the assignment key[v] = w (u, v)) can be implemented in heap in $O(\log_2 V)$ time. Thus total running time of Prim's algorithm is $\mathbf{O(V \log_2 V + E \log_2 V) = O(E \log_2 V)}$.

Algorithm of Prim's Algorithm

Algorithm 11.1. Prim's Algorithm for Finding Minimum Spanning Tree ....

1. Create two sets *i.e.* V & V′, with the following initial conditions:
   (a) Set V′ is empty
   (b) Set V contains all the vertices of a graph
   (c) Select the minimum weighted edge (i,j)from the graph G and insert both its end-points *i.e.* i & j into the set V′
2. Repeat step 3 while set V is not equal to set V'
3. Find all the neighbors (***i.e.*** outgoing edges) of all the vertices which are in set V′ such that one-end point of the neighboring edge is in set V′ and another is not in V′
4. Sum up all the selected edge's weight
5. Exit

11.5.2. Kruskál's Algorithm For Minimum Spanning Tree

Similar to Prim's algorithm, Kruskal's algorithm builds the minimum spanning tree of a graph by adding edges to the spanning tree one-by-one. Initially all vertices in the graph corresponds to one tree *i.e.* initially we have N trees (where N is number of vertices), this can be thought of a forest with N trees. The edges are selected and added to the spanning tree in increasing order of their weights *i.e.* an edge with minimum weight is selected during every round. An edge is added to the tree only if it does not create a cycle. If the edge connects two different trees, they are formed as one. The general working of algorithm is given below:

1. Create a forest F (a set of trees), where each vertex in the graph is a separate tree.
2. Create a set S containing all the edges in the graph (Usually a priority queue with edge with minimum weight is at front of the queue)
3. Till S does not become empty
   (*i*) Remove an edge with minimum weight from S
   (*ii*) if that edge connects two different trees, then add it to the forest, combining two trees into a single tree
   (*iii*) else discard that edge

If the vertices u and v of an edge (u, v) under consideration belong to the same tree then the edge forms a cycle, and is not part of the minimum-cost spanning tree. On the other hand, suppose u and v are belong to two different trees then this must be an edge (u, v) in the minimum-cost spanning tree. In this case, the two tree are merged (*i.e.* their union is done) and now both vertices u and v belong to one tree.

DO YOU KNOW

JOSEPH BERNARD KRUSKAL (BORN 1928) Joseph Kruskal, born in New York City, attended the University of Chicago and received his Ph.D from Princeton University of Wisconsin, and later he was an assistant professor at the University of Michigan. In 1959 he became a member of the technical staff at Bell laboratories, a position he continues to hold. His current research interests include statistical linguistics and psychometrics. Besides his work on minimum spanning tress, Kruskal is also known for contribution to multidimensional scaling. Kruskal discovered his algorithm for producing minimum spanning trees when he was a second year graduate student. He was not sure his 2- page paper on this subject was worthy of publication, but was convinced by others to submit it.

The Pseudo Code for the algorithm is given below:

Kruskal(G)

1. for each vertex v in G do
2. Create a set (actually tree) with element v
3. Initialize a priority queue Queue that contains all edges in G in descending order of their weights
4. Define Forest ← Ø (it will ultimately contain the edges of the MST)
5. while T has edges < = N — 1 do (N = number of vertices)
   (*a*) Select an edge with minimum weight by removing it from the Queue
   (*b*) Check if u and v does not belong to same tree. (find operation)
   (*c*) Let T(v) be the tree containing v, and let T(u) be the Tree containing u
   (*d*) If T (v) ≠ T (u) then
   Add edge (u, v) to Forest
   Merge T(v) and T(u) into one tree, that is, union T(v) and T(u).
6. return tree T

To understand the algorithm, consider the figure given below where we find the MST of the graph using Kruskal's algorithm. The figure also shows the contents of the forest being grown:

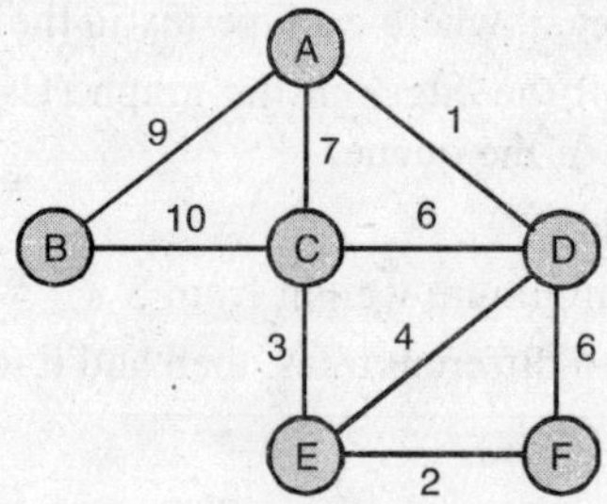

(i) Forest – {{A}, {B}, {C}, {D}, {E}, {F}}

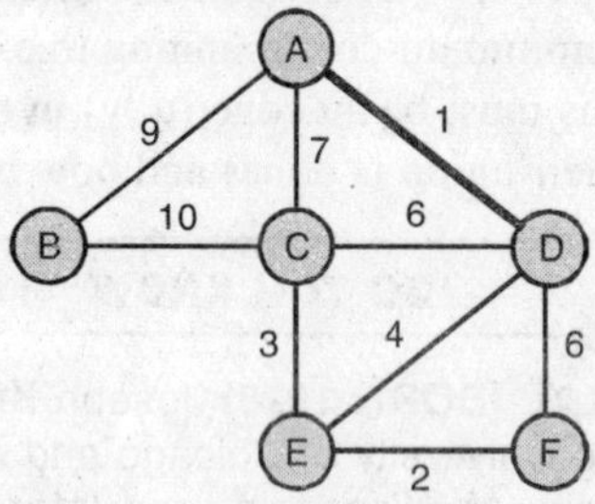

(ii) Forest = {{A, D}, {B}, {C}, {E}, {F}}

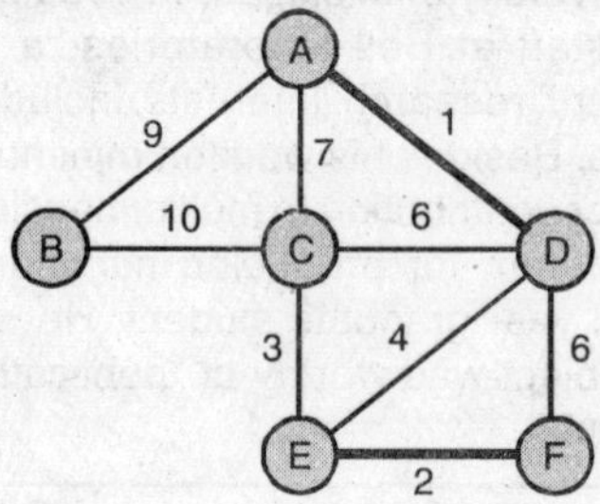

(iii) Forest = {{A, D}, {B}, {C}, {E, F}}

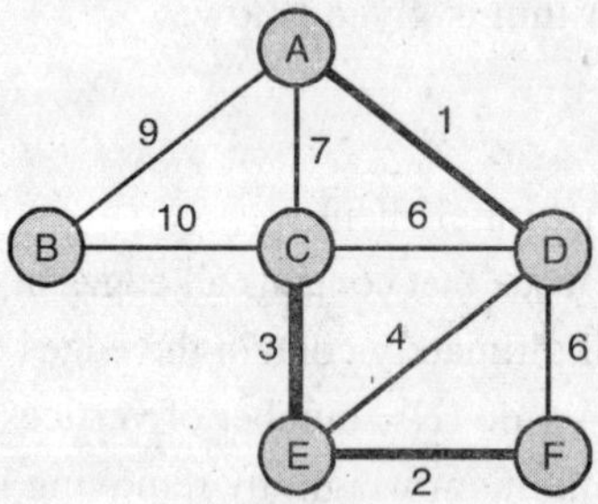

(iv) Forest = {{A, D}, {B}, {C, E, F}}

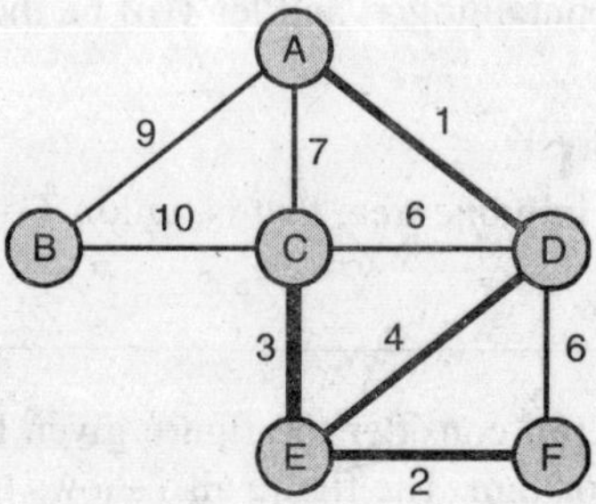

(v) Forest = {{B} {A, C, D, E, F}}

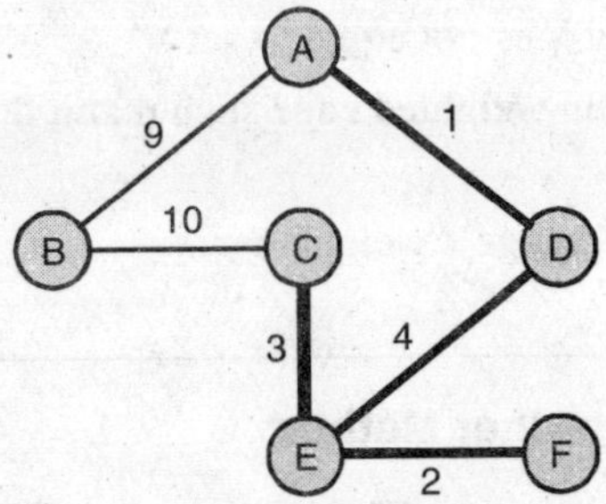

(vi) Edges (D, F), (C, D) and (A, C) are not considered
(they are discarded) as they form a cycle (vertices of edges belong to same tree)

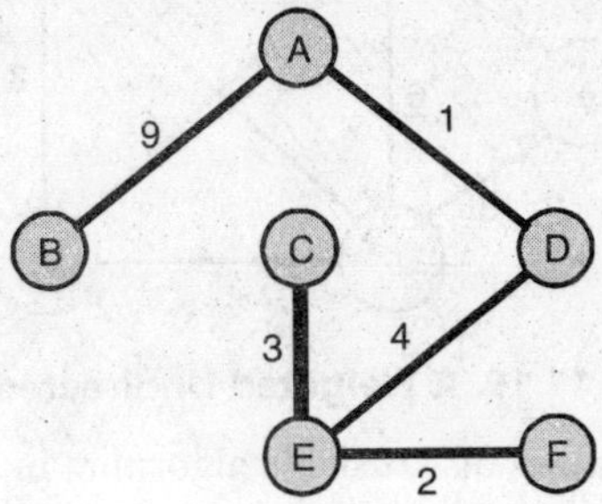

(vii) Forest = {<A, B, C, D, E, F>}
Weight = 19

Fig. 11.18. Kruskal's MST.

Performance

Considering number of edges and vertices in the graph as E and V respectively the running time of Kruskal's algorithm can be O (E log E) time, or equivalently, O (E log V) time, all with simple data structures. This equivalence is valid because maximum number of edges E in a graph is V^2 *i.e.* $E=V^2$

Taking logarithm both sides we have log E = 2 log V or O (log E) = O(log V)

Line number 2 and 3 causes O(V) time; line number 4 takes O(E $\log_2$ E) time (sorting time). There are O(E) operations on the forest (actually disjoint set forest), because all the E edges will be checked whether any of the them will be part of the MST. In totality the running time of all the operations is O(E α (E, V)) (where α is 1/A, A is Ackermann's function). As α (E, V) = O($\log_2$ E) (Can't show here how?). Total running time of Kruskal's algorithm is O(E $\log_2$ E)

If you compare the algorithm with Prim's you will find that during its execution the set of edges selected by Prim's algorithm forms exactly one tree. On the other hand, the set of edges selected by Kruskal's algorithm forms a forest of trees.

Algorithm of Kruksal's Algorithm

Algorithm 11.2. Kruksal's Algorithm for finding Minimum Spanning Tree ....

1. Create two sets *i.e.* V and V′, with the following initial conditions:

   (a) Set V′ is empty

   (b) Set V contains all the vertices of a graph

   (c) Select the minimum weighted edge (i,j)from the graph G and insert both its end-points *i.e.* i and j into the set V′

2. Repeat step 3 while set V is not equal to set V′
3. Select another minimum weighted edge such that it does not create a cycle and add its end-points in V′
4. Sum up all the selected edge's weight
5. Exit

Kruskal's Execution Using Another Method

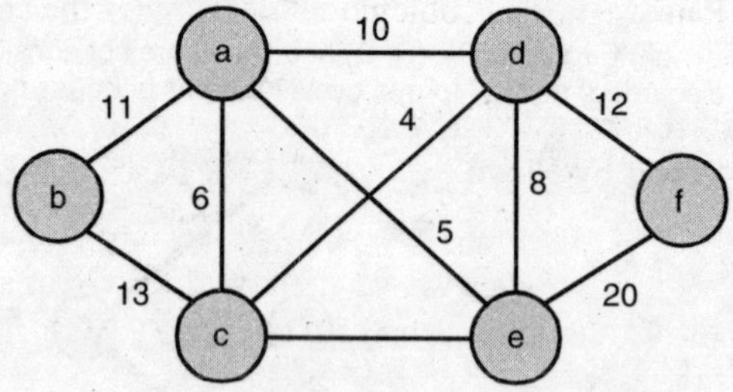

Fig. 11.19. A Weighted Undirected Graph.

The following shows the execution of Kruskal's algorithm in solving the MST problem of the graph above :

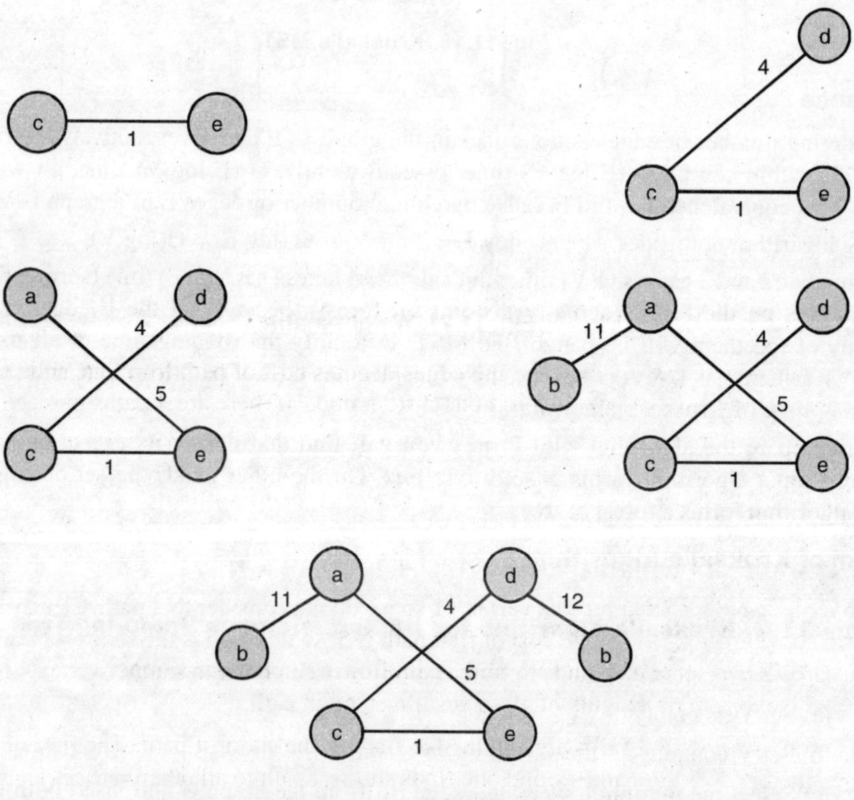

Fig. 11.20. Result of Applying Kruskal's Algorithm on above Graph.

11.6. SHORTEST PATH ALGORITHMS

Another classic set of problems in graphs is finding the shortest path given a weighted graph. In finding the shortest path, there is a need to get the *length* which, in this case, is the sum of the nonnegative cost of each edge in the path.

There are two path problems on weighted graphs:

- Single Source Shortest Paths (SSSP) Problem that determines the cost of the shortest path from a source vertex u to a destination vertex v, where u and v are elements of V.
- All-Pairs Shortest Paths (APSP) Problem that determines the cost of the shortest path from each vertex to every other vertex in V.

We will discuss the algorithm created by Dijkstra to solve the SSSP problem, and for the APSP, we will use the algorithm developed by Floyd.

In order to understand the notion of shortest path, we need to understand weight matrix. Consider a directed graph G with N vertices ordered as v1, v2 .... VN. For any directed edge from vi to vj (i, j from 1 to N) we have a positive number defined for the edge (v_i, v_j). This positive number is known as weight of length of the edge (v_i, v_j). An example of directed weighted graph and its corresponding weight matrix is shown below:

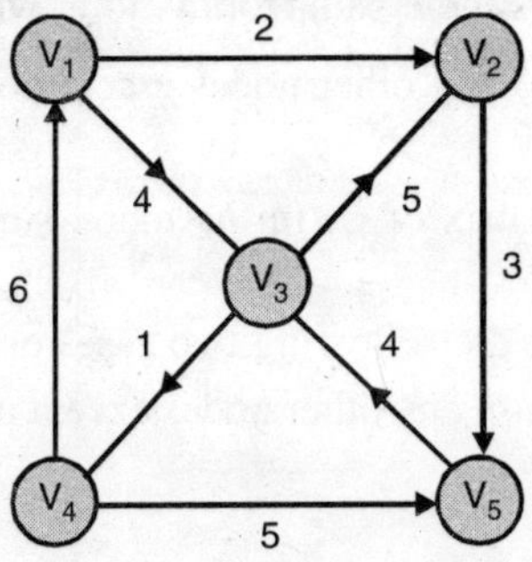

(*a*) Directed Weighted Graph

	V_1	V_2	V_3	V_4	V_5
V_1	0	2	4	0	0
V_2	0	0	0	0	3
V_3	0	5	0	1	0
V_4	6	0	0	0	5
V_5	0	0	4	0	0

(*b*) Corresponding weight matrix

Fig. 11.21. Weight matrix.

In the graph shown above weight on the edges denotes cost of path from one node to another node. Now consider the problems of reaching to node 5 from node 1.There are 3 paths to reach to node 5 from node 1. They are

(*i*) V_1—>V_2—>V_5. Its cost is 2+3 =5

(*ii*) v1—>v3-—>v2—>v5.It cost is4+5+3=12

(*iii*) v1 —> v3 —> v4 —> v5. Its cost is 4 + 1 + 5 = 10

There is one more path that repeats vertex v1 so we do not consider that path. Clearly among the three paths shown above from v1 to v5, the cost of first path is 5, so this path is the shortest path among all three paths. Thus problem of shortest path is to find a path from one vertex to another vertex with the minimum cost. The cost is the sum of weights of all edges traversed in path.

In this section we will see two algorithms for finding the shortest path. The first one finds shortest path among all pair of vertices and second one finds shortest path to all other vertices from a given source vertex.

11.6.1. Warshall's Algorithm : Shortest Paths (All Pair Shortest Path Algorithm)

Let G be a dircted graph with m nodes, $v_1, v_2, \dots v_m$. Suppose we want to find the path matrix P of the graph G. Warshall gave an algorithm for this purpose that is much more efficient than calculating the powers of the adjacency matrix A and using 8.3. This algorithm is described in this section, and similar algorithm is used to find shortest paths in G when G is weighted.

First we define m–squqare Boolean matrices $P_0, P_1, \dots, P_m$ as follows. Let $P_k(i, j)$ denote the ij entry of the matrix P_k. Then we cefine:

$$P_k[i, j] = \begin{cases} 1 \text{ if there is a simple path from } v_I \text{ to } v_J \\ \quad \text{which does not use any other nodes} \\ \quad \text{except possibly } v_1, v_2, \dots v_k \\ 0 \text{ otherwise} \end{cases}$$

In other words,

$P_0[i, j] = 1$ if there is an edge from v_i to v_j

$P_1[i, j] = 1$ If there is a simple path from v_i to v_j which does not use any other nodes escept posssibly v_1

$P_2[i, j] = 1$ If there is a simple path from v_1 to v_2 which does not use any other nodes escept posssibly v_1 and v_2

....

First observe that the matriox $P_0 = A$, the adjacency matrix of G. furthermore, since G has only m nodes, the last matrix $P_m = P$, the path matrix of G.

Warshall observed that $P_k[i, j] = 1$ can occur only if one the following two cases occurs :

(1) There is a simple path from v_1 to v_2 which does not use any other nodes except possibly $v_1, v_2, \dots, v_{k-1}$: hence

$$P_{k-1}[i, j] = 1$$

(2) There is a simple path from v_i to v_k and a simple path from v_k to v_j where each path does not use any other nodes except possible $v_1, v_2, \dots v_{k-1}$; Hence

$$P_{k-1}[i, k] = 1 = P_{k-1}[k, j] = 1$$

These two cases are pictured, respectively, in Fig. 11.22 (*a*) and (*b*) where

$$\rightarrow \cdots \rightarrow$$

denotes part of a simple path which does not use any nodes except possibly $v_1, v_2 \dots v_{k-1}$.

$v_i \rightarrow \cdots \rightarrow v_j$	$v_i \rightarrow \cdots \rightarrow v_k \rightarrow \cdots \rightarrow v_j$
(*a*)	(*b*)

Fig. 11.22.

Accordingly, the elements of the matrix P_k can be obtained by

$$P_k[i, j] = P_k[i, j] \vee (P_{k-1}[i, k] \wedge P_{k-1}[k, j])$$

where we use the logical operations of $\wedge$ (AND) and $\vee$ (OR). In other words we can obtain each entry in the matrix P_k by looking at only tjree ectries in the matrix P_{k-1}. Warshall's algorithm follows.

Algorithm 11.3. Warshall's Algorithm ....

A directed graph G with M nodes is maintained in memory by its adjacency matrix A. This algorithm finds the (Boolean) path matrix P of the graph. G.

1. Repeat for I, J = 1, 2, . . . , M: [Initializes P.]
   If A[I, J] = 0, then: Set P[I, J]: = 0;
   [End of loop.]
2. Repeat Steps 3 and 4 for K = 1, 2, . . . , M: [Updates P.]
3. Repeat Step 4 for I = 1, 2, . . . , M
4. Repeat for J = 1, 2, . . ., M:
   Set P[I, J]: = P[I, J] ∪ (P[I, K] ∩ P[K, J]).
   [End of loop.]
   [End of Step 3 loop.]
   [End of Step 2 loop.]
5. Exit.

Shortest-Path Algorithm

Let G me a directed graph with m nodes, $v_1, v_2, \ldots, v_m$, Suppose G is weighted: that is suppose eacn edge e in G is assigned a nonnegative number $w(e)$ called the weight or length of the edge e. Then G may be maintained in memory by its weight matrix W - (w_{ij}), defined as follows.

$$w_{ij} = \begin{cases} w(e) & \text{if there is an edge } e \text{ from } v_i \text{ to } v_j \\ 0 & \text{if there is no edge from } v_i \text{ to } v_j \end{cases}$$

The path matrix P tells us whether or not there are paths between the nodes. Now we want to find a matrix Q which will tell us the lengths of the shortest paths between the noodes or, more exactly, a matrix $Q = (q_{ij})$ where

$$q_{ij} = \text{Length of a shortest path from } v_i \text{ to } v_j$$

Next we describe a modification of Warshall's algorithm which finds us the matrix Q.

Here we define a sequence of matrices $Q_0, Q_1, \ldots, Q_m$ (analogous to the above matrices $P_0, P_1, \ldots, P_m$) whose entries are defined as follows:

$$Q_k[i,j] = \text{the smaller of the length of the preceding path from } v_i \text{ to } v_j \text{ or the sum of the lengths of preceding paths from } v_i\ v_k \text{ and from } v_k \text{ to } v_j$$

More exacly,

$$Q_k[i,j] = \text{MIN}(Q_{k-1}[i,j],\ Q_{k-1}[\text{i, k}] + Q_{k-1}[k,j])$$

The initial matrix Q_0 is the same as the weight matrix W except that each 0 in W is replaced by ∞ (or a very, very large number). The final matix Q_m will be the desired matrix Q.

Example 11.3.

Consider the weighted graph G in FIg 8.6. Assume v1 = R, v2 = S, v3 = T and v4 = U. Then the weight matrix W of G is as follows:

$$W = \begin{pmatrix} 7 & 5 & 0 & 0 \\ 7 & 0 & 0 & 2 \\ 0 & 3 & 0 & 0 \\ 4 & 0 & 1 & 0 \end{pmatrix}$$

Applying the modified Warshall's algorithm, we obtain the following matrices Q_0, Q_1, Q_2, Q_3 and $Q_4 = Q$. To the righ of each matrix Q_k, we show the matrix of paths which correspond to the lengths in the matrix Q_k.

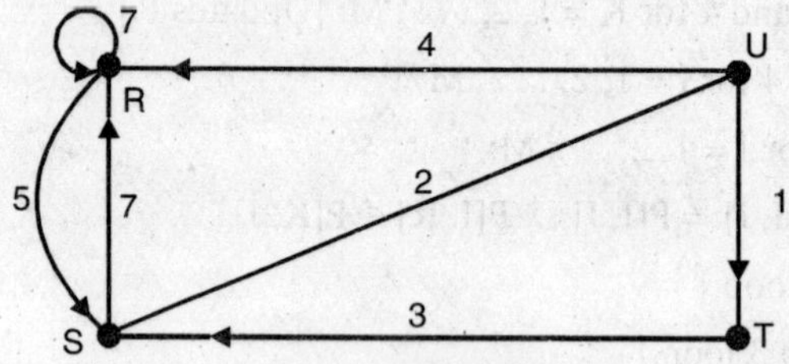

Fig. 11.23.

$$Q_0 = \begin{pmatrix} 7 & 5 & \infty & \infty \\ 7 & \infty & \infty & 2 \\ \infty & 3 & \infty & \infty \\ 4 & \infty & 1 & \infty \end{pmatrix} \qquad \begin{pmatrix} RS & RS & - & - \\ SR & - & - & SU \\ - & TS & - & - \\ UR & - & UT & - \end{pmatrix}$$

$$Q_1 = \begin{pmatrix} 7 & 5 & \infty & \infty \\ 7 & 12 & \infty & 2 \\ \infty & 3 & \infty & \infty \\ 4 & 9 & 1 & \infty \end{pmatrix} \qquad \begin{pmatrix} RS & RS & - & - \\ SR & SRS & - & SU \\ - & TS & - & - \\ UR & URS & UT & - \end{pmatrix}$$

$$Q_2 = \begin{pmatrix} 7 & 5 & \infty & 7 \\ 7 & 12 & \infty & 2 \\ 10 & 3 & \infty & 5 \\ 4 & 9 & 1 & 11 \end{pmatrix} \qquad \begin{pmatrix} RR & RS & - & RSU \\ SR & SRS & - & SU \\ TSR & TS & - & TSU \\ UR & URS & UT & URS \end{pmatrix}$$

$$Q_3 = \begin{pmatrix} 7 & 5 & \infty & 7 \\ 7 & 12 & \infty & 2 \\ 10 & 3 & \infty & 5 \\ 4 & 4 & 1 & 6 \end{pmatrix} \qquad \begin{pmatrix} RR & RS & - & RSU \\ SR & SRS & - & SU \\ TSR & TS & - & TSU \\ UR & UTS & UT & UTSU \end{pmatrix}$$

$$Q_4 = \begin{pmatrix} 7 & 5 & 8 & 7 \\ 7 & 11 & 3 & 2 \\ 9 & 3 & 6 & 5 \\ 4 & 4 & 1 & 6 \end{pmatrix} \qquad \begin{pmatrix} RR & RS & RSUT & RSU \\ SR & SURS & SUT & SU \\ TSUR & TS & TSUT & TSU \\ UR & UTS & UT & UTSU \end{pmatrix}$$

We indicate how the cicled entries are obtained:

$$Q_1[4, 2] = \text{MIN}(Q_0[4, 2], Q_0[4, 1] + Q_0[1, 2]) = \text{MIN}(\infty, 4 + 5) = 9$$

$$Q_2[1, 3] = \text{MIN}(Q_1[1, 3], Q_1[1, 2] + Q_1[2, 3]) = \text{MIN}(\infty, 5 + \infty) = \infty$$

$$Q_3[4, 2] = \text{MIN}(Q_2[4, 2], Q_3[4, 3] + Q_3[3, 2]) = \text{MIN}(9, 3 + 1) = 4$$

$$Q_4[3, 1] = \text{MIN}Q_3[3, 1], Q_3[3, 4] + Q_3[4, 1]) = \text{MIN}(10, 5 + 4) = 9$$

The formal statement of the algorithm follows:

Algorithm 11.4. Shortest-Path Algorithm ....

A weighted graph G with M nodes is maintained is memory by its weight matrix W. This algorithm finds a matrix Q such that O[I, J] is the lenth of a shortest path from VI to nodes VJ. INFINTY is very large number, and MIN is the minimum value function.

1. Repeat for I, J = 1, 2, . . ., M: [Intializes Q.]
   W[I, J] = 0, then: Set Q[I, J]: = INFINITY;
   Else : Set Q[I, J]: = W[I, J].
   [End of loop.]
2. Repeat Steps 3 and 4 for K = 1, 2, . . . , M: [Updates Q.]
3. Repeat Step 4 for I = 1, 2, . . ., M:
4. Reapeat for J = 1, 2, . . ., M:
   Set Q[I, J]: = MIN (Q[I, J], Q[I, K] + Q[K, J]
   [End of loop.]
   [End of Step 3 loop.]
   [End of Step 2 loop.]
5. Exit.

Algorithm 11.5 can also be used for a graph G without weights by simply assigning the weight $w(e) = 1$ to each edge e in G.

11.6.2. Dijkstra's Algorithm (Single Source Shortest Path Algorithm)

Dijkstra's algorithm is an algorithm for solving the single-source, shortest-path problem on an edge-weighted graph in which all the weights are non-negative. It finds the shortest paths from some initial vertex, say vs, to all the other vertices one-by-one. The essential feature of Dijkstra's algorithm is the order in which the paths are determined : The paths are discovered in the order of their weighted lengths, starting with the shortest, proceeding to the longest.

For each vertex v, Dijkstra's algorithm keeps track of three pieces of information, k_u, d_u and p_u.

k_u :

The Boolean-valued flag k_u indicates that the shortest path to vertex v is known. Initially, K_u = false for all $v \in V$.

d_u :

The quantity d_u is the length of the shortest known path from v_s to v. When the algorithm begins, no shortest paths are known. The distance d_u is a tentative distance. During the course of the algorithm candidate paths are examined and the tentative distances are modified.

Initially, $d_u = \infty$ for all $v \in V$ such that $v \neq v_s$, while $du = 0$.

p_u:

The predecessor of vertex v on the shortest path from v_s to v, *i.e.*, the shortest path from v_s to v has the form $\{v_s, p_u, v\}$.

Initially, p_i is unknown for all $v \in V$.

DO YOU KNOW

EDSGAR WYBE DIJKSTRA (1930-2002) Edsgar Dijkstra, born in the Nehtherland, began programming computers in the early 1950s while studying theoretical physics at the University of Lieden. In 1952, realizing that he was more interested in programming than in physics, he quickly completed the requirements for his physics degree and began his career as a programmer, even though programming is not a recognized profession.(In 1957, the authorities in Amsterdam refused to accept "programming" as his profession on his marriage license. However, they did accept "theoretical physics" when he changed his entry to this). Dijkstra has been one of the most forceful proponents of programming as a scientist discipline. He has made fundamental contribution to the areas of operating system, including deadlock avoidance; programming languages, including the notion of structured programming and algorithms. In 1972, Dijkstra received the Turing Award from the Association for Computing Machinery, one of the most prestigious award in computer science. Dijkstra became a Burroughs research Fellow in 1973, and in 1984 he was appointed to a chair in Computer Science at the university of Texas, Austin.

Algorithm 11.5. Dijkstra's algorithm ....

Dijkstra's algorithm proceeds in phases. The following steps are performed in each pass :

1. From the set of vertices for with k_u = false, select the vertex v having the smallest tentative distance d_u.
2. Set $k_u \leftarrow$ true.
3. For each vertex w adjacent to v for which $k_u \neq$ true, test whether the tentative distance d_w is greater than $d_u + C(v, w)$. If it is, set $d_w \leftarrow d_u + C(v, w)$ and set $p_w = v$.
4. Exit.

In each pass exactly one vertex has its k_u set to true. The algorithm terminates after | v | passes are completed at which time all the shortest paths are known.

Table illustrates the operation of Dijkstra's algorithm as it finds the shortest paths starting from vertex b in graph G_1 shown in Fig. 11.24.

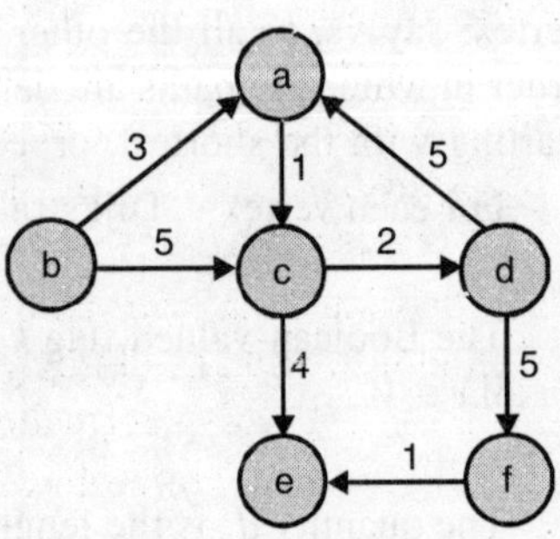

Fig. 11.24.

Initial

	a	*b*	*c*	*d*	*e*	*f*
Ku	F	F	F	F	F	F
du	∞	0	∞	∞	∞	∞
Pu	–	–	–	–	–	–

Pass 1

	a	*b*	*c*	*d*	*e*	*f*
Ku	F	T	F	F	F	F
du	3	0	5	∞	∞	∞
Pu	*b*	–	*b*	–	–	–

Pass 2

	a	*b*	*c*	*d*	*e*	*f*
Ku	T	T	T	F	F	F
du	3	0	4	∞	∞	∞
Pu	*b*	–	*a*	–	–	–

Pass 3

	a	*b*	*c*	*d*	*e*	*f*
Ku	T	T	T	F	F	F
du	∞	0	∞	6	8	∞
Pu	*b*	–	*a*	*c*	*c*	–

Pass 4

	a	*b*	*c*	*d*	*e*	*f*
Ku	T	T	T	T	F	F
du	3	0	4	6	8	9
Pu	*b*	–	*a*	*c*	*c*	*e*

Pass 5

	a	*b*	*c*	*d*	*e*	*f*
Ku	T	T	T	T	T	F
du	3	0	4	6	8	9
Pu	*b*	–	*a*	*c*	*c*	*e*

Pass 6

	a	*b*	*c*	*d*	*e*	*f*
Ku	T	T	T	T	T	T
du	3	0	4	6	8	9
Pu	*b*	–	*a*	*c*	*c*	*e*

Initially all the tentative distances are ∞, except for vertex *b* which has tentative distance zero. Therefore, vertex *b* is selected in the first pass. The mark 'T' an entry in Table indicates that the shortest path is known (k_u = true).

Next we follow the edges emanating from vertex *b*, $b \to a$ and $b \to c$, and update the distances accordingly. The new tentative distances for a becomes 3 and the new tentative distance for *c* is 5. In both cases, the next-to-last vertex on the shortest path is vertex *b*.

In the second pass, vertex a is selected and its entry is 'T', indicating the shortest path is known. There is one edge originating from *a, i.e.* $a \to c$. The distance to *c* via *a* is 4. Since this is less than the tentative distance to *c*, vertex c is given the new tentative distance 4 and its predecessor on the shortest-path is set to *a*. The algorithm continues in this fashion for a total of | *v* | passes until all the shortest paths have been found.

The shortest-path information contained in the last part of the Table can be represented in the form of a vertex-weighted graph as shown in Fig. 11.25.

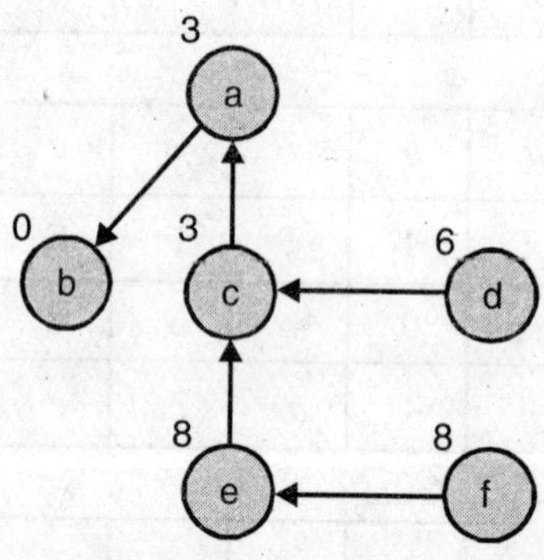

Fig. 11.25. The Shortest-Path Graph for G_1.

This graph contains the same set of vertices as the problem graph G_1. Each vertex *v* is labelled with the length *dv* of the shortest path from *b* to *v*. Each vertex (except *b*) has a single emanating edge that connects the vertex to the next-to-last vertex on the shortest-path. By following the edges in this graph from any vertex *v* to vertex *b*, we can construct the shortest path from *b* to *v* in reverse.

Performance

Like Prim's algorithm, Dijkstra's algorithm runs in O(|E| $\log_2$ |V|) time. Assuming priority queue is implemented as linear array. Line number 6 takes O(V) time and there are |V| such operations. Therefore, total time for extracting the vertex with minimum key in while-loop is $O(V^2)$. Since the total number of edges in all the adjacency list is | E |. Therefore for-loop iterates |E| times with each iteration taking O(1) time. Hence, the running time of the algorithm with array implementation is $O(V^2 + E) = 0(V^2)$.

If priority queue is implemented as heap, then line number 6 (extracting the vertex with minimum key) operations takes $O(\log_2 V)$ time and there are |V| such operations. The binary heap can be build in O(V) time.

Operation in line number 9 takes $O(\log_2 V)$ time and there are at most |E| such operations. Hence, the running time of the algorithm with binary heap is $O((V + E) \log_2 V)$. Note that this time becomes $O(E \log_2 V)$ if all vertices in the graph are reachable from the source vertices.

11.7. TOPOLOGICAL SORTING

A topological sorting is a sorting process of items over which a partial ordering is defined, *i.e.*, where an ordering is given over some pairs of items but not between all of them. The following are examples of partial orderings. For examples

1. In a dictionary or glossary, words are defined in terms of other words. If a word v is defined in terms of a word w, we denote this by v < w. Topological sorting of the words in a dictionary means arranging them in an order such that there will be no forward references.
2. A task (e.g, , an engineering project) is broken up into subtasks. Completion of certain subtasks must usually precede the execution of other subtasks. If a subtask v must precede a subtask w, we write v —> w. Topological sorting means their arrangement in an order such that upon initiation of each subtask all its prerequisite subtasks have been completed.
3. In a university curriculum, certain courses must be taken before others since they rely on the material presented in their prerequisites. If a course v is a prerequisite for course w, we write v —> w. Topological sorting means arranging the courses in such an order that no course lists a later course as prerequisite.
4. In a program, some procedures may contain calls of other procedures. If a procedure v is called by a procedure w, we write v -> w. Topological sorting implies the arrangement of procedure declarations in such a way that there are no forward references.

In general, a partial ordering of a set S is a relation between the elements of S. It is denoted by the symbol "a", verbalized by precedes, and satisfies the following three properties (axioms) for any distinct elements x, y, z of S:

1. if x < y and y < z, then x < z (transitivity)

2. if x <y, then not y < x (asymmetry)

3. not z < z (ir-reflexivity)

The sets S to be topologically sorted by an algorithm are usually finite. Hence, a partial ordering can be illustrated by drawing a diagram or graph in which the vertices denote the elements of S and the directed edges represent ordering relationship. An example is shown in Fig.

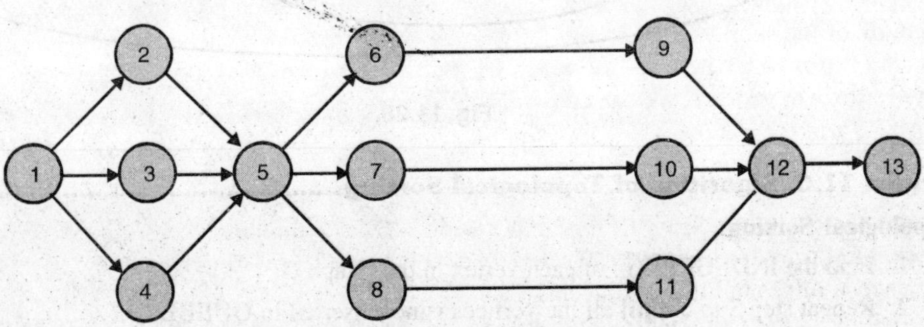

Fig. 11.26. Topological sort (dependency of task)

An -> from i to j means task i has been finish first before starting task j.

Topological sorting in graph

A topological sort of a directed acyclic graph (DAG) G is an ordering of the vertices of G such that for every edge (e_i, e_j) of G we have $i < j$. That is, a topological sort is a linear ordering of all its vertices such that if DAG G contains an edge (e_i, e_j), then e_i appears before e_j in the ordering. DAG is cyclic then no linear ordering is possible.

In simple words, a topological ordering is an ordering such that any directed path in DAG G traverses vertices in increasing order.

It is important to note that if the graph is not acyclic, then no linear ordering is possible. That is, we must not have circularities in the directed graph. For example, in order to get a job you need to have work experience, but in order to get work experience you need to have a job.

Consider the following graph:

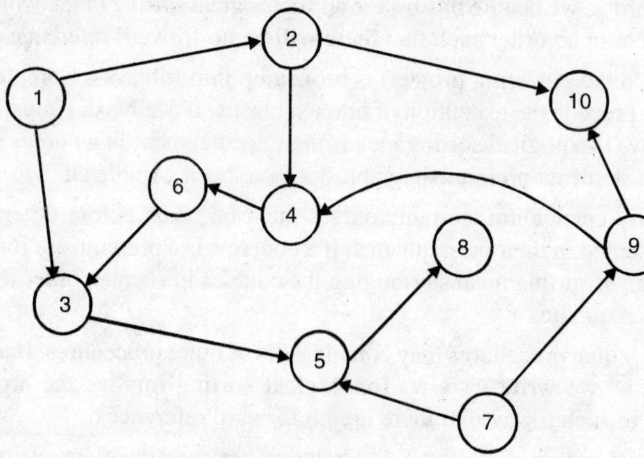

Fig. 11.27.

It topological sorting is shown linearly by placing all vertices horizontally on some distances and drawing a line as per the definition given above.

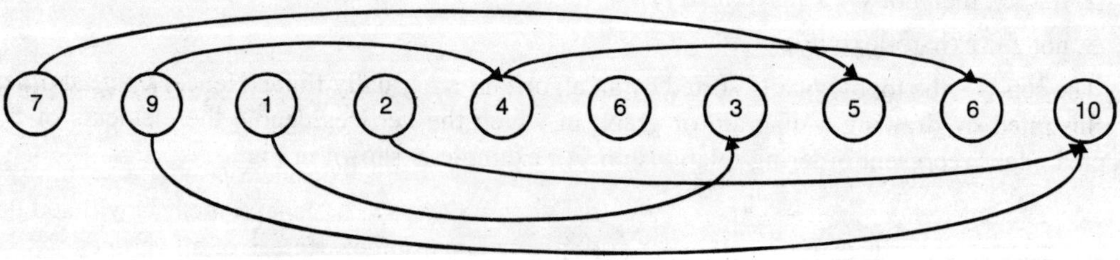

Fig. 11.28.

Algorithm 11.6. Algorithm of Topological Sorting..

Topological Sorting:

1. Find the INDEGREE(v) of each vertex in the Graph G
2. Repeat step 3 to 5 until all the vertices is not inserted in QUEUE Q
3. Insert all the vertices in a QUEUE Q which have its INDEGREE(v) = 0

4. Delete a vertex 'u' from the FRONT of the queue Q
5. Repeat the following for each neighbor 'v' of the deleted vertex 'u'
   (a) Set INDEGREE(v) := INDEGREE(v) – 1
   (b) If INDEGREE(v) = = 0, then insert 'v' to the REAR of QUEUE Q.
6. Join all the edges in the QUEUE Q in the same order as in given graph
7. If all the edges having the direction from left to right it indicates that the graph is topologically sorted.
8. EXIT

SOLVED PROBLEMS

1. What is a regular graph?

Ans. A graph where all vertices have the same degree is called regular graph. The graph on left is a regular graph where as on right is not the regular one.

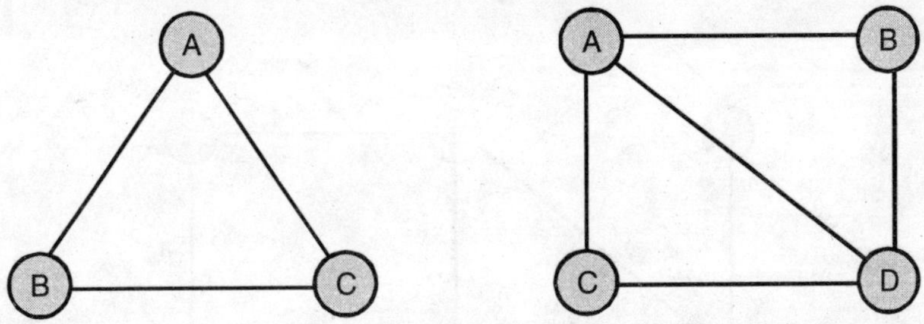

Fig. 11.29.

2. Is a complete graph also a regular graph?

Ans. Yes, in a complete graph each vertex is of same degree ($n-1$, n = number of vertices) so it is also a regular graph.

3. Prove "the sum of the degrees of all the vertices in a finite graph is equal to twice the number of edges in it."

Ans. Let G = (V, E) is a graph with n number of edges, *i.e.* $n = |E|$. Consider one edge $e_1 = (u, v)$ of graph. This one edge e_1 will add one degree to vertex u one to v. In general n edges will add degree of one set of vertices & n to other set of vertices. Thus sum of degrees of all vertices

$$\sum_{i=1}^{|v|} \text{degree}(v_i) = n + n = 2 \times \text{Number of edges}$$

4. What is the difference between a closed path & a cycle?

Ans. In a closed path only the start & end vertex is same & no other vertices are repeated where as in a cycle many vertices can be repeated.

5. For the following graph, draw all possible spanning trees.

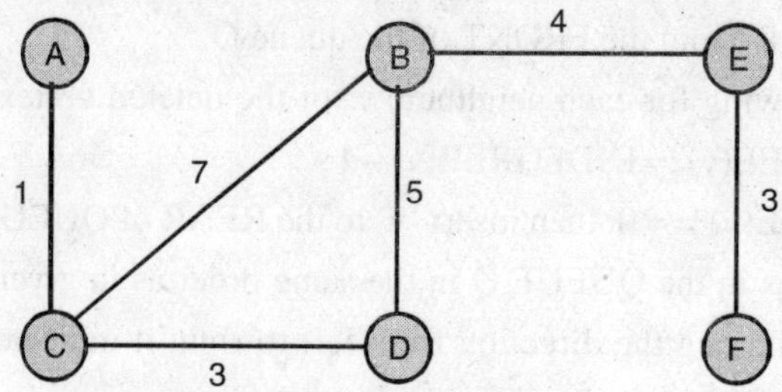

Fig. 11.30.

Ans.

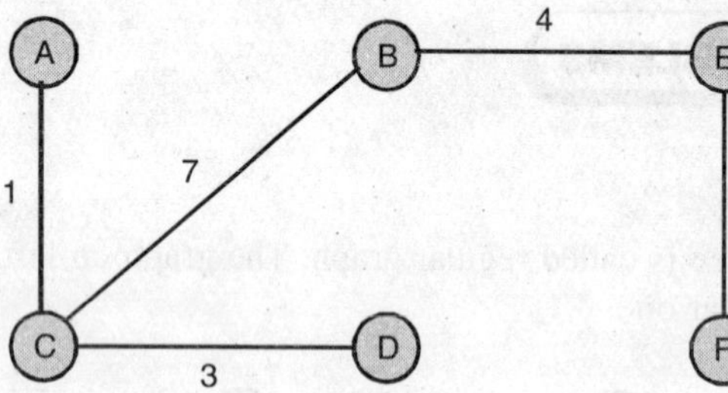

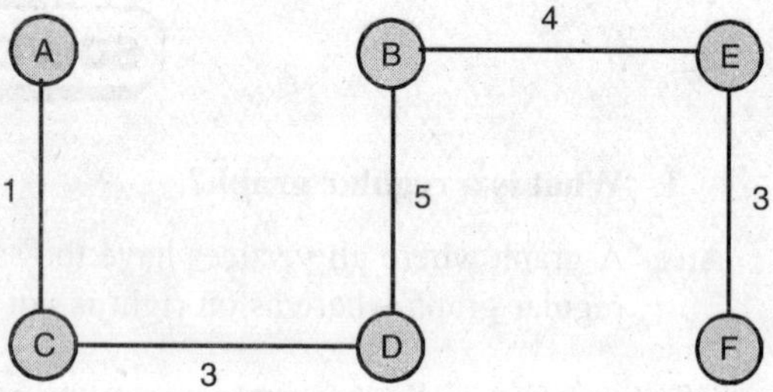

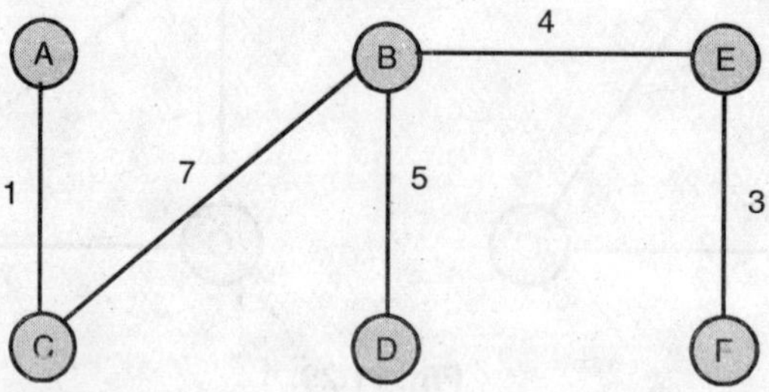

Fig. 11.31.

6. **What is the definition of tree in terms of graph?**

Ans. A tree is a connected graph without any circuits.

7. What is analogy of DFS & BFS with traversal of tree?

Ans. DFS traversal of graph is analogous to preorder traversal and BFS traversal is analogous to level order traversal.

8. For the following graph with v_1 as source apply dijkstra's algo.

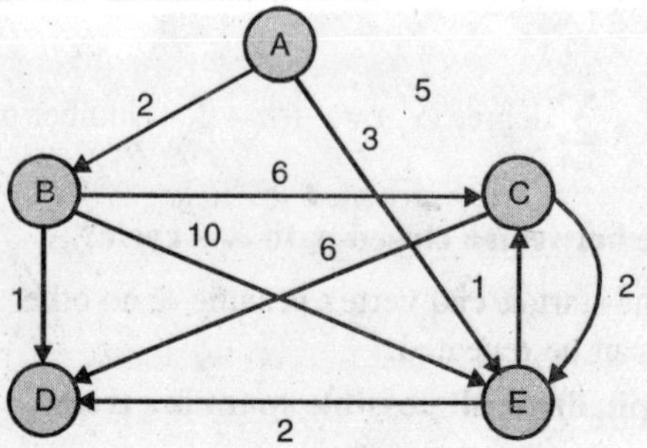

Fig. 11.32.

Ans.

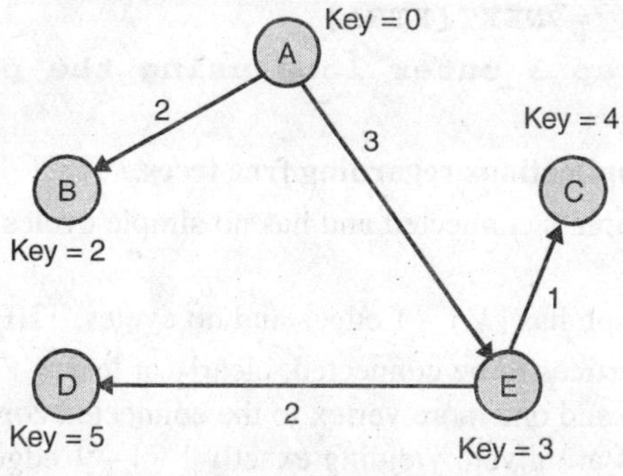

Fig. 11.33.

9. Suppose a graph G is maintained in memory in the form
GRAPH (NODE, NEXT, ADJ, START, DEST, LINK)
Write a procedure which finds the indegree INDEG and the outdegree OUTDEG of each node of G.

Ans. First we traverse the node list, using the pointer PTR in order to initialize the arrays INDEG and OUTDEG to zero. Then we traverse the node list, using the pointer PTRA, and for each value of PTRA, we traverse the list of neighbors of NODE[PTRA], using the pointer PTRB. Each time an edge is encountered, PTRA gives the location of its initial node and DEST[PTRB] gives the location of its terminal node. Accordingly, each edge updates the arrays INDEG and OUTDEG as follows :

```
OUTDEG[PTRA] : = OUTDEG[PTRA] + 1
```

and

```
INDEG[DEST [PTRB]] := INDEG[DEST[PTRB]] + 1
```

The formal procedure follows :

```
    Procedure P8.15 : DEGREE(NODE, NEXT, ADJ, START, DEST,
LINK, INDEG, OUTDEG)
```

This procedure finds the indegree INDEG and outdegree OUTDEG of each node in the graph G in memory.

```
1. [Initialize arrays INDEG and OUTDEG.]
   (a) Set PTR : = START
   (b) Repeat while PTR ≠ NULL : [Traverses node list.]
      (i)  Set INDEG[PTR] : = 0 and OUTDEG [PTR] : = 0
      (ii) Set PTR : = NEXT[PTR].
   [End of loop.]
2. Set PTRA : = START.
3. Repeat Steps 4 to 6 while PTRA ≠ NULL : [Traverses node list.]
4. Set PTRB : = ADJ [PTRA].
5. Repeat while PTRB ¹ NULL : [Traverses list of neighbors.]
   (a)     Set OUTDEG[PTRA] : = OUTDEG[PTRA] + 1 and
      INDEG[DEST[PTRB]] : = INDEG[DEST[PTRB]] + 1.
   (b)     Set PTRB : = LINK [PTRB].
```

```
    [End of inner loop using pointer PTRB.]
6. Set PTRA : = NEXT[PTRA].
    [End of step 3 outer loop using the pointer PTRA.]
7. Return.
```

10. Prove the following implications regarding free trees.

(*a*) If an undirected graph is connected and has no simple cycles, THEN the graph has | *V* | – 1 edges.

(*b*) If an undirected graph has | *V* | – 1 edges and no cycles, THEN the graph is connected.

Ans. (*a*) For a graph of *n* vertices to be connected, clearly at least | *V* | – 1 edges are required since each edge serves to add one more vertex to the connected component. No cycles means that no additional edges are given, yielding exactly | *V* | – 1 edges.

(*b*) Proof by contradiction. If the graph is not connected, then by definition there are at least two components. At least one of these components has *i* vertices with *i* or more edges (by the pigeonhole principle). Given *i* – 1 edges to connect the component, the *i*th edge must then directly connect two of the vertices already connected through the other edges. The result is a cycle. Thus, to avoid a cycle, the graph must be connected.

11. Explain why, in the worst case, Dijkstra's algorithm is (asymptotically) as efficient as any algorithm for finding the shortest path from some vertex I to another vertex J.

Ans. In the worst case, and algorithm for finding the shortest path between a given pair of vertices *i* and *j* will have to visit every node in the graph. In the process of visiting every node in the graph, we can determine the shortest paths from the start vertex to all the nodes. Thus, in the worst case, the cost for finding all of the shortest paths is no worse than the cost to find the shortest path between a specified pair of vertices.

12. Write an algorithm to determine whether an undirected graph of | *V* | vertices contains a cycle. Your algorithm should run in –(| *V* |) time.

Ans. To solve this problem, simply run the standard DFS algorithm, returning the fact that there is a cycle if any already visited vertex is encountered (ignoring the edge returning to the vertex that you just came from). This algorithm has has time Θ(|*V* |) because the graph can only have Θ(|*V* |- 1) edges if it does not contain a cycle, and any existing cycle will be detected in at most |*V* | edge visits.

13. The single-destination shortest-paths problem for a directed graph is to find the shortest path from every vertex to a specified vertex V. Write an algorithm to solve the single-destination shortest-paths problem.

Ans. Simply reverse the direction of all the edges, then run the standard algorithm for Single Source Shortest Paths.

14. Write an algorithm to find a maximum cost spanning tree, that is, the spanning tree with highest possible cost.

Ans. Simply use any Minimal Cost Spanning Tree algorithm, but always pick the greatest edge among the available choices instead of the least edge.

15. What is/are the limitation of Dijkstra's Algorithm?

Ans. Dijkstra's Algorithm does not support negative edges as well as negative edge cycles.

16. What is a spanning Tree?

Ans. A spanning tree is a tree associated with a network. All the nodes of the graph appear on the tree once. A minimum spanning tree is a spanning tree organized so that the total edge weight between nodes is minimized.

17. Does the minimum spanning tree of a graph give the shortest distance between any 2 specified nodes?

Ans. No.

Minimal spanning tree assures that the total weight of the tree is kept at its minimum. But it doesn't mean that the distance between any two nodes involved in the minimum-spanning tree is minimum.

18. Write the procedure to traverse a graph using DFS. Also workout the DFS to find minimum spanning tree for the following graph:

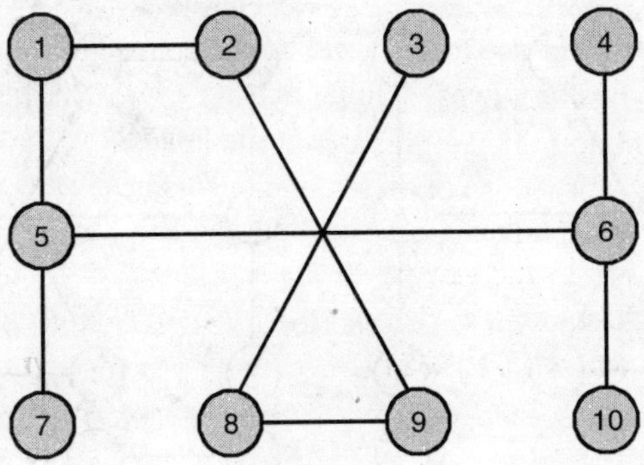

Fig. 11.34.

Ans. With depth first search (DFS), the graph is traversed as deeply as possible. Given a graph with vertices marked ***unvisited*****, the traversal is performed as follows:**

1. Select a start, unvisited vertex. If no vertex can be found, DFS terminates.
2. Mark start vertex visited.
3. Process adjacent vertex :
   (*a*) Select an unvisited vertex, say *u*, adjacent to (or from) the start vertex
   (*b*) Mark the adjacent vertex visited
   (*c*) Initiate a DFS with u as start vertex. If no such vertex can be found, go to step (1)
4. If more unvisited, adjacent vertices are encountered, go to step (*c*)

Starting at vertex 1, the DFS listing of elements of the following graph is

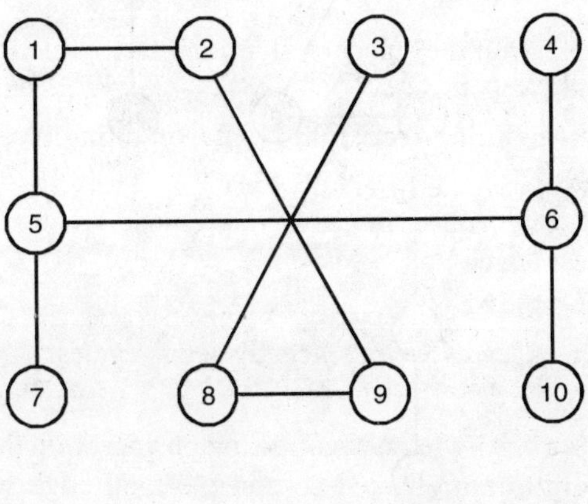

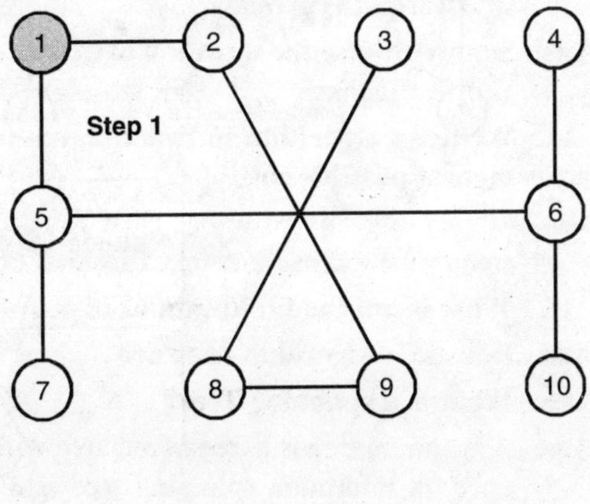

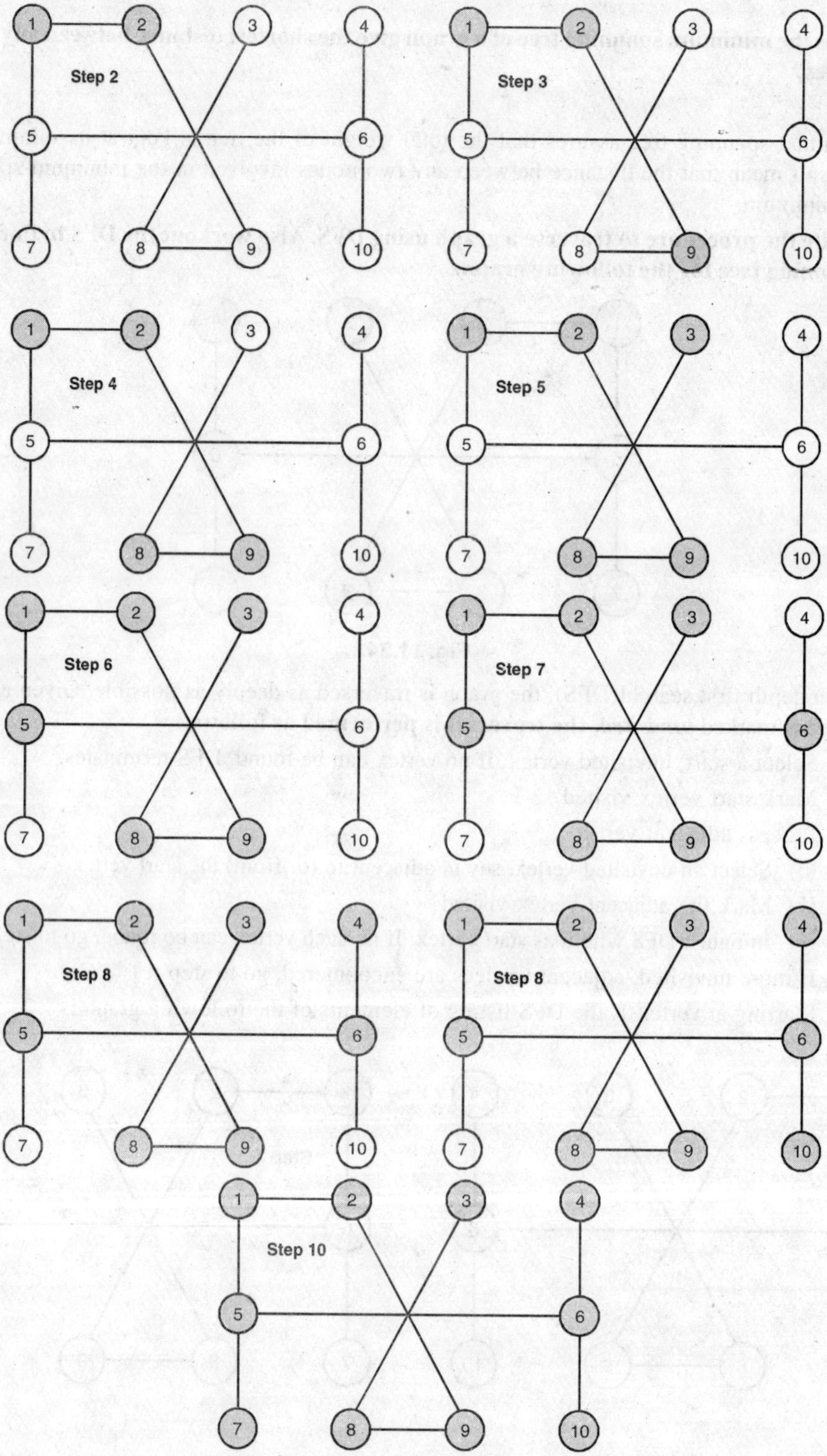

1 2 9 8 3 5 6 4 10 7

Fig. 11.35. Required Spanning Tree using Depth First Search.

19. Write the procedure to traverse a graph using BFS. Also workout the BFS to find minimum spanning tree for the following graph :

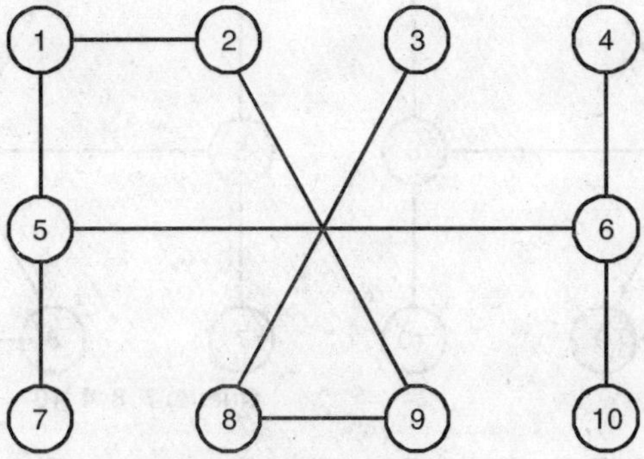

Fig. 11.36.

Ans. Breadth first search (BFS) traverses as broadly as possible. Given a graph with vertices marked unvisited, the traversal is performed as follows:

1. Select an unvisited vertex, say vertex *i*, and mark it visited. Then, search as broadly as possible from *i* by visiting all the vertices adjacent to it.
2. Repeat (1) until all the vertices in the graph are visited.

   Just like in DFS, conflict on which node to visit next is resolved via visiting in increasing order of numeric label.

Starting at vertex 1, the BFS listing of elements of the previous graph is

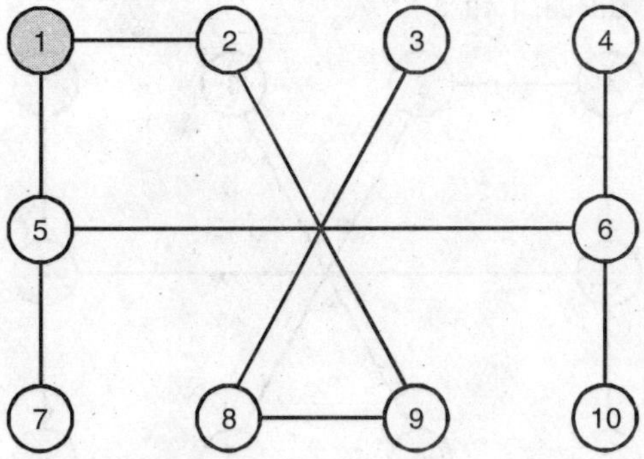

Queue: 1 2 5

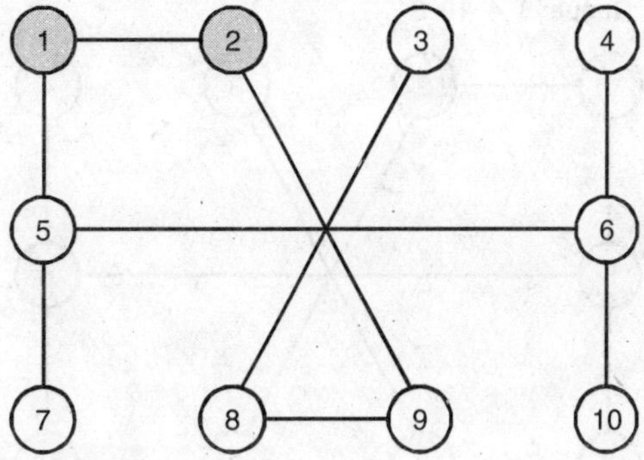

Queue: 2 5 9

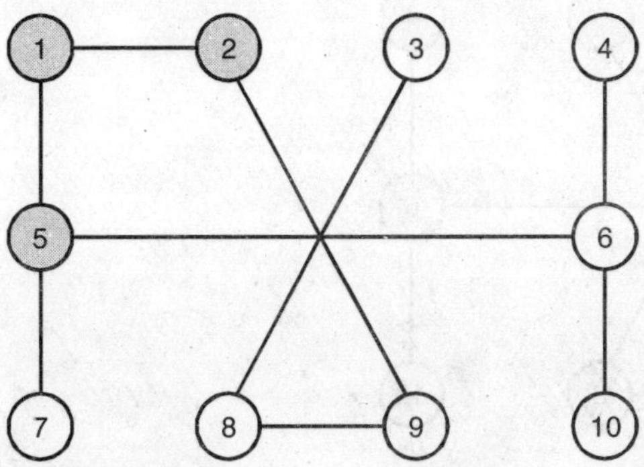

Queue: 5 9 6 7

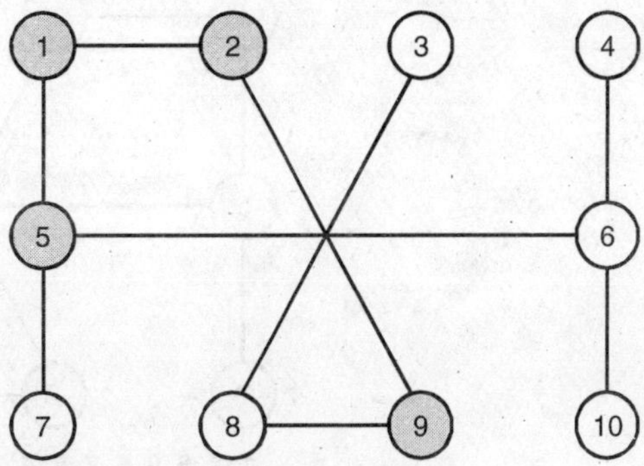

Queue: 9 6 7 8

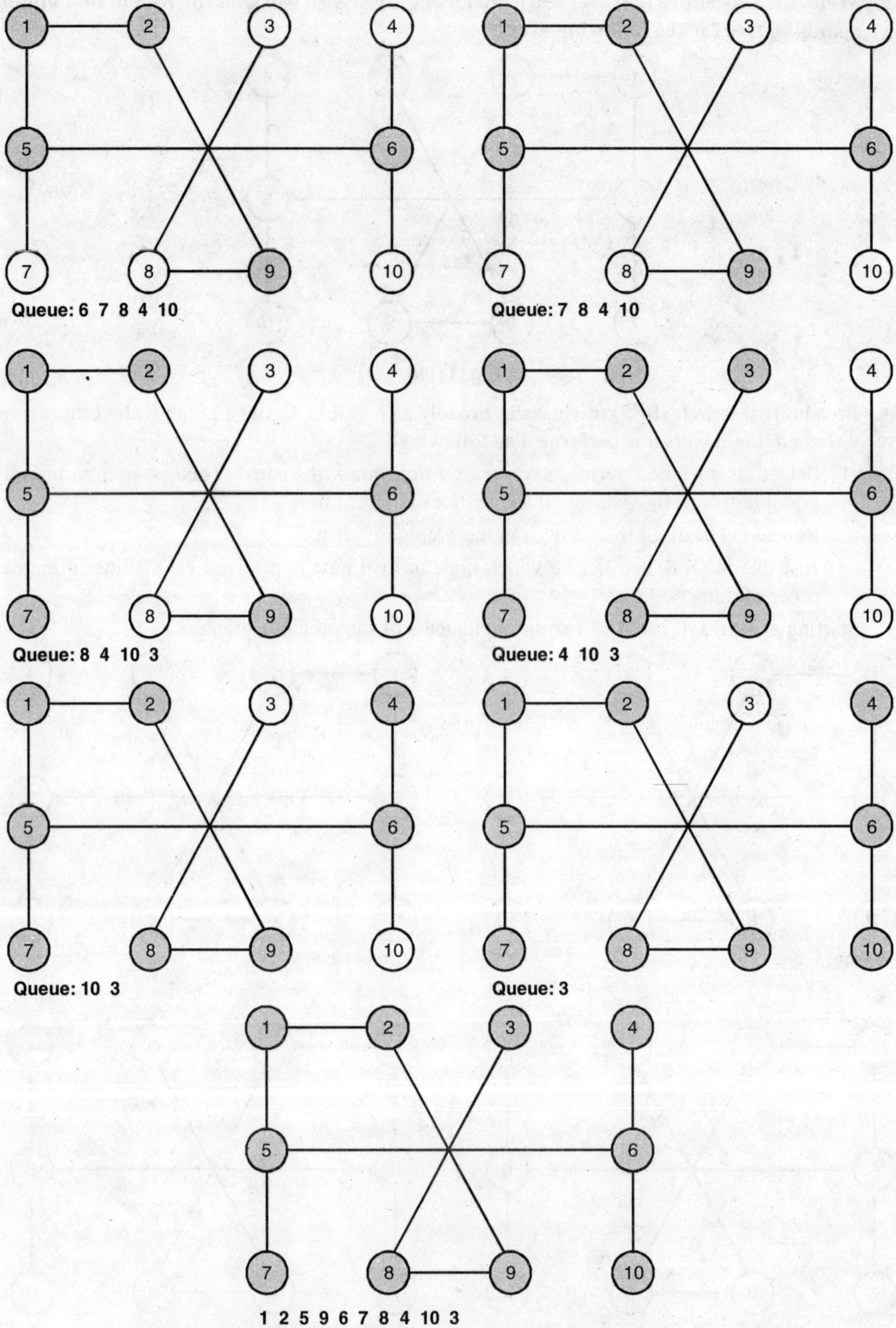

Fig. 11.37. Required Spanning Tree using Breadth First search

20. What is connected graphs and strongly connected component?

Ans. Two vertices i and j are **connected** if there is a path from vertex i to vertex j. If for every pair of distinct vertices i and j there is a directed path to and from both vertices, it is said to be **strongly connected.** A maximal connected subgraph of an undirected graph is known as **connected component in an undirected graph.** In a directed graph G, **strongly connected component** refers to the maximal strongly connected component in G.

21. What is minimum spanning tree theorem?

Ans. Let $G = (V, E)$ be a connected, weighted, undirected graph. Let U be some proper subset of V and (u, v) be an edge of least cost such that $u \in U$ and $v \in (V - U)$. There exists a minimum cost spanning tree T such that (u, v) is an edge in T.

22. Give the adjency list representation of following graph :

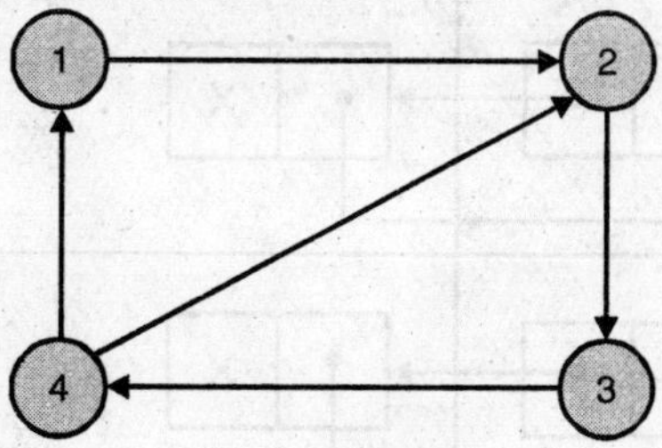

Fig. 11.38.

Ans.

Vertices	Neighbour (Adjacent nodes)
1	2, 4
2	3
3	4
4	2

The number of vertices in the graph is equivalent to number of node list and each node list comprises of :

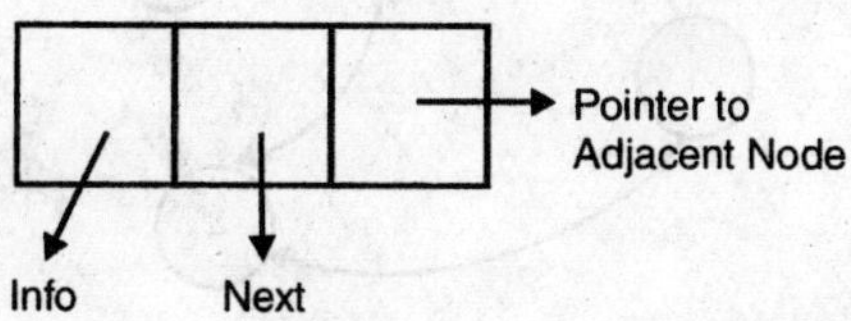

And the number of neighbours of each vertext is the number of edge list for each vertex, which comprises of

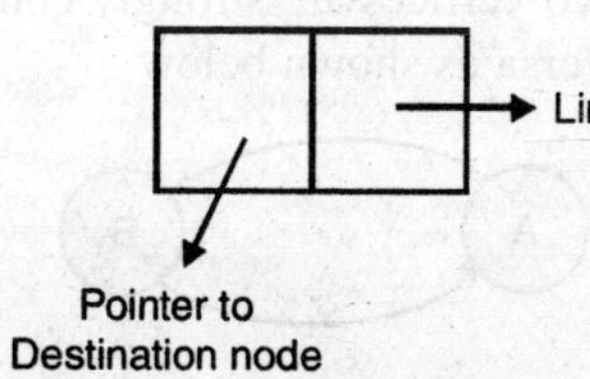

The adjacency list is shown below :

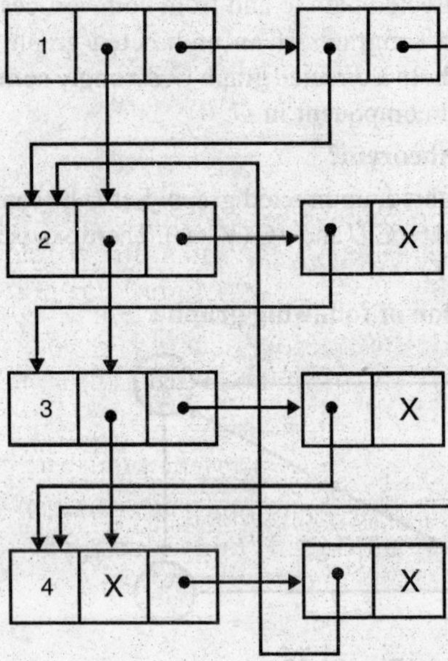

Fig. 11.39.

23. Is the directed graph below is strongly connected ? Show how this graph would look if represented by

(*i*) Adjacency matrix

(*ii*) Adjacency list

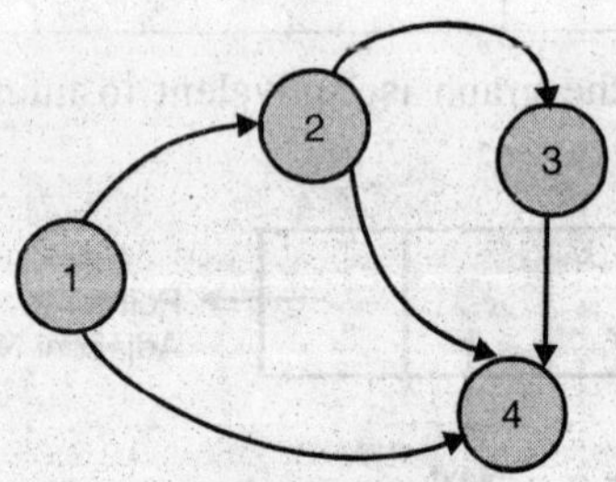

Fig. 11.40.

Ans. No the above given graph is not strongly connected as all the vertices don't have the edges connected to one another *i.e.* two vertices are strongly connected, if for two vertices there is an edge between *A* to *B* and vice versa as shown below

Fig. 11.41. Strongly connected graph.

A graph is strongly connected if it consists of cycles.

(i) Using *Adjacency matrix*

$$A = \begin{array}{c} \\ 1 \\ 2 \\ 3 \\ 4 \end{array}\begin{array}{c} \begin{array}{cccc} 1 & 2 & 3 & 4 \end{array} \\ \begin{pmatrix} 0 & 1 & 0 & 1 \\ 0 & 0 & 1 & 1 \\ 0 & 0 & 0 & 1 \\ 0 & 0 & 0 & 0 \end{pmatrix}_{\substack{4\times 4 \\ |V|\times|V|}} \end{array}$$

$$|V| = 4$$
$$V = \{1, 2, 3, 4\}$$

$$A[i,j] = \begin{cases} 1, & \text{if there is a direct edge between } i \text{ to } j \\ 0, & \text{otherwise} \end{cases}$$

(ii) Using *Adjacency List* : There are 4 number of node list and each node list have some edge list, which is equivalent to number of adjacent nodes.

Vertex	Adjacent nodes
1	2, 4
2	3, 4
3	4
4	–

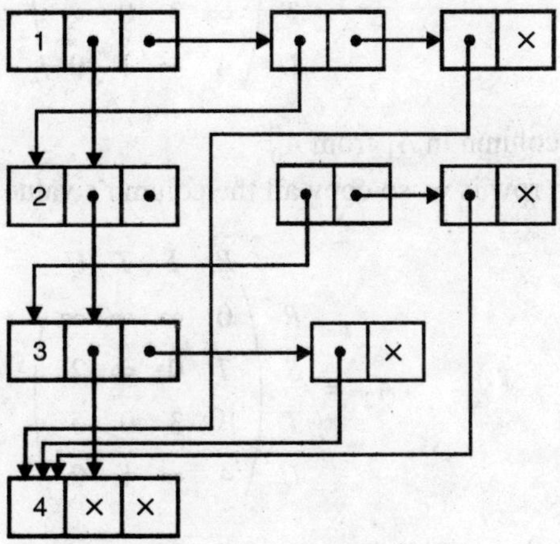

Fig. 11.42.

24. Find all pairs shortest path for the following graph using Floyd's Warshall's algorithm ?

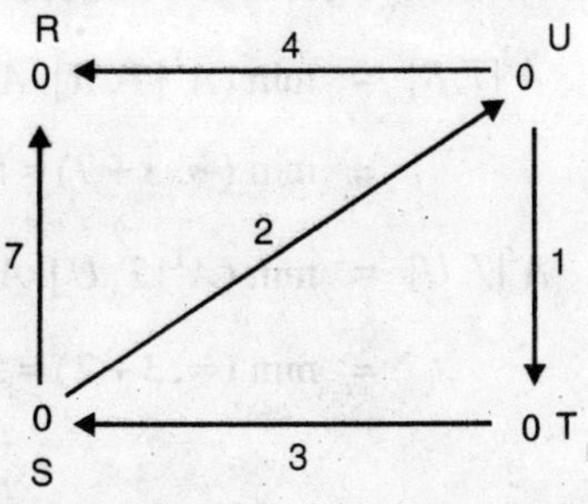

Ans. To apply Floyd's Warshall algorithm :

First of all there is total of $|V| + 1$ number of matrices and each matrix is filled according the following rules :

(*i*) To create/fill the matrix A^K, fill or copy the Kth row and Kth column contents of A^{K-1} matrix.

(*ii*) Copy the contents of rows in A^K from A^{K-1}, at which there is ∞ entry at column Kth in A^{K-1} matrix.

(*iii*) Copy the contents of columns in A^K. From A^{K-1}, at which there is ∞ entry at row Kth in A^{K-1} matrix.

(*iv*) All remaining entry in A^{K-1}, if any, is filled using the following formula :

$$A^K[i,j] = \min(A^{K-1}[i,j], A^{K-1}[i,k] + A^{K-1}[k,j])$$

$$A_0 = \begin{array}{c} \\ R \\ S \\ T \\ U \end{array} \begin{array}{c} \begin{array}{cccc} R & S & T & U \end{array} \\ \begin{pmatrix} 0 & \infty & \infty & \infty \\ 7 & 0 & \infty & 2 \\ \infty & 3 & 0 & \infty \\ 4 & \infty & 1 & 0 \end{pmatrix} \end{array}$$

$$A_1 = \begin{array}{c} \\ R \\ S \\ T \\ U \end{array} \begin{array}{c} \begin{array}{cccc} R & S & T & U \end{array} \\ \begin{pmatrix} 0 & \infty & \infty & \infty \\ 7 & 0 & \infty & 2 \\ \infty & 3 & 0 & \infty \\ 4 & \infty & 1 & 0 \end{pmatrix} \end{array}$$

Copy Rth row and Rth column in A_1 from A_0

As all the entries in Rth row is ∞, so copy all the column's value in A_1 from A_0.

$$A_2 = \begin{array}{c} \\ R \\ S \\ T \\ U \end{array} \begin{array}{c} \begin{array}{cccc} R & S & T & U \end{array} \\ \begin{pmatrix} 0 & \infty & \infty & \infty \\ 7 & 0 & \infty & 2 \\ 10 & 3 & 0 & 5 \\ 4 & \infty & 1 & 0 \end{pmatrix} \end{array}$$

Copy Sth row and Sth column in A^2 from A^1

As third entry in Sth row is ∞ hence copy Tth column in A^2 from A^1

Generally, first and fourth entry in Sth column is ∞, hence copy Rth and Uth row.

$$A^2[T,R] = \min(A^1[T,R], A^1[T,S] + A^1[S,R])$$

$$= \min(\infty, 3+7) = \min(\infty, 10) = 10$$

$$A^2[T,U] = \min(A^1[T,U], A^1[T,S] + A^1[S,U])$$

$$= \min(\infty, 3+2) = 5$$

$$A^3 = \begin{array}{c} \\ R \\ S \\ T \\ U \end{array} \begin{array}{c} \begin{array}{cccc} R & S & T & U \end{array} \\ \begin{pmatrix} 0 & \infty & \infty & \infty \\ 7 & 0 & \infty & 2 \\ 10 & 3 & 0 & 5 \\ 4 & 4 & 1 & 0 \end{pmatrix} \end{array}$$

Copy Tth row and Tth column in A^3 from.

As there is no entry *i.e.* ∞ in Tth row. NO COPY.

As there are ∞ at first and second entry in Tth column, hence copy First and Second row.

$$\begin{aligned} A^3[U,R] &= \min(A^2[U,R], A^2[U,T] + A^2[T,R]) \\ &= \min(4, 1+10) = 4 \\ A^3[U,S] &= \min(A^2[U,S], A^2[U,T] + A^2[T,S]) \\ &= \min(\infty, 1+3) = 4 \\ A^3[U,U] &= \min(A^2[U,U], A^2[U,T] + A^2[T,T]) \\ &= \min(0, 1+5) = 0 \end{aligned}$$

Parallely the matrix A^4 is,

$$A^4 = \begin{array}{c} \\ R \\ S \\ T \\ U \end{array} \begin{array}{c} \begin{array}{cccc} R & S & T & U \end{array} \\ \begin{pmatrix} 0 & \infty & \infty & \infty \\ 6 & 0 & 3 & 2 \\ 9 & 3 & 0 & 5 \\ 4 & 4 & 1 & 0 \end{pmatrix} \end{array}$$

25. What is the difference between Adjacent matrix and path matrix ?

Ans. Both of these matrices are used to represent a graph in memory. Using arrays.

Adjacency matrix

$$A^K[i,j] = \begin{cases} 1, & \text{if there is an direct edge between vertex } i \text{ to vertex } j \\ 0, & \text{otherwise} \end{cases}$$

Path Matrix

$$P^K[i,j] = \begin{cases} 1, & \text{if there is a path between vertex } i \text{ to vertex } j \\ & \text{with either } 0, 1, \ldots \text{ or } k \text{ vertices in between} \\ 0, & \text{otherwise} \end{cases}$$

$$\begin{aligned} P^0 &= A^0 \\ P_1 &= A^0 + A^1 \\ P^2 &= A^0 + A^1 + A^2 \end{aligned}$$

Parallely $\quad P^n = A^0 + A^1 + \ldots A^n$

26. We are given two graphs, one is directed and other is undirected. Suggest which technique do you prefer to store the graph in memory ?

Ans. There are two techniques to represent a graph (directed or undirected) in memory :

(*a*) Adjacency matrix using ARRAYS

(*b*) Adjacency list using LINKED LIST.

If a graph is un-directed, then we will prefer adjacency matrix to represent it in memory. While if a graph is directed, then we will prefer adjacency list to represent it in memory.

27. List the difference between breadth-first-search and depth-first-search ?

Ans.

Breadth-first	**Depth-first**
1. The starting vertex is given to traverse a graph	No starting vertex given
2. Uses queues to traverse a graph	Uses STACK to traverse a graph
3. A parent vertex is traversed first compared to its child vertex	A child vertex is traversed first compared to its parent vertex
4. BFS does not uses recursion and back-tracking.	DFS uses back-tracking
5. Its application is to find spanning tree as well as shortest path from starting vertex	Its applicatio is to sort a graph using topological sorting.
6. In BFS, we will find the distance of every vertex from the starting vertex.	In DFS, we will find the vertex's discovery as well as finishing time using time-stamps.

28. List the difference between Prim's and Kruskal to find minimum spanning tree ?

Ans.

PRIM's	**KRUSKAL**
1. The starting vertex is given to find MST.	No such vertext is initially known.
2. It uses priority queues to find MST.	It uses MAX-Heap Queue to find MST.
3. It doesn't uses any disjoint-set operations.	It uses some disjoint-set operations like make-set, find-set, UNION etc.
4. In PRIMs, there is only one subtree and each time or iteration a vertex and edge is added to it.	In Kroskal, there are different subtrees and all the subtrees are connected together to form a MST.
5. PRIMS algorithm is similar to Dijkstra's algorithm and hence its used to find shortest distance from the starting vertex.	Kruskal's algorith is only used to find maximum spanning tree.

29. List the limitations of topological sorting ?

Ans. The topological sorting can't be applied on the graph which is

(1) REFLEXIVE (HAVING SELF LOOP)

(2) Symmetric (Having Cycles)

(3) NOT TRANSITIVE

(4) Un-directed graph.

30. Find the weight matrix and reachability for the following graph from source vertex A to destination vertex H. Show all steps ?

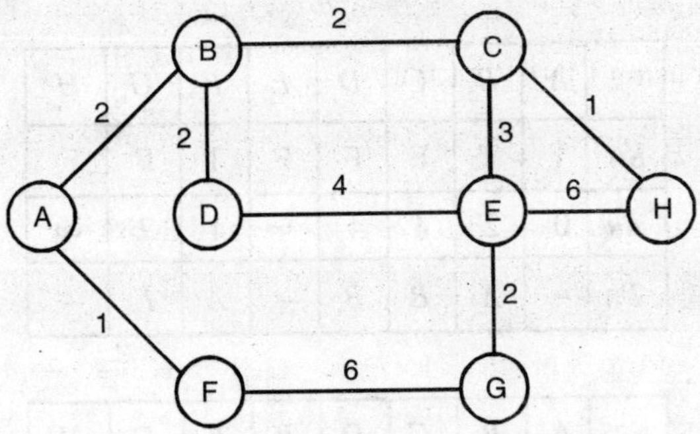

Fig. 11.43.

Ans. Dijkstra's algorithm is used here as we are given a single source vertex *A*, the algorithm proceeds in phases :

(1) From the set of vertices for with K_u = false, select a vertex *v* having the smallest tentative distance *dk*

(2) Set $K_u \leftarrow$ true

(3) For each vertex '*w*' adjacent to *v* for which $K_u \neq$ true, test whether the tentative distance *dw* is greater than $du = C(u, w)$. If it is , set $dw \leftarrow du + C(u, w)$ and set $Pw \leftarrow V$.

(4) Exit

Initial

	A	*B*	*C*	*D*	*E*	*F*	*G*	*H*
Ku	F	F	F	F	F	F	F	F
du	0	∞	∞	∞	∞	∞	∞	∞
Pu	–	–	–	–	–	–	–	–

Pass I

	A	*B*	*C*	*D*	*E*	*F*	*G*	*H*
Ku	T	F	F	F	F	F	F	F
du	0	2	∞	∞	∞	1	∞	∞
Pu	–	*A*	–	–	–	*A*	–	–

Pass II

	A	*B*	*C*	*D*	*E*	*F*	*G*	*H*
Ku	T	F	F	F	F	T	F	F
du	0	2	∞	∞	∞	1	7	∞
Pu	–	*A*	–	–	–	*A*	*F*	–

Pass III

	A	*B*	*C*	*D*	*E*	*F*	*G*	*H*
Ku	T	T	F	F	F	T	F	F
du	0	2	4	4	∞	1	7	∞
Pu	–	*A*	*B*	*B*	–	*A*	*F*	–

Pass IV

	A	*B*	*C*	*D*	*E*	*F*	*G*	*H*
Ku	T	T	T	F	F	T	F	F
du	0	2	4	4	7	1	7	5
Pu	–	*A*	*B*	*B*	*C*	*A*	*F*	*C*

Pass V

	A	*B*	*C*	*D*	*E*	*F*	*G*	*H*
Ku	T	T	T	T	F	T	F	F
du	0	2	4	4	7	1	7	5
Pu	–	*A*	*B*	*B*	*C*	*A*	*F*	*C*

Pass VI

	A	*B*	*C*	*D*	*E*	*F*	*G*	*H*
Ku	T	T	T	T	F	T	F	T
du	0	2	4	4	7	1	7	5
Pu	–	*A*	*B*	*B*	*C*	*A*	*F*	*C*

Pass VII

	A	*B*	*C*	*D*	*E*	*F*	*G*	*H*
Ku	T	T	T	T	T	T	F	T
du	0	2	4	4	7	1	7	5
Pu	–	*A*	*B*	*B*	*C*	*A*	*F*	*C*

Pass VIII

	A	*B*	*C*	*D*	*E*	*F*	*G*	*H*
Ku	T	T	T	T	T	T	T	T
du	0	2	4	4	7	1	7	5
Pu	–	*A*	*B*	*B*	*C*	*A*	*F*	*C*

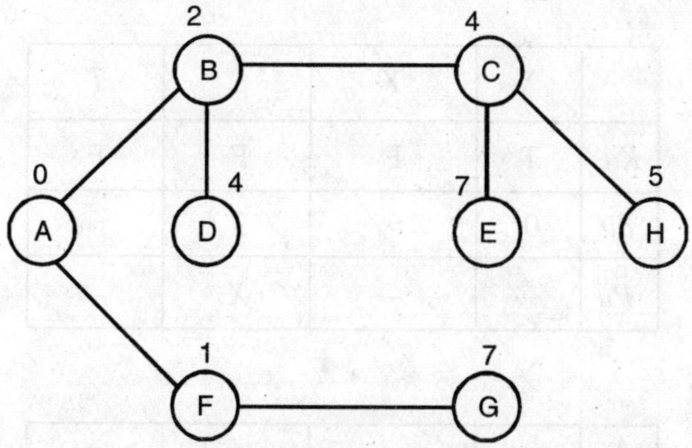

Fig. 11.44.

Vertex *A* to Vertex *H*

$$A \to B \to C \to H$$

Distance $dH = 5$.

31. A weighted graph *G* is maintained in memory by a node array *A* and weight matrix *W* as *A* : [*X*, *Y*, *S*, *T*]

$$W = \begin{bmatrix} 0 & 0 & 3 & 0 \\ 5 & 0 & 1 & 6 \\ 2 & 0 & 0 & 4 \\ 0 & 5 & 4 & 0 \end{bmatrix}$$

Draw the graph and find the shortest path from vertex *X* ?

Ans. The graph for the given weighted matrix is :

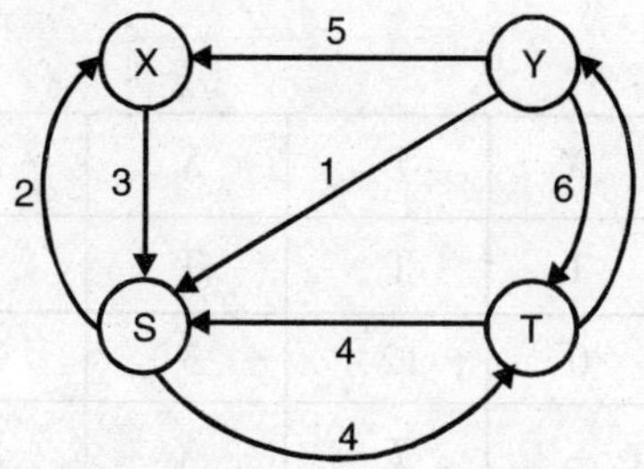

Apply Dijkstra's algorithm, which proceeds as follows :

Initial

	X	*Y*	*S*	*T*
Ku	F	F	F	F
du	0	∞	∞	∞
Pu	–	–	–	–

Pass I

	X	Y	S	T
Ku	T	F	F	F
du	0	∞	3	∞
Pu	–	–	X	–

Pass II

	X	Y	S	T
Ku	T	F	T	F
du	0	∞	3	7
Pu	–	–	X	S

Pass III

	X	Y	S	T
Ku	T	F	T	F
du	0	12	3	7
Pu	–	T	X	S

Pass IV

	X	Y	S	T
Ku	T	T	T	T
du	0	12	3	7
Pu	–	T	X	S

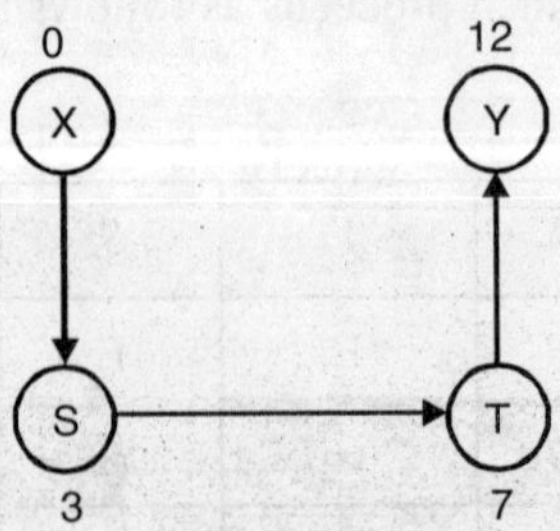

32. Find shortest path for the following graph using Dijkstra's single source shortest algorith taking V_0 as starting node.

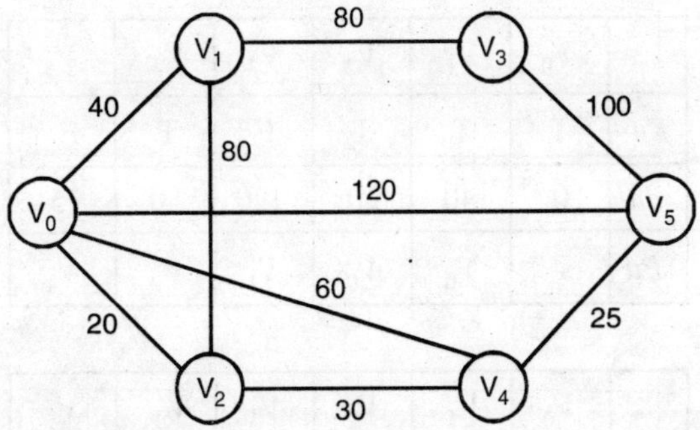

Ans. Initial

	V_0	V_1	V_2	V_3	V_4	V_5
Ku	F	F	F	F	F	F
du	0	∞	∞	∞	∞	∞
Pu	–	–	–	–	–	–

Pass I

	V_0	V_1	V_2	V_3	V_4	V_5
Ku	T	F	F	F	F	F
du	0	40	20	∞	60	120
Pu	–	V_0	V_0	–	V_0	V_0

Pass II

	V_0	V_1	V_2	V_3	V_4	V_5
Ku	T	F	F	F	F	F
du	0	40	20	∞	50	120
Pu	–	V_0	V_0	–	V_2	V_0

Pass III

	V_0	V_1	V_2	V_3	V_4	V_5
Ku	T	T	T	F	F	F
du	0	40	20	120	50	120
Pu	–	V_0	V_0	V_1	V_2	V_0

Pass IV

	V_0	V_1	V_2	V_3	V_4	V_5
Ku	T	T	T	F	T	F
du	0	40	20	120	50	75
Pu	–	V_0	V_0	V_1	V_2	V_4

Pass V

	V_0	V_1	V_2	V_3	V_4	V_5
Ku	T	T	T	F	T	T
du	0	40	20	120	50	75
Pu	–	V_0	V_0	V_1	V_2	V_4

Pass VI

	V_0	V_1	V_2	V_3	V_4	V_5
Ku	T	T	T	T	T	T
du	0	40	20	120	50	75
Pu	–	V_0	V_0	V_1	V_2	V_4

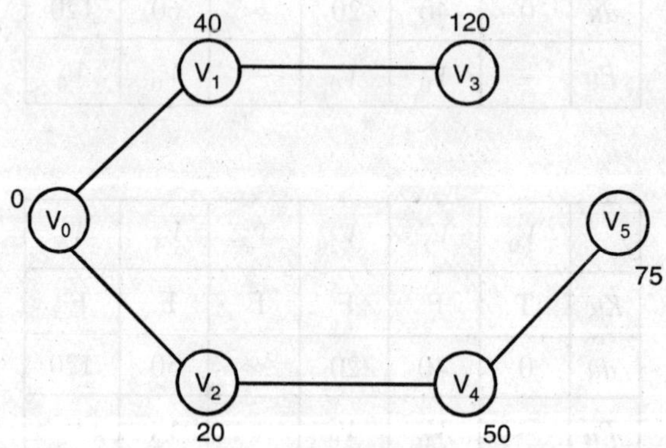

33. Consider the following specification of a graph *G*.

$$V(G) = \{1, 2, 3, 4\}$$

$$E(G) = \{(1, 2), (1, 3), (3, 3), (3, 4), (4, 1)\}$$

(*a*) Draw a picture of an undirected graph.

(*b*) Draw its adjacency matrix.

Ans. (*a*)

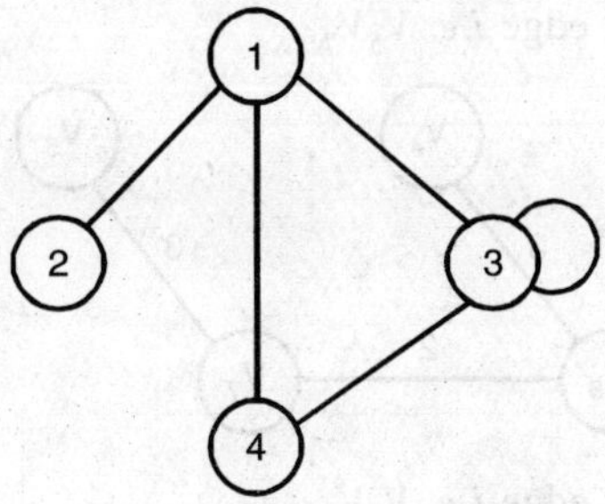

(*b*) Adjacency matrix of *G*,

$$\begin{array}{c} \\ 1 \\ 2 \\ 3 \\ 4 \end{array} \begin{array}{c} \begin{array}{cccc} 1 & 2 & 3 & 4 \end{array} \\ \begin{bmatrix} 0 & 1 & 1 & 1 \\ 1 & 0 & 0 & 0 \\ 1 & 0 & 1 & 1 \\ 1 & 0 & 1 & 0 \end{bmatrix} \end{array}$$

34. Work out the Kruskal's algorithm to find the minimum spanning tree of the following graph :

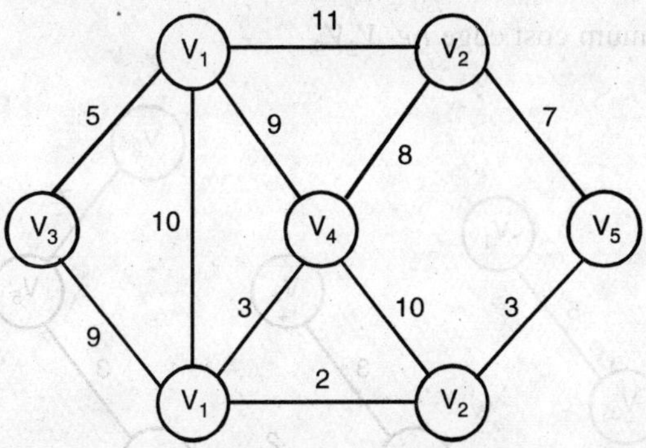

Fig. 11.45.

Ans. In Kruskal algorithm, we will find a minimum-cost spanning tree by adding an edge each time in a subtree such that the subtree doesn't have cycle and if in a spanning subtree, there are total of $|V| - 1$ number of edges then the subtree prepared is the minimum cost spanning tree as shown below :

(*i*) insert minimum cost edge *i.e.* V_6V_7

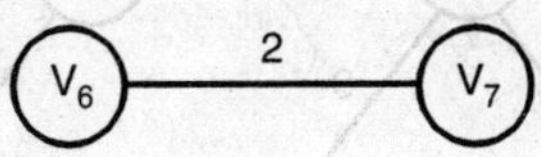

(*ii*) insert next minimum cost edge *i.e.* V_6V_4

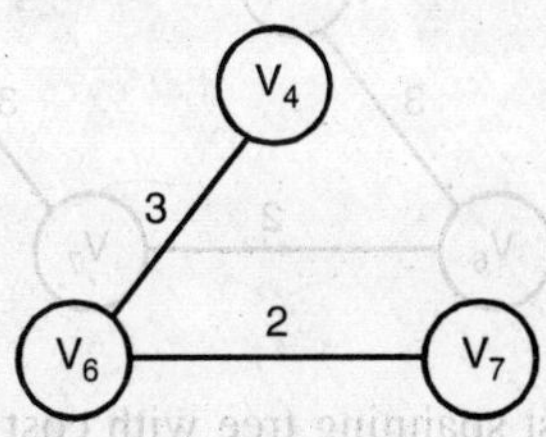

(*iii*) insert next minimum cost edge *i.e.* V_5V_7

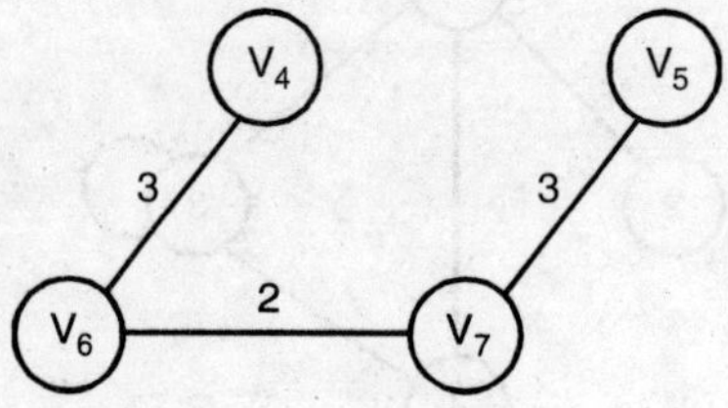

(*iv*) insert next minimum cost edge *i.e.* V_3V_1

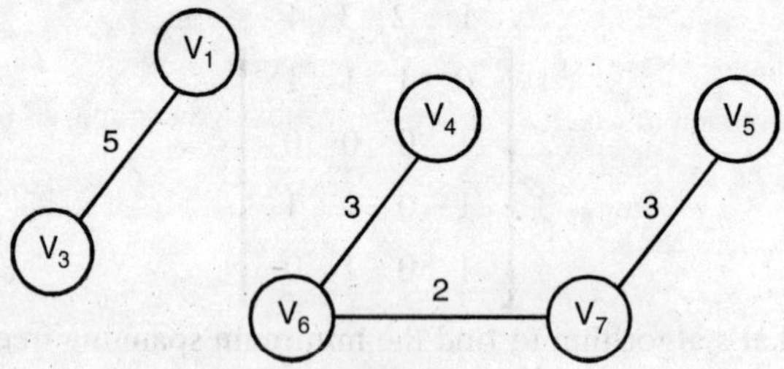

(*v*) insert next minimum cost edge *i.e.* V_2V_5

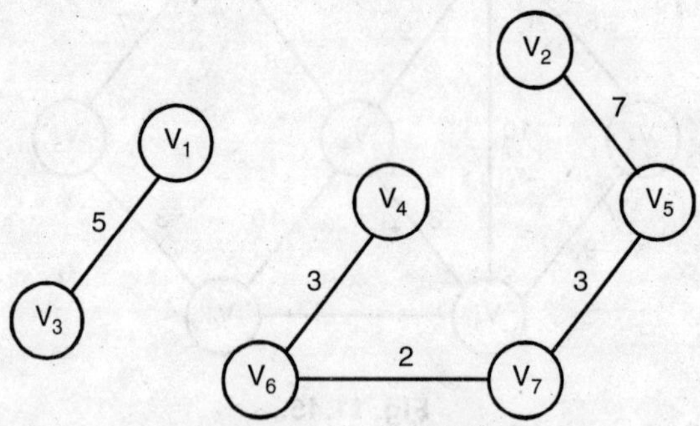

(*vi*) next minimum cost edge is V_4V_2 but it creates a cycle so don't insert it. Hence insert either V_3V_6 or V_1V_4 *i.e.*

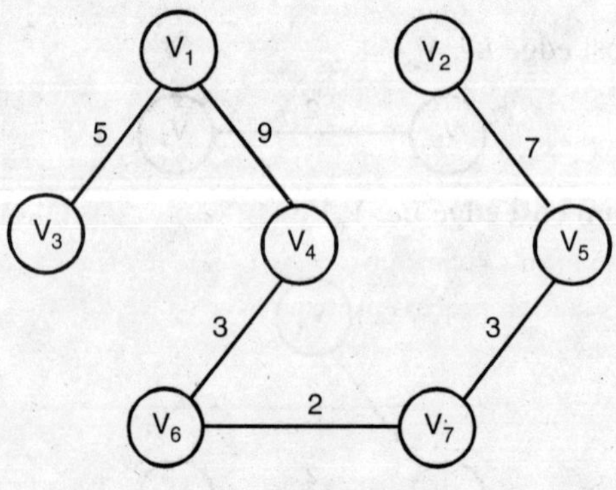

The required minimum-cost spanning tree with cost = 29.

SUMMARY

1. A graph G is composed of a finite set of vertices and a finite set of edges. A graph is denoted as G = (V, E). The elements of V are the vertices and those of E are the edges. The vertex set of V are denoted by V_G and edge set is denoted by E_G. Thus G = (V_G, E_G)
2. An edge associated with one vertex is called a loop.
3. An edge is called incident on each of its endpoints. Two edges incident on the same endpoints are called adjacent.
4. The degree of a graph is the sum of all the degrees of the vertices in the graph.
5. A vertex with no edges incident on it is isolated
6. A path is said to be closed when first and last vertex is same *i.e.* $V_0 = V_N$.
7. A connected graph is a graph in which for any given vertex in the graph, all the other vertices are reachable from it that is there is a path between any two vertices.
8. Complete graph is simple graph in which every pair of distinct vertices are adjacent. A complete graph will always be connected.
9. A directed graph or digraph G (written as DAG)is a finite set of vertices and a finite set of directed edges, where each edge is associated with an ordered pair of vertices called its endpoints.
10. In a directed graph indegree (written as indeg (v)) of a vertex is the number of edges coming to it and outdegree (written as outdeg (v)) of a vertex is number of edges going out from it.
11. A circuit is a trail that begins and ends on the same vertex. Another definition of circuit is that a circuit is a closed walk that does not contain a repeated edge.
12. Traversing a graph means accessing every vertex of graph and edge and performing some processing if desired (for example to display it) For that we have two popular graph traversal algorithms. The first one is known as Breadth First Search (BFS) and another is Depth First Search (DFS). We will examine both one by one.
13. In BFS we take a source vertex or root vertex from which we have to start the traversal. For keeping track of the vertices, we take a queue implemented as linked list.
14. In the DFS we start from source vertex and reach to one of its adjacent node; we then move to node adjacent to adjacent to root and so on *i.e.* in DFS we do not process all adjacent nodes first instead we go deeper inside the graph by following one adjacent node to another.
15. A Spanning tree of a graph consists of all vertices and some of the edges so that graph does not contain a cycle
16. A minimum spanning tree (MST) is defined for a weighted graph. It is the spanning tree for a weighted graph whose sum of weight of edges in the spanning tree is minimum.
17. For finding a MST for a graph, there are two most popular algorithms given by Prims and Kruskal's.
18. The shortest path is to find a path from one vertex to another vertex with the minimum cost. The cost is the sum of weights of all edges traversed in path.

    We have two algorithms for finding the shortest path. The first one finds shortest path among all pair of vertices and second one finds shortest path to all other vertices from a given source vertex.
19. Dijkstra's algorithm finds solution of shortest path to any vertex in the graph where source vertex is given. Dijkstra's algorithm solves the single-source shortest- path problem when all edges have positive weights. The algorithm is similar to Prim's algorithm.
20. A topological sorting is a sorting process of items over which a partial ordering is defined, *i.e.*, where an ordering is given over some pairs of items but not between all of them.

EXERCISES

1. What is a graph? Explain with example.
2. How graph is said to be a non-linear data structure?

3. How a graph is represented in computer's memory?
4. With respect to graph define the followings:
   (*i*) Simple graph (*ii*) multigraph
   (*iii*) cycle (*iv*) path
   (*v*) circuit (*vi*) walk
   (*vii*) degree (*viii*) weighted graph
   (*ix*) complete graph (*x*) DAG
5. How to compute path matrix for a given graph?
6. Compare and contrast the adjacency matrix and linked list representation for graph.
7. What is spanning tree? Where it is useful?
8. For a given graph with V vertices and E edges how many spanning trees are possible?
9. What is minimum spanning tree? Explain with example.
10. Describe algorithm of Kruskal to find a minimum spanning tree T of a weighted graph G.
11. Repeat the problem 10 using Prims algorithm.
12. Compare and contrast algorithms of Prims and kruskals for finding minimum spanning tree of a graph.
13. What are the two popular methods of graph traversal ?
14. How BFS is different from DFS?
15. Show that depth first search starting from vertex V visits all vertices reachable from V.
16. Show that the sum of degrees of all vertices in an undirected graph is twice the number of edges.
17. Show that the number of vertices of odd degree in a finite graph is even.
18. How many edges will be in a complete graph of N vertices?
19. Explain how existence of a cycle can be detected in an undirected graph using the DFS traversal?

❑❑❑

CHAPTER **12**

SORTING

INSIDE THIS CHAPTER

12.1. INTRODUCTION

Sorting refers to arranging some set of elements (may be number, alphabet or strings) in ascending (increasing) or descending (decreasing) order. There are number of methods for performing sorting. Each sorting method differs from each other by the technique it uses to place the element in the proper place. Apart from this, the difference lies in the running time taken by a particular method in sorting number of elements (hence complexity), and various other data structures that a method employ other than the set of elements. In general the main task involved in sorting is basically comparison and interchanging position of elements but there are sorting methods that employs other techniques other than comparisons for the sorting purpose. In this section, we are going to study number of sorting methods, will present C code for each of them and will do the analysis of each of the sorting method.

12.2. BUBBLE SORT

The bubble sort is the easiest and frequently used sorting algorithm among all the sorting algorithms. The algorithm has got its name as after every pass, the largest element bubbles up and move to end of the array. To understand it lets sort the following numbers using bubble sort:

```
2  1  9  0  7  4  12  3
```

In every pass of the algorithm, we have number of steps which decreases as the number of pass increases. Let's start with pass 1:

```
S1 swap if A[0]>A[1]
                     1  2  9  0  7  4  12  3
S2 swap if A[1]>A[2]
Changed array:       1  2  9  0  7  4  12  3
```

```
S3 swap if A[2]>A[3]
Changed array:     1  2  0  9  7  4  12  3
S4 swap if A[3]>A[4]
Changed array:     1  2  0  7  9  4  12  3
S5 swap if A[4]>A[5]
Changed array:     1  2  0  7  4  9  12  3
S6 swap if A[5]>A[6]
No change:         1  2  0  7  4  9  12  3
S7 swap if A[6] >A[7]
Changed array:     1  2  0  7  4  9  3  12
```

Note at end of pass 1, largest element has reached up to the last position in the array. For the subsequent passes, we show the processed elements of the array only.

Pass 2

S1: 1 2 0 7 4 9 3 12	**S2:** 1 0 2 7 4 9 3 12
S3: 1 0 2 7 4 9 3 12	**S4:** 1 0 2 4 7 9 3 12
S5: 1 0 2 4 7 9 3 12	**S6:** 1 0 2 4 7 3 9 12

Pass 3

S1: 0 1 2 4 7 3 9 12	**S2:** 0 1 2 4 7 3 9 12
S3: 0 0 2 4 7 3 9 12	**S4:** 0 1 2 4 7 3 9 12
S5: 0 0 2 7 4 3 9 12	

Pass 4

S1: 0 1 2 4 3 7 9 12	**S2:** 0 1 2 4 3 7 9 12
S3: 0 1 2 4 3 7 9 12	**S4:** 0 1 2 4 3 7 9 12

Pass 5

S1: 0 1 2 3 4 7 9 12	**S2:** 0 1 2 3 4 7 9 12
S3: 0 1 2 3 4 7 9 12	

Pass 6

S1: 0 1 2 3 4 7 9 12	**S2:** 0 1 2 3 4 7 9 12

Pass 7

S1: 0 1 2 3 4 7 9 12	

At the end of pass 7, the array is sorted.

Notice for an array of 8 elements, sorting finishes after 7th pass i.e. for an array of N elements, sorting will be done after N-1 pass.

Implementation of bubble sort is now easy to understand. Outer for loop runs for every pass and inner for loop runs for steps for a single pass. If if condition turns out to be true, swapping of array elements at position *j* and *j*+1 is done.

```
void bubblesort (int arr[ ], int N)
{
int i, j, t;
for (i = 0; i< = N-1; i++)
{
```

```
for(j = 0; j< = N-i-1;j++)
{
If  (arr[j]>arr[j+1])
{
t = arr[j];
arr[j]  =  arr[j+1];
arr[j+1]  =  t;
}
}
}
}
```

12.2.1. Complexity of Bubble Sort

For an array of N elements there are N-1 passes (outer for loop) and in each pass there are N-1 comparisons. Therefore total number of comparisons are (N-1) * (N-1) i.e. N^2 - 2N+1. This is of the order of $O(N^2)$. Thus complexity of bubble sort is $O(N^2)$. But this complexity is in general not fixed. This is because number of comparisons may differ if the input array is already sorted or sorted in reverse order. Further as we have 3 statements for swapping, these statements also contribute to running time of bubble sort. We now consider only number of interchanges (number of times if condition will be true) for the following array: 1 2 3 4 5 6 7 8 as input:

It is clearly visible from the code that no interchange of elements will take place as if condition will never be true for the above input. From this we can say that for a sorted input bubble sort runs in linear time i.e. complexity is O(N). This will be the best case for bubble sort where array is already sorted.

Now if the array is sorted in reverse order i.e. our array is: 8 7 6 5 4 3 2 1, then for every iteration of the outer for loop, if condition will turn out to be true. That is in this case maximum number of interchanges ((N-1)*(N-l)) will take place and complexity will be $O(N^2)$. This will be the worst case for bubble sort where array is sorted in reverse order.

If the array elements are in the arbitrarily order (not in any of the above order) then complexity will also be $O(N^2)$.

12.2.2. Improved Bubble Sort

The efficiency of the bubble sort can be increased by reducing the number of passes in which the array will be sorted. We would like not to run any further pass when elements have been sorted in fewer than N-1 pass or array is already sorted. This information can be obtained by knowing whether any interchange has taken place in the inner loop. It interchange has taken place then pass will continue otherwise it won't.

The modified bubble sort algorithm is given below:

```
void bubblesort (int arr[ ], int N)
{
int i, j, t;
int swap=1;
for (i = 0; i< N-1 && swap = = 1;i++)
{
swap = 0;
```

```
for (j = 0 ; j< N-i-1;j++)
{
if(arr[j]>arr[j+1])
{
swap = 1;
t = arr[j];
arr[j] = arr[j+1];
arr[j+1] = t;
}
}
}
}
```

We have taken one integer variable by the name swap. The initial value of this variable is 1. The outer for loop runs till i< = N-1 and swap = = 1. In the outer for loop, value of this variable is set to 0. Now if any interchange takes place, swap will have value 1, else it will remain 0. Because of this condition in the outer for loop will turned out to be false and next pass will not run.

12.3. INSERTION SORT

In the insertion sort, elements are inserted into their proper location such that array will be sorted. Assume we have array ARR with 5 elements.

(*i*) In the first round ARR[0] will be itself sorted.

(*ii*) In the second round ARR[0] and ARR[1] will be compared and they will be sorted such that two elements are in order as: ARR[0] < ARR[1].

(*iii*) In the third round ARR[2] will be inserted among ARR[0] and ARR[l] such that the three will appear in order as: ARR[0] <ARR[1] <ARR[2].

(*iv*) In the fourth round ARR[3] will be inserted among ARR[0] to ARR[2] such that the four will appear in order as: ARR[0] <ARR[1] <ARR[2] < ARR[3].

(*v*) Finally in the last round ARR [4] will be inserted among elements ARR [0] to ARR[3] such that the five will appear in order as: ARR[0] <ARR[1] <ARR[2] <ARR[3] <ARR[4].

Thus in every pass proper location of the element to be inserted is found and element is inserted at that location. To better understand lets sort the following numbers using insertion sort.

45 67 4 2 5 66 89 0 23

At the end of step 10 the array is sorted.

```
void inssort(int arr[ ])
{
int i, j, k, t;
for (i = 1;i< = MAX-1;i++)
{
for (j = 0; j<i ; j++)
{
if (arr [j] >arr [i])
{
```

```
t = arr[j];
arr [j] = arr[i];
for(k = i; k>j ; k-)
arr[k] = arr[k-1];
arr[k+1] = t;
}
}
}
}
```

Steps	Elements in Array										Remarks
1	12	45	67	4	2	5	66	89	0	23	A[0] is at proper place
2	12	45	67	4	2	5	66	89	0	23	A[0] is and A[1]1 are sorted
3	12	45	67	4	2	2	66	89	0	23	A[0] is and A[2]1 are sorted
4	4	12	45	67	2	5	66	89	0	23	A[3] is inserted at A[0]
5	2	4	12	45	67	5	66	89	0	23	A[4] is inserted at A[0]
6	2	4	5	12	45	67	66	89	0	23	A[5] is inserted at A[0]
7	2	4	5	12	45	66	67	89	0	23	A[6] is inserted at A[0]
8	2	4	5	12	45	66	67	89	0	23	A[7] is at proper place
9	0	2	4	5	12	45	66	67	89	23	A[8] is inserted at A[0]
10	0	2	4	5	12	23	45	66	67	89	A[9] is inserted at A[5]

For the implementation of insertion sort, we have two for loops. The outer for loop runs from i = 1 to 9 and inner for loop runs from j = 0 to i. This is because at any step in the insertion sort we have to take into account 0 to i elements from the array. To understand the working of the two for loops, we show few steps of the algorithm.

For i = 1, j = 0, if condition is false. Outer for loop runs again. This time i = 2 and j = 0; if condition is false again. For i = 2, j = 1, once again if condition is false. Outer for loop runs again. This time i = 3, j = 0, if condition is true as A [0] > A[3]; control enters into the if block:

```
t = p[j] = p[0] = 12; p[j] = p[i] = 4; i.e. p[0] = p[3] = 4;
```

The following for loop runs for shifting the array elements to find a proper location where t can be inserted.

```
for (k = i ; k>j ; k-)
    p[k] = p[k-1];
```

For the first iteration of for loop we have p[3] = p[2]; first four elements in the array are:

```
4 45 67 67
```

Next time we have p[2] = p[l]; this time first four elements in the array are:

```
4   45   45   67
```

Next time we have p[1] = p[0]; this time first four elements in the array are:

```
4   4   45   67
```

Next time k becomes 0 and for loop terminates. After the for loop t is inserted as position k+1 in the array as: p[k+1] = t i.e. p[1] = 12.

Now the first four elements of the array are sorted. Continuing in the above manner, the whole array is sorted.

12.3.1. Algorithm For Insertion Sort

Algorithm 12.1. INSERTION-SORT (A) ....

```
for j ← A[j]
    do key ← A[j]
        do insert A[j] into the sorted squence A[1 .. j – 1]
        while i > 0 and A[i] > key
            do A[i + 1] ← A[i]
            i ← i – 1
        A[i + 1] ← key
```

Example 12.1.

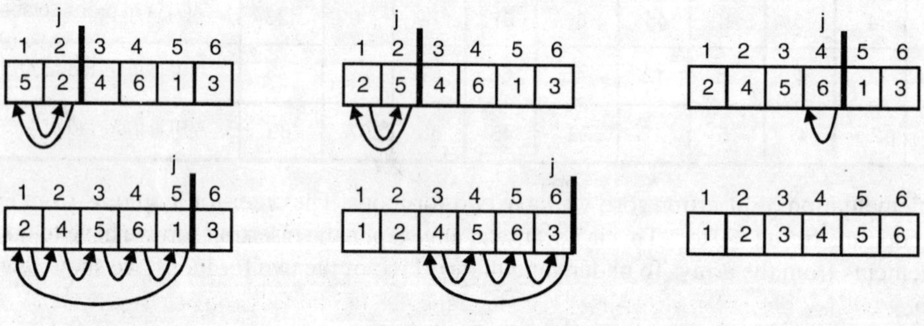

[Read this figure row by row. Each part shows what happens for a particular iteration with the value of j indicated. *j* indexes the "current card" being inserted into the hand. Elements to the left of *A*[*j*] that are agreater than *A*[*j*] move one position to the right, and A[j] moves into the evacuated position. The heavy vertical lines separate the part of the array in which an iteration works—*A*[1.. *j*] – from the part of the array that is unaffected by this iteration—*A*[*j* + 1 ... *n*]. The last part of the figure shows the final sorted array.]

12.3.2. Complexity Of Insertion Sort

The outer for loop runs for N-1 number of times. When the array is already sorted in reverse then there will be maximum number of comparisons (N-1) (maximum number of interchanges also, as if condition will turn out to be true in every case). In the first iteration number of comparisons will be 1, in the second it will be 2, upto N-1, so total number of comparisons will be

```
1+2+3+4+  (N-1)=N(N-1)/2
```

Therefore complexity of insertion sort in worst case: $O(N^2)$

Now if the array is already sorted then comparison will be done but no interchange will be taking place and insertion sort will run in linear time i. e. complexity of insertion sort when array already sorted is : $O(N)$. This is the best case for insertion sort.

After the discussion of worst and best case, we consider average case. On an average there will be (N-1)/2 comparisons, so total number of comparisons will be :

```
1+2+3+4+ ... (N-1)/2= N(N-1)/4
```

Therefore complexity of insertion sort in average case: $O(N^2)$

Search time for finding a proper location within the array or any segment of array for any of the array element under consideration can be minimized when instead of linear search we perform binary search. In this case number of comparisons will be log N instead of $(N - 1)/2$. But still we have to shift $(N - 1)/2$ elements forward (The inner for loop within inner for loop). Thus the choice of linear search or binary search does not affect the complexity of the insertion sort algorithm. This comes out from this fact also that insertion sort is generally employed for sorting when N is small and for small values of N, linear search and binary search gives almost same results.

12.4. SELECTION SORT

The logic for selection sort is very simple. Given a set of elements N, we find out the location of the minimum element among the set and replace it with the location of first element. Next we do the same task for N-1 elements and find out the location of the minimum element among $N - 1$ elements and replace is with the location of the first element (of set $N - 1$). After the first step minimum element will be at the first location. After second step second minimum element among the remaining list will be at second position and so on. This process is repeated and finally after $N - 1$ steps the list or set will be sorted. That is in every step location of the minimum element is selected hence name selection sort. This is shown below for the given set of numbers, assuming array name is A :

2 5 6 7 8 12 34 20 1 67

Steps	Elements in the array										Remarks
1	**2**	5	6	7	8	12	34	20	1	67	Lmin = 8, swap A[0] ←→ A[8]
2	2	**5**	6	7	8	12	34	20	**2**	67	Lmin = 8, swap A[1] ←→ A[8]
3	1	2	**6**	7	8	12	34	20	**5**	67	Lmin = 8, swap A[2] ←→ A[8]
4	1	2	5	**7**	8	12	34	20	**6**	67	Lmin = 8, swap A[3] ←→ A[8]
5	1	2	5	6	**8**	12	34	20	**7**	67	Lmin = 8, swap A[4] ←→ A[8]
6	1	2	5	6	7	**12**	34	20	**8**	67	Lmin = 8, swap A[5] ←→ A[8]
7	1	2	5	6	7	8	**34**	20	**12**	67	Lmin = 8, swap A[6] ←→ A[8]
8	1	2	5	6	7	8	12	**20**	34	67	Lmin = 8, swap A[8] ←→ A[8]
9	1	2	5	6	7	8	12	20	**34**	67	Lmin = 8, swap A[9] ←→ A[8]
	1	2	5	6	7	8	12	20	34	67	Array is now sorted

The code for the selection sort is given below :

```
void min(int arr[ ], int k, int*loc)
{
int i, m;
m = arr[k];
```

```
*loc = k;
for(i = k+1;i<MAX;i++)
if (m>arr [i])
}
m = arr [i];
*Loc = i;
}
}
void selsort (int arr[ ])
{
int i, loc, t;
for(i = 0; i<MAX;i++)
{
min (arr, i, & loc);
t = arr [i];
arr [i] = arr [loc];
arr[loc] = t;
}
}
```

For finding minimum of set of array elements during every pass of the for loop we have written the function min. Inside the function we pass the array, starting index say k of the subset of the array and a pointer loc. When the function returns loc will have location of the smallest element among the array elements p[k] to p[MAX-l]. Inside the function selsort within for loop element at loc is swapped with the current element at index i. This continues for every iteration of the for loop. When the loop exhausts and function returns array will be sorted.

12.4.1. Algorithm For Selection Sort

Algorithm 12.2. Selection-Sort (A)..

1. n ← length[A]
2. for j ← 1 to n – 1
3. do smallest ← j
4. for i ← j + 1 to n
5. do if A[i] < A[smallest]
6. then smallest ← i
7. exchange A[j] ↔ A[smallest]

The algorithm maintains the loop invariant that at the start of each iteration of the outer **for** loop, the subarray $A[1 .. j-1]$ consists of the $j-1$ smallest elements in the array $A[1 ... n]$, and this subarray is in sorted order. After the first $n-1$ elements, the subarray $A[1 ... n-1]$ contains the smallest $n-1$ elements, sorted and therefore element $A[n]$ must be the largest element.

The running time of the algorithm is $\Theta(n^2)$ for all cases.

12.4.2. Complexity of Selection Sort

The for loop runs $N - 1$ times in the function selsort. For every minimum selected for the array or segment of array just one swapping operation is performed. This is because it is not placed in any inner for loop. The function mm requires $n - 1$ comparison during first iteration of the for loop, $n - 2$ in second iteration and so on. Thus total number of comparisons:

$$N - 1 + N - 2 + 3 + 2 + 1 = N(N - 1)/2$$

Thus complexity of selection sort is $O(N^2)$. If we consider array in sorted order or in reverse order then this does not affect number of comparisons in function min. Thus number of comparisons in selection sort is independent of the original order of the elements in the array. Moreover, the complexity of bubble sort and selection sort is same but selection sort executes faster than bubble sort. This is because of less number of swapping operations.

12.5. MERGE SORT (Ver 1)

Merging of two arrays means combining them in such manner so that resultant array is sorted. For that both the source array need to be sorted first. If the size of first and second array is N_1 and N_2 then size of merged array will be $N_1 + N_2$. Consider two arrays Al and A2 with the elements as given below. Note the arrays are sorted. In case they are not sorted, they need to be sorted first.

```
0  2  6  8(A1)          1  3  5  9  (A2)
```

For merging of two arrays we compare first element of array *Al* with first element of *A*2. The lower of two is assigned to the destination array and the corresponding array index is incremented. We assume index *i* for *A*l and *j* for *A*2 and *k* for the array *M* (the merged array using *A*l and *A*2).

*S*1 compare 0 and 1, as 0<1 assign 0 to *M*, increment *i*, *k* by 1.

*S*2 compare 2 and 1, as 1<2 assign 1 to *M*, increment *j*, *k* by 1.

*S*3 compare 2 and 3, as 2<3 assign 2 to *M*, increment *i*, *k* by 1.

*S*4 compare 6 and 3, as 3<6 assign 3 to *M*, increment *j*, *k* by 1.

*S*5 compare 6 and 5, as 5<6 assign 5 to *M*, increment *j*, *k* by 1.

*S*6 compare 6 and 9, as 6<9 assign 6 to *M*, increment *i*, *k* by 1.

*S*7 compare 8 and 9, as 8<9 assign 8 to *M*, increment *i*, *k* by 1.

If any of the arrays exhausts, rest of the elements of other array is assigned to *M*. Here first array exhausts and in the second array only element 9 remains. This element is assigned to *M*.

For the implementation we have written the following *C* function. The macro HALF is MAX/2 and MAX is the size of the merged array. Size of both source arrays is HALF.

```
void mergesort (int p[ ], int q[ ], int r[ ] )
{
int i, j, k, t;
for(i = 0;i<HALF;i++)
{
for (j = i+1;j<HALF;j++)
{
if(p[i]>p[j] )
{
t = p[i];
```

```
    p[i] = p[j];
    p[j] = t;
    }
    if(q[i]>q[j])
    {
    t = q[i];
    q[i] = q[j];
    q[j] = t;
    }
    }
    }
    for (i = j = k = 0; i< = MAX;)
    {
    if(p [j] < = q[k])
        r[i++] = p [j++];
    else
        r[i++] = q[k++];
    if (j = = HALF| | k = = HALF)
        break;
    }
    for (;j<HALF;)
        r[i++] = p [j ++];
    for (;k<HALF;)
        r[i++] = q [k++];
    }
```

In the program, the first for loop does the sorting of the two array elements. Next for loop assigns the elements to destination array by comparing array elements of the two source arrays. When this for loop terminates, remaining elements from any of the arrays are assigned to destination array.

12.5.1. Merge Sort (Ver 2)

In the previous function we considered two separate arrays and we did the merging of two arrays to get the merged sorted array. Here we have just one array and we apply the merge sort method over this single array. The logic is simple and similar to previous one but here some subset of array is treated as the source array for merging. To understand it practically, let's take the following set of numbers in an array:

```
6  2  8  3  9  7  1  0
```

In the first pass every individual array element is treated as one array. Each of the array element A[0] to A[7] is treated as one array; so in the first round we have 8 subarrays.

```
{A[0]}, {A[1]}, {A[2]}, {A[3]}, {A[4]}, {A[5]}, {A[6]}, {A[7]}
```

Writing the actual values of A[i]'s we have

```
{6}, {2}, {8}, {3}, {9}, {7}, {1}, {0}
```

Each of them will itself be sorted. In the second pass taking two adjacent array elements in sorted manner (*i.e.* merging them) as one subarrays we have 4 subarrays :

```
{A[1], A[0]}, {A[3], A[2]}, {A[5], A[4]}, {A[7], A[6]},
```

Writing the actual values of A[i]'s we have

```
{2, 6}, {3, 8}, {7, 9}, {0, 1}
```

In the third pass taking two subarrays at a time & from previous pass and merging them, we have 2 subarrays of 4 elements each. We show only the actual values of the array elements and not their subscripted form:

```
{2, 3, 6, 8} {0, 1, 7, 9}
```

Finally the above two arrays are merged to get the final sorted array:

```
{0, 1, 2, 3, 6, 7, 8, 9)
```

Now we see the implementation of mergesort sorting method. We take the same array elements as discussed above so MAX will be 8. The code is given below :

```
void mergesort (int*p)
{
int temp[MAX];
int i, j, k, L1, L2, S, UP1, UP2;
S = 1;
while (S<MAX)
{
L1 = 0;k = 0;
while (L1+S<MAX)
{
L2 = L1+S;
UP1 = L2-1;
UP2 = (L2+S-1<MAX) ? L2+S-1 : MAX-1;
for(i = L1, j = L2;i< = UP1 && j< = UP2;k++)
     if(p[i]< p[j])
          temp[k] = p[i++];
     else
          temp[k] = p[j++];
for(;i< = UP1;k++)
          temp[k] = p[i++];
for(;j< = UP2;k++)
          temp[k] = p[j++];
L1 = UP2+1;
}
for(i = 0; i<MAX;i++)
        p[i] = temp[i];
```

```
S = S*2;
}
}
```

Initially S = 1 denotes the size of the subarrays. L1, L2 are the lower bounds and UP1, UP2 are the upper bounds for the two subarrays under consideration. For holding the elements from the subarrays temporarily we have taken array temp. In the inner while loop as condition is true we have L1 = 0, UP1 = 0, L2 = 1, UP2 = 1;

After the first for loop temp[0] is 2, i = 0, j = 2, k = 1. The first for loop runs and temp[1] will be 6; In the second iteration of inner while loop, L1 = 2, L2 = 3, UP1 = 2, UP2 = 3. After the first for loop temp[2] is 3, i = 2, j 4, k = 3. The first for loop runs and temp[3] will be 8; In the third iteration of inner while loop, L1 = 4, L2 = 5, UP1 = 4, UP2 = 5. After the first for loop temp[4] is 7, i = 4, j 6, k = 5. The first for loop runs and temp[5] will be 9; In the fourth iteration of inner while loop, L1 = 6, L2 = 7, UP1 = 6, UP2 = 7. After the first for loop temp [6] is 0, i = 6, j = 8, k = 7. The first for loop runs and temp [7] will be 1;

At this instance, inner while loop ends as L1+S (8+1) not less than MAX (8). After the inner while loop the elements from temp array are copied back to original array and S becomes 2 i.e. size of each subarray becomes 2. Processing continues as discussed above.

12.5.2. Algorithm For Merge Sort

Algorithm 12.3. Merge-Sort (A, p, r) ....

```
if p < r
    then q ← [(p + r)/2]
        MERGE-SORT (A, p, q)
        MERGE-SORT (A, q + 1, r)
        MERGE (A, p, q, r)
Exit
```

Algorithm 12.4. Merge-Sort (A, p, q, r) ....

```
n_1 ← q – p + 1
n_2 ← r – q
create arrays L[1...n_1 + 1] and R[1...n_2 + 1]
for i ← 1 to n_1
    do L[i] ¬ A[p + i – 1]
for j ← 1 to n_2
    do R[j] ← A[q + j]
L[n_1 + 1] ← ∞
R[n_2 + 1] ← ∞
i ← 1
j ← 1
for k ← p to r
    do if L[i] ≤ R[j]
        then A[k] ← L[i]
            i ← i + 1
        else A[k] ← R[j]
            j ← j + 1
```

12.5.3. Complexity of Merge Sort

It is clearly visible from the code that value of S increases by a power of 2, so there cannot be more than $\log_2$ N pass in the merge sort and in each pass we can have maximum N comparisons. Therefore complexity of merge sort is O(N $\log_2$ N). However, the algorithm requires additional space in the form of auxiliary array temp so in terms of space complexity it is) O(N).

12.5.4. Recursive Merge Sort

```
void merge (int arr[ ], int L1, int U1, int L2, int U2)
{
int p, q, j, i;
int temp[100];
p = L1;q = L2;i = 0;
while(p< = U1 && q< = U2)
      if(arr[p]<arr[q])
            temp[i++] = arr[p++];
      else
            temp[i++] = arr[q++];
for(;p< = U1;)
      temp[i++] = arr[p++];
for(;q< = U2;)
      temp[i++] = arr[q++];
for(q = L1, i = 0;q< = U1;q++, i++)
      arr[q] = temp[i];
for (q = L2, j = i ; q< = U2;q++, j++)
      arr[q] = temp[j];
}
void mergesort(int arr [ ], int LB, int UB)
{
int m;
if (UB>LB)
{
m = (UB+LB)/2;
mergesort (arr, LB, m);
mergesort (arr, m+1, UB);
merge (arr, LB, m, m+1, UB);
}
}
```

The code is same as the earlier ones but here we have shown you the recursive version of the merge sort algorithm. In the code we have two functions, first one is the merge and second one is the mergesort. In the mergesort function we make two recursive call to mergesort function and later we call merge function for the merging operation. We leave it for the reader to understand the code as a lot of discussion has been done in the earlier two versions.

In the mergesort we recursively solves two subarrays using call to mergesort. This taken 2 T(N/2) time where T(N) is time taken by the mergesort on N numbers. The merge operation taken O(N) time for combining N elements of array. Thus total running time can be expressed using the equation:

```
T(N) = 2 T (N/2) + O(N)
```

Solution of the above is $O(N \log_2 N)$ which is the worst case running time for mergesort.

12.6. QUICK SORT

In quick sort we find the proper position of an element in the array. Assume proper position is pos, now this pos will be such that all the elements that are lower in index then pos, all such elements will be lower than element at pos and all the elements that are higher in index then pos, all such elements will be higher then element at that pos. This proper position is returned and array is divided into two partitions: One from L to pos and other from pos+1 to U (where L and U are lower and upper bound in array). To understand this consider the following set of numbers:

```
23  3  5  45  6  78  11  13  1
```

We want to find out the proper position of the array element 23 (say index is L). Starting from the element at L+1 moving left to right we search for an element that is > than 23. First such element is 45 (say this index is p). Now starting from element at position q (it is U in the beginning) moving from right to left we search for an element that is < 23(decrement q). The first such element is 1. Now element at position p and q are swapped; Array elements now at:

```
23  3  5  1  6  78  11  13  45  (p=3, q=8)
```

From position p searching for an element that is > than 23, we have 78 at position 5 (p = 5). Starting from q searching for an element that is < 23, we have 13 at position (q = 7). Now elements at position q and p are swapped; Array elements now at:

```
23  3  5  1  6  13  11  78  45  (p=5,q=7)
```

Continuing in the above manner p will be at position 8 and q will be at position 6. As q crosses p during this search, we stop this processing and element at L (i. e. 23) is swapped with element at position q (6). Array elements now:

```
11  3  5  1  6  13  23  78  45
```

You can see that element 23 has got its proper position in the array and all elements before 23 are smaller than 23 and all elements that come after 23 are greater than 23.

Now all the elements from L to q constitute one partition of the array and elements from q+1 to U constitute another partition of the array. Quick sort is called again over these two subarray. The value q is returned from the function that is responsible for doing the partition. The main purpose of this function is to find a proper position of the element in the array. This element is known as pivot element.

```
int partition(int arr[ ], int L, int U)
{
int i, p, q, t;
p=L+1;
q=U;
i = arr[L];
while(q> = p)
```

```
{
while (arr [p]<i)
      p++;
while(arr[q]>i)
     q–;
if(q>p)
{
t = arr[p];
arr[p] = arr[q];
arr[q] = t;
}
}
t = arr[L];
arr[L] = arr[q];
arr[q] = t;
return q;
}
void Qsort (int p[ ], int L, int U)
{
int i;
if (U>L)
{
i = partition(p, L, U);
Qsort(p, L, i-1);
Qsort(p, i+1, U);
}
}
```

For the implementation of quick sort we have written two functions: one for the quick sort and another for partitioning the arrays. After determining value of q for the original array, function Qsort is called again for both the partitions. At the end when recursion stops, array will be sorted.

12.6.1. Algorithm For Quick Sort

Algorithm 12.5. QUICKSORT (A, p, r) ....

1. if p < r
2.   then q ← PARTITION (a, p, r)
3.   QUICKSORT(A, p, q – 1)
4.   QUICKSORT(A, q + 1, r)

Algorithm 12.6. PARTITION(A, p, r) ....

1. x ← A[r]
2. i ← p – 1

```
for j ← p to r – 1
      do if A[j] ≤ x
            then i ← i + 1
                    exchange A[i] ↔ A[j]
exchange A[i + 1] ↔ A[r]
return i + 1
```

12.6.2. Complexity of Quick Sort

Consider size of array N is some power of 2 say $N = 2^M$. Taking $\log_2$ both sides we have $M = \log_2 N$. Now assume that the partition method always gives position of the pivot element as middle position in any of the subarray. If this happens then there will be nearly N comparisons in the first pass, $N/2$ comparisons in the second pass (for each of the two sub array), $N/4$ comparison in the third pass (for each of the 4 sub arrays) and so on up to pass number M. Thus total number of comparisons:

$$N + 2 * (N/2) + 4*(N/4) + \ldots\ldots\ldots\ldots N * (N/N)$$

Or $\quad N + N + \ldots M$ times

So total number of comparisons $O(N * M)$ as we have $M = \log_2 N$, so complexity of Quick sort in best case is $O(N \log_2 N)$

The worst case of Quick sort occurs when the input array is already sorted. In this case the first pivot element will be at position 0, so two subarrays will be of size o and N. Next time pivot element will be at position 1, so two sub arrays of size 1 and N-1 will be there. Continuing in this manner we will be having sub arrays of size N, $N - 1$, $N - 2$....... If for sorting an array of size N, requires N comparisons then total number of comparisons will be:

$$N + N - 1 + N - 2 + N - 3 \ldots\ldots\ldots\ldots 1 = N(N + 1)/2$$

So worst case time complexity of quick sort is $O(N^2)$

12.7. SHELL SORT

In the shell sort, we make a shell of the elements at position i and $i + m$, where m is any power of 2. If element at position i is greater than an element at position $i + m$ we do the swapping of the elements. The value of m changes in every pass of the method (in fact it decreases in every pass and in last pass it must be 1) and different value of m gives us shells of different size. The value m is also known as increment. The value of m decreases by a power of 2 after one pass is over. When value of m is 1, it means shell of size 1 and array will be sorted. If p is the number of pass, then one must choose increment m_i such that

$$N > m_1 > m_2 > \ldots\ldots\ldots\ldots > m_p = 1$$

In general initial value m_1 is chosen as $m_1 = N$ div 2 and successive increment are chosen as: $m_{i+1} = m_i$ div 2.

Let's understand by applying it over elements of array given below :

1 56 4 5 2 90 0 3

Pass 1 : *m* = 8/2 = 4

Steps	Elements in the array								Remarks
1	1	56	4	5	2	90	0	3	A[0] < A[0 + 4] no swapping
2	1	56	4	5	2	90	0	3	A[1] < A[1 + 4] no swapping
3	1	56	4	5	2	90	0	3	A[2] > A[1 + 4] do swapping
4	1	56	0	5	2	90	4	3	A[3] > A[3 + 4] do swapping
5	1	56	0	3	2	90	4	5	

Pass 2 : *m* = 4/2 = 2

Steps	Elements in the array								Remarks
1	1	56	0	3	2	90	4	5	A[0] > A[0 + 2] do swapping
2	0	56	1	3	2	90	4	5	A[1] > A[1 + 2] do swapping
3	0	3	1	56	2	90	4	5	A[2] < A[2 + 2] no swapping
4	0	3	1	56	2	90	4	5	A[3] < A[3 + 2] no swapping
5	0	3	1	56	2	90	5	5	A[4] < A[4 + 2] no swapping
6	0	3	1	56	2	90	4	5	A[5] > A[5 + 2] do swapping
7	0	3	1	56	2	5	4	90	A[3] > A[3 + 2] do swapping
8	0	3	1	5	2	56	4	90	

Pass 3 : *m* = 2/2 = 1

Steps	Elements in the array								Remarks
1	0	3	1	5	2	56	4	90	A[0] < A[0 + 1] no swapping
2	0	3	1	56	2	5	4	90	A[1] > A[1 + 1] do swapping
3	0	1	3	56	2	5	4	90	A[2] < A[2 + 1] no swapping
4	0	1	3	56	2	5	4	90	A[3] > A[3 + 1] do swapping
5	0	1	3	2	56	5	4	90	A[2] > A[2 + 1] do swapping
6	0	1	2	3	56	5	4	90	A[4] > A[4 + 1] do swapping
7	0	1	2	3	5	5	4	90	A[5] > A[5 + 1] do swapping
8	0	1	2	3	5	56	56	90	A[6] < A[6 + 1] no swapping
9	0	1	2	3	5	4	56	90	A[4] > A[4 + 1] do swapping
	0	1	2	3	4	5	56	90	Array is now sorted

```
void shellsort (int*a)
{
int i, m, j, t;
m = MAX/2;
while (m > 0)
{
for (j = m;j<MAX;j++)
    for(i = j-m;i> = 0;i = i-m)
        if (a[i]>a[i+m])
```

```
            {
                t = a[i];
                a[i] = a[i+m];
                a[i+m] = t;
            }
            else
                break;
        m = m/2;
        }
        }
```

For the implementation we have written a for loop that runs till $m>0$. Initial value of m is MAX/2 and successive value of *rn* is calculated from the expression $m = m/2$. For every value of m, we have a for loop that runs from $j = m$ to MAX – 1. The next for loop does the swapping of elements at position i and $i + m$ if $a[i] > a[i + m]$. Note the decrement $i = i - m$ will be useful only when after swapping $a[i]$ and $a[i + m]$ causes arrangement of $a[i]$ and $a[i + m]$ such that $a[i] > a[i + m]$ for any value of i even after they were satisfying the condition earlier ($a[i]$ were less than $a[i + m]$). To understand it consider the following snapshot of elements of arrays. The value of m is 2 here :

0	3	1	56	2	90	4	5
0	3	1	56	2	5	4	90
0	3	1	5	2	56	4	90

Take case where i is 5. We compare element A[5] and A [7] in the first row, as A [5] > [7], we swap them. Row second shows the swapped array elements. In the second row as A [4] > A [6], they are swapped and resultant array is shown in the third row.

12.7.1. Complexity of Shell Sort

Till date it has not been possible to evaluate the time complexity of the shell sort correctly on an arbitrary array and for any general choice of increment. The actual time complexity is a factor of value of m chosen. If an approximate values of m is chosen in sequence then time complexity of shell sort is approximately given by $O(N (\log_2 N)^2)$

12.8. RADIX SORT

In the radix sort number are sorted on the basis of radix of a number system. For decimal number system radix is 10, this is because we have digits 0 to 9 in the decimal number system. Now 10 temporary buckets are used for holding the numbers; numbered 0 to 9. Which number goes to which bucket will be determined by examining the unit most digits in the first pass. For example in the first pass number 234 will go to bucket number 4 and number 032 will go to bucket number 2. In the second pass numbers from all the buckets are collected back (in order) into the original array and this process repeats. This time numbers are placed onto the bucket by checking the 10th digit in the number. For example, number 234 will go to bucket number 3 and number 032 will go to bucket number 3. This continues for the subsequent passes.

The number of passes depends upon the number of digits in the maximum number in the given array. To better understand this we sort the following number using radix sort:

1 234 456 654 697 874 243 385 902 23

Elements in array	B0	B1	B2	B3	B4	B5	B6	B7	B8	B9
001		001	902	243	234	385	456	697		
234				023	654					
456					874					
654										
697										
874										
243										
385										
902										
023										

Pass 1, placing number in bucket on the basis of unit digit

Elements in array	B0	B1	B2	B3	B4	B5	B6	B7	B8	B9
001	001		023	234	243	654		874	385	697
902	902					456				
243										
023										
234										
654										
874										
385										
456										
697										

Pass 2, placing number in bucket on the basis of 10th digit

Elements in array	B0	B1	B2	B3	B4	B5	B6	B7	B8	B9
001	001		234	385	456		654		874	902
902	023		243				697			
023										
234										
243										
654										
456										
874										
385										
697										

Pass 3, placing number in bucket on the basis of 100th digit

Now at the end of pass 3, we take out elements from all the buckets in order and place back onto the array. The sorting has been done, as can be seen from the contents of the array

001 023 234 243 385 456 654 697 874 902

At this moment, you have understood the working of the radix sort. Now understand it we have implemented it in the function radix sort, we examine the function for the same set of numbers.

```
void radixsort (int arr[ ])
{
int i, j, k, max, r, pass;
int dC = 0, div = 1;
int bucket[10] [10], bC[10];
max = arr[0];
for(i = 1;i<MAX;i++)
        if (arr[i] >max)
                max = arr[i];
while (max>0)
{
dC++;
max = max/10;
}
for(pass = 1;pass< = dC;pass++)
{
for(i = 0;i<10;i++)
        bC[i] = 0;
for(i = 0;i<MAX;i++)
{
r = (arr[i] / div) % 10;
bucket[r] [bC[r]] = arr[i];
bC [r] ++;
}
i = 0;
for (k = 0;k<10;k++)
        for(j = 0;j<bC[k];j++)
                arr[i++] = bucket[k] [j];
div = div*10;
}
}
```

Initially using first for loop we have find out the maximum element of the array i. e. 902. Next, while loop gives us number of digits in the maximum element i. e. dC stores 3. For denoting 10 buckets we have declared a 2-D array bucket as: int bucket[10][10]; The second dimension 10 shows that we can have at most 10 elements in any of the bucket in every pass, first dimension is due to radix 10. For storing the count that how many elements are in any of the bucket in every pass we have taken a 1-D array bC. For example say in pass 1 we have 3 elements in the bucket 2 so bC[2] = 3. The array is initialized to 0 in every pass of the function radix sort in the for loop (discussion follows next)

The crux of the function is the for loop that runs from pass = 1 to pass 3.We show just one iteration of this for loop. Initially the bC is set to 0 for all elements. The next inner for loop finds out the least significant digit (right most) and places it into the corresponding bucket. For every r, bC[r] is incremented by 1 as one more element has been added to the bucket r. The value of div is 1 in the outer for loop and is multiplied by 10 in every pass.

```
for(i =0;i<MAX;i++)
{
r = (arr [i]/div)%10;
bucket[r] [bC[r]] = arr[i];
bc [r] ++;
}
```

When the inner for loop terminates, all array elements have been placed in their respective buckets and count of every bucket is stored in the array bC.

The next for loop takes out all elements from the buckets and place them back into the original array. For taking into account the numbers of elements in each bucket we have used array bC in the inner for loop. The whole coding is listed here again:

```
i = 0;
for (k = 0;k<10;k++)
        for(j = 0;j<bC[k];j++)
                arr[i++] = bucket[k] [j];
```

Before moving onto the next pass div is multiplied by 10. This is necessary as next time we have to take out the 10th digit from all array elements.

12.8.1. Algorithm For Radix Sort

Key idea : Sort least significant digits first

To sort d digits :

Algorithm 12.7. RADIX-SORT (A, d)

1. for i ← 1 to d
2. do use a stable sort to srt array A on digit i

Example :

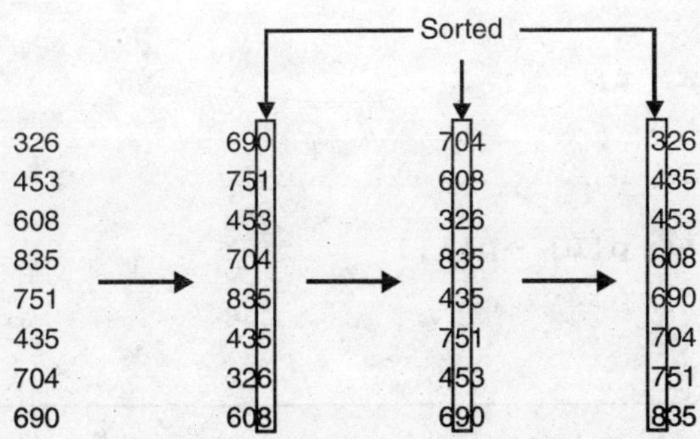

12.8.2. Complexity of Radix Sort

The complexity of radix sort depends upon number of elements (N) and number of digits in the maximum element of the array (say M). It is clear then number of passes will be equal to M. As outer for loop runs for M times and inner second for loop runs for N times, the running time of radix sort can be approximated by $O(N * M * d)$, where d is the radix (in case of decimal number system as we have used it is 10) . As radix is a constant term, the general approximated running time of radix sort is $O(N*M)$. Now in the worst case if $M = N$ then running time will be $O(N^2)$ and in case $M = \log_{10} N$, then running time will be $O(N \log_{10} N)$.

12.9. HEAP SORT

A heap is type of data structure that is similar to a binary tree. Heap is of two types: max heap and min heap. In max heap the parent node is greater than from both of its child nodes and in min heap the parent node is smaller than from any of its child node. In the function we have used max heap and when we refer heap it means we are talking about max heap. For example the following two figures shown below are example of max heap (left) and of min heap (right)

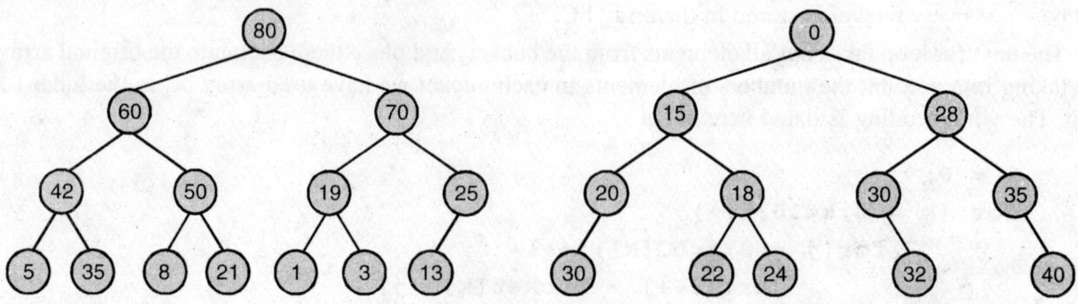

Fig. 12.1. Max and Min Heap.

The basic idea behind heap sort is to first create a heap from the given array. For that we have written the function BuildHeap. After the heap has been built, the maximum element of the array will be at the root of the heap and minimum element will be at the last right most leaf node of the heap(not necessary). We swap these two elements and decrease the size of heap by 1. Processing in this manner the whole array will be sorted as maximum element will be out from the heap in every iteration. This we have written in the function HeapSort. But the most important thing we have to perform is to maintain the heap property as we do the swapping. For that we have written the function Heapify. The function takes the array and an index into array. The function of Heapify is to let value $A[i]$ be at proper place in the heap so that a sub tree rooted at $A[i]$ is a heap. For index i the left child is at $2*i+1$ and right child is at $2*i+2$.

```
void Heapify (int p[ ], int n, int i)
{
int L, R, max, t;
L = 2*i+1;
R = 2*i+2;
if(L< = n-1 && p[L] >p[i])
    max = L;
else
```

```
    max = i;
if (R< = n-1 && p[R]>p[max])
     max = R;
if (max! = i)
{
t = p[i];
p[i] = p[max];
p[max] = t;
Heapify(p, n, max);
}
}
void BuildHeap (int p[ ], int n)
{
int i;
for (i = n/2-1 ; i> 0; i—)
Heapify (p, n, i);
}
void HeapSort (int p[ ], int n)
{
int i, t;
BuildHeap (p, MAX);
for (i = n-1;i> = 1;i—)
{
t = p[0];
p[0] = p[i];
p[i]  = t;
n—;
Heapify (p, n, 0);
}
}
```

To understand the Heapify function, consider the following set of numbers for the array

0	1	2	3	4	5	6	7	8	9
23	56	7	1	2	90	0	3	21	73

For the creation of heap, we have written the function BuildHeap and in this function Heapify is called MAX/2 times. The shown above is shown again but now in the form of tree (Which is not a heap).

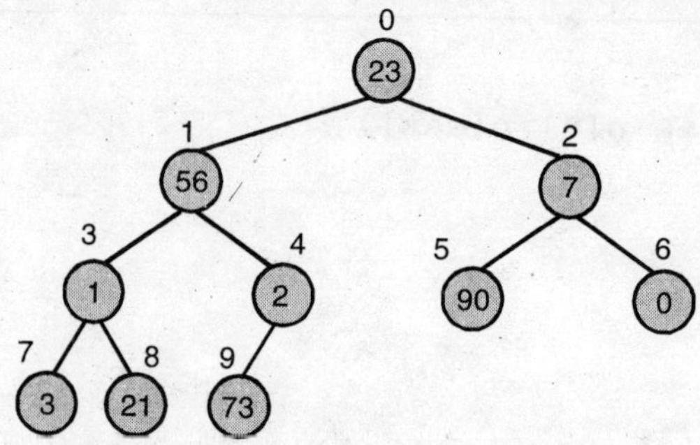

Fig. 12.2. Heap array in tree form.

In the function BuildHeap, the initial value of n is 4, function Heapify is called as:

Heapify (p,10,4). Inside the function value of L is 9; element $p[4]$ and $p[9]$ are compared; as $p[9] > p[4]$, max will be 9. As there is no element at $A[10]$ and $R <= n-1$ is false, second if condition will be false. As max and i are not same, $A[4]$ and $A[9]$ will be swapped and recursive call to Heapify will be made with $i = 9$. For $i = 9$, both the if condition will be false and function will terminate. You can check that subtree rooted at $A[4]$ is a heap. Next time when Heapfy is called from BuildHeap. value of i is 3, L is 7 and R is 8. Element $A[3]$ is compare with $A[7]$, max will be 7 as ($A[7]>A[3]$). Next if condition is true as $A[R]>A[\text{max}]$ ($A[8]>A[7]$) so max will be 8. The last if condition is true (max is 8 and i is 3) so $A[3]$ and $A[8]$ are swapped. Now the subtree rooted at $A[4]$ is heap. Continuing in this manner the heap built from array will be :

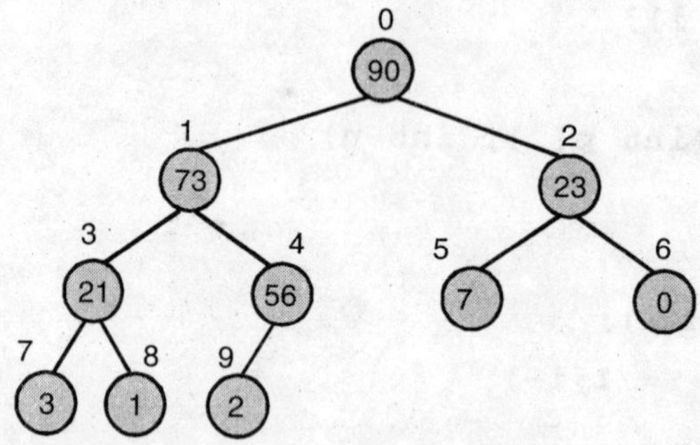

Fig. 12.3. Heap buit from array after applying Build Heap.

After the heap has been built, the following code is responsible for the sorting operation:

```
for(i = n-1;i> = 1;i–)
{
t = p[0];
p[0] = p[i];
p[i] = t;
n--;
Heapify (p, n, 0);
}
```

The initial value of i is 9, element p[0] and p[9] are swapped. Value of n changes from 10 to 9, Because of last element has now become the root of the heap, the heap no longer remains a heap. To make it heap again we call the Heapify function with index 0. After the heap has re-built again (when Heapify returns) root contains the maximum element from the remaining 9 elements. This is swapped with the last element and processing continues in this manner. The steps are shown below. Note the resultant heap is shown in each step, operation of Heapify has not shown again:

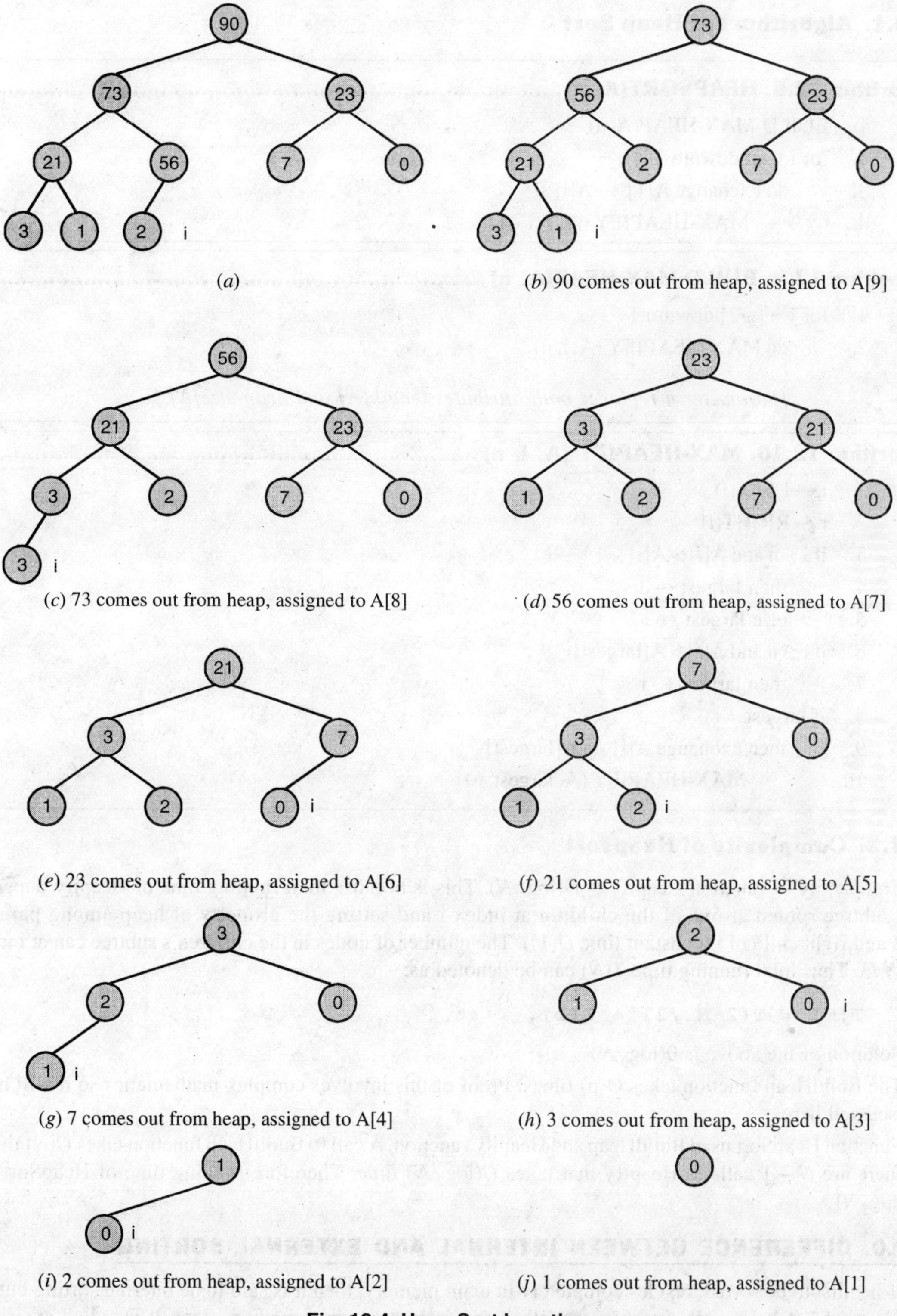

(*a*)

(*b*) 90 comes out from heap, assigned to A[9]

(*c*) 73 comes out from heap, assigned to A[8]

(*d*) 56 comes out from heap, assigned to A[7]

(*e*) 23 comes out from heap, assigned to A[6]

(*f*) 21 comes out from heap, assigned to A[5]

(*g*) 7 comes out from heap, assigned to A[4]

(*h*) 3 comes out from heap, assigned to A[3]

(*i*) 2 comes out from heap, assigned to A[2]

(*j*) 1 comes out from heap, assigned to A[1]

Fig. 12.4. Heap Sort in action.

12.9.1. Algorithm For Heap Sort

Algorithm 12.8. HEAPSORT(A, n) ....

```
BUILD-MAX-HEAP(A, n)
for i ← n downto 2
    do exchange A[1] ↔ A[i]
        MAX-HEAPIFY (A, 1, i – 1)
```

Algorithm 12.9. BUILD-MAX-HEAP(A, n) ....

```
for i ¬ [n/2] downto 1
    do MAX-HEAPIFY (A, i, n)
```

[Parameter n replaces both attributes length[A] and heap-size[A].]

Algorithm 12.10. MAX-HEAPIFY (A, i, n) ....

```
i ← LEFT(i)
r ← RIGHT(i)
if l ≤ n and A[l] > A[i]
    then largest ← l
    else largest ← i
if r ≤ n and A[r] > A[largest]
    then largest ← r
if largest ≠ i
    then exchange A[i] ↔ A[largest]
        MAX-HEAPIFY (A, largest, n)
```

12.9.2. Complexity of Heapsort

Complexity of function Heapify is $O(\log_2 N)$. This is because total running time of Heapify depend upon subtree rooted at one of the children at index i and setting the property of heap among parent i, left and right child of i (constant time $O(1)$). The number of nodes in the children's subtree can at most be $2N/3$. Thus total running time $T(N)$ can be denoted as:

$$T(N) = T(2N/3) + O(1).$$

Solution of the above is $0(\log_2 N)$

The BuildHeap function takes O(n) times, Proof of this involves complex mathematics so it will not be discussed here.

Function HeapSort used BuildHeap and Heapify function. A call to BuildHeap function takes $O(N)$ time and there are $N - 1$ calls to Heapify that takes $O(\log_2 N)$ time. Therefore running time of HeapSort is $O(N \log_2 N)$

12.10. DIFFERENCE BETWEEN INTERNAL AND EXTERNAL SORTING

When list to be sorted, resides completely in main memory, then it is said to be internal sorting but if some part of list is in main memory and other part is in secondary memory, then it is called external sorting. External sorting is applied to files due to their large size.

File Sorting are of two types :

(*a*) **Sorting with disk :** File resides on disk.

(*b*) **Sorting with tape :** File resides on tape

12.11. EXTERNAL SORT/MERGE ALGORITHMS

This topic discusses external sort/merge algorithms used for sorting large data files that could not fit into the main memory of the computer. The algorithms presented are two-way sort/merge, the balanced two-way sort/merge, the balanced k-way sort/merge, and the polyphase sort/merge. The advantages of sorting records include the following: sorting transaction to match the key order of the master file can reduce the time required to locate a matching master record; sorted transactions can improve the maintenance of master files with other types of file organization; and many reports generated from files require that the file be sorted, as it is more presentable and readable when sorted.

12.11.1. Two-Way Sort/Merge

A sort/merge algorithm involves two steps :

1. The records in the file to be sorted are divided into several groups, called run, and each run fits into main memory. An internal sort is applied to each run, and the resulting sorted runs are distributed to two external files.
2. One run at a time from each of the external files created in step 1 merge into larger runs of sorted records. The result is stored in a third external file. The data are distributed from the third file back into the first two files, and the merge continues until all records are in one large run.

Suppose we have a file of records with the following keys to be sorted in a ascending order:

50 110 95 10 100 36 153 40 120 60 70 130 22 140 80

Let us assume that the size of the run is 3 records/run. By using step 1, we can divide the keys into groups of 3 as follows: (50, 110, 95), (10, 100, 36), (153, 40, 120), (60, 70, 130), (22,140, 80). The number of groups indicates that we have 5 runs. These 5 runs are then loaded into 2 files: File 1 contains run 1 (50, 110, 95), run 3 (153, 40, 120), and run 5 (22, 140, 80); File 2 contains run 2 (10, 100, 36), and run 4 (60, 70, 130). All in sorted order. The distribution of runs in each file is alternated in order to have the same amount of runs in each file, at worst case one file will have one more run than the other, just like our example above.

In step 2 we will be merging the two files together one run at a time into a third file, say File3. Run 1 in File 1 is merged with run 2 in File 2 to produce the first run in File 3 with 6 sorted records. Run 3 in File 1 is merged with run 4 in File 2 to produce second run in File 3 also with 6 sorted records. Now File 2 is empty and File 1 contains run 5, Run 5 is merged into File 3 to make the third run of 3 sorted records.

Now we are going repeat step 1 by distributing each run in file 3 into file 1 and file 2, and then merge the two files into file 3 again. These steps are repeated until we have only one run in file 3. The number of passes through the two-way sort/merge is the ceiling of *lg* NR, where NR means number of runs and *lg means log base 2.*

12.11.2. Balanced Two-Way Sort/Merge

Balanced two-way sort/merge algorithm is nothing more than an improvement to the two-way sort/ merge algorithm. By increasing the number files used, the balanced two-way sort/merge algorithm merges two input files and stores the merged runs on two output files alternatively. As a result, the algorithm eliminates the need to redistribute the runs to the two input files before it can start merging. This algorithm merges runs from file 1 and file 2 and stores the merged runs into file 3 and file 4, it then merges the runs the runs in file 3 and file 4 and stores the merged runs into file 1 and file 2, again the algorithm continues until all the records are merged into one run.

The number of passes through the two-way sort/merge algorithm, that is ceiling of *lg* NR.

12.11.3. Balanced k-way sort/Merge

This also is an improvement on the previous algorithm. By increasing the number of input and output files, the number of runs in each file is decreased, with the result that fewer merges have to be performed. However, the merging phase is more complicated with *k* input files because a *k-way* merge needs to distribute merged runs into *k* output files.

Suppose we have the same file that was described above with 15 records and we were to sort the file using the balanced *k-way* sort/merge algorithm. The distribution of the records into runs is similar to the one used in the *two-way* sort/merge algorithm. However the merge phase will be merging into 3 files, that is, run 1 from file 1 and run 2 from file 2 and run 3 from file 3 are merged into file 4 to produce 1 run of sorted records. Run 4 from file 1 and run 5 from file 2 are then merged into file 5 to produce 1 run of sorted records. Now, the two output files (file 4 & file 5) will now become input files (just like in balanced two-way sort/merge) and the two runs will be merged into file 1 to produce 1 run of sorted records and the sort/merge is done. The example just described can be called the balanced 3-way sort/merge as we have 3 input files and 3 output files, even through file 6 was never used, as we did not have enough runs to use file 6. The advantages of this algorithm are that the sort/merge phase is faster than the algorithm that were discussed above, but the disadvantage is the excess of use of files (disk space). The number of passes required in this algorithm is ceiling lg_k NR of *k* is the number of files to be merged and it is also the number of output files.

12.11.4. Polyphase Sort/Merge

The polyphase sort/merge algorithm is again an improvement to the other algorithm described. This algorithms works just like the Balanced *k-way* sort/merge algorithm but it does not use as many files. The algorithm simply make use of the empty input file by turning that file into output file. Suppose we have the same file used earlier with 15 records and we were to sort the records. For now the distribution can be done just like we have been doing before into two files input files, file 1 and file 2. Run 1 from file 1 is merged with run 2 of file 2 to produce first run of file 3. Run 3 from file 1 is merged with run 4 of file 2 to produce second run of file 3. At this moment file 2 is empty as the last run in the file has been used, but file 1 still have 1 more run with 3 sorted records and file 3 has two runs with 6 sorted records each. The algorithm will then begin to merge file 1 and file 3 into file 2, making file 2 the output file. As soon as file 1 becomes empty, file 1 then become the next output file. This process is repeated until we have 1 run with sorted records in one of the files. Also note that this algorithm is only efficient if we have unequal distributions of runs in each input file.

SOLVED PROBLEMS

1. The Bubble Sort implementation has the following inner for loop :

```
for (int j=n-1; j>i; j–)
```

Consider the effect of replacing this with the following statement :

```
for (int j=n-1; j>0; j–)
```

Would the new implementation work correctly? Would the change affect the asymptotic complexity of the algorithm? How would the change affect the running time of the algorithm?

Ans. The revised algorithm will work correctly, and its asymptotic complexity will remain $\Theta(n^2)$. However, it will do about twice as many comparisons, since it will compare adjacent elements within the portion of the list already known to be sorted. These additional comparisons are unproductive.

2. When implementing Insertion Sort, a binary search could be used to locate the position within the first *i* – 1 elements of the array into which element *i* should be inserted. How would this affect the number of comparisons required? How would using such a binary search affect the asymptotic running time for Insertion Sort?

Ans. While binary search will find the proper place to locate the next element, it will still be necessary to move the intervening elements down one position in the array. This requires the same number of operations as a sequential search. However, it does reduce the number of element/element comparisons, and may be somewhat faster by a constant factor since shifting several elements may be more efficient than an equal number of swap operations.

3. Recall that a sorting algorithm is said to be stable if the original ordering for duplicate keys is preserved. Of the sorting algorithms Insertion Sort, Bubble Sort, Selection Sort, Shellsort, Quicksort, Mergesort, Heapsort, Binsort, and Radix Sort, which of these are stable, and which are not? For each one, describe either why it is or is not stable. If a minor change to the implementation would make it stable, describe the change.

Ans.

- Insertion Sort is stable. A swap is done only if the lower element's value is LESS.
- Bubble Sort is stable. A swap is done only if the lower element's value is LESS.
- Selection Sort is NOT stable. The new low value is set only if it is actually less than the previous one, but the direction of the search is from the bottom of the array. The algorithm will be stable if "less than" in the check becomes "less than or equal to" for selecting the low key position.
- Shell Sort is NOT stable. The sublist sorts are done independently, and it is quite possible to swap an element in one sublist ahead of its equal value in another sublist. Once they are in the same sublist, they will retain this (incorrect) relationship.
- Quick-sort is NOT stable. After selecting the pivot, it is swapped with the last element. This action can easily put equal records out of place.
- Conceptually (in particular, the linked list version) Mergesort is stable.

  The array implementations are NOT stable, since, given that the sublists are stable, the merge operation will pick the element from the lower list before the upper list if they are equal. This is easily modified to replace "less than" with "less than or equal to."
- Heapsort is NOT stable. Elements in separate sides of the heap are processed independently, and could easily become out of relative order.
- Binsort is stable. Equal values that come later are appended to the list.
- Radix Sort is stable. While the processing is from bottom to top, the bins are also filled from bottom to top, preserving relative order.

4. Consider a recursive Mergesort implementation that calls Insertion Sort on sublists smaller than some threshold. If there are n calls to Mergesort, how many calls will there be to Insertion Sort? Why?

Ans. Consider Mergesort in terms of a full binary tree. Each call to Mergesort either results in two new calls to Mergesort, or else a single call to Insertion Sort. Thus, the calls to Insertion Sort are equivalent to the leaf nodes of a full binary tree. We know from the Full Binary Tree Theorem that the number of leaf nodes in a full binary tree of n nodes is $n/2$. Thus, if there are n calls to Mergesort, there will be $n/2$ calls to Insertion Sort.

5. **Suppose that a record is 32 bytes, a block is 1024 bytes (thus, there are 32 records per block), and that working memory is 1MB (there is also additional space available for I/O buffers, program variables, etc.). What is the expected size for the largest file that can be merged using replacement selection followed by a single pass of multiway merge? Explain how you got your answer.**

Ans. Since working memory is 1MB and the block size is 1KB, the number of blocks in working memory is 1024. The expected runlength is 2MB, since replacement selection will, on average, produce runs that are twice the mem- ory size. 1024 runs can be merged in a single multiway merge operation. Thus, the largest expected file size for a single pass of multiway merge is 2 Gigabytes.

6. **Assume that working memory size is 256KB broken into blocks of 8192 bytes (there is also additional space available for I/O buffers, program variables, etc.). What is the expected size for the largest file that can be merged using replacement selection followed by two passes of multiway merge? Explain how you got your answer.**

Ans. Since working memory is 256KB and the blocksize is 8KB, the working memory holds 32 blocks. The expected runlength is 512KB, so a single pass of multiway merge forms runs of length 16MB. The second pass then forms a run as large as 512MB.

7. **Assume that a company has a computer configuration satisfactory for processing their monthly payroll. Further assume that the bottleneck in payroll processing is a sorting operation on all of the employee records, and that an external sorting algorithm is used. The company's payroll program is so good that it plans to hire out its services to do payroll processing for other companies. The president has an offer from a second company with 100 times as many employees. She realizes that her computer is not up to the job of sorting 100 times as many records in an acceptable amount of time. Describe what impact each of the following modifications to the computing system is likely to have in terms of reducing the time required to process the larger payroll database.**
   (*a*) A factor of two speedup to the CPU.
   (*b*) A factor of two speedup to disk I/O time.
   (*c*) A factor of two speedup to main memory access time.
   (*d*) A factor of two increase to main memory size.

Ans. (*a*) Speeding up the CPU will have little effect on an external sorting operation.

(*b*) Cutting the disk I/O time will substantially improve the external sorting time. A reasonable estimate is that cutting disk I/O time in half will cut the sorting time by around 1/3.

(*c*) Main memory access time will not help a great deal, since disk I/O is the probable bottleneck. However, for the sorting operation, main memory access time is in fact more of a bottleneck than CPU speed, so it should help more to speed the memory than to speed the CPU.

(*d*) Increasing the memory size by a factor of two will increase the file size that can be processed by a single pass of multi-way merge by a factor of four, in two passes by a factor of eight, and so no. If this leads to a reduction in the number of passes need to process the file, then a substantial time savings will be realized. This could easily cut the processing time by 1/3 or 1/4 since 2 or 3 passes of multiway merge under the initial conditions are reasonable to expect.

8. **Assume that a computer system has disk blocks of 1024 bytes, and that you are storing records that have 4-byte keys and 4-byte data fields. The records are sorted and packed sequentially into the disk file.**

(*a*) **Assume that a linear index uses 4 bytes to store the key and 4 bytes to store the block ID for the associated records. What is the greatest number of records that can be stored in the file if a linear index of size 256KB is used?**

(*b*) **What is the greatest number of records that can be stored in the file if the linear index is also stored on disk (and thus its size is limited only by the second-level index) when using a second-level index of 1024 bytes (i.e., 256 key values) as illustrated by Figure 10.2? Each element of the second-level index references the smallest key value for a disk block of the linear index.**

Ans. (*a*) A record in the linear index refers to a block of sorted data records. Assuming that the linear index stores a key and a 4 byte block number, the index can hold information for 32K blocks, for a total file size of 32MB, or 4M records.

(*b*) This second level index allows the first level index to be 128 blocks, or 16K records long. Thus, the record file can contain 16K blocks, or 16MB, which is 2M records. While this is smaller than the situation in (*a*), there is only a very small amount of main memory in use.

9. What is advantage and disadvantage of a selection sort?

Ans. Advantages : Easy to implement In-place sort (requires no additional storage space).

Disadvantages : Doesn't scale well: $O(n^2)$

10. What is advantage and disadvantage of Quick sort and compare it with Heap sort and Merge sort ?

Ans. Advantages :

1. One of the fastest algorithms on average
2. Does not need additional memory (the sorting takes place in the array - this is called in-place processing)

Disadvantages :

1. The worst-case complexity is $O(N^2)$

Applications :

1. Quick Sort generally runs fast
2. No additional memory
3. The above advantages compensate for the rare occasions when it runs with $O(N^2)$

Comparison with Heap sort :

1. $O(N \log N)$ complexity
2. Quick sort runs faster, (does not support a heap tree)
3. The speed of quick sort is not guaranteed

Comparison with Merge sort :

1. Merge sort guarantees $O(N \log N)$ time
2. Merge sort requires additional memory with size N
3. Usually Merge sort is not used for main memory sorting, only for external memory sorting

11. Is bubble sort more efficient technique than insertion sort?

Ans. Bubble sort and insertion sort both have the same time complexity (and space complexity) in the best, worst, and average cases. However, these are purely theoretical comparisons. In practical real-world scenarios, insertion sort (or any other sort, for that matter) will almost always be the better choice over a bubble sort.

12. List the fastest to slowest sorting algorithms?

Ans. The simplest sorting algorithm that most beginners learn is the **bubble sort**. It uses a simple logic - swapping adjacent entries repeatly in running order until the whole list is sorted. However, this method is one of the slowest sorting algorithms.

Insertion sort is faster - it works like how we would pick the next entry and insert it in the correct place in a list of sorted entries.

Selection sort ranks the same as insertion sort - it works by choosing the next smallest number in the remaining non-sorted list and adds it to the sorted list. However, its way of choosing the next smallest number is naive - looking through all the entries in the remaining non-sorted list.

Heap sort is an optimised selection sort whereby the smallest number is maintained in a heap - a data structure that allows more efficient management of numbers. This puts heap sort a mark above the others mentioned above.

Merge sort ranks about the same as heap sort. It uses the idea of divide and conquer, splitting up the whole job into smaller manageable jobs and then combining the results of the smaller jobs.

However, under most conditions, **Quick sort** is still the fastest sorting algorithm. It uses a divide and conquer approach as well using its famous partition algorithm. However, the downside is that its worst case running time is still of the magnitude of bubble sort's.

13. Write the classification of sorting algorithms ?

Ans. Sorting algorithms are often classified using different metrics :

1. **Computational complexity :** Classification is based on worst, average and best behavior of sorting a list of size (*n*).
   - For typical sorting algorithms acceptable/good behavior is $O(n \log n)$ and unacceptable/bad behavior is Ω (n^2).
   - Ideal behavior for a sort is $O(n)$.
2. **Memory usage** (and use of other computer resources) :
   - Some sorting algorithms are "in place", such that only $O(1)$ or $O(\log n)$ memory is needed beyond the items being sorted.
   - Others need to create auxiliary data structures for data to be temporarily stored. We've seen in class that merge sort needs more memory resources as it is not an "in place" algorithm, while quick sort and heap sort are "in place". Radix and bucket sorts are not "in place".
3. **Recursion :** Some algorithms are either recursive or non-recursive. (e.g., merge sort is recursive).
4. **Stability** : Stable sorting algorithms maintain the relative order of elements/records with equal keys/values. Radix and bucket sorts are stable.
5. **General method :** Classification is based on how sort functions internally.
   - Methods used internally include insertion, exchange, selection, merging, distribution etc. Bubble sort and quick sort are exchange sorts. Heap sort is a selection sort.
6. **Comparison sorts** : A comparison sort examines elements with a comparison operator, which usually is the less than or equal to operator (=). Comparison sorts include :
   - Bubble sort
   - Insertion sort
   - Selection sort
   - Shell sort
   - Heapsort

- Mergesort
- Quicksort.

7. **Non-Comparison sorts :** These use other techniques to sort data, rather than using comparison operations. These include :
   - Radix sort (examines individual bits of keys)
   - Bucket sort (examines bits of keys)
   - Counting sort (indexes using key values)

14. What is advantage and disadvantage of radix and bucket sort ?

Ans. Advantages :

- Radix and bucket sorts are stable, preserving existing order of equal keys.
- They work in linear time, unlike most other sorts. In other words, they do not bog down when large numbers of items need to be sorted. Most sorts run in $O(n \log n)$ or $O(n\^2)$ time.
- The time to sort per item is constant, as no comparisons among items are made. With other sorts, the time to sort per time increases with the number of items.
- Radix sort is particularly efficient when you have large numbers of records to sort with short keys.

Drawbacks :

- Radix and bucket sorts do not work well when keys are very long, as the total sorting time is proportional to key length and to the number of items to sort.
- They are not "in-place", using more working memory than a traditional sort.

15. List Different types of sorting algorithms?

Ans.

Problem Definition	Sorting Algorithms
The data to be sorted is small enough to fit into a processor's main memory and can be randomly accessed *i.e.* no extra space is required to sort the records (Internal Sorting).	Insertion Sort, Selection Sort, Bubble Sort
Source data and final result are both sorted in hard disks (too large to fit into main memory), data is brought into memory for processing a portion at a time and can be accessed sequentially only (External Sorting).	Merge Sort (business application, database application)
Constant alphabet (ordered alphabet of constant size, multi set of characters can be stored in linear time), sort records that are of multiple fields.	Radix Sort
The input elements are small.	Insertion Sort
The input elements are too large.	Merge Sort, Shell Sort, Quick Sort, Heap Sort
The input elements are repeated in the list and sorted the list in order to maintain the relative order of record with equal keys.	Bubble Sort, Merge Sort, Insertion Sort, counting sort
The input elements are repeated in the list and sorted the list so that their relative orders are not maintain with equal keys.	Quick Sort, Heap Sort, Selection Sort, Shell Sort

16. List all comparison based sorting algorithms with their complexities, advantages and disadvantages?

Ans.

Name	Average Case	Worst Case	Method	Advantage/Disadvantage
Bubble Sort	$O(n^2)$	$O(n^2)$	Exchange	1. Straightforward, simple and slow. 2. Stable. 3. Inefficient on large tables.
Insertion Sort	$O(n^2)$	$O(n^2)$	Insertion	1. Efficient for small list and mostly sorted list. 2. Sort big array slowly. 3. Save memory
Selection Sort	$O(n^2)$	$O(n^2)$	Selection	1. Improves the performance of bubble sort and also slow. 2. Unstable but can be implemented as a stable sort. 3. Quite slow for large amount of data.
Heap Sort	$O(n \log n)$	$O(n \log n)$	Selection	1. More efficient version of selection sort. 2. No need extra buffer. 3. Its does not require recursion. 4. Slower than Quick and Merge sorts.
Merge Sort	$O(n \log n)$	$O(n \log n)$	Merge	1. Well for very large list, stable sort. 2. A fast recursive sorting. 3. Both useful for internal and external sorting. 4. It requires an auxiliary array that is as large as the original array to be sorted.
In place-merge Sort	$O(n \log n)$	$O(n \log n)$	Merge	1. Unstable sort. 2. Slower than heap sort. 3. Needs only a constant amount of extra space in addition to that needed to store keys.
Shell Sort	$O(n \log n)$	$O(n \log^2 n)$	Insertion	1. Efficient for large list. 2. It requires relative small amount of memory, extension of insertion sort. 3. Fastest algorithm for small list of elements. 4. More constraints, not stable.
Quick Sort	$O(n \log n)$	$O(n^2)$	Partition	1. Fastest sorting algorithm in practice but sometime Unbalanced partition can lead to very slow sort. 2. Available in many slandered libraries. 3. $O(\log n)$ space usage. 4. Unstable sort and complex for choosing a good pivot element.

17. List all non-comparison based sorting algorithms with their complexities, advantages and disadvantages?

Ans.

Name	Average Case	Worst Case	$n << 2^K$	Advantage/disadvantage
Bucket Sort	$O(n.k)$	$O(n^2.k)$	No	1. Stable, fast. 2. Valid only in range o to some maximum value M. 3. Used in special cases when the key can be used to calculate the address of buckets.
Counting Sort	$O(n+2^k)$	$O(n+2^k)$	Yes	1. Stable, used for repeated value. 2. Often used as a subroutine in radix sort. 3. Valid in the rang(o k] Where k is some integer.
Radix Sort	$O(n._s^k)$	$kO(n._s^k)$	No	1. Stable, straight forward code. 2. Used by the card-sorting machines. 3. Used to sort records that are keyed by multiple fields like date (Year, month and day).

18. What are the methods available in storing sequential files?

Ans. 1. Straight merging,
2. Natural merging,
3. Polyphase sort,
4. Distribution of Initial runs.

19. Explain Heap and Build a Heap H step-by-step for the following data

44, 30, 50, 22, 60, 55, 77, 55

Ans. A Heap H is a binary tree, which can be represented using two ways :

(*i*) **MAX-HEAP :** A HEAP tree in which the root node's value is larger than both it's child's value.

(*ii*) **MIN-HEAP :** A Heap tree formed in which the root node's value is smaller than both it's child's value.

Construction of MAX-HEAP H for the sequence :

(*a*) 

(*b*) insert 30

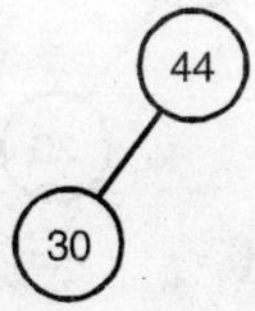

(*c*) insert 50

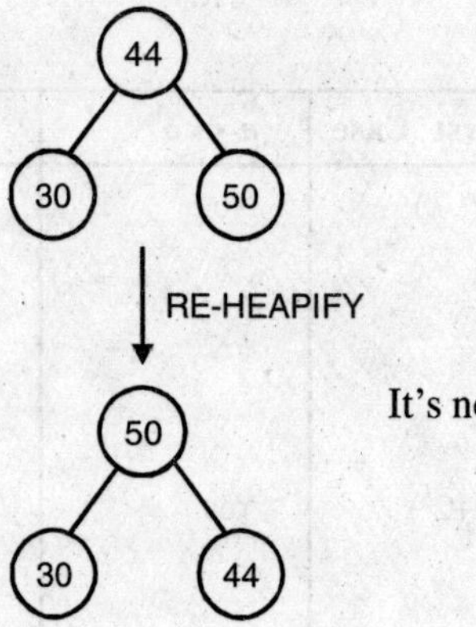

It's not a MAX-HEAP TREE. So re-heapify it.

(*d*) insert 22

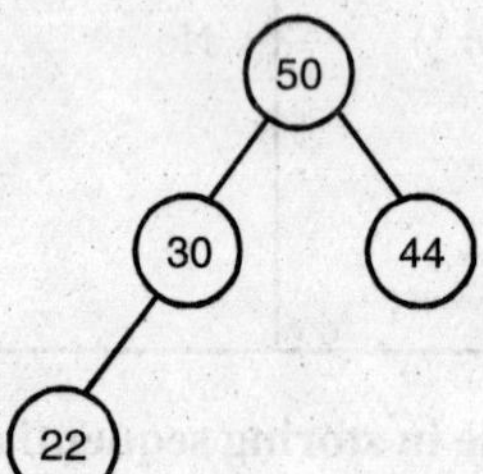

(*e*) insert 60

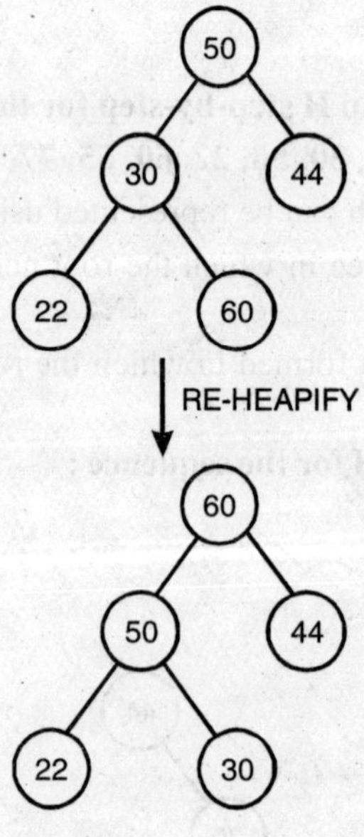

(*f*) insert 55

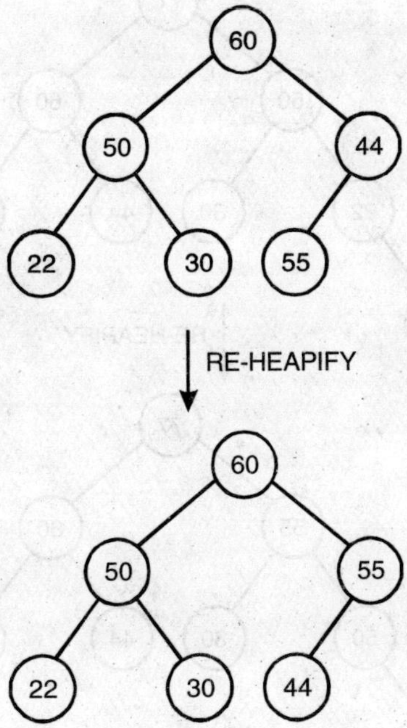

(*g*) insert 77

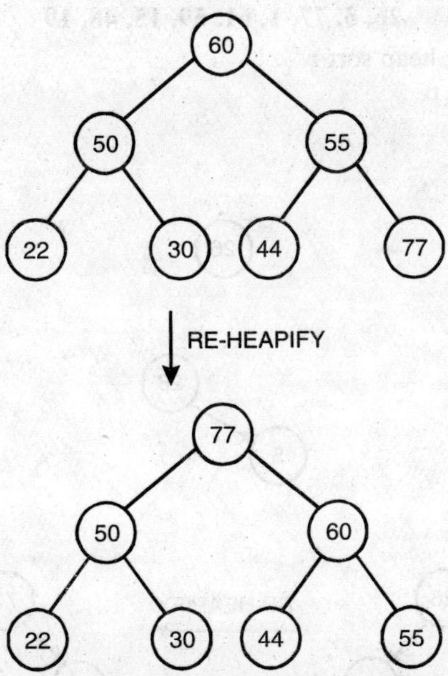

(*h*) insert 55

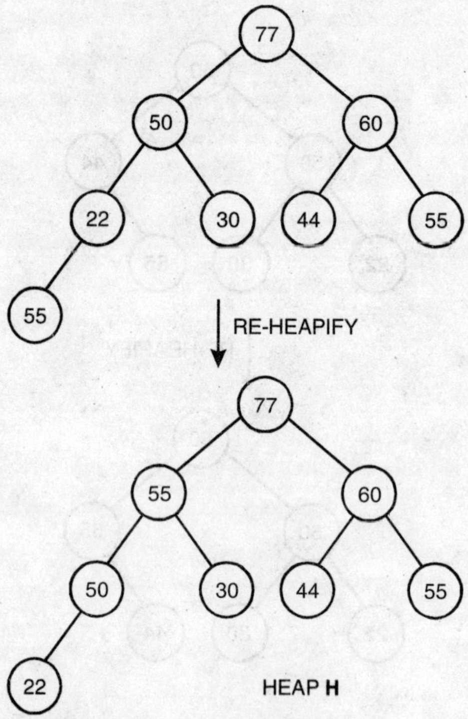

20. Trace the steps of Heap Sort of the following data :

26, 5, 77, 1, 61, 59, 15, 48, 19

Ans. To sort a sequence using heap sort :

(*a*) BUILD A MAX HEAP

(*b*) APPLY SORTING

(*a*) (I) Insert 26

(II) Insert 5

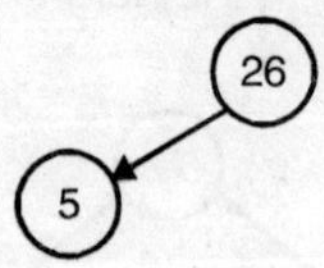

(III) Insert 77

(IV) Insert 1

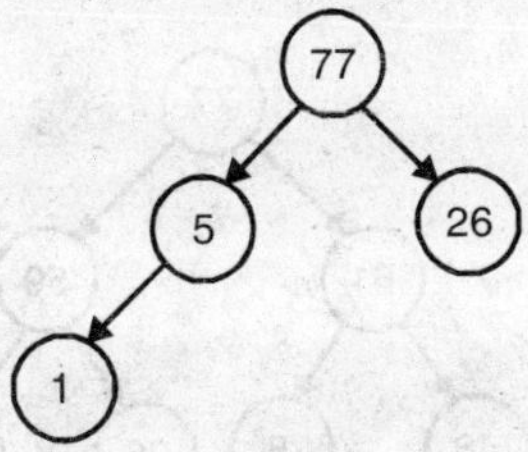

(V) Insert 61

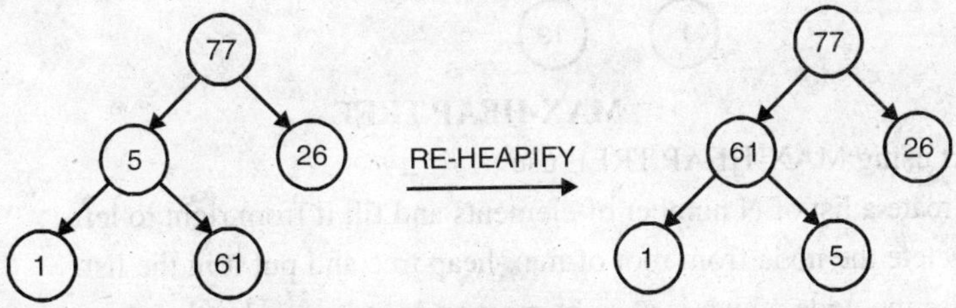

(VI) Insert 59

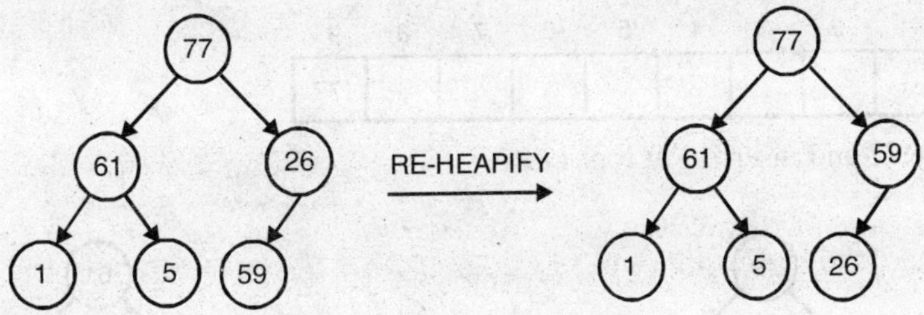

(VII) Insert 15

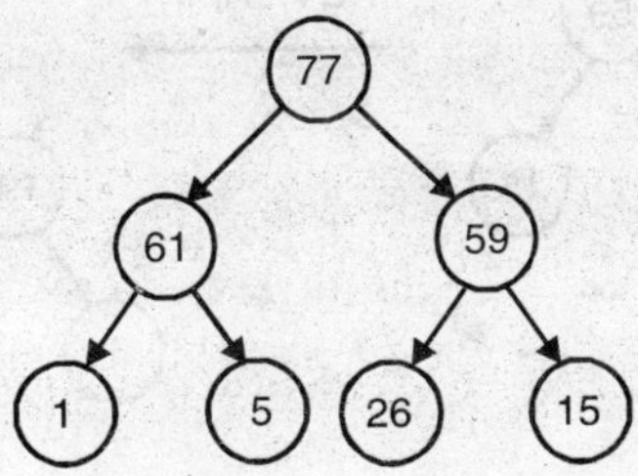

(VIII) Insert 48

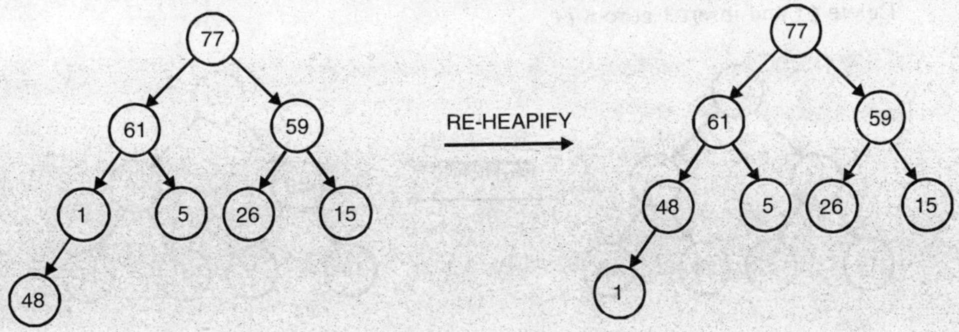

(IX) Insert 19

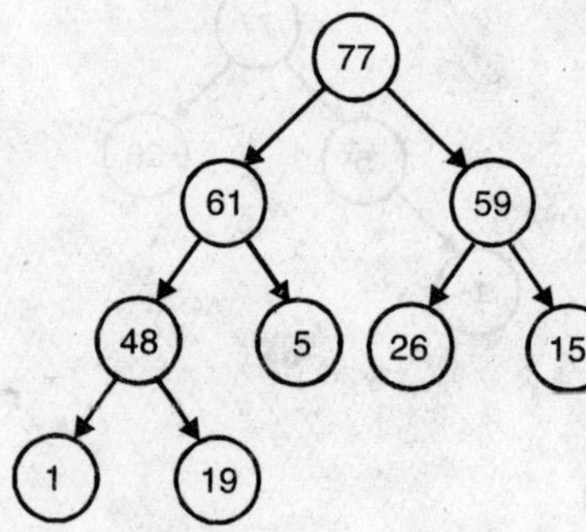

MAX-HEAP TREE

(*b*) To sort using MAX-HEAP TREE use

(*i*) Create a list of N number of elements and fill it from right to left.

(*ii*) Delete the node from root of max-heap tree and put it in the list.

(*iii*) Use the node which is at right-most node at lowest level and insert it at root.

(*iv*) Repeat the above 2 steps until the list is completely filled.

	1	2	3	4	5	6	7	8	9
(I)									77

Delete 77 and insert 19 at root *i.e.*

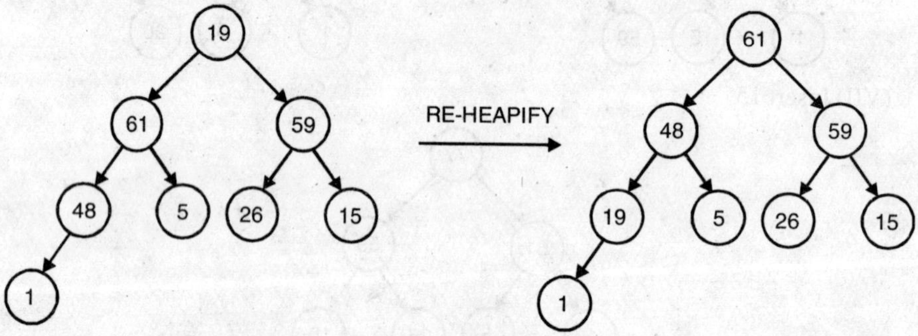

(II)								61	77

Delete 61 and insert 1 at root *i.e.*

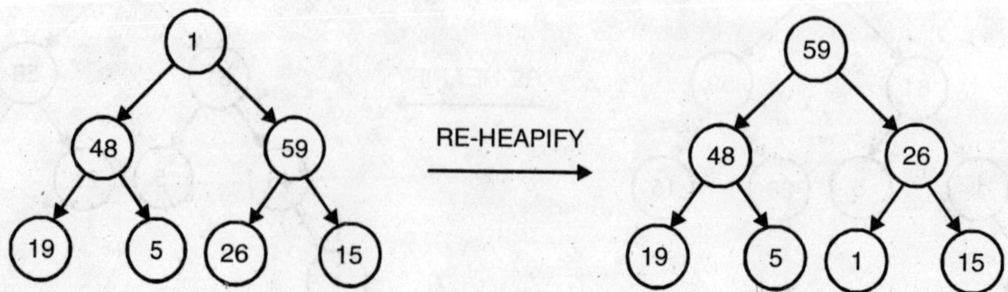

(III)							59	61	77

Delete 59 and insert 15 at root node *i.e.*

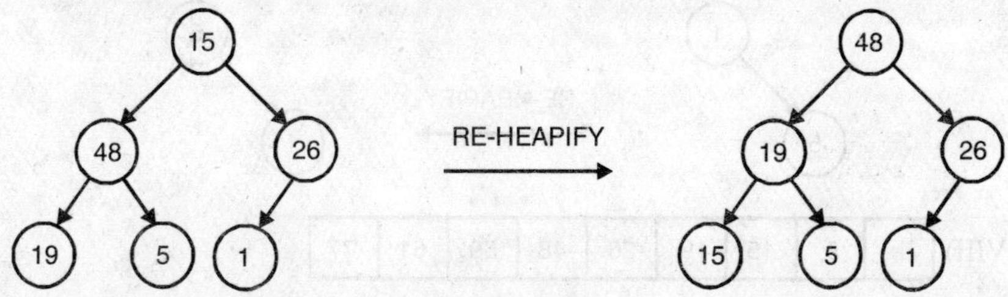

(IV)						48	59	61	77

Delete 48 and insert 1 at root node *i.e.*

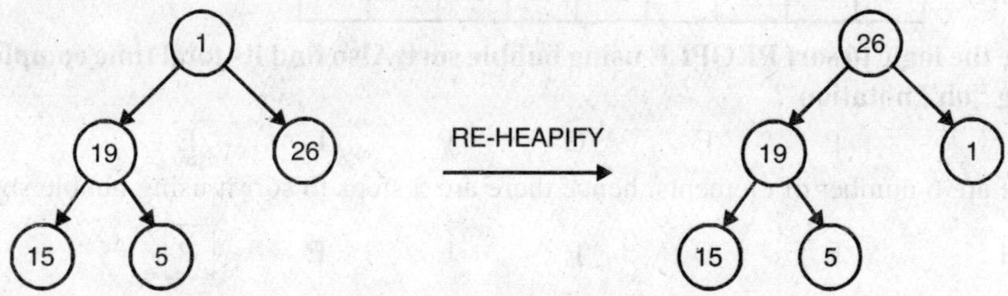

(V)					26	48	59	61	77

Delete 26 and insert 5 at root node *i.e.*

(VI)				19	26	48	59	61	77

Delete 19 and insert 5 at root node *i.e.*

(VII)

		15	19	26	48	59	61	77

Delete 15 and insert 1 at root node *i.e.*

(VIII)

	5	15	19	26	48	59	61	77

Delete 5 and insert 1 at root node *i.e.*

(IX)

1	5	15	19	26	48	59	61	77

21. Trace the logic to sort PEOPLE using bubble sort. Also find its total time complexity in terms of big "oh" notation ?

Ans. P E O P L E

There are 6 number of elements, hence there are 5 steps to sort it using bubble sort

Step I : E O P L E [P]

$(n-1)$ comparison

Step II : E O L E [P] [P]

$(n-2)$ comparisons

Step III : E L E [O] [P] [P]

$(n-3)$ comparisons

Step IV : E E [L] [O] [P] [P]

$(n-4)$ comparisons

Step V : E [E] [L] [O] [P] [P]

$(n-5)$ comparisons

The complexity is the sum of all the iteration of each step and starting from $(n-1)$ iteration upto 1 iteration at last step therefore,

$$T(n) = (n-1) + (n-2) + \ldots + 1$$

$$= \frac{n(n-1)}{2}$$

$$= \frac{n^2}{2} = \frac{n}{2}$$

$$T(n) = O(n^2)$$

22. Trace the quicksort algorithm for the following data :

6, 5, 4, 7, 12, 11, 9, 10, 8, 2

Ans. Take $x[0]$ as pivot and place it at its proper position by swapping the elements which are on the wrong side to get two subarrays and repeat the process on each subarray.

Initial array is

6, 5, 4, 7, 12, 11, 9, 9, 10, 8, 2

Take 6 as pivot and place it at its proper place by swapping 2 with 7 and finally 2 with 6 we get

(2, 5, 4) 6(12, 11, 9, 10, 8, 7)

We get two subarrays. We first sort the first subarray

2 and (5, 4)

2 4 5

Now, we sort the second subarray

12, 11, 9, 10, 8, 7

(11, 9, 10, 8, 7) and 12

The second subarray is a single element and is therefore soreted. We will sort the first subarray

11, 9, 10, 8, 7

(9, 10, 8, 7) and 11

(7, 8) 9 (10) 11

Now both subarrays are sorted. Placing all subarrays together, we get the final sorted output as

2, 4, 5, 7, 8, 9, 10, 11, 12

23. Consider the following array and draw the heap that this array represents

90	80	40	50	60	10	20	30

Ans. The heap for the given array will be as follows :

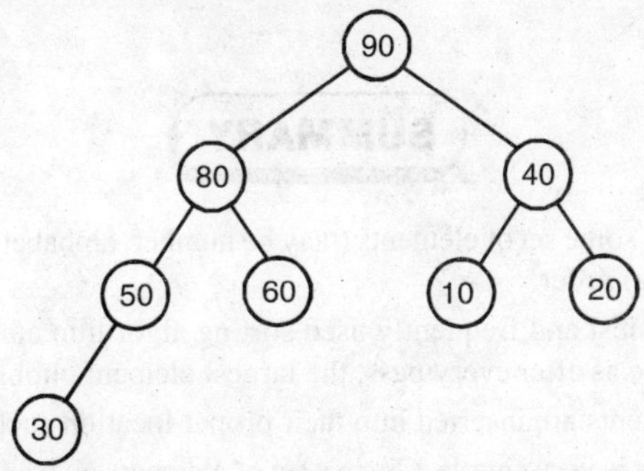

24. Apply merge sort algorithm for the following list of numbers (run the algorithm and clearly. Show the value of variables in each step) :

85, 76, 46, 92, 30, 41, 12

Ans. Divide the given list of numbers into two subarrays or sublist using a middle index i.e.

$$\text{mid} = \left\lfloor \frac{LB + UB}{2} \right\rfloor$$

where LB and UB are lower and upper bound respectively and $\lfloor \ \rfloor$ is the floor function after finding the mid. The given list is divided into two subarray *i.e.* from LB to mid and from mid + 1 to UB as shown below :

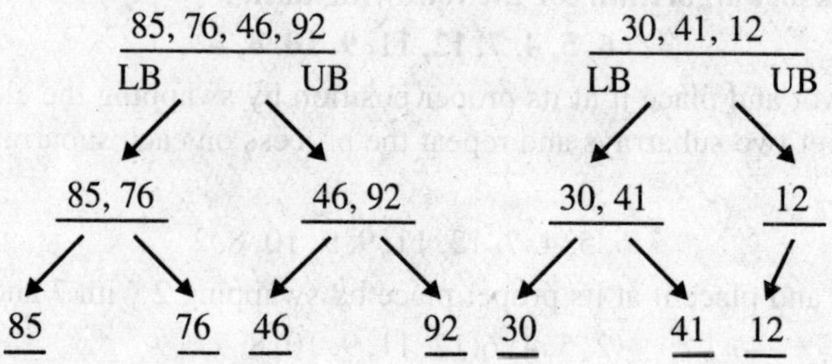

After dividing a given array of *n* number of elements int *n* array each containing only one element.

Now after getting n different arrays, we have to recursively merge two subarrays at a time to get the sorted sequence of original array as shown below :

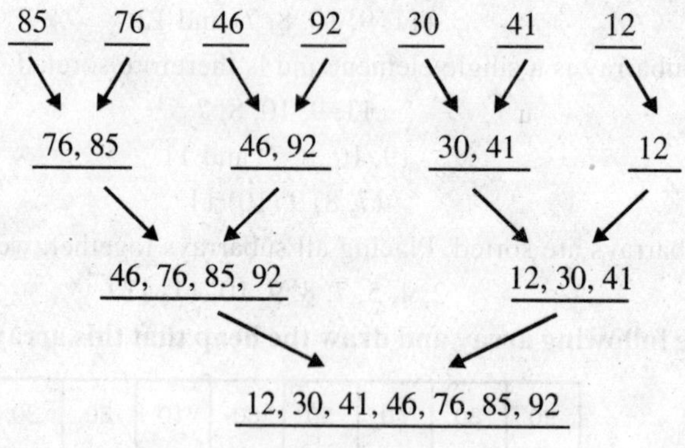

On merging, we are having two given sorted sub-arrays and we have to merge them together to get the sorted array.

SUMMARY

1. Sorting refers to arranging some set of elements (may be number, alphabet or strings) in ascending (increasing) or descending (decreasing) order.
2. The bubble sort is the easiest and frequently used sorting algorithm among all the sorting algorithms. The algorithm has got its name as after every pass, the largest element bubbles up and move to end of the array.
3. In the insertion sort, elements are inserted into their proper location such that array will be sorted.
4. The logic for selection sort is very simple. Given a set of elements N, we find out the location of the minimum element among the set and replace it with the location of first element. Next we do the same task for N-1 elements and find out the location of the minimum element among N-1 elements and replace is with the location of the first element (of set N-1).
5. Merging of two arrays means combining them in such manner so that resultant array is sorted.
6. In quick sort we find the proper position of an element in the array. Assume proper position is pos, now this pos will be such that all the elements that are lower in index then pos, all such elements will be lower than element at pos and all the elements that are higher in index then pos, all such elements will be higher then element at that pos.
7. In the shell sort, we make a shell of the elements at position i and i+m, where m is any power of 2. If element at position i is greater than an element at position i+m we do the swapping of the elements. The value of m

changes in every pass of the method (in fact it decreases in every pass and in last pass it must be 1) and different value of m gives us shells of different size.

8. In the radix sort number are sorted on the basis of radix of a number system. For decimal number system radix is 10, this is because we have digits 0 to 9 in the decimal number system. Now 10 temporary buckets are used for holding the numbers; numbered 0 to 9.
9. A heap is type of data structure that is similar to a binary tree. Heap is of two types: max heap and min heap. In max heap the parent node is greater than from both of its child nodes and in min heap the parent node is smaller than from any of its child node.
10. When list to be sorted, resides completely in main memory, then it is said to be internal sorting but if some part of list is in main memory and other part is in secondary memory, then it is called external sorting.

EXERCISES

1. How insertion sort is different than selection sort?
2. Consider a list of numbers sorted in ascending order. Which of the following sorting method result in best case
   (*i*) Selection (*ii*) Insertión sort (*iii*) Bubble Sort
3. In the question 2, what about worst case
4. Write name of all sorting algorithms whose average case and worst case time complexity for an input of N numbers is same.
5. Write name of all sorting algorithms whose time complexity for average case is $O(N^2)$.
6. Sort the following numbers using radix sort:
   123, 456, 342, 789, 90, 2, 875, 453, 290, 998, 223
7. Show first 5 steps of shell sort ever the numbers
   45, 78, 2, 12, 34, 9, 89, 4
8. Sort the following string "DATA STRUCTURES" considering the ascii value for each of the character, including space.
9. Try to sort characters of your name using radix sort? What problem you will be facing as compare to soting decimal numbers?
10. Describe the best sorting algorithm when input list of N numbers is very small and very large.
11. Construct a heap from the following set of characters, use alphabetic comparison:
    'B','U''I', 'L', 'D','H','E','A','P'.
12. Is Quick sort is an application of stack?

❑❑❑

CHAPTER 13

SEARCHING

INSIDE THIS CHAPTER

13.1. INTRODUCTION

Searching refers to finding a given data item in a set, list or array. In general searching is of two type: Linear search and Binary search. There is another searching technique known as "Hashing" but it will be taken in detail later as separate topic.

13.2. LINEAR SEARCH/SEQUENTIAL SEARCH

In sequential search we search for a key in a sequential manner, accessing each element only once from beginning of the data structure. These include arrays, lists, and sequential files. This search will take O(n) worst case (average case will approach O(n) if there are many keys not found). We would generally do a sequential search when:

- There are very few items or search is rare compared to other operations.
- The items are ordered by frequency and there is a strong likelihood that only the top few items will be needed
- Items cannot be ordered (keys can only be equal or not equal to other keys; no <, >)
- It is too expensive to convert the data structure to a more flexible one.

If element is found then search is said to be successful else unsuccessful. The C code for the same is given below. Here array is assumed to be unsorted.

```
void Lsearch (int arr [ ],
              int MAX,
              int num)
{
int   i;
int   found    =0;
for(i = 0;    i<MAX;     i++)
```

```
if (arr [i]  = = num)
{
found = 1;
break;
}
if(found = = 1)
printf("Element exist at %d position\n", i);
else
printf("Element does not exist in array\n");
}
```

For searching an element in the array using linear search, we search the entire array from the beginning and comparing every array element with the data to be searched, As soon as matching occurs we set the found variable to 1 (which was 0 in the beginning) and break from loop. Outside the loop it is checked whether found is 1 or not and depending upon result is displayed.

13.2.1. Linear Search In Sorted Array

For writing code that performs linear search in a sorted array, the logic is simple and it depends upon whether array is sorted in ascending order or descending order. In case array is sorted in increasing order we check only those elements which are less then the desired elements. For that we write the condition as:

```
if (arr [i]> = num)
```

In case the array elements are sorted in descending order then condition will be different as given below:

```
if(arr[MAX-1]<num)
```

And if you want to combine conditions for both types of arrays then you can write the condition as:

```
if(arr[i]> = num || arr[MAX-1]<num)
```

The code for this is given below:

```
void Lsearch (int arr[ ], int MAX, int num)
{
int i;
int found = 0;
for(i = 0;i<MAX; i++)
{
if(arr [i] > = num)
{
if(arr [i] = = num)
{
found = 1;
break;
}
}
```

```
}
if(found = = 1)
printf("Element exist at %d position\n",i);
else
printf("Element does not exist in array\n");
}
```

13.2.2. Complexity of Linear Search

Complexity of any searching method depends on the number of comparison performed to search for a specific element in the array of N elements. In case element is found within the very first comparison, it will be the best case for searching (least probable) and time complexity will be $O(1)$. In the worst case element may be found at the end of the list so that number of comparisons will be N. Time complexity in this case will be $O(N)$. On an average case the number of comparisons will be approximately $(N+1)/2$. But still complexity will be $O(N)$. This result is obtained from the probabilistic theory.

Assume $p(i)$ is the probability that element at index i will be found. The value of $p(i)$ will be between 0 and 1 and $p(0) + p(1) + p(2) + ... + p(n-1) = 1$. The expression means that the data item that we are searching will indeed be found. In sequential search when data item is searched in the array, the comparison will depend on i where element appears in arr[i]. Considering this, the average number of comparisons will be:

$$= 1*p(0) + 2*p(1) + 3*p(2) + \ldots\ldots\ldots\ldots\ldots + n*p(n-1)$$

assume p(i) is 1/N *i.e.* all elements are having equal probability that they will be retrieved, then

$$= 1*1/N + 2*1/N + 3*1/N + \ldots + N*1/N$$

$$= 1/N(1 + 2 + 3 \ldots + N)$$

$$= 1/N\,(N\,(N+1)/2)$$

$$= (N+1)/2$$

From the result obtained it is clear that average number of comparisons required for finding an element in the array is nearly equal to half of the elements in the array.

13.3. BINARY SEARCH

In the binary search over the array of N elements, we first find out the middle position of the array. We then compare the middle element of the array with the data to be searched. If data is equal to the middle element, it is found. If data is less then the middle element, it resides into the lower half of the array else it resides into the upper half of the array. The process is repeated for the other half of the array (lower or upper) and finally the element will be found as the middle element or search will be unsuccessful.

To understand it lets apply the binary search over the following set of numbers:

```
1   3   6   12   34   56   67   78   89   90
```

Say we want to search for the element 78. Initially L = 0, U = 9, M = (0+9)/ 2=4 arr[4] not equal to 78; if condition is false. Next if condition is false as arr[4]<78 so L = M+1 = 5 and U = 9; New value of M is (5+9)/2 = 7. Now arr[7] is equal to 78. Element is found at position 7.

Next consider searching element 2 in the array. Initially L = 0, U = 9, M = (0+9)/2 = 4 arr[4] is not equal to 2; if condition is false. Next if condition is true as arr[4] (34)> 2 so U = M-1 = 3 and L = 0; New value of M is (0+3)/2 = 1. Now arr[1] is compared with 2, the if condition is false; Next if condition is true as arr[1] (3) > 2, so U = M-1 = 0 and L = 0. Now arr[0] is compared with 2, the if condition is false and next if condition is also false as arr[0] (1) < 2, so L = M+1 = 1. As L is now greater than U, condition in the while will be false and loop terminates. The search results in failure.

The function given below implements binary search:

```
void Bsearch(int arr[ ], int MAX, int num)
{
int L = 0, U = MAX-1, M;
int found = 0;
while(L< = U)
{
M = (U+L)/2;
if(arr[M] = = num)
{
found = 1;
break;
}
if (arr [M] >num)
U = M-1;
else
L = M+1;
}
if (found)
    printf("Element exist at %d position\n", M);
else
printf ("Element does not exist in array\n");
 }
```

13.3.1. Iterative and Recursive Binary Search Algorithm

Procedure BINARY-SEARCH takes a sorted array A, a value v, and a range [low ... high] of the array, in which we search for the value v. The procedure compares v to the array entry at the midpoint of the range and decides to eliminate half the range from further consideration. We give both iterative and recursive versions, each of which returns either an index i such that A[i] = v, or NIL if no entry of A[low ... high] contains the value v. the initial call to either versions should have the parameters A, v, l, n.

Algorithm 13.1. ITERATIVE-BINARY-SEARCH (A, v, low, high)

```
1. while low ≤ high
2.    do mid ← [(low + high)/2]
3.       if v = A[mid]
4.          then return mid
5.       if v > A[mid]
6.          then low ← mid + 1
7.          else high ← mid – 1
8. return NIL
```

Algorithm 13.2. RECURSIVE-BINARY-SEARCH (A, v, low, high) ....

1. if low > high
   then return NIL
2. mid ← ⌊(low + high)/2⌋
3. if v = A[mid]
   then return mid
4. if v > A[mid]
   then return RECURSIVE-BINARY-SEARCH (A, v, mid+1, high)
   else return RECURSIVE-BINARY-SEARCH (A,v, low, mid −1)

Both procedures terminate the search unsuccessfully when the range is empty (*i.e.* low > high) and terminate it successfully if the value *v* has been found. based on the comparison of *v* to the middle element in the searched range, the search continues with the range halves. The recurrence for these procedures is therefore $T(n) = T(n/2) + \Theta(1)$, whose solution is $T(n) = \Theta(lg\ n)$.

13.3.2. Complexity of Binary Search

It is clearly visible from the code for binary search that each comparison divides the sub list into two halves. For example if original list is of say size 8, then in the first iteration we will have two list of size 4 each. In the second iteration either of the two list will be selected and will further be divided into two list of size 2. In the final iteration either of the two lists will be selected and will further be divided into two lists of size 1. Finally either the search will be successful or not. Thus for an array of size N, at most $\log_2 N + 1$ comparisons will be required (+1 is required when size is not the power of 2). Therefore the complexity of binary search will be $O(\log_2 N)$. Due to this complexity, binary search is considered as one of the most efficient searching technique. Consider size N of list is 1000 (close to 2^{10} which is 1024), thus approximately 10 comparisons will be required to find an element in an array of size 1000 (as $\log_2 1000 \approx 10$).

But there are some limitations of binary search. The first one is that original list must be sorted. In case list is not sorted, we need to apply some sorting algorithm to sort the list first. Second there must be a way to access the middle element directly. In case of array, it is ok, but if there are frequent insertion and deletion along with searching the item in the sorted list, then this may not be the efficient method. However despite these limitations binary search remains the most efficient searching algorithm.

13.4. HASHING

Hashing is a searching technique that relies on the concept of hash table, In hashing we have a set of key values and each key value is mapped to some address. In computer science, a hash table, or a hash map, is a data structure that associates keys with values. The primary operation it supports efficiently is a lookup: given a key (e.g. a person's name), find the corresponding value (e.g. that person's telephone number). It works by transforming the key using a hash function into a hash, a number that is used as an index in an array to locate the desired location ("bucket") where the values should be. So a hash table is a data structure that stores key/value pairs and can be quickly searched by the key. Because insertion and removal are operations dependent on the speed of the search, they tend to be fast as well.

13.4.1. Hash Functions

Hashing is the application of a mathematical function (called a **hash function**) to the key values that results in mapping the possible range of key values into a smaller range of relative addresses. The hash function is something like a locked box that needs a **key** to get the output which, in this case, is the **address** where the key is stored:

key ====> | Hash FunctionH(k) | ===> address

With hashing, there is no obvious connection between the key and the address generated since the function "randomly selects" an address for a specific key value, without regard to the physical sequence of the records in the file. Hence, it is also known as **randomizing scheme**.

Two or more input keys, say k_1 and k_2, when applied to a hash function, may *hash* to the same address, an accident known as **collision**. Collision can be reduced by allocating more file space than the minimum required to store the number of keys. However, this approach leads to wasted space. There are several ways to handle collisions, which will be discussed later.

In hashing, there is a need to choose a good hash function and consequently, select a method to resolve, if not eliminate, collisions. A good hash function performs fast computation, in time complexity of $O(1)$, and produces less (or no) collisions.

DO YOU KNOW

AAN MATHISON TURING (1912-1954) Aan Turing was born in London, although he was conceived in India, where his father was employed in the Indian Civil Service. As a boy, he was fascinated by chemistry, performing a wide variety of experiments, and by machinery. After completing his dissertation, which include a rediscovery of the central limit theorem, a famous theorem in statistics, he was elected a fellow of his college.. In 1953 Turing became fascinated with the decision problem, a problem posed by great germen mathematics *Hilbert*, Which asked whether there is a general method that can be applied to any assertion to determine whether the assertion is true. Turing enjoyed running and one day, while resting after a aru7n, he discovered the key ideas needed to solve the decision problem. Using these machines he found a anew problem, involving what he called computable numbers that could not be decided using a general method.

From 1936 to 1938 Turing visited Princeton University to work with Alonzo Church, who had also solved Hilbert's decision problem. In 1939 Turing returned to King's college. However, at the outbreak of World War II, he joined the Foreign Office, performing cryptanalysis of German ciphers. His contribution to the breaking of the code of the Enigma, a mechanical German cipher machine, played an important role in winning the war.

After the war, Turing worked on the development of early computers. He was interested in the ability of machines to think, proposing that if a computer could not be distinguished from a person based on written replies to questions, it should be considered to be "thinking". He was also interested in biology, having written on morphogenesis, the development of form in organisms. In 1954 Turing committed suicide by taking cyanide, without leaving a clear explanation. Legal troubles related to a homosexual relationship and hormonal treatments mandated by the court to lesson his sex drive may have been factors in his decision to end his life.

To perform searching using hashing we need to devise a function to convert the key directly to an address, that is we need a function $H(k)$ that is a one-to-one mapping to integers in $(0, L - 1)$. (where L is the size of our hash table). This is a helper function that converts any object into an integral index suitable for subscripting the hashtable. These functions are known as hashing function. Thus a hash function is any function that transforms a given key into an index into the hash table. The H in $H(k)$ is known as hash function, $H(k)$ is hash of the key and the result obtained after applying H to k will be known as hash value of key k. Thus in general any hash function H is a mapping from set of keys K to set of locations or indexes L. Therefore we can write a hash function H as

```
H : K->L
```

It is very much possible when calculating the hash value for two keys $k1$ and $k2$ to be same *i.e.* $H(kl) = H(k2)$. This situation is termed as collision. That is $H(k1)$ will give us some index I into the hash table and key $k1$ will be placed at that index into thc hash table. Next time when $H(k2)$ will be calculated, it will give us the same index I !. Now it is quite obvious that two keys cannot be stored into the same location. This is collision. We will take this topic in detail later on.

A good hash function is essential for good hash table performance. A good hash function has the following properties :

- The hash value is fully determined by the data being hashed. For instance, it should not have a 'memory'.
- The hash should be fast to compute.
- The hash function "uniformly" distributes the data across the entire set of possible hash values.
- The hash function generates very different hash values for similar strings. Variables like key l, key2, etc. should not be mapped into a cluster.

After discussing about hash function we now discuss some of the most popular hash functions :

1. The Division Method or mod Function

This method is one of the most common randomizing scheme. If the key value is divided by a number $\boldsymbol{n}$, the range of address generated will be of the form $\mathbf{0}$ to $\boldsymbol{n} - \mathbf{1}$. The formula is :

$$h(k) = k \bmod n$$

where k is the integer key value and n is a prime number.

If n is the total number of relative locations in the file, this method can be used to map the keys into n record locations. n should be chosen to attempt to reduce number of collisions. If n is even, the result of the hash function has revealed an even value and if n is odd, the resulting value is also odd. Division with a prime number will not result to much collision, that is why it is the best choice for the divisor in this method. We could choose a prime number that is close to the number of record positions in the file.

However, this method may be used even if n is not a prime, but be prepared to handle more collisions. For example, let $n = 13$

Key Value k	Hash Value $h(k)$	Key Value k	Hash Value $h(k)$
125	8	234	0
845	0	431	2
444	2	947	11
256	9	981	6
345	7	792	12
745	4	459	4
902	5	725	10
569	10	652	2
254	7	421	5
382	5	458	3

2. The Multiplicative Method

In this method follow the steps given below :

- Multiply the key by a constant A, $0 < A < 1$,
- Extract the fractional part of the product,
- Multiply this value by m.

Thus the hash function is :

```
h(k) = floor(m * (kA - floor(kA)))
```

That is multiply the key by any real number between 0 and 1; take out the fractional part of the product thus obtained (kA- floor (kA)) and multiply it by m. Finally take the floor of this value. This will be the final value (between 0 and m- 1) where key k will be stored.

In this case, the value of m is not critical and we typically choose a power of 2 so that we can get the following efficient procedure on most digital computers:

(*i*) Choose $m = 2p$.

(*ii*) Multiply the w bits of k by floor(A * 2w) to obtain a 2w bit product.

(*iii*) Extract the p most significant bits of the lower half of this product.

It seems that :

```
A = (sqrt(5)-1)/2 = 0. 6180339887 is a good choice
```

3. The Midsquare Method

In the midsquare method the key is squared (*i.e.* multiplied by itself), from the product thus obtained, few digits from the left, right or from middle are chosen (depend upon one's choice and on size of table). This value is used an index into the table. In case value thus obtained is not between 0 and m-1 then it may be divided by the table size m and remainder may be chosen as index value. The problem with this method is that it does not distribute the keys uniformly as compared to other two methods discussed earlier.

4. The Folding Method

Another simple hashing technique is folding. In this technique, the key value is split into two or more parts and then *added*, *ANDed*, or *XORed* to get a hash address. If the resulting address has more digits than the highest address in the file, the excess high- order digits are truncated.

There are different ways of folding. A key value can be ***folded in half***. This is ideal for relatively small key values since they would easily fit in the available addresses. If in any case, the key would be unevenly split, the left fold must be greater than the right fold. A key value can also be ***folded in thirds*** this is ideal for somewhat large key values. We can also ***fold alternate digits***. The digits in the odd positions form one part, and the digits in the even positions form another. Folding in half and in thirds can be done in yet another two ways. One is ***boundary folding*** where some parts of the folded keys are reversed (imitating the way we fold paper) and then summed. Last is ***shift folding*** where no parts of the folded keys are reversed.

The following are some examples of **shift folding** :

1. Even digits, folding in half

   125758 => 125+758 => 883

2. Folding in thirds

   125758 => 12+57+58 => 127

3. Odd digits , folding in half

7453212 => 7453+212 => 7665

4. Different digits, folding in thirds

74532123 => 745+32+123 => 900

5. Using XOR, folding in half

100101110 => 10010+1110 => 11100

6. Alternate digits

125758 => 155+278 => 433

The following are some examples of **boundary folding :**

1. Even digits, folding in half

125758 => 125+857 => 982

2. Folding in thirds

125758 => 21+57+85 => 163

3. Odd digits, folding in half

7453212 => 7453+212 => 7665

4. Different digits, folding in thirds

74532123 => 547+32+321 => 900

5. Using XOR, folding in half

100100110 => 10010+0110 => 10100

6. Alternate digits

125758 => 155+872 => 1027

This method is useful for converting keys with large number of digits to a smaller number of digits so the address fits into a word memory. It is also easier to store since the keys do not require much space to be stored.

Example 13.1.

Suppose we have to assign a 4 digit student id to 45 students studying in an institution. To store information about the students we have 80 locations. We just take 4 students here and assume their student ids are: 2345, 5378, 7321, and 6911. As the numbers of locations are 80, we assume a prime number close to 80 as value of m. We chose value as 79. We apply the above 4 methods to find an index to place the above 4 ids (actually student information) into some index into the hash table.

1. Division/Mod method

(*i*) *H*(2345) = 2345 mod 79 = 54, information of student with id 2345 is stored at index 54 into the hash table.

(*ii*) *H*(5378) = 5378 mod 79 = 6, information of student with id 5378 is stored at index 06 into the hash table.

(*iii*) *H*(7321) = 7321 mod 79 = 53, information of student with id 7321 is stored at index 53 into the hash table.

(*iv*) *H*(6911) = 6911 mod 79 = 38, information of student with id 6911 is stored at index 38 into the hash table.

2. The Multiplicative Method

(*i*)

H(2345) = floor(79 * (2345*0. 618033 – floor(2345*0. 618033))

= floor(79 * (1426. 071885 – floor(1426. 071885))

= floor(79 * (1426. 071885 – 1426))
= floor(79 * 0. 071885)
= floor(5. 678915)
= 5

Information of student with id 2345 is stored at index 05 into the hash table.

(*ii*) H(5378) = floor(79 * (5378*0. 618033 - floor(5378*0. 618033))
= floor(79 * (3323. 781474 -floor(3323. 781474))
= floor(79 * (3323. 781474 -3323))
= floor(79 *0. 781474)
= floor(61. 736446) = 61

Information of student with id 5378 is stored at index 61 into the hash table.

(*iii*) H(7321) = floor(79 * (7321*0. 618033 - floor(7321*0. 618033))
= floor(79 * (4524. 619593 -floor(4524. 619593))
= floor(79 * (4524. 619593 - 4524))
= floor(79 *0. 619593)
= floor(48. 947847)
= 48

Information of student with id 7321 is stored at index 48 into the hash table.

(*iv*) *H*(6911) = floor(79 * (6911*0 618033 - floor(6911*0. 618033))
= floor(79 * (4271. 226063 - floor(4271. 226063))
= floor(79 * (4271. 226063 -4271))
= floor(79 *0. 226063)
= floor(17. 858977)
= 17

Information of student with id 6911 is stored at index 17 into the hash table.

3. The midsquare method

K	K^2	H(k)
2345	54**99**025	09
5378	289**22**884	22
7321	535**97**041	97
6911	477**61**921	61

For finding the value of $H(k)$ from k^2, first and last 3 digits are removed. Note in the case of key 7321, index obtained is 97 which is outside the locations 0 to 79. Therefore 97 mod 79 (it will be 18) is chosen as index value and key 7321 is stored at index 18.

4. The Folding Method

(A) Without reversing the digits of second key

(*i*) H(k) = k1 + k2, where k1 = 23 and k2 = 45

H(*k*) = 23 +45 = 68, key 2345 is stored at location 68

(*ii*) *H*(*k*) = k1 +k2, where k1 = 53 and k2 = 78

$H(k)$ = 53 +78 = 31 (ignore carry), key 5378 is stored at location 31

(*iii*) $H(k)$ = k1 +k2, where k1 = 73 and k2 =21

$H(k)$ = 73 +21 = 94, key 7321 is stored at location 94 mod79 = 15

(*iv*) $H(k)$ = k1 +k2, where k1= 69 and k2 = 11

$H(k)$ = 69 +11 = 80, key 6911 is stored at location 80 mod 79 = 01

(B) With reversing the digits of second key

(*i*) $H(k)$ = k1 +k2, where kl = 23 and k2 = 54

$H(k)$ = 23 +54 = 77, key 2345 is stored at location 77

(*ii*) $H(k)$ = k1+ k2, where k1 = 53 and k2 = 87

$H(k)$ = 53 +87 = 40 (ignore carry), key 5378 is stored at location 40

(*iii*) $H(k)$ = kl +k2, where k1 = 73 and k2 = 12

$H(k)$ = 73 +12 = 85, key 7321 is stored at location 85 mod 79 = 06

(*iv*) $H(k)$ = kl +k2, where kl = 69 and k2 = 11

$H(k)$ = 69 +11 = 80, key 6911 is stored at location 80 mod 79 = 01

13.4.2. Resolving Collision

Choosing a good hashing algorithm based on how few collisions are likely to occur is the first step in avoiding collision. However, this will just minimize and will not eradicate the problem. To avoid collision, we could :

- *spread out records, i.e.*, finding a hashing algorithm that distributes the records fairly randomly among the available addresses. However, it is hard to find a hashing algorithm that distributes the records evenly.
- *use extra memory,* if there are memory addresses to distribute the records into, it is easier to find a hashing algorithm than if we have almost equal number of addresses and records. One advantage is that the records are spread out evenly, eventually reducing collisions. However, this method wastes space.
- *use buckets, i.e.*, put more than one record at a single address.

We know that when two keys have got the same index into the hash table after applying some hash function, then this situation is termed collision. In this section we will see, how to handle the collision during hashing. First of all note that collision cannot be avoided at all, no matter which hash function you choose. It can only be minimized using various techniques that we are going to discuss in this section.

There are mainly two techniques for collision resolution:

(*i*) Open addressing or Rehashing

(*ii*) Chaining

Both will be discussed one by one.

1. Open Addressing

In open addressing, when the address produced by a hash function $h(k)$ has no more space for insertion, an empty slot other than $h(k)$ in the hash table is located. This process is called **probing**. In this empty slot, the new record that has collided with the old one, known as the **colliding key**, could be safely placed. In this method, we introduce the permutation of the table addresses, say 0, 1, ... $m - 1$, This permutation is called **probe sequence**.

In this section, we will cover two techniques: *linear probing* and *double hashing*.

Linear Probing

Linear probing is one of the simplest techniques in handling collisions in which the file is scanned or probed sequentially as a circular file and the colliding key is stored in the nearest available space to the address. This is used in systems where a "first come, first served" basis is followed. An example is an airline reservation system where seats for chance passengers are given away when passengers for reserved seats don't arrive. Whenever a collision occurs at a certain address, the succeeding addresses are probed, that is, searched sequentially until an empty one is found. The key then uses this address. Array should be considered circular, so that when the last location is reached, the search proceeds to the first location of the array.

In this method, it is possible that the entire file could be searched, starting at position $i+1$, and the colliding keys are distributed throughout the file. If a key hashes to position i, which is occupied, positions $i+1, \ldots, n$ are searched for an empty location.

Double Hashing

The slots in a hash table may contain only one key or could also be a bucket. In this example, buckets of capacity 2 are used to store the following keys :

Key Value k	Hash Value $h(k)$	Key Value k	Hash Value $h(k)$
125	8	234	0
845	0	431	2
444	2	947	11
256	9	981	6
345	7	792	12
745	4	459	4

Key Value k	Hash Value h(k)	Key Value k	Hash Value h(k)
902	5	725	10
569	10	652	2
254	7	421	5
382	5	458	3

For keys 125, 845, 444, 256, 345, 745, 902, 569, 254, 382, 234, 431, 947, 981, 792, 459 and 725, storage is pretty straightforward – no overflow happened.

	KEY1	KEY2
0	845	234
1		
2	444	431
3		
4	745	459
5	902	382
6	981	
7	345	254
8	125	
9	256	
10	569	725
11	947	
12	792	

Inserting 652 into the hash table results to an overflow at address 2 so we rehash :

h2(652) = 652 mod 11 = 3

rh(2, 652) = (2 + 3) mod 13 = 5, but address 5 is already full so rehash :

rh(5, 652) = (5 + 3) mod 13 = 8, with space - so store here.

	KEY1	KEY2
0	845	234
1		
2	444	431
3		
4	745	459
5	902	382
6	981	
7	345	254
8	125	652
9	256	
10	569	725
11	947	
12	792	

Hashing 421 also results in a collision, so we rehash:

h2(421) = 421 mod 11 = 3

rh(5, 421) = (5 + 3) mod 13 = 8, but address 8 is already full so rehash:

rh(8, 421) = (8 + 3) mod 13 = 11, with space - so store here.

	KEY1	KEY2
0	845	234
1		
2	444	431
3		
4	745	459
5	902	382
6	981	
7	345	254
8	125	652
9	256	
10	569	725
11	947	421
12	792	

Lastly, key 458 is stored at address 3.

This method is an improvement of linear probing in terms of performance, that is, the new address is computed using another hash function instead of sequentially searching the hash table for an empty space. However, just like linear probing, this method also suffers from the displacement problem.

2. Chaining

Chaining is another method for resolving collision. In chaining no rehashing method is used instead links or pointer are used to link together record with similar hash addresses. To understand this considers 10 records (A to J) with the following hash addresses: 5, 7, 8, 5, 6, 9, 3, 9, 2, 3. This is shown in the table given below :

Record	A	B	C	D	E	F	G	H	I	J
H(K)	5	7	8	5	6	9	3	9	2	3

We keep a hash table whose indexes represents $H(k)$ values and their contents are the pointer to the linked list. For every hash value we have a separate linked list. Whenever a new record with the same hash value is inserted a new node is added at the end in the corresponding linked list. For the above information of record and $H(k)$ values we have the following figure:

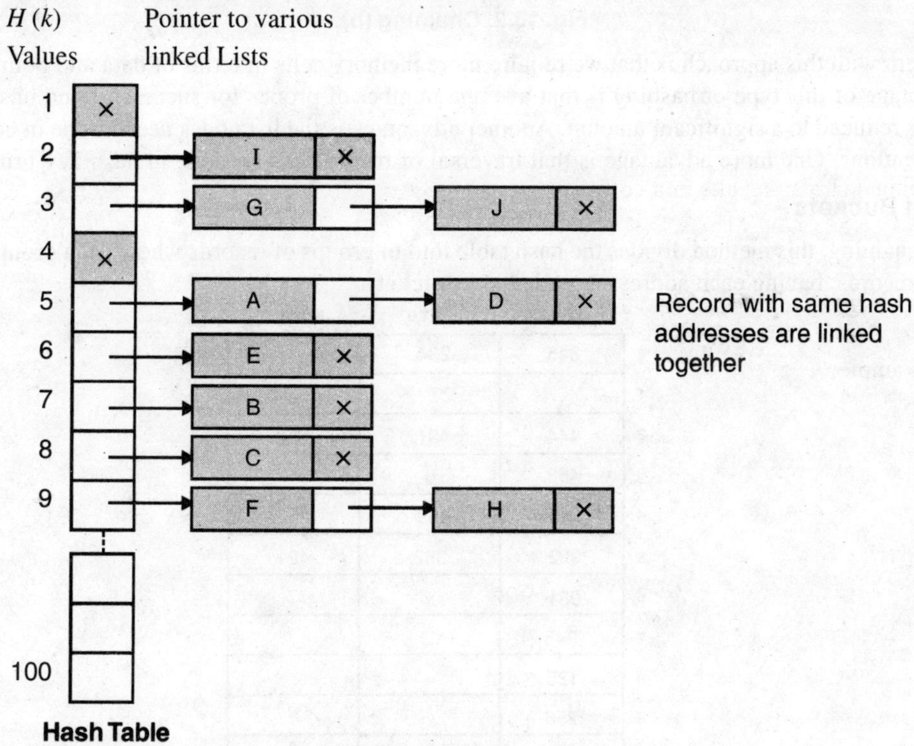

Fig. 13.1. Chaining (a).

Another figure for chaining is shown below. In the following figure we have assumed that instead of separate linked lists in different memory locations, we have a big table of free nodes with fields data and next. The next pointer field links together various nodes of a specific linked list. We have another table that is similar to the hash table used in previous figure but here we show the addresses of the first node in every linked list. Thus the hash table can be thought as array/ table of pointers.

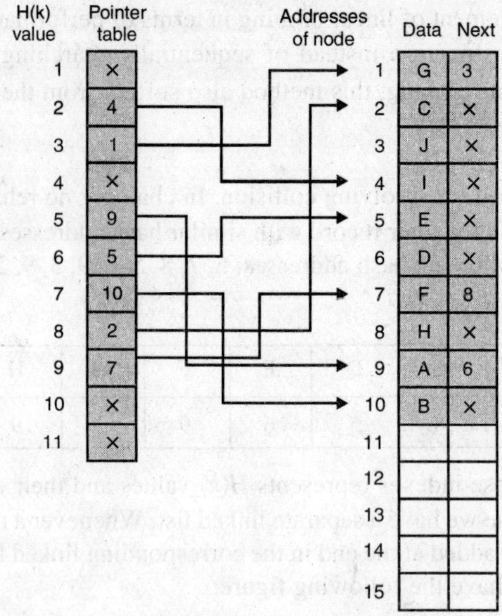

Fig. 13.2. Chaining (b).

Problem with this approach is that we require more memory cells in terms of data and pointer fields. The advantage of this type of hashing is that average number of probes for successful and unsuccessful searches is reduced to a significant amount. Another advantage is the list nodes need not be in contiguous storage locations. One more advantage is that traversal of records can be done in hash-key order.

3. Use of Buckets

Like chaining, this method divides the hash table into **m** groups of records where each group contains exactly **b** records, having each address regarded as a **bucket**.

For example:

	KEY 1	KEY 2	KEY 3
0	845	234	
1			
2	444	431	652
3	458		
4	745	459	
5	902	382	421
6	981		
7	345		
8	125		
9	256		
10	569	725	
11	947		

Collision is redefined in this approach. It happens when a bucket overflows – that is, when an insertion is attempted on a full bucket. Hence, there is a significant reduction in the number of collisions. However, this method wastes some space and is not free from overflowing further, in which case an overflow policy must be invoked. In the above example, there are three slots in each address. Being static in size, problem arises when more than three values hash to a single address.

13.4.3. Efficiency of Hash Functions

Efficiency of hash function is measured in terms of average number of probes needed to find a suitable empty position or searching for a record with a given key in the hash table. The efficiency depends on the load factor (LF) that is defined as N/ L where N is number of keys and L number of locations in the hash table *i.e.* size of table. Usually we are interested in average number of probes for successful and unsuccessful search. This is usually denoted by SS(LF) and US(LF). For linear probing, for tables with high value of L it has been proved mathematically that average number of probes for successful search is given by the followingformula :

```
SS(LF)    = 2-LF / (2-2LF)
          = 0. 5 (2-LF) / (1-LF)
          = 0. 5 + 0. 5(1-LF)
```

And average number of probes for unsuccessful search is given by following formula:

```
US (LF) = 0. 5 (1+ 1/ (1-LF) 2)
```

In case of chaining, average number of probes for a successful and unsuccessful search is given approximately by the formula:

SS (LF) = 1+ 0. 5 LF and US(LF) = e^{-LF} **+ LF**

13.4.4. Hash Table Organization

Organization	Advantages	Disadvantages
Chaining	■ Unlimited number of elements ■ Unlimited number of collisions	■ Overhead of multiple linked lists
Re-hashing	■ Fast re-hashing ■ Fast access through use of main table space	■ Maximum number of elements must be known ■ Multiple collisions may become probable
Overflow area	■ Fast access ■ Collisions don't use primary table space	■ Two parameters which govern performance need to be estimated

13.5. DYNAMIC FILES AND HASHING

The hashing techniques discussed so far make use of fixed address space (hash table with **n** addresses). With static address space, overflow is possible. If the data to store is dynamic in nature, that is, a lot of deletion and insertions are possible, it is not advisable to use static hash tables. This is where dynamic hash tables come inuseful. In this section, we will discuss two methods: *extendible hashing* and *dynamic hashing*.

13.5.1. Extendible Hashing

Extendible hashing uses a self-adjusting structure with unlimited bucket size. This method of hashing is built around the concept of a ***trie***.

Trie

The basic idea is to build an index based on the binary number representation of the hash value. We use a minimal set of binary digits and add more digits as needed. For example, assume each hashed key is a sequence of three digits, and in the beginning, we need only three buckets :

Bucket	Address
A	00
B	10
C	11

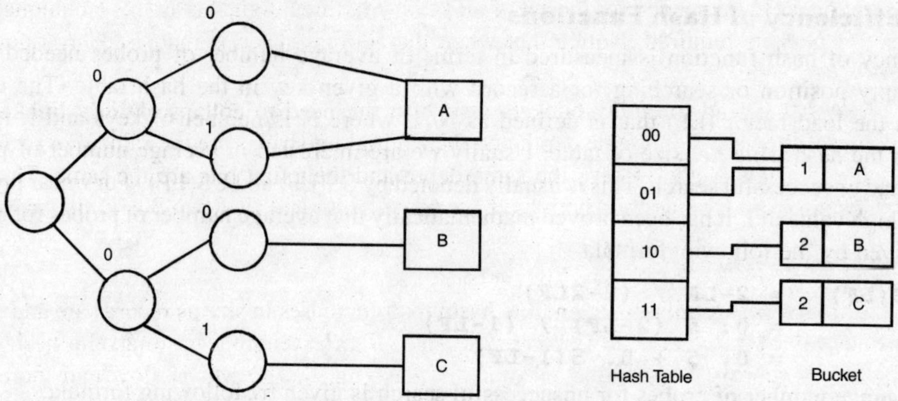

Fig. 13.3. Trie and Hash Table.

The figure on the left shows the trie. The figure on the right shows the hash table and the pointers to the actual buckets. For bucket A, since only one digit is considered, both the addresses 00 and 01 point to it. The bucket has structure (DEPTH, DATA) where **DEPTH** is the number of digits considered in addressing or the number of digits considered in the trie. In the example, the depth of bucket A is 1, while for buckets B and C, it is 2.

When A overflows, a new bucket is created, say D. This will make more room for insertion. Hence the address of **A** is expanded into two addresses:

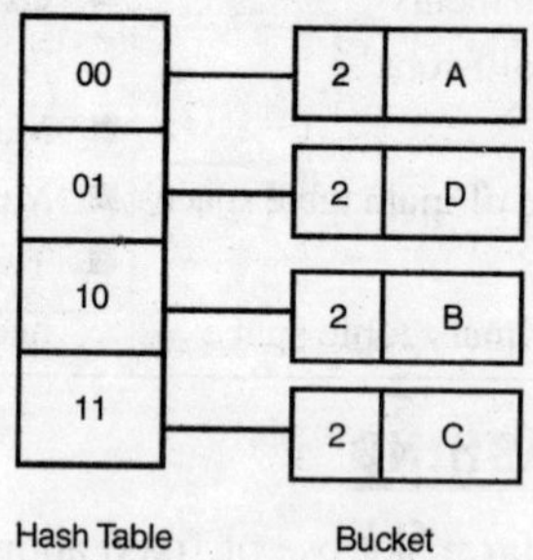

Fig. 13.4. Hash Table.

When B overflows, a new bucket D is then created and the address of B is expanded into three digits:

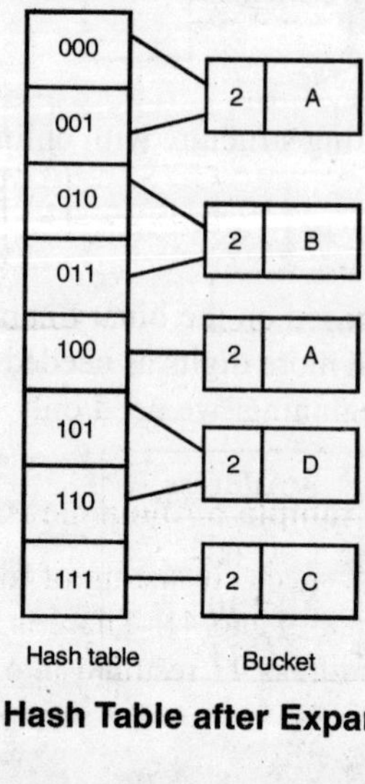

Fig. 13.5. Hash Table after Expansion.

Usually, eight buckets are needed for the set of all keys with three digits but in this technique, we use the minimal set of buckets required. Notice that when the address space is increased, it is double the original size.

Deletion of keys that lead to empty buckets results to the need to collapse **buddy buckets**. Two buckets are buddies if they are at the same depth and their initial bit strings are the same. For example, in the previous figure, buckets A and D have the same depth and their first bits are the same. The case is similar with buckets B and E since they both have the same depth and first two bits in the address.

13.5.2. Dynamic Hashing

Dynamic hashing is very similar to extendible hashing, it increases in size as records are added. They differ only in the way they dynamically change the size. If in extendible hashing, the hash table is doubled in size every time it is expanded, in dynamic hashing, the growth is slow and incremental. Dynamic hashing starts with a fixed address size similar to static hashing and then grows as needed. Usually, two hash functions are used. The first one is to see if the bucket is in the original address space and the second one is used if it is not. The second hash function is used to guide the search through the *trie*.

The following figure shows an example:

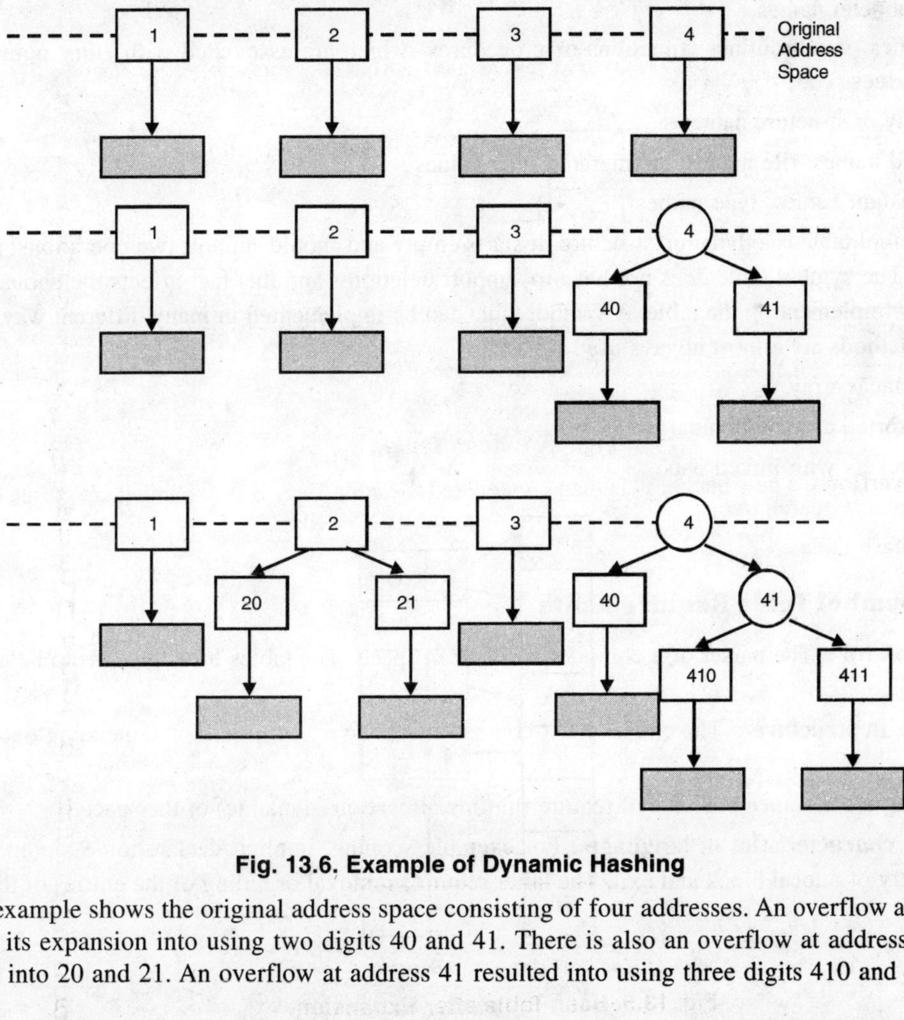

Fig. 13.6. Example of Dynamic Hashing

The example shows the original address space consisting of four addresses. An overflow at address 4 results to its expansion into using two digits 40 and 41. There is also an overflow at address 2, so it is expanded into 20 and 21. An overflow at address 41 resulted into using three digits 410 and 411.

13.6. SYMBOL TABLE

A symbol table is a data structure used by a language translator such as a compiler, assembler or interpreter, where each identifier in a program's source code is associated with information relating to its declaration or appearance in the source, such as its type, scope level and sometimes its location.

The symbol table is accessed by most phases of a language translator (LT), beginning within the lexical analysis scanning. The symbol table carries the collected information about each named object in the program to other phases of the LT. As soon as a named token is found, depending on the token type, a call to the symbol table routines is made.

The organization of the symbol table is the key to fast language translation (assembly or compilation). Even when working on a small program, the LT may use the symbol table hundreds of times and, consequently, an efficient implementation of the table can cut the language translation time significantly even for short programs.

The symbols for the symbol table are usually tokens. A token is the smallest unit for a program. They may be: Keywords of the language, special characters/ punctuation and user defined identifiers used for different meanings, constants, operators etc. Identifiers are user defined

- variables, memory location labels
- parameter names
- names of subroutines (functions or procedures) which are associated with entry points of the routines' code
- array or structure names
- field names; file names; enumerated type values
- constant names; type names

The symbol table is a dynamic structure. It starts empty and should support two operations, insertion and search. The symbol table does not have to support deletions, and this fact affects the choice of data structure for implementing the table. A symbol table can be implemented in many different ways but the following methods are almost always used :

- A linear array.
- A sorted array with binary search.
- Buckets with linked lists.
- A binary search tree.
- A hash table.

13.6.1. Symbol Table Requirements

Fast look up : The parser of a compiler is linear in speed. The tables lookup needs to be as fast as possible.

Flexible in structure : The entries must contain all necessary information, depending on usage of identifier

Efficient use of space : This will require runtime allocation (dynamic) of the space.

Handle characteristics of language : For example scoping, implicit declaration Scoping requires handling entry of a local block and exit. The latter requires removal or hiding of the entries of the block.

Example 13.2.

Consider the following code snippet in C:

```
int a=10,b=20, c=30;
float x = 2. 3, y = 4. 5;
char ch1 = 'A';
```

The symbol table will look like as (we consider only some of the fields) assuming arbitrary addresses.

ID	NAME	TYPE	SIZE	VALUE	ADDRESS
ID1	a	int	2	10	0x2030122
ID2	b	int	2	20	0x2030124
ID3	c	int	2	30	0x2030126
ID4	x	float	4	2.3	0x2030130
ID5	y	float	4	4.5	0x2030134
ID6	Ch1	char	1	'A'	0x2030135

SOLVED PROBLEMS

1. Write the C code so that binary search routine performs insertion of the element when element does not found in the array?

Ans. The code is given below:

```
void Bsearch(int arr[],int num,int N)
{
int L=0,U=N-1,M;
int found=0;
whle (L<=U)
{
  M=(U+L)/2;

if (arr[M] = = num)
{
found=1;
break ;
}
if(arr[M] > num)
  U=M-1;
else
        L=M+1;
}
if (found = =0)
{
if (arr[M] < num)
 arr [M] = num;
else
```

```
        arr [M+1] = num;
    }
}
```

Where N is location where you want to put the element in the array. The function can be used as

```
void main ( )
{
int a[10]={1,2,3,4,5};
Bsearch (a,7,6);
Bsearch(a,8,7);
for(int i=0; i<10;i++)
   printf("%d",a[i]);
}
```

OUTPUT

```
1  2  3  4  5  7  8  0  0  0
```

2. Suppose an array A has 100 elements. What is the worst case running time for searching an element x in the array. Assuming x has equal probability to be found at any index in the array.

Ans. From the time complexity of binary search we can write $2^7 > 100$, so total number of comparisons are 7.

3. What is the average case running time using the probabilistic method in the above problem?

Ans. Using the probabilistic method we first need to find out how many elements N_k in A will require k comparisons. Using the complexity of binary search *i.e.* $\log_2 N$ we note that for 1 element we have 1 comparison, for 2, it will be 2, for 4 it will be 3 and so on. This is given below:

N_k	1	2	4	8	16	32	37
k	1	2	3	4	5	6	7

The last column contains 37 elements and 7 comparisons 'coz of fact that after the comparisons of elements (1 + 2 + 4 + 8 + 16 + 32 = 63), 100 – 63 = 37 elements remains. Now using the probabilistic method, we have the formula :

$$f(N) = \frac{1}{N}\sum_{k=1}^{7} k.N_k$$

$$= \frac{1\times1+2\times2+3\times4+4\times8+5\times16+6\times32+7\times37}{100}$$

$$= \frac{1+4+12+32+80+192+259}{100}$$

$$= \frac{580}{100} = 5.8$$

4. Name condition under which sequential search of a list is preferable to binary search?

Ans. If the given list is not sorted than we will prefer sequential search compared to binary search as the time for linear search is $O(n)$ for the sorted as well as unsorted list, while the time for binary search is $O(\log_2 n)$ if the list is sorted otherwise binary search is not applicable.

5. What is the bucket size, when the overlapping and collision occur at same time?

Ans. One. If there is only one entry possible in the bucket, when the collision occurs, there is no way to accommodate the colliding value. This results in the overlapping of values.

6. Suppose that a file contains N = 11 entries indexed from 0 through 10 and the following keys are mapped into the file.

10, 100, 32, 45, 58, 126, 3, 29, 200, 400, 0.

(*i*) Determine the hash addresses and find how many collision occur when the hash function used key mode *N*.

(*ii*) Find a hash function that will produce no collision for these keys.

Ans. (*i*)

$$\frac{10}{11} = 10 \quad \text{(collision)-II}$$

$$\frac{100}{11} = 1 \quad \text{(collision)-I}$$

$$\frac{32}{11} = 10 \quad \text{(collision)-II}$$

$$\frac{45}{11} = 1 \quad \text{(collision)-I}$$

$$\frac{58}{11} = 3 \quad \text{(collision)-III}$$

$$\frac{126}{11} = 5$$

$$\frac{3}{11} = 3 \quad \text{(collision)-III}$$

$$\frac{29}{11} = 7$$

$$\frac{200}{11} = 2$$

$$\frac{400}{11} = 4$$

$$\frac{0}{11} = 0$$

$$h(k) = 10, 1, 10, 1, 3, 5, 3, 7, 2, 4, 0$$

There are total of three collisions as shown above.

(*ii*) To avoid the collision of the given keys, there are several methods.

Here we will use linear probing, here, if there is a collision, then we will search the next empty location in the memory *i.e.*

(*a*) When there is a first collision for keys 10 and 32, the key 32 is stored at next free location which is location 6.

(*b*) For collision of keys 100 and 45 the key 45 is stored at location 8.

(*c*) For collision of keys 58 and 3 the key 3 is stored at location 9.

The hash function, for which there is no collision is key Mod N alongwith linear probing.

7. How many comparisons are necessary to (*i*) determine that the value is not in the array, (*ii*) to find a value in the array in case of sequential unordered search, sequential ordered search

and binary search

20	40	90	100	110	160	190	200	300
1	2	3	4	5	6	7	8	9

The key values to be searched are :

(*a*) 20
(*b*) 10
(*c*) 90
(*d*) 95
(*e*) 115
(*f*) 300
(*g*) 305

Ans. The number of comparisons required to either determine the value in the array or to determine that the value is not in the array are given in the following table.

Key value	Sequential unordered search	Sequential ordered search	Binary search
20	1	1	3
10	9	1	3
90	3	3	3
95	9	4	4
115	9	6	3
300	9	9	4
305	9	9	4

SUMMARY

1. Searching refers to finding a given data item in a set, list or array.
2. In sequential search we search for a key in a sequential manner, accessing each element only once from beginning of the data structure.
3. In the binary search over the array of N elements, we first find out the middle position of the array. We then compare the middle element of the array with the data to be searched. If data is equal to the middle element, it is found. If data is less then the middle element, it resides into the lower half of the array else it resides into the upper half of the array.
4. Hashing is a searching technique that relies on the concept of hash table, In hashing we have a set of key values and each key value is mapped to some address
5. Thus a hash function is any function that transforms a given key into an index into the hash table. The H in $H(k)$ is known as hash function, $H(k)$ is hash of the key and the result obtained after applying H to k will be known as hash value of key k.
6. We know that when two keys have got the same index into the hash table after applying some hash function, then this situation is termed collision
7. A symbol table is a data structure used by a language translator such as a compiler, assembler or interpreter, where each identifier in a program's source code is associated with information relating to its declaration or appearance in the source, such as its type, scope level and sometimes its location.

8. **Fast look up :** The parser of a compiler is linear in speed. The tables lookup needs to be as fast as possible.
9. Hashing maps the possible range of key values into a smaller range of relative addresses.
   - Simple hash techniques include the prime number division method and folding.
   - Chaining makes use of m linked lists, one for each possible address in the hash table.
   - In the use of buckets method, collision happens when a bucket overflows.
   - Linear probing and double hashing are two open addressing techniques.
   - Dynamic hash tables are useful if data to be stored is dynamic in nature. Two dynamic hashing techniques are extendible hashing and dynamic hashing.

EXERCISES

1. What do you understand by searching?
2. Discuss various types of searching methods.
3. How searching for a number in sorted list affects the execution time?
4. How binary search is different than linear search?
5. For searching a number in a list of 1234 element, how many comparisons will be required in binary search?
6. Can we perform binary search in a sorted linked list?
7. What is hashing? How it is different from linear and binary search?
8. What are hash functions? What is the basic requirement to be a good hash function?
9. Discuss all different types of hash function applying over the following keys:

   2987, 6509, 7127, 6528
10. Write separate C routine for all 4 hash function discussed in this chapter? Your C function should take the key as integer and return hash address.
11. What is collision? Can it be eliminated completely?
12. What are the measures for resolving collision in hashing?
13. Discuss the concept of chaining alone with some suitable example.
14. What is double hashing? How it is different from linear and quadratic probing?
15. Write a program that simulates chaining using separate linked list for each hash address?
16. What is symbol table? Where it is used?
17. Name three condition under which sequential search of a list is preferable to binary search.
18. Suppose we are hashing integers with a hash table using the hash function

    $$h(i) = i \bmod 7$$

    (*i*) Show the resulting open hash table if the perfect cubes 1, 8, 27, 64, 125,216, 343 are inserted.

    (*ii*) Repeat part (i) using a closed hash table with linear resolution of collisions.

❑❑❑

CHAPTER 14 FILES

14.1. LIMITATIONS OF STORING DATA ON MAIN MEMORY

1. If data is too large to be stored in main memory, then it may not fit into main memory
2. Main memory is volatile memory, hence data stored in it lost as soon as program quits.
3. **Expensive** : Compared to secondary and tertiary storage
4. **Fast** : operations in memory are fast

INSIDE THIS CHAPTER

14.1.1. Limitations of Storing Data on Secondary Memory or Storage (Hard Disk)

1. **Less expensive** – compared to main memory
2. **Slower** – compared to main memory, faster compared to tapes
3. **Persistent** – data from one run can be saved to the disk to be used in the next run

14.1.2. Limitations of Storing Data on Tertiary Storage (Tapes)

1. Slowest – sequential data access
2. Used for data archives

To overcome the problems mentioned above we can use an external storage as special data holding entities called files.

14.2. SOME BASIC DEFINITIONS

File : A file is a collection of logically related records.

Record : A record is a collection of logically related fields of information about a particular entity.

Entity : An Entity is something that has certain attribute or properties.

Attribute : Each attribute of an entity set has a range of values, the set of all possible values that can be assigned to a particular attribute.

Field : It is a singe elementary unit of information representing an attribute of an entity.

Primary Key : It is that field which contains only unique values for all records.

Key Values : The values in the primary key field are called key field values.

14.3. QUERIES

Key fields : These are those fields in the file which helps in retrieving records of a file. The records may be retrieved by specifying values for some or all of these keys.

Query : A combination of key values specified for retrieval will be termed as query

Example : Suppose in a employee file the fields emp number, name, designation, sex & salary have been designated as key fields.

Employee File

E #	Name	Designation	Sex	Salary
100	Andrews	Programmer	M	$ 10,000
200	Tames	Analyst	F	$15,000
300	Williams	Analyst	M	$ 12,000
400	Seymur	Programmer	F	$ 12,000
500	Armstrong	Analyst	M	$ 9,000

14.3.1. Types of fields Query

1. Simple Query

Query : Retrieve the record with sex M.

Response to query

E #	Name	Designation	Sex	Salary
100	Andrews	Programmer	M	$ 10,000
300	Williams	Analyst	M	$ 12,000
500	Armstrong	Analyst	M	$ 9,000

2. Range Query

Query : Retrieve record with salary > $10,000

Response to query

E #	Name	Designation	Sex	Salary
200	Tames	Analyst	F	$15,000
300	Williams	Analyst	M	$ 12,000
400	Seymur	Programmer	F	$ 12,000

3. Functional Query

Query : Retrieve all records with salary greater than average salary of all employees

Response to query

E #	Name	Designation	Sex	Salary
200	Tames	Analyst	F	$15,000
300	Williams	Analyst	M	$ 12,000
400	Seymur	Programmer	F	$ 12,000

4. Boolean Query

Query: Retrieve all records with (sex = F & designation = programmer) or (sex =M & salary > 12000)

Response to query

E #	Name	Designation	Sex	Salary
400	Seymur	Programmer	F	$ 12,000

14.4. INDEXING

A data structure that allows the DBMS to locate particular records in a file more quickly & very similar to the index at the end of a book to locate various topics covered in the book.

14.4.1. Types of Index

1. **Primary index** – one primary index per file
2. **Secondary index** – many secondary indexes per file
3. **Clustering index**

14.4.2. Pros and Cons of indexing

1. **Faster** access to rows where the indexed column is searched on.
2. Inserts & deletes **slower**
3. Updates to indexed columns **slower**
4. Increases **storage** used

14.5. TYPES OF FILE SYSTEMS

I. On the basis of response time

(*i*) Real-time files

(*ii*) Batch files

II. On the basis of data storage format

(*i*) Text files

(*ii*) Binary files

14.5.1. Batch Files

A **batch file** is a text file containing a series of commands intended to be executed by the command interpreter. When a batch file is run, the shell program (usually COMMAND.COM or cmd.exe) reads the file and executes its commands, normally line-by-line. Batch files are useful for running a sequence of executables automatically and are often used by system administrators to automate tedious processes. Unix-like operating systems (such as Linux) have a similar type of file called a shell script.

14.5.2. Text Files

A **text file** (sometimes spelled "textfile": an old alternate name is "flatfile") is a kind of computer file that is structured as a sequence of lines. A text file exists within a computer file system. The end of a text file is often denoted by placing one or more special characters, known as an end-of-file marker, after the last line in a text file.

"Text file" refers to a type of container, while plain text refers to a type of content. Text files can contain plain text, but they are not limited to such.

14.5.3. Binary Files

A **binary file** (extension **.bin** unless otherwise specified) is a computer file which may contain any type of data, encoded in binary form for computer storage and processing purposes; for example, computer document files containing formatted text. Many binary file formats contain parts that can be interpreted as text; binary files that contain only textual data - without, for example, any formatting information - are called plain text files. In many cases, plain text files are considered to be different from binary files because binary files are made up of more than just plain text. When downloading, a completely functional program without any installer is also often called program binary, or binaries (as opposed to the source code).

14.6. TYPES OF STORAGE DEVICES

Secondary storage devices are of two types:

1. Sequential access storage devices (SASD)
2. Direct access storage devices (DASD)

14.6.1. Sequential Access Storage Devices (SASD)

In these devices data is stored and retrieved sequentially. Accessing a particular record on these devices requires accessing of all previous records in the file.

For Examples: magnetic tape, paper tape, card reader etc.

14.6.2. Direct Access Storage Devices (DASD)

In these devices data can be retrieved both sequentially & randomly. For example: magnetic disks (HDD, Floppy disks), CDs, DVD's etc

All DASD's are made up of cylinder surface, tracks & sectors. A surface of disk consists of tracks & sectors.

Tracks are consecutive circular areas on surface of disks. Sectors are angular areas on surface of disks.

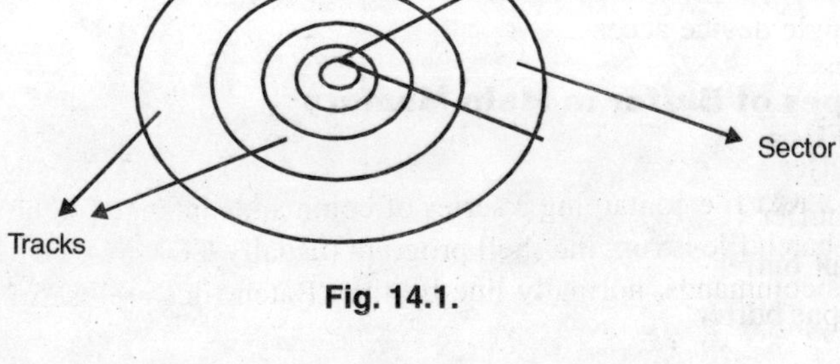

Fig. 14.1.

A cylinder is a set of vertically aligned tracks, which resides on platters of disk

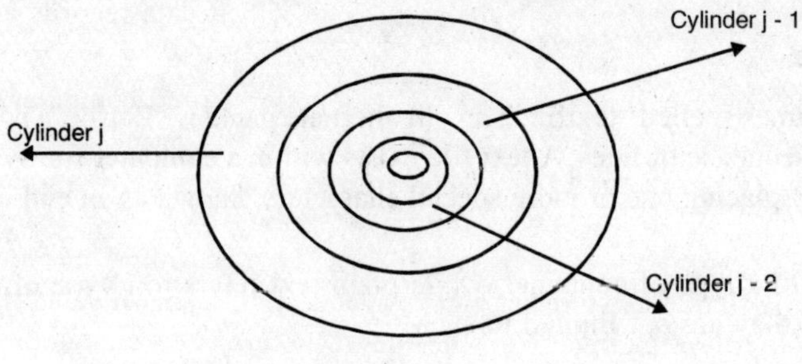

Fig. 14.2.

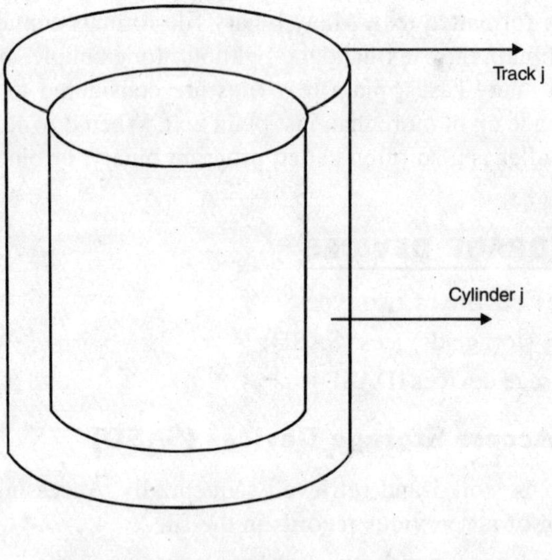

Fig. 14.3.

Note : For a track j, the corresponding cylinder number will be cylinder j because a cylinder is made up of jth tracks of all 's' surfaces, combined together on a disk if there are 'c' tracks on any of its surface then it means these are 'c' cylinders on that disk

14.7. MAIN MEMORY V/S FILE

I.Buffer – It is an area of main memory reserved to hold data for file. It is used to resolve differences in read/write commands of a program.

II.Block - To reduce program's device access records may be blocked such that multiple records are read / written in a single device access.

14.7.1. Four Types of Buffer in Main Memory

I. file input buffer
II. file output buffer
III. console input buffer
IV. console output buffer

14.8. FILE ORGANIZATION

The concept of file organization refers to the manner in which data records are arranged on file storage medium.

File organization is the methodology which is applied to structured computer files. Files contain computer records which can be documents or information which is stored in a certain way for later retrieval. File organization refers primarily to the logical arrangement of data (which can itself be organized in a system of records with correlation between the fields/columns) in a file system. It should not be confused with the physical storage of the file in some types of storage media. There are certain basic types of computer file, which can include files stored as blocks of data and streams of data, where the information streams out of the file while it is being read until the end of the file is encountered.

DO YOU KNOW

MAURICE KARNAUGH (BORN 1924) Maurice Karnaugh, born in New York city , received his B.S. from the city College of New York and his Ph.D from Yale University. He was a member of the technical staff at bell laboratories from 1952 until 1966 and manager of research and development at the Federal system Division of AT&T from 1966 to 1970. In 1970 he joined IBM as a member of the research staff. Karnaugh has made fundamental contribution to the application of digital techniques in both computing and telecommunication. His current interests include knowledge- based system in computers and heuristic search method.

We will look at two components of file organization here:

1. The way the internal file structure is arranged and
2. The external file as it is presented to the O/S or program that calls it. Here we will also examine the concept of file extensions.

We will examine various ways that files can be stored and organized. Files are presented to the application as a stream of bytes and then an EOF (end of file) condition.

A program that uses a file needs to know the structure of the file and needs to interpret its contents.

Another Definition of File Organization

Data in files is organized as a collection of records. A record is a grouping of related information that has a predefined size and a predefined number and layout of fields. Each record in a file has the same number and layout of fields, which hold specific parts of the record's information. For example, each record can contain information about a bank account, with fields for the account number, a name, the balance, and other information. The way the records in a file are arranged is called the file organization.

A file has one primary index, which defines the physical sequence of records in the file. A file can also have any number of secondary indexes that provide alternative sequences in which the file's records can be accessed. An index can be seen as a list of pointers to records in a file, where the primary index lists the actual sequence of the records, and each secondary index lists the pointers in a different sequence. Application programs can read, update, add, delete, and browse data in local or remote files.

14.8.1. Factors that Affect File Organization

1. Nature of operations that are to be performed on the file.
2. Volume of file access & frequency of file access
3. Required response time to a query
4. External storage media on which file is to reside.

14.8.2. Responsibilities of a File System

1. Maintaining directory for file identification, this directory keeps information about location of file
2. To facilitate data flow between main memory & secondary storage
3. To coordinate communication between CPU & secondary storage device
4. To handle asynchronous data transfer
5. To prepare file for input & output use
6. To handle the files when their use is terminated.

Three file organizations and their equivalent VSAM terms are as follows :

1. **Entry-sequenced**

   Entry-sequenced data set (ESDS)
2. **Relative**

   Relative record data set (RRDS)
3. **Clustered**

   Key-sequenced data set (KSDS)

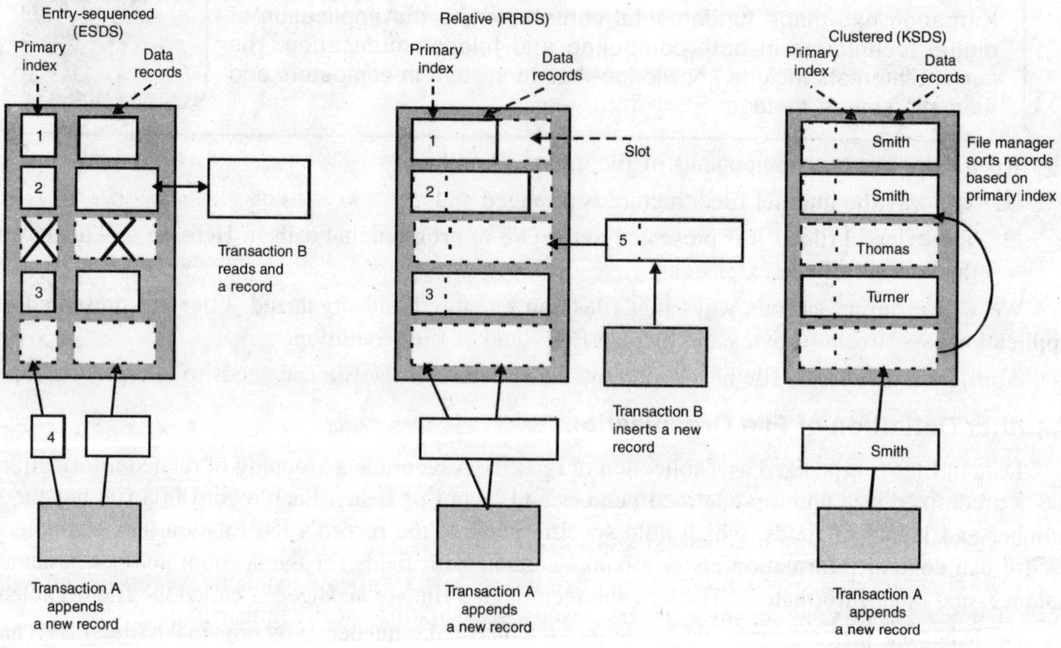

Fig. 14.4.

Entry-Sequenced File (ESDS)

Records in an entry-sequenced file are stored in the sequence in which they are written into the file. New records are always appended to the end of the file. When records are deleted, the disk space that they used is not automatically reclaimed or reused. You can reclaim it by reorganizing the file.

Entry-sequenced files are often used when the records are accessed in the chronological sequence in which they were written; for example, for log files and audit trail files. The primary index, which is stored separately from the file, lists records in the sequence in which they were added to the file.

Records can be fixed or variable length. When an existing record is updated, the updated record cannot be longer than its original length.

Relative File (RRDS)

A relative file is an array of fixed-length slots in which records can be stored. A record can be added into the first free slot from the beginning of the file, at the end of the file, or in a specified slot in the file. Records can be fixed or variable length up to the slot size that was predefined to the file manager. You can update or delete any record. Any slot that is freed when a record is deleted can be reused by the insertion of another record in the free slot. The primary index is physically part of the data in the record.

Clustered (KSDS)

A clustered file has each of its records identified by a key field in a predefined position within the record. The keys need not be unique. The records in a file are sorted automatically according to the value of their keys, to cluster records with the same and adjacent keys. This reduces the cost of searching for ranges of records.

Records that are in a clustered file do not have a numerical index as do entry-sequenced and relative files. You can base the primary index on any field or combination of fields. The records that are in a clustered file are sequenced as specified by the primary index. When you add or delete records, the file manager automatically updates the primary index and, if needed, moves records to maintain clustering. Disk space that is freed by deleting records is automatically reused.

Methods and Design Paradigm

It is a high-level design decision to specify a system of file organization for a computer software program or a computer system designed for a particular purpose. Performance is high on the list of priorities for this design process, depending on how the file is being used. The design of the file organization usually depends mainly on the system environment. For instance, factors such as whether the file is going to be used for transaction-oriented processes like OLTP or Data Warehousing, or whether the file is shared among various processes like those found in a typical distributed system or standalone. It must also be asked whether the file is on a network and used by a number of users and whether it may be accessed internally or remotely and how often it is accessed.

However, all things considered the most important considerations might be:

1. Rapid access to a record or a number of records which are related to each other.
2. The Adding, modification, or deletion of records.
3. Efficiency of storage and retrieval of records.
4. Redundancy, being the method of ensuring data integrity.

A file should be organized in such a way that the records are always available for processing with no delay. This should be done in line with the activity and volatility of the information.

14.8.3. Operations Performed on Files

(*i*) Creation of new fie
(*ii*) Opening a existing fie
(*iii*) Reading from a fie
(*iv*) Writing to a file
(*v*) Moving to a specific location in a file
(*vi*) Closing a file

14.9. TYPES OF FILE ORGANIZATION

Organizing a file depends on what kind of file it happens to be: a file in the simplest form can be a text file, (in other words a file which is composed of ASCII (American Standard Code for Information Interchange) text.) Files can also be created as binary or executable types (containing elements other than plain text.) Also, files are keyed with attributes which help determine their use by the host operating system.

14.9.1. Techniques of File Organization

The three techniques of file organization are:

1. Unordered or Heap (unordered)
2. Ordered or sequential Sorted
   (*i*) Sequential (SAM)
   (*ii*) Line Sequential (LSAM)
   (*iii*) Indexed Sequential (ISAM)
3. Hashed or Direct

In addition to the three techniques, there are four methods of organizing files. They are **sequential, line-sequential, indexed-sequential, inverted list** and **direct or hashed access** organization.

14.10. FILE OPERATIONS PERFORMED ON DIFFERENT TYPES OF FILES

1. **Unordered Or Heap File**

   Records are stored in the same order in which they are created

   (a) Insert operation
   - **Fast** – because the incoming record is written at the end of the last page of the file

   (b) Search (or update) operation
   - **Slow** – because linear search is performed on pages

   (c) Delete Operation
   - **Slow** – because the record to be deleted is first searched for
   - Deleting the record creates a **hole** in the page
   - Periodic file compacting work required to **reclaim the wasted space**

2. **Ordered or Sequential File**

   Records are sorted on the values of one or more fields

   (*a*) **Search** (or update) Operation
   - **Fast** –binary search is performed on sorted records

   (*b*) **Delete** Operation
   - **Fast** – because searching the record is fast
   - Periodic **file compacting** work is required

   (*c*) **Insert** Operation
   - **Poor** – because if we insert the new record in the correct position we need to **shift** all the subsequent records in the file

     Alternatively an **'overflow file'** is created
   - contains all the new records as a **heap**
   - Periodically overflow file is merged with the main file
   - If overflow file is created, search and delete operations for records in the overflow file have to be **linear**

3. Hash Files

   Hash file is an array of 'buckets'

   (*a*) Insert Operation is **Fast**
   - hash function computes the index of the bucket to which the record belongs
   - If that bucket is full you go to the next free one

   (*b*) Search Operation is **Fast**
   - hash function computes the index of the bucket
   - Performance may degrade if the record is not found in the bucket suggested by hash function

   c) Delete Operation is **Fast**
   - hash function computes the index of the bucket
   - function able to locate the record quickly

14.11. FUNDAMENTAL FILE ORGANIZATION TECHNIQUES

1. Sequential file organization
2. Random file organization
3. Indexed file organization
4. Multi-key file organization

14.11.1. Sequential File Organization

In sequential file organization, records are written consecutively when a file is created and must be accessed consecutively when file is used.

A sequential file contains records organized in the order they were entered. The order of the records is fixed. The records are stored and sorted in physical, contiguous blocks within each block the records are in sequence.

Records in these files can only be read or written sequentially.

Once stored in the file, the record cannot be made shorter, or longer, or deleted. However, the record can be updated if the length does not change. (This is done by replacing the records by creating a new file.) New records will always appear at the end of the file.

If the **order of the records** in a file is not important, **sequential organization** will suffice, no matter how many records you may have. Sequential output is also useful for report printing or sequential reads which some programs prefer to do.

A sequential file may be stored on SASD or DASD. The time required to access two records on SASD is a function of space between their locations on the device whereas time required to access two records on DASD is not necessarily dependent on their relative locations.

Insertion/deletion of Records in Sequential Files

Records are generally added at the end of the file. When a record is deleted, it is marked deleted in the file by tagging it. But physically the record is still present in the file as a dummy record. If there are large number of deletions, file will contains many dummy records and lot of storage and processing time will be wasted. Records are removed physically from a file by creating a new file & by copying only active (undeleted) records. This operation takes a long time for large files, so it must be done only when necessary.

Advantages of Sequential files

1. Simple
2. can be stored on inexpensive media like tapes
3. suitable for batch processing

Limitation

1. Not a very good response time i.e. not suitable for real time applications
2. When number of keys is more than one, this technique is not adequate.

Performance can be determined using three factors :

1. **Block Size** – It should be as large as possible
2. **File Length** – performance of this technique is inversely proportional to file length
3. **Key selection** – key should be selected in such a way that it is the field on which searches are made most often.

14.11.2. Random File Organization

In this technique, records are stored at random locations on the disk. Here individual records are accessed directly without accessing are previous records.

DASD supports this technique but SASD does not. In this technique, the records key value is transformed into an address using mapping function 'f'. The record is then stored at that address i.e.

```
f (key value) = address
```

Randomization can be achieved by:

(*a*) Direct addressing

(*b*) Directory look-up

(*c*) Hashing.

14.11.2.1. Direct Addressing

In this technique, if we have equal sized records, the available disk space is divided into nodes large enough to hold a record. The numeric value of the key is used to determine the node which will hold the particular record; it means mapping function is –

```
Key value = address
```

For example – if primary key is student roll number, then record for student number = 100 will be stored in node 100; this mapping function is called absolute addressing.

Limitation of Direct Addressing

(*a*) It is not always possible to supply appropriate type of key values.

(*b*) Absolute addressing is device dependent if storage device holding the file is changed, the key value will also be required to change.

(*c*) Leads to costage of memory

14.11.2.2. Directory look up

It keeps a table or directory of key values & address pair. To retrieve a record on a random file the table is searched for key value & address is used to access the record itself.

Example: Directory table

Key	Address		Random File
5	4	1	25
10	2	2	10
15	5	3	20
20	3	4	5
25	1	5	15

Directory tree (BST form)

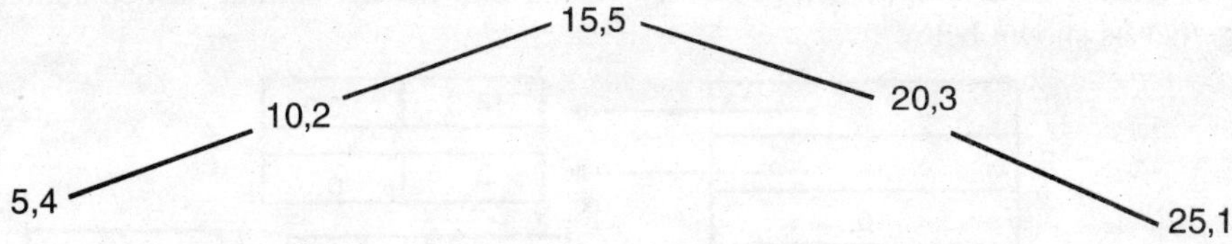

Fig. 14.5.

Limitation of Directory look-up

If a random file is large, directory may also be large. So lots of additional space is required to hold the directory.

14.11.2.3. Hashing

Hashing is a technique in which record's key value is transformed into an address using a mapping function H called as hash function.

H (key value) = hash address

Or

Fig. 14.6.

Types of Hash Function

(1) Division Remainder method

(2) Mid square method

(3) Folding with reversing

(4) Folding with out addressing or reversing

AS DISCUSSED IN CHAPTER 13

Collision Resolution

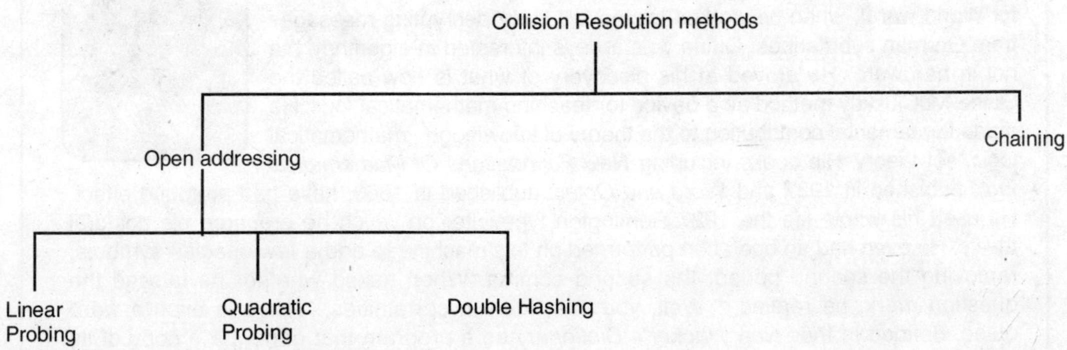

Fig. 14.7.

Chaining

Involves a table to be maintained in memory using link list i.e. similar hashed address records are linked together as shown below

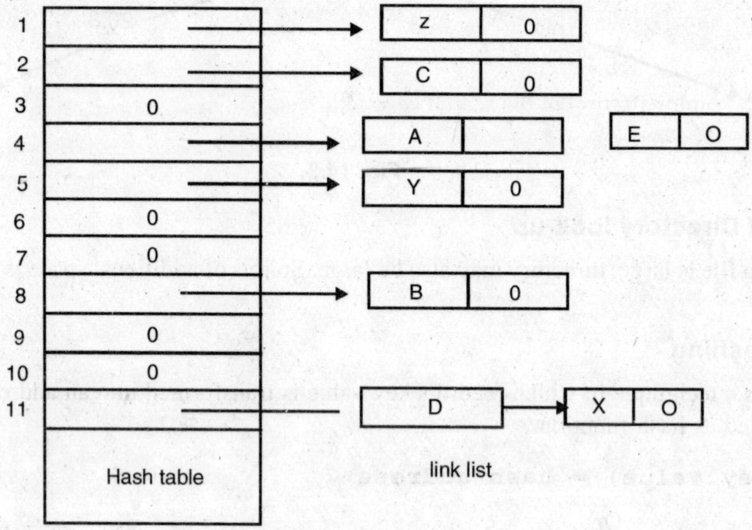

Fig. 14.8.

14.11.3. Indexed File Organization (IFO)

An indexed file organization allows both types of accesses i.e. sequential access & direct access. DASD support IFO but SASD doesn't.

Deuse index: They contain an entry for every record

Non deuse index : Contains an entry for a group of records

DO YOU KNOW

WILARD VAN ORMAN QUINE(1908-2000) Williard quine, born in Akron, Ohio, attended Oberlin college and later Harvard University, where he received his Ph.D in philosophy in 1932. He became a junior fellow at Harvard in 1933 and was appointed to a position on the faculty there in 1936. He remained at Harvard his entire professional life, except for World war II, when he worked for the US Navy decrypting messages from German submarines. Quine was always interested in algorithm, but not in hardware. He arrived at his discovery of what is now called the Quine-McCluskey method as a device for teaching mathematical logic.He made fundamental contribution to the theory of knowledge , mathematical logic, set theory. His books including *New Foundations Of Mathematical logic* pubished in 1937 and *Word and Object* published in 1960, have had profound effect. He used his whole life the 1927 Remington typewriter on which he prepared his doctoral thesis. He even had an operation performed on this machine to add a few specias symbols, removing the second period, the second comma. When asked whether he missed the question mark, he replied, " Well, you see, Idea in certainities". There is even a word quine, defined in the *New Hacker's Dictionary* as a program that generate a copy of its own source code as its complete output. Producing the shortest possible quine in a given programming language is a apopular puzzle for hacker.

Key searches are improved by this system too. The single-level indexing structure is the simplest one where a file, whose records are pairs, contains a key pointer. This pointer is the position in the data file of the record with the given key. A subset of the records, which are evenly spaced along the data file, is indexed, in order to mark intervals of data records.

This is how a key search is performed: the search key is compared with the index keys to find the highest index key coming in front of the search key, while a linear search is performed from the record that the index key points to, until the search key is matched or until the record pointed to by the next index entry is reached. Regardless of double file access (index + data) required by this sort of search, the access time reduction is significant compared with sequential file searches.

Let's examine, for sake of example, a simple linear search on a 1,000 record **sequentially organized** file. An average of 500 key comparisons is needed (and this assumes the search keys are uniformly distributed among the data keys). However, using an index evenly spaced with 100 entries, the total number of comparisons is reduced to 50 in the index file plus 50 in the data file: a five to one reduction in the operations count!

Hierarchical extension of this scheme is possible since an index is a sequential file in itself, capable of indexing in turn by another second-level index, and so forth and so on. And the exploit of the hierarchical decomposition of the searches more and more, to decrease the access time will pay increasing dividends in the reduction of processing time. There is however a point when this advantage starts to be reduced by the increased cost of storage and this in turn will increase the index access time.

Hardware for Index-Sequential Organization is usually Disk-based, rather than tape. Records are physically ordered by primary key. And the index gives the physical location of each record. Records can be accessed sequentially or directly, via the index. The index is stored in a file and read into memory at the point when the file is opened. Also, indexes must be maintained.

Life sequential organization the data is stored in physical contiguous box. How ever the difference is in the use of indexes. There are three areas in the disc storage:

- **Primary Area :** Contains file records stored by key or ID numbers.
- **Overflow Area :** Contains records area that cannot be placed in primary area.
- **Index Area :** It contains keys of records and there locations on the disc.

Cylindrical surface indexing (ISAM)

It is also called as indexed sequential access method.

It assumes that file is sorted by key & records are stored in sequential order onto disk. Now assuming that file size is very large & hence it is stored on more than one disk i.e. on'd' disks & on each disks, there are cylinders, surfaces & tracks whose number may be same on all disks or different on different disks.

Now accessing a particular record of the file requires the disk to be identified. Once disk is identified as ith disk that the cylinder on the disk that may have the record is searched for say the cylinder is jth, then surface is searched say it is kth & finally jth track.

Let us assume an Example

Cylinder 1 has indexes ranging from 1 to 400

CYLINDER 1					
	50	60	70	80	Track 1
	90	110	120	130	Track 2
	140	160	168	180	Track 3
	190	210	220	230	Track 4
	270	300	350	400	Track 5

Cylinder 2 has indexes ranging from 401 to 800.

CYLINDER 2					
	450	460	470	480	Track 1
	490	510	520	530	Track 2
	540	560	568	580	Track 3
	590	610	620	630	Track 4
	670	700	750	800	Track 5

To search a record using cylinder surface indexing, we have the following steps viz:

1. First we have to identify the cylinder number, in which the record resides
2. After identifying the cylinder number, we have to identify the track number in the identified cylinder.

Cylinder Index : It is maintained for all the cylinders as shown below.

Cylinder number with maximum index.

1	400	2	800

SIMILARLY **TRACK INDEX** is maintained for each cylinder individually as shown below

TRACK INDEX FOR CYLINDER 1

1	80	2	130	3	180	4	230	5	400

TRACK INDEX FOR CYLINDER 2

1	480	2	530	3	580	4	630	5	800

Suppose now if we want to search a record with index 570 then the records is in cylinder number 2 with track number 3

Hashed indexing

Explained before in chapter 11

Tree indexing

M-way search trees : They are generalized version of BST. The goal of m-way search tree is to minimize the accesses while retrieving a key from a file.

Definition : An m-way search tree T may be an empty tree. But if T is non empty, it satisfies the following property.

(1) For some integer m, known as order of tree each node is of degree which can reach a maximum of m.

(2) If a node has k Childs where $K \le m$, then node can have only (K-1) keys $K_1, K_2, \ldots K_{K-1}$ Contained in node

(3) For a node $A_0, (k_1, A_1), (k_2, A_2), \ldots (K_{m-1}, A_{m-1})$ all (2) key values in the Subtree pointed to by A_i are less than key K_{i+1}, $0 \le i \le m-2$, and all key values in the subtree pointed to by A_{m-1} are greater than k_{m-1}.

(4) Each of subtrees A_i, are also m-way search tree

Example : 5-way search tree

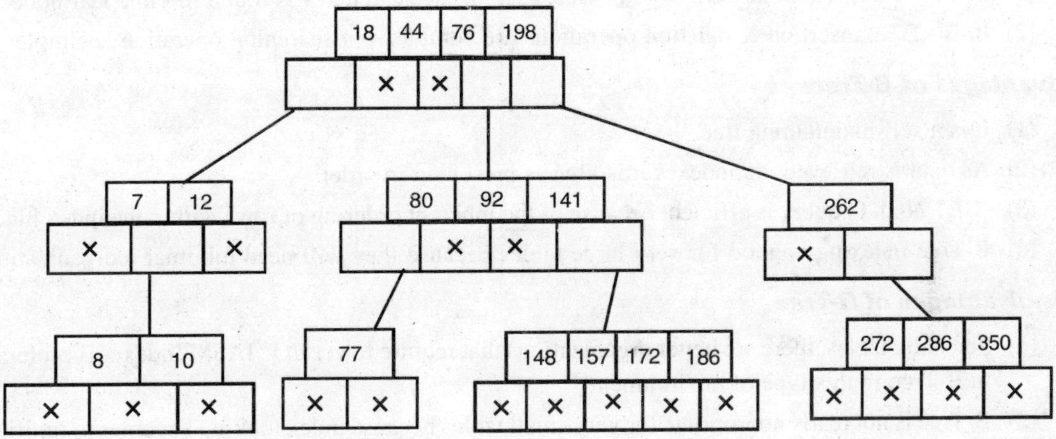

Fig. 14.9.

B-trees

M-way search tree have the advantage of minimizing file access due to their restricted height. However, it is essential that height of tree be kept as low as possible & therefore there arises a need to maintain balanced m-way search tree such a balanced m-way search tree is defined as a B-Tree.

Definition : A B-tree of order m, if non-empty is a m-way search tree in which

(*i*) The root has at least two child nodes and at most m child nodes

(*ii*) The internal nodes except the root have at least [m/2] child nodes and at most m child nodes

(*iii*) The number of keys in each internal nodes is one less that number of child nodes & these keys partition the keys in the subtrees of nodes in a manner similar to that of m-way search tree.

(*iv*) All leaf nodes are on same level.

A B-tree of order 3 is referred to as 2-3 trees since the internal nodes are of degree 2 or 3 only

Example : B-Tree of order 5

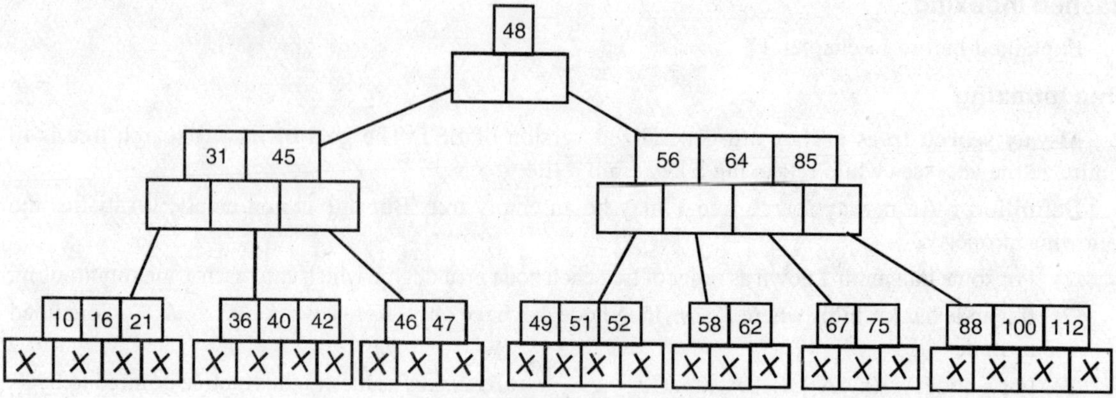

Fig. 14.10.

There are several variations of B-Tree, including B^+-Tree and B*-Tree. The B^+-Tree indices are similar to B-Tree indices but the main differences are:

(1) In a B^+-Tree, search key are stored twice, each of the search key is found in same leaf node.

(2) In B^+-Tree, insertion & deletion operations are complex but searching operation is simple.

Advantages of B-Trees

(1) It is a self-maintaining tree

(2) As data is retrieved via indexes it is always presented in order.

(3) 'GET NEXT' query is efficient because of the inherent ordering of rows within the index blocks.

(4) B-Tree indexing is good for very large tables because they will need minimal reorganization.

Disadvantages of B-Tree

(1) For static tables, there are better organizations that require fewer I/O. IASM Indexes is preferable to B-Tree in this type of environment.

(2) B-Tree is not really appropriate for very small tables because index look-up becomes a significant part of overall access time.

(3) The index can use considerable disk space, especially in products which allow different users to create separate indexes on same table / column combinations.

Trie indexing

A trie is a tree of degree m>=2 in which the branching at any level is determined by only the portion of the key instead of the entire key values.

The trie indexing contains two types of nodes viz:

1. First type is called a branch node
2. & second type is called information node.

Example : Key value: 088, 100, 218, 910, 999

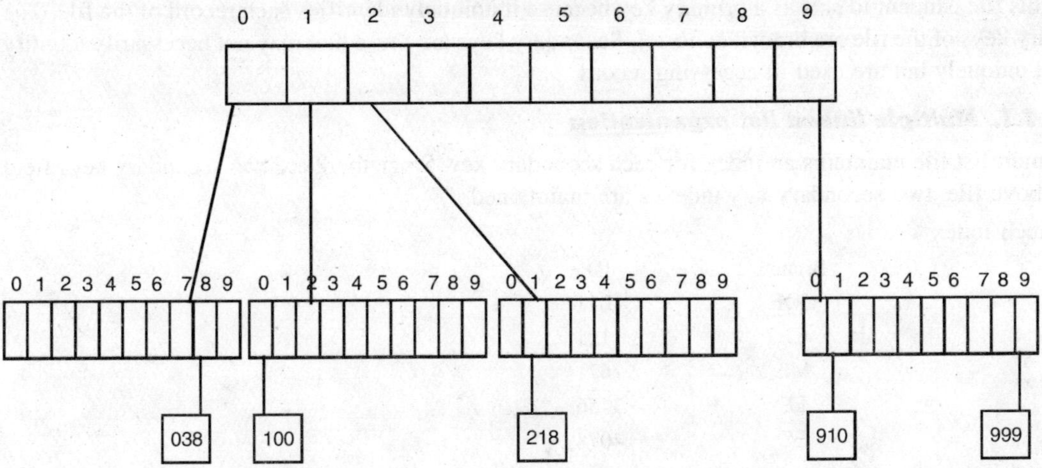

Fig. 14.11.

14.11.4. Multi key file organization

In all the file organization, we have studied till now; allow records to be accessed by one key only. But in reality, records in a file may be required to be accessed via more than one key.

To access the records using more than one key, we have many solutions viz

(*i*) One solution could be to duplicate data across files & access each file by a different key. But the major problem is storage space requirement.

(*ii*) Another solution to this problem is to have one data file with multiple across paths. This approach is called multi key file organization.

Two types of multi key file organization are :

(*a*) Multiple linked list or multi list

(*b*) Inverted files.

To explain both techniques consider the following **example:**

Student File

Record id	ID	Name	Branch	Sem	Hostel	Room no
1	1011	Rajan	COE	1	A	100
2	1121	Kamal	CE	2	B	25
3	1671	Shikha	ME	1	C	110
4	2981	Amit	COE	4	A	60
5	3456	Meena	EE	2	D	43
6	4078	Rajesh	EC	3	C	50
7	4567	Rama	EE	3	B	45
8	5025	Ajay	ME	5	A	80
9	5617	Nikhil	CE	4	C	65
10	6231	Ravi	COE	7	D	40
11	6710	Pran	EE	1	B	35
12	7002	Illa	ME	2	C	10
13	7108	Ishu	EC	5	B	70
14	8096	Ashu	COE	7	A	5
15	8300	Prayan	CE	1	D	20
16	8001	rohit	COE	7	C	55

In this file, student id acts as a primary key because it uniquely identifies each record of the file. The secondary keys of the file are branch & hostel. Secondary keys are those that may not necessarily identify a record uniquely but are used in accessing record.

14.11.4.1. Multiple linked list organization

A multi list file maintains an index for each secondary key. Since there are two secondary keys field in the above file, two secondary key indexes are maintained.

Branch index

Branch	ID
COE	1011
CE	1121
ME	1671
EE	3456
EC	4078

Hostel index

Hostel	ID
A	1011
B	1211
C	1671
D	3456

Now records which have equivalent values for a given secondary key are linked together to form a link list.

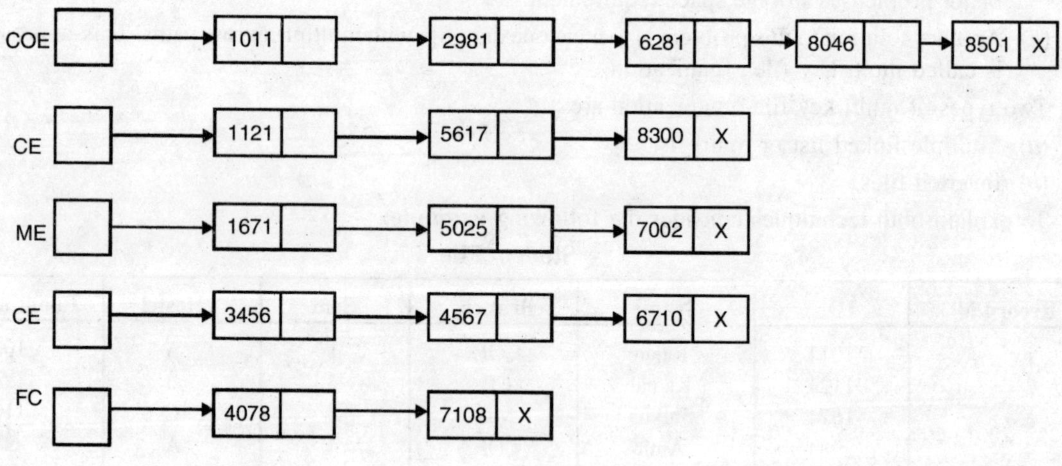

Fig. 14.12.

Similarly hostel list is maintained.

14.11.4.2. Inverted files

Conceptually, inverted files are similar to multi lists. The difference is that in multi list, the records with same keys values are linked together with link info being kept in individual records, while in the case of inverted files, this link info is kept in index itself. Hence inversion does not affect data files.

In file organization, this is a file that is indexed on many of the attributes of the data itself. The inverted list method has a single index for each key type. The records are not necessarily stored in a sequence. They are placed in the are data storage area, but indexes are updated for the record keys and location.

Here's an example, in a company file, an index could be maintained for all products, another one might be maintained for product types. Thus, it is faster to search the indexes than every record. These types of file are also known as **"inverted indexes."** Nevertheless, **inverted list files** use more media space and the storage devices get full quickly with this type of organization. The benefits are apparent immediately because searching is fast. However, updating is much slower.

Content-based queries in text retrieval systems use **inverted indexes** as their preferred mechanism. Data items in these systems are usually stored compressed which would normally slow the retrieval process, but the compression algorithm will be chosen to support this technique.

When querying a file there are certain circumstances when the query is designed to be **modal** which means that rules are set which require that different information be held in the index. Here's an example of this modality: when phrase querying is undertaken, the particular algorithm requires that offsets to word classifications are held in addition to document numbers.

When a record is added or deleted to/from the data file, we must update the inversion index & maintain it in sorted order.

Example : inverted list for student's hostel with variable length entries.

Hostel	Primary key
A	1011, 2981, 5025, 8096
B	1121, 4567, 6710, 7108
C	1671, 4078, 5617, 7002, 8501
D	3456, 6231, 8300

An inverted list can be more efficiently stored as a tree instead of a table.

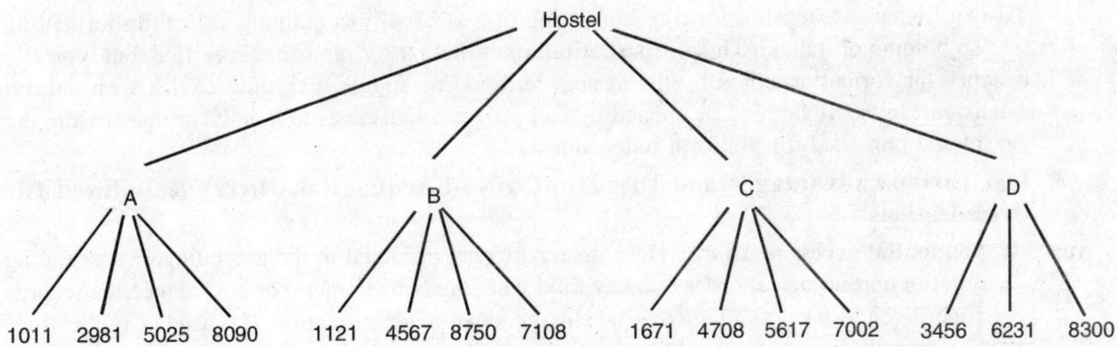

Fig. 14.13.

14.11.5. Cellular Partition

In order to reduce file search time the storage media may be divided into cells. A cell may be an entire disk space or it may be a cylinder lists are localized to lie within a cell. Thus if we had a multilist organization in which list for KEY1 = PROG list included records on several different cylinders. When we would break this list into several smaller lists, where each PROG list included only those records, which are in same cylinder. The index entry for PROG will now contains several entries of the type (addr, length), where 'addr' is pointer to start of list of record with KEY1 = PROG & 'length' is the number of records on this list. By implementing this, all records on same cell may be accessed without moving read/write head.

In case, a cell is a disk pack, then using cellular partition, it is possible to search different cells in parallel.

In real situation, a cellular partition resembles dividing a hard disk into disk drives like C: D: E: etc. a file searched inside a particular drive.

SOLVED PROBLEMS

1. Explain the indexing structures?

Ans. Indexes structure speed up the retrieval of records under certain search conditions. Indexes also called as secondary access paths and it do not affect the placement of records. They provide fast access for searches based on the indexing field. Some types of indexes work only in conjunction with a certain file organization. Single level ordered indexes allow us to search for a record by searching the index file using binary search. The index is typically defined on a single field of the file called the indexing field.

There are several kinds of ordered index structures:

- A primary index
- clustering index
- secondary index

2. What is Text File? How do text files differ from other type of files?

Ans. A text file also known as flatfile is a kind of computer file that is structured as a sequence of lines. A text file exists within a computer file system. The end of a text file is often denoted by placing one or more special characters, known as an end-of-file marker, after the last line in a text file. "Text file" refers to a type of container, while plain text refers to a type of content. Text files can contain plain text.

.txt is a filename extension for files consisting of text usually containing very little formatting (ex: no bolding or italics). The precise definition of the .txt format is not specified, but typically matches the format accepted by the system terminal or simple text editor. Files with the .txt extension can easily be read or opened by any program that reads text and, for that reason, are considered universal (or platform independent).

3. List various advantages and Disadvantages of Sequential, Direct & Indexed file Organizations?

Ans. 1. **Sequential access method :** Here the records are arranged in the ascending or descending order or chronological order of a key field which may be numeric or both. Since the records are ordered by a key field, there is no storage location identification. It is used in applications like payroll management where the file is to be processed in entirety, i.e. each record is processed. Here, to have an access to a particular record, each record must be examined until we get the desired record. Sequential files are normally created and stored on magnetic tape using batch processing method.

Advantages :

- Simple to understand.
- Easy to maintain and organize
- Loading a record requires only the record key.
- Relatively inexpensive I/O media and devices can be used.
- Easy to reconstruct the files.
- The proportion of file records to be processed is high.

Disadvantages :

- Entire file must be processed, to get specific information.
- Very low activity rate stored.
- Transactions must be stored and placed in sequence prior to processing.
- Data redundancy is high, as same data can be stored at different places with different keys.
- Impossible to handle random enquiries.

2. **Direct access files organization :** (Random or relative organization). Files in his type are stored in direct access storage devices such as magnetic disk, using an identifying key. The identifying key relates to tits actual storage position in the file. The computer can directly locate the key to find the desired record without having to search through any other record first. Here the records are stored randomly, hence the name random file. It uses online system where the response and updation are fast.

Advantages :

- Records can be immediately accessed for updation.
- Several files can be simultaneously updated during transaction processing.
- Transaction need not be sorted.
- Existing records can be amended or modified.
- Very easy to handle random enquiries.
- Most suitable for interactive online applications.

Disadvantages :

- Data may be accidentally erased or over written unless special precautions are taken.
- Risk of loss of accuracy and breach of security. Special backup and reconstruction procedures must be established.
- Less efficient use of storage space.
- Expensive hardware and software are required.
- High complexity in programming.
- File updation is more difficult when compared to that of sequential method.

3. **Indexed sequential access organization**: Here the records are stored sequentially on a direct access device i.e. magnetic disk and the data is accessible randomly and sequentially. It covers the positive aspects of both sequential and direct access files. The type of file organization is suitable for both batch processing and online processing. Here, the records are organized in sequence foe efficient processing of large batch jobs but an index is also used to speed up access to the records. Indexing permit access to selected records without searching thc entire file.

Advantages :

- Permits efficient and economic use of sequential processing technique when the activity rate is high.
- Permits quick access to records, in a relatively efficient way when this activity is a fraction of the work load.

Disadvantages :

- Slow retrieval, when compared to other methods.
- Does not use the storage space efficiently.
- Hardware and software used are relatively expensive.

SUMMARY

1. **File** - A file is a collection of logically related records.
2. **Record** – A record is a collection of logically related fields of information about a particular entity.
3. **Field** – It is a singe elementary unit of information representing an attribute of an entity.
4. **Key values** - The values in the primary key field are called key field values
5. **Query** – A combination of key values specified for retrieval will be termed as query
6. A **batch file** is a text file containing a series of commands intended to be executed by the command interpreter.
7. A **text file** (sometimes spelled "textfile": an old alternate name is "flatfile") is a kind of computer file that is structured as a sequence of lines. A text file exists within a computer file system
8. A **binary file** (extension **.bin** unless otherwise specified) is a computer file which may contain any type of data, encoded in binary form for computer storage and processing purposes; for example, computer document files containing formatted text
9. File organization is the methodology which is applied to structured computer files. Files contain computer records which can be documents or information which is stored in a certain way for later retrieval.
10. In sequential file organization, records are written consecutively when a file is created and must be accessed consecutively when file is used.
11. In Random file organization, records are stored at random locations on the disk. Here individual records are accessed directly without accessing are previous records.
12. An indexed file organization allows both types of accesses i.e. sequential access & direct access. DASD support IFO but SASD doesn't.
13. M-way search tree have the advantage of minimizing file access due to their restricted height. However, it is essential that height of tree be kept as low as possible & therefore there arises a need to maintain balanced m-way search tree such a balanced m-way search tree is defined as a B-Tree.
14. A trie is a tree of degree m>=2 in which the branching at any level is determined by only the portion of the key instead of the entire key values.
15. A multi list file maintains an index for each secondary key. Since there are two secondary keys field in the above file, two secondary key indexes are maintained.
16. Conceptually, inverted files are similar to multi lists. The difference is that in multi list, the records with same keys values are linked together with link info being kept in individual records, while in the case of inverted files, this link info is kept in index itself. Hence inversion does not affect data files.

❑❑❑

APPENDICES

A. Some Selected Problems
B. List of Practicals
C. Great Founders in the Field of Computer Engineering

APPENDIX

A

SOME SELECTED PROBLEMS

1. **What is data structure?**

Ans. A data structure is a way of organizing data that considers not only the items stored, but also their relationship to each other. Advance knowledge about the relationship between data items allows designing of efficient algorithms for the manipulation of data.

2. **List out the areas in which data structures are applied extensively?**

Ans. Compiler Design,Operating System, Database Management System, Statistical analysis package, Numerical Analysis, Graphics, Artificial Intelligence, Simulation

3. **What are the major data structures used in the following areas : RDBMS, Network**

Ans. data model & Hierarchical data model.

RDBMS	–	Array (i.e. Array of structures)
Network data model	–	Graph
Hierarchical data model	–	Trees

4. **If you are using C language to implement the heterogeneous linked list, what pointer type will you use?**

Ans. The heterogeneous linked list contains different data types in its nodes and we need a link, pointer to connect them. It is not possible to use ordinary pointers for this. So we go for void pointer. Void pointer is capable of storing pointer to any type as it is a. generic pointer type

5. **Minimum number of queues needed to implement the priority queue?**

Ans. Two. One queue is used for actual storing of data and another for storing priorities.

6. **What is the data structures used to perform recursion ?**

Ans. Stack. Because of its LIFO (Last In First Out) property it remembers its 'caller' so knows whom to return when the function has to return. Recursion makes use of system stack for storing the return addresses of the function calls.

Every recursive function has its equivalent iterative (non-recursive) function. Even when such equivalent iterative procedures are written, explicit stack is to be used.

7. **What are the notations used in Evaluation of Arithmetic Expressions using prefix and postfix forms?**

Ans. Polish and Reverse Polish notations.

8. Convert the expression ((A + B) * C – (D – E) ^ (F + G)) to equivalent Prefix and Postfix notations.

Ans. Prefix Notation:

^ - * +ABC - DE + FG

Postfix Notation:

AB + C * DE - - FG + ^

9. Sorting is not possible by using which of the following methods?

(a) Insertion

(b) Selection

(c) Exchange

(d) Deletion

Ans. Using insertion we can perform insertion sort, using selection we can perform selection sort, using exchange we can perform the bubble sort (and other similar sorting methods). But no sorting method can be done just using deletion.

10. A binary tree with 20 nodes has null branches?

Ans. Let us take a tree with 5 nodes (n=5)

It will have only 6 (ie,5+1) null branches. In general,

A binary tree with n nodes has exactly n+1 null nodes.

11. What are the methods available in storing sequential files ?

Ans. Straight merging,

Natural merging,

Polyphase sort,

Distribution of Initial runs.

12. How many different trees are possible with 10 nodes ?

For example, consider a tree with 3 nodes (n=3), it will have the maximum combination of 5 different (i.e., 2^3 - 3 = 5) trees.

In general:

If there are n nodes, there exist 2^n – n different trees.

13. List out few of the Application of tree data-structure?

Ans. The manipulation of Arithmetic expression,

Symbol Table construction, Syntax analysis.

14. List out few of the applications that make use of Multilinked Structures?

Ans. Sparse matrix,

Index generation.

15. In tree construction which is the suitable efficient data structure?

(a) Array (b) Linked list (c) Stack (d) Queue (e) none

Ans. (b) Linked list

16. What is the type of the algorithm used in solving the 8 Queens problem?

Ans. Backtracking

17. In an AVL tree, at what condition the balancing is to be done?

Ans. If the 'pivotal value' (or the 'Height factor') is greater than 1 or less than –1.

18. What is the bucket size, when the overlapping and collision occur at same time?

Ans. One. If there is only one entry possible in the bucket, when the collision occurs, there is no way to accommodate the colliding value. This results in the overlapping of values.

19. Traverse the given tree using Inorder, Preorder and Postorder traversals.

Ans. Given tree:

Inorder : D H B E A F C I G J

Preorder: A B D H E C F G I J

Postorder: H D E B F I J G C A

20. There are 8, 15, 13, 14 nodes were there in 4 different trees. Which of them could have formed a full binary tree?

Ans. In general:

There are $2^n - 1$ nodes in a full binary tree.

By the method of elimination:

Full binary trees contain odd number of nodes. So there cannot be full binary trees with 8 or 14 nodes, so rejected. With 13 nodes you can form a complete binary tree but not a full binary tree. So the correct answer is 15.

Note: Full and Complete binary trees are different. All full binary trees are complete binary trees but not vice versa.

21. In the given binary tree, using array you can store the node 4 at which location?

Ans. At location 6

Root	LC1	RC1	LC2	RC2	LC3	RC3	LC4	RC4
1	2	3	-	-	4	-	-	5

where LCn means Left Child of node n and RCn means Right Child of node n

22. Sort the given values using Quick Sort?

65 70 75 80 85 60 55 50 45

Ans. Sorting takes place from the pivot value, which is the first value of the given elements, this is marked bold. The values at the left pointer and right pointer are indicated using L and R respectively.

65 70^L 75 80 85 60 55 50 45^R

Since pivot is not yet changed the same process is continued after interchanging the values at ^{L} and ^{R} positions

65 45 75^L 80 85 60 55 50^R 70

65 45 50 80^L 85 60 55^R 75 70

65 45 50 55 85 65 45 50 55

When the L and R pointers cross each other the pivot value is interchanged with the value at right pointer. If the pivot is changed it means that the pivot has occupied its original position in the sorted order (shown in bold italics) and hence two different arrays are formed, one from start of the original array to the pivot position-1 and the other from pivot position+1 to end.

60^L 45 50 55^R **65** **85^L** 80 75 70^R

55^L 45 50^R **60** **65** **70^R** 80^L 75 **85**

50^L 45^R 55 **60** **65** **70** 80^L 75^R **85**

In the next pass we get the sorted form of the array.

45 50 55 60 65 70 75 80 85

23. For the given graph, draw the DFS and BFS?

Ans. BFS: A X G H P E M Y J

DFS: A X H P E Y M J G

24. Classify the Hashing Functions based on the various methods by which the key value is found.

Ans. Direct method,

Subtraction method,

Modulo-Division method,

Digit-Extraction method,

Mid-Square method,

Folding method,

Pseudo-random method.

25. What are the types of Collision Resolution Techniques and the methods used in each of the type?

Ans. Open addressing (closed hashing),

The methods used include:

Overflow block,

Closed addressing (open hashing)

The methods used include:

Linked list,

Binary tree…

26. In RDBMS, what is the efficient data structure used in the internal storage representation?

Ans. B+ tree. Because in B+ tree, all the data is stored only in leaf nodes, that makes searching easier. This corresponds to the records that shall be stored in leaf nodes.

27. Of the following tree structure, which is, efficient considering space and time complexities?

(*a*) Incomplete Binary Tree

(*b*) Complete Binary Tree

(*c*) Full Binary Tree

Ans. (*b*) Complete Binary Tree.

By the method of elimination:

Full binary tree loses its nature when operations of insertions and deletions are done. For incomplete binary trees, extra storage is required and overhead of NULL node checking takes place. So complete binary tree is the better one since the property of complete binary tree is maintained even after operations like additions and deletions are done on it.

28. What is a spanning Tree?

Ans. A spanning tree is a tree associated with a network. All the nodes of the graph appear on the tree once. A minimum spanning tree is a spanning tree organized so that the total edge weight between nodes is minimized.

29. Does the minimum spanning tree of a graph give the shortest distance between any 2 specified nodes?

Ans. No.

Minimal spanning tree assures that the total weight of the tree is kept at its minimum. But it doesn't mean that the distance between any two nodes involved in the minimum-spanning tree is minimum.

30. Which is the simplest file structure?

(a) Sequential

(b) Indexed

(c) Random

Ans. (a) Sequential

31. Whether Linked List is linear or Non-linear data structure?

Ans. According to Access strategies Linked list is a linear one.

According to Storage Linked List is a Non-linear one.

32. Draw a binary Tree for the expression:

$$A * B - (C + D) * (P / Q)$$

33. What is space time tradeoff?

Ans. It refers to a algorithm solution of data processing in which one is allowed to reduce the time of execution by increment in space or vice versa.

Complexity: complexity of an algorithm is a function f(x) which measures the total time & space used by an algorithm to input size x.

Now, as time is given more importance compared to space complexity.

Therefore, complexity is nearly equal to time complexity.

Now the complexity comprises of compile & run time and runtime is more than the compile time of an algorithm hence we can neglect the space complexity of the algorithm &

Total complexity = runtime complexity

Which in tern depends on total number of comparisons in an algorithm?

34. What is Data Structure?

Ans. Data structure is the study of logical and mathematical model of particular organization of data.

Study of Data Structure involves the following steps:

(1) Logical and mathematical description of structure

(2) Implementation of structure on computer

(3) Quantitative analysis of structure, which involves determining the amount of memory needed to store structure and time required to process the structure.

35. List the disadvantages of using an array?

Ans. (1) Relatively expensive to insert and delete elements in array.

(2) Problem in memory management.

36. List out the advantages of using a linked list?

Ans. (1) It is easier to insert & delete elements.

(2) Uses pointers (*i.e.* link) which reduces the total complexity.

37. List out the advantages of BST (Binary Search Tree)?

Ans. (1) BST enables one to easily insert & delete elements.

(2) BST enables one to search for and find an element with a running time complexity.

$F(x) = O(\log_2 x)$ but is expensive.

38. (*a*) What is the difference between automatic & static storage of variables ? Explain through examples.

(*b*) What will be the output of following C program ?

```
#include < stdio.h>
main ( )
```

```
{       int a, b=0;
        satic int c [1 0] = {1,2,3,4,5,6,7,8,9,0};
        for (a=0; a< 10 ; ++a)
        if ((c[a]%2)== 0)  b+c[a] ;
        printf ("%d", b);
  }
```

39. (*a*) Write a program to calculate *n* terms of a Fibonacci series using Recursive function.

(*b*) Write a program to arrange n integer numbers in decreasing order.

40. (*a*) How an array is passed to a function ? How it differs from call by value ? Explain through an example.

(*b*) Write a function (like strcpy ( ) library function) in C which copies a string to another string.

41. (*a*) Create a structure to specify data of customers in a bank. The data to be stored is: Account no., Name, Balance in account. Assuming maximum of 200 customers in the bank. Write a program to print the Account number & Name of customers with Balance below Rs.100.

(*b*) What is the difference between a structure & a union ?

(c) How a structure variable can be initialized ? Explain through an example.

42. (*a*) How the pointers are passed to function ? Explain through an example.

(*b*) What will be the output of the following program :

```
# include< stdio.h>
int funct 1(int n);
main ( )
{       int n = 10;
print f ("%d", funct1(n));}
int funct 1(int n)
{       if (n>0) return (n+funct1(n-2));
}
```

43. (*a*) What is an array ? How does it differ from ordinary variable ?

(*b*) WAP for Binary search.

(c) Explain difference between passing an array to a function and passing a single valued data to a function ?

44. (*a*) Explain the difference between function definition and function declaration ?

(*b*) Differentiate between call by reference and call by value.

(*c*) Write a function program to find smallest number in a matrix and location of row and column in which element is available ?

45. (*a*) Explain the various storage classes in C ?

(*b*) What is a pointer ? How is a pointer initialized ?

(*c*) Write a program that uses recursive function to find factorial of number.

46. (*a*) Explain purpose of following functions with examples.

1. strcpy ()
2. strcat ()
3. strlen ()

(*b*) Write a program to reverse the character of a string using pointers.

(*c*) Explain what will be the output of following programme :

```
#include < stdio.h>

main ()
{
int a, b, * ptr ;
a =9;
ptr= & a;
b= * ptr ;
a++;
}
```

What is value of a, b, * ptr ?

47. (*a*) Write a program to multiply two 3*3 matrices.

(*b*) Explain the following functions

strlen(), strcpy(), strcmp(), strcat(), getch()

48. (*a*) Explain the difference between for, do-while and while loops.

(*b*) Write a program to cheek whether a given string is a palindrome or not.

49. (*a*) Write a program to draw a pyramid.

(*b*) Explain the following

(*i*) int * p;

(*ii*) int ** p;

(*iii*) int * p();

(*iv*) a =& b;

(*v*) Arithmetic operations on pointers.

50. (*a*) Differentiate between pass-by-reference and pass-by- value in function parameter passing in C.

(*b*) Differentiate between UNION and STRUCTURE in C with syntax.

51. (*a*) Draw a flow chart to find whether the given number is prime or not.

(*b*) Write a program to calculate the sum of following series:

$$\frac{1}{1^2} + \frac{1}{2^2} + \frac{1}{3^2} + \dots \frac{1}{n^2}$$

52. (*a*) Write an algorithm to find the roots of a quadratic equation.

(*b*) Write a program that calculates & prints the sum & product of the numbers between 1 and 20 that are divisible by 2 or 3.

53. (*a*) What will be the output of the following program :

```
# include < stdio.h>
main ( )
    {  int i=0 , x=0 ;
    do{   if ( i%5= =0)
    {  x ++ ;
            print f (" %d " , x) :
    }
    ++ i ;
```

```
}
while (i < 20) ;
print f(‘ “\nx = % d” , x) ;
}
```

(*b*) Differentiate between break and continue statements. Write syntax of ‘for’ statement.

54. (*a*) What is recursion ? Write a program to calculate the n terms of Fibonacci series ?

(*b*) Draw a flow chart to find the given number is prime or not

(*c*) Explain the difference between passing an array to a functions ϑ passing a single valued data to a function ?

55. What will be the output of the following programs (if any) ?

(*a*)
```
#include <stdio .h>
int funcl (int count)
main ()
{ int a,count;
for (count = 1; count < = 5; + + count)
{ a=func1 (count);
print (“ % d”, a)
}
}
int funcl (int x)
{ int y;
y=x* x;
return(y) ;
}
```

(*b*)
```
# include < stdio.h >
main()
{ int i = 0 , x=0;
for (i=0; i <=5 ; i+ +)
{ switch(i.1)
{ case 0;
case (1); x+=1: break;
case (3): x+=2: break;
default: x+=3 ;
}
print (“%d” ,x);
}}
```

(*c*)
```
# include<stdio.h >
int func1 (int count);
main()
{ int a , count;
for (count = 1; count <=5; + + count)
{ printf (“%d “, a );
```

```
}}
func1 (int x)
{static int y =0 ;
 y+ = x;
return (x) : }
```

56. (*a*) Explain the various storage classes in C

(*b*) What is the difference between structure and union

(*c*) Write a program to read names, roll no's and total marks of 50 students in a class from key board generate a merit list.

57. (*a*) Write a function in 'C' Which copies one string to another string ?

(*b*) What is pointer ? How is a pointer initialized ?

(*c*) Write a program using pointers to need in a array of integers and print its elements in reverse order ?

58. (*a*) Write a program that reads a file & count number of characters, spaces tabs and new line characters present in the file ?

(*b*) Compare fread and fwrite functions with fscanf and fprintf functions .

(*c*) What is the purpose of library function feof ?

59. (*a*) What will be the output of the following program:

```
# include< stdio. h>
main ()
{ int i = 0, x=0 ;
 for (i=0;i< 10; + + i)
 {  if(i% 2 = i)
x+ = i;
else
x--;
print(" % d", x);
continue;:
 }
print("/nx=%d",x):
 }
```

(*b*) What is except sequence ? Explain its purpose

(*c*) What is the scope of preprocessor directive Within the program file ?

(*d*) Differentiate the following with example :

(*i*) while vs. do while (ii) break vs. continue

60. (*a*) Write a program to reverse a number ?

(*b*) Draw a flow chat to find out the roots of quadratic equating.

61. (*a*) A C program contains the following declarations and initial assignment int i=10, j=2; char c='c',d='d'; float x=0 .005 , y=0.01,z;

(*i*) z=(c=0) ? X:0 (ii) (i-3*j)%(c+2*d)/(x-y)

(*b*) Write a program to display the following output

A

A B

A B C
A B C D

62. (*a*) What do you mean by algorithm ? Give an example.
(*b*) What is the difference between a flowchart & data flow diagram ?
(*c*) What is the difference between the declaration & definition of a variable ?
(*d*) What will be the output of the following program

```
#include <stdio.h>
main()
{
static int a[10], i;
for(i=0;i<10;i++)
printf("\n %d",i[a]);
}
```

(*e*) Why it is necessary to use '&' sign before a variable name in scanf function ?
(*f*) What is the difference between a while loop and do-while loop ?
(*g*) What is the comma operator. Give an example ?
(*h*) Differentiate between a 2 Dimensional array and an array of pointers ?
(*i*) What do you mean by recursion. Give an example ?
(*j*) Differentiate between a pointer to constant and a constant pointer ?

63. (*a*) What are various Data-types in 'C' Language ? Discuss in brief ?

OR

Write a program to print your college name without using semicolon anywhere in your program ?

(*b*) Interpret the meaning of the control string in each of the following scanf function.
(*i*) scanf ("12ld %hd %15lf %15le", &a, &b, &c, &d);
(*ii*) scanf ("%10lx 56hx %5hu %14lu", &a, &b, &c, &d);
(*iii*) scanf ("%12D %hd %15f %15e",&a, &b, &c, &d);
(*iv*) scanf ("%8d %*d %12lf", &a, &b, &c);

OR

A C program contains the following variable declaration int i= 12345, j = 0xabcd9, k = 077777; Show the output resulting from each printf statement
(*i*) printf ("%d %x % 0", i, j, k);
(*ii*) printf ("%3d %3x % 30", i, j, k);
(*iii*) printf ("%8d %8x % 80", i, j, k);

64. In what way is the initialization of an external variable more restricted than the initialization of an automatic variable ? Also explain the initialization dangers in the use of an external variable ?

OR

Describe the output of the following program

(*a*)
```
#include <stdio.h>
main()
{
int i, j, x = 0;
```

```
for (i = 0; i<5;i++)
for(j = 0;j<5;j++)
{
x = x + i +j;
printf("\n %d",x);
}
printf("\n %d",x);
}
```

(*b*)

```
#include <stdio.h>
main()
{
    int i,j,x=0;
for(i=0;i<38;i++)
for(j=0;j<5;j++)
{
if( (i+j)%2 == 0)
continue;
switch (i+j-1)
{
case -1:
case 0: x = x + 1;
case 1: break;
case 2:
case 3:x+=2;
break;
default: x+=3;
}
printf("%d",x);
}
printf("%d,x);
}
```

Describe the output of the code writtern above ?

OR

Describe the output of the program

```
#include<stdio.h>
int a = 100, b = 200;
int func1(int count);
int func2(int c);

main()
{
int count;
```

```
for (count=1;count<=5;++count)
printf(%d",func1(count));
}

int func1 (int x)
{
int c, d;
c = func2(x);
d = (c<100) ?(a+c):b;
return (d);
}
int func2(int x)
{
static int prod = 1;
prod * = x;
return (prod);
}
```

65. (*a*) Write an algorithm to generate Fibonacci series and draw the flow chart and give correspond program ?

(*b*) Explain in brief different modes of Vi editor and list the commands ?

(*c*) Why does data flow diagram is required in modeling software system. Give some data flow diagram at different levels for any example software system ?

(*d*) Explain in brief all bitwise operators in 'C' ?

66. (*a*) State whether true or false from the following: (support your answers with relevant explanation)

(*i*)
```
i = 0;
while( i++ < 10)
printf("\n the world");
printf("\n is beautiful");
```
The above code will result in an infinite loop (T / F)

(*ii*)
```
int a[50];
int *ptr;
ptr = a;
```
To access the 7th element of an array *(*ptr+6) is the correct statement (T / F)

(*iii*)
```
int i=3, a=1, b=1;
void func()
{
    int a=0;
    static int b = 0;
    a++;
    b++;
}
main()
{
```

```
        for (i=0;i<5;i++)
        {
            func();
            a++;
            b++;
        }
        printf(" a = %d, b = %d \n",a,b);
    }
```

The output of the above code will be a = 0, b = 6. (T / F)

(*b*) Write short notes on the following :

(*i*) Recursion (*ii*) Passing array as an argument.

67. (*a*) Can the following code be compiled ? If not why ?

```
#include<stdio.h>
struct xyz
{
int a;
struct xyz b;
}
struct xyz c;
main()
{
printf("IP University");
}
```

OR

How is the member of a 'Union' variable assigned an initial value ? in what ways does the initialization of a 'Union' variable differ from the initialization of a structure variable ?

(*a*) What do you mean by dynamic memory allocation ? Allocate a 2 dimemsional integer array of size MXN dynamically.

OR

Interpret the meaning of 'P' in the following declaration

(*i*) int (*P (char *a)) [10];

(*ii*) int (* P (char *a[]));

(*iii*) int (*P [10]) (char a);

(*iv*) int *(*P)(char *a[]);

68. (*a*) Draw a flowc hart that reads two numbers 'x' & 'y' and prints the bigger of them. It gives an appropriate message if both the numbers are same.

(*b*) Write an algorithm to swap two numbers without using third variable and also write the program in C for the same.

(*c*) What is the difference between a flow chart & Data flow Diagram ?

69. (*a*) Draw flowchart for the following

(*i*) to sort a list of 'N' numbers

(*ii*) to find the sum of digits of a number.

OR

Draw flow chart for the following

(*i*) to find the smallest number from a list of 'N' numbers

(*ii*) to print the digits of a number in reverse order.

(*b*) Briefly discuss the features of VI editor.

70. (*a*) What is the purpose of keyword void in function declaration ?

(*b*) Why does not the code of a[i] = i++ work ?

(*c*) What is an exit function ?

(*d*) What is bounds checking ?

(*e*) What type of array is also known as vectors ?

(*f*) What is the output of the following code:

(*i*)
```
void main()
{
int a[] = ( 15, 16, 17, 18};
clrscr();
printf("%d",a[2.9999]);
getch();
}
```

(*ii*)
```
void main()
{
int a[10];
printf("%d %d",a[0], a[12]);
}
```

(*g*) What are the advantages of modular programming ?

(*h*) Distinguish between actual and formal parameters with an examples ?

71. (*a*) What do you mean by function prototyping and what are the advantages of function prototyping in 'c' ?

(*b*) How many times the following loops be executed:

(*i*)
```
for (a=35; a>20;a--)
{
S2;
}
```

(ii)
```
for(a=0;a>=14;a++)
{
S4;
}
```

(iii)
```
for(;a<=10;a++)
{
S5;
}
```

(iv)
```
for(;;)
{
```

```
S2;
}
```

(c) Write a program using functions to check wheather the diagonal elements of 4*4 matrix are all 0.

72. (a) Write a function int is_prime(int i) has return 1 if n is a prime and 0 otherwise ?

(b) Find the output of following

(i)
```
#include<stdio.h>
main()
{
int x;
for(x=0;x<=99;x++)
{
printf("%d\n",x++);
}
}
```

(ii)
```
#include<stdio.h>
main()
{
int x=0
for(x=99;x>0;--x)
{printf("%d\n",--x);
}
}
```

73. Write short notres on following:

(a) function definition & declaration

(b) global & local variables

(c) comma operator

(d) 3D array with example

74. (a) What is a preprocessor ?

(b) What are the advantages of 'C' language ?

(c) (i) 'C' has only _____________ keywords ?

(ii) Bit wise operators are applicable to only char & int. State true/false.

(d) How do I write printf() so that width of the files can be specified at run time.

(e) If A = 1, b=-2, c=3, d=4, e='S' and f='J', then state wheather the following conditions are true or false:-

(i) A == B

(ii) (A>B || C<D)

(iii) A>B

(iv) ((A+C)>(B-D)) && ((B+C)<(D-A))

(v) (A<C && B<D)

(f) Explain what happens when the following statement is executed:

if (abs(x) < min) x = (x>0) ?xmin:-xmin;

75. (*a*) A 'C' program contains the following statements:

```
#include<stdio.h>
int a=0177, b=055, c=0xa8, d=0x1ff
```

Write a printf() function for each of the following groups of variables of expressions

(*i*) a,b,c & d

(*ii*) (a+b), (c-d)

(*b*) What are the different data types available in 'C'

(*c*) A fibonnacci sequence is defined as follows – first and second term are 0 and 1, subsequent terms are found by adding the preceding 2 terms in the sequence. A part of sequence is 0,1,1,2,3,5,8,13,21,34………. Obtain a flowchart to find all numbers in sequence less than 200.

76. (*a*) Differentiate between the following with the help of suitable examples:

(*i*) Call by Value & Call by Reference

(*ii*) While & Do-While loop

(*iii*) Continue & break

(*b*) WAP to find average of 'n' numbers entered by the user without using arrays. The value of 'n' is entered by the user ?

(*c*) What will be the output of the following code:

```
main()
{
        static int a[10], i;
        for (i=0; i<=10; printf(" \n %d",i[a]));
}
```

77. (*a*) Write a program to input a letter of alphabet in uppercase and display both octal and hexa decimal ASCII code for each character.

(*b*) Trace through the following segment of code for each new value of A, B, C and state the output in each case

(*i*) A = 16, b = 16, C = 32

(*ii*) A=16, B=-18, C = 32

(*iii*) A=-2, B=-4, C=16

```
if(a>0)
{ if (B<0)
putchar ('x');
else
if(C>20)
putchar(Y');
}
else putchar ('z');
```

(*c*) What is a syntactic error. Name some common syntax error ?

78. (*a*) Find the errors in the following statements :

(*i*) name = 'Arun';

(*ii*) area = 314 * r ** 2;

(*iii*) int num (15)

(*b*) What is the output of the following programs ?

(*i*) main()

```
        {
            int i;
            for (i=1;i<=5;printf("\n%d",i));
            i++;
        }
```

(*ii*)
```
main()
        {
            char ch = 'x';
            switch (ch)
            {
                case 'x':
                case 'y':
                    printf("\n You entered y");
                case 'x':
                    printf("\n a as in ashar");
            }
        }
```

(*c*) Write a function which receive five integers and returns the sum, average and standard deviation of these numbers. Call this function from main() and prints the result in main() ?

(*d*) What is the difference between the following two # include directives ?

#include "conio.h"

#include <conio.h>

(*e*) What is header files ?

(*f*) Write macro definition for obtaining the bigger of two numbers ?

(*g*) Add the missing statement for the following program to print 80,

```
main()
{
    int j, *ptr;
    *ptr = 80;
    printf("\n %d",j);
}
```

(*h*) Write short notes for the given below :

(*a*) Formatted console I/O functions and unformatted console I/O functions.

(*b*) Macros versus functions

(*c*) Structures and Unions

79. (*a*) Differentiate between the following with the help of suitable examples:

(*i*) An identifier and a keyword

(*ii*) Constant and variable

(*iii*) Assignment & relational operator

(*iv*) Logical & Syntax error

(*b*) What are the characteristics of an algorithm ?

(*c*) Explain the functions getche() and putch() with the help of suitable example.

80. Write short notes on following:

(*a*) DFD with an example

(*b*) Screen editor & line editor

(*c*) Limitations of conditional operator

(*d*) Can we get the remainder of a floating point division. Explain it with an example ?
(*e*) Difference between formatted & unformatted inout & output functions ?

81. (*a*) What is the use of keyword 'Void' in functions ?
(*b*) Explain the concept of nested in structures
(*c*) Differentiate between call by value & call by reference ?
(*d*) Give three conditions for infinite loop in while& do while loop
(*e*) What will happen if break is used in default part of switch statrement ?

82. (*a*) Give the output
```
int arr[5] = { 1, 2, 3, 4, 5}
printf("%d",arr);
printf("%d",arr[1]);
printf("%d",arr[0]);
printf("%d",&arr[0]);
```
(*b*) Write a program that will printout all rotation of a string typed into it. For example rotation of word 'SPACE' are space, paces, acesp,cespa,esape etc.
(*c*) Define a structure Employee having the members as emp no, name, department & joining date. Write a program to print the names of those employees who have completed atleast one year service.

83. (*a*) Write short notes on the following :
(*i*) Vi editor
(ii) Ternary & bitwise operator used in C
(*b*) int i=10, j=15, k;
float c;
Determine the values of each of the following expression. Use the above values given for the variables.
(*i*) k = (j&i) && (i!=0);
(ii) k = i > 8 ? j < 20 : printf(" see") : 30 : printf(" world");
(*c*) You have to display " Best of luck" on the output screen. Do this by writing a program in C in such a way that there is not a single semicolon used in the program ?

84. (*a*) Generate the following pyramid of digits using nested loops:
```
            1
        2   3   2
    3   4   5   4   3
4   5   6   7   6   5   4
```
(*b*) if a1 & a2 are declared as char a1[10] & a2[10] and a1 has been initialized, what will be the effect of statement a2=a1 ?
(*c*) differentiate between structures & unions with examples ?

85. Write short notes on :
(*a*) Recursion
(*b*) While & do while
(*c*) Break & continue
(*d*) Actual & formal arguments

86. (*a*) What is the significance of '&' in scanf() function ?
(*b*) What are unary operators ?Also give their precedence level ?
(*c*) Write an algorithm to find sum of the series:
$\Sigma X \wedge r$ (x raised to the power r), r varying from 0 to 100.

(*d*) a*b = c
Is this the legal statement in c ?

(*e*) What is the meaning of the statement #include<stdio.h> >

(*f*) What are relational operators ?

87. (*a*) Given the four sides of a rectangle. Write a program to find out wheather its area is greater than its perimeter ?

(*b*) Given three numbers A, B, C. write a program to print their values in an ascending order. For example if A = 10, B = 11, C = 6, your result should be:
Smallest no. = 6
Next higher no = 10
Highest no = 11

(*c*) Find out the error:

(*i*)
```
main()
{ int x=10;
if x>=2
printf("%d",x);
}
```

(ii)
```
main()
{int I = 2, j=5;
if (I ==2 && j==5)
printf("satisfied at last");
}
```

88. What will be the output of the following codes (also explain your answer):

(*i*)
```
main()
{
    int k=-1, j=2;
    switch(k%=j*k)
    {
            case -1: printf(" %d",k%=5);
            case 2:  printf(" %d",j*k);
            default: printf(" %d",j/=3);
            case 3:  break;
            case 4:  printf(" hello world");
    }
}
```

(*ii*)
```
main()
{
    int c=0, d=5, e=10, a;
    a=c<1 ? d>1 || e>1 ? 100 : 200 : 300;
    if(a)
    printf(" \n the value of a is : %d",a);
    printf(" \n world");
}
```

(iii)
```
main()
{
```

```
        if('A' < 'a')
        printf(" value of 'a' is greater");
    else
    printf(" value of 'A' is greater");
                    }
(iv) main()
    {
        int x=5, y=0, z;
        printf(" the value of x is : %d %d ",x++,++x);

        z = y++ + --x + x++;

        if( (y | x) && z )
        printf(" \n value of y is : %d",++y);
        else
        printf(" \n value of y is : %d",++y);
    }
```

89. (*a*) What is an array and explain how is the memory allocated to an array ?

(*b*) What are the rules regarding the relationship between actual parameters passed to the function and its formal parameters ?

(*c*) WAP to find whether a number entered by the user is a palindrome or not (using while loop).

(*d*) WAP to display all the even numbers from 1 to 100 (using for loop).

90. (*a*) With the help of example explain the use of an algorithms ?

(*b*) What is the output of the following code

```
int x;
x = 1;
x = 2;
x = 3;
printf("%d %d %d",x,x,x);
```

(*c*) Write a program to find the roots of a quatradic equation. Find the roots for the cases when D<0, D=0, D>0.

(*d*) What is a conditional operator ?

91. (*a*) Discuss in brief data types in 'C' with suitable example programs ?

(*b*) Explain different forms of decision control structures and give some example programs ?

(*c*) Explain the given below with suitable example programs:

(*i*) putchar and getchar

(*ii*) switch and break statements

(*iii*) scanf & printf

(*iv*) assignment operator

(*v*) storage classes in 'C'

92. Write short notes on the following:

(*a*) input & output statements of 'C'

(*b*) DFD

(*c*) VI editor

(*d*) Flowcharts

93. *(a)* **C is a Middle level language. Comment**

(b) **differerentiate between scanf() & getchar()**

(c) **What is logical error ? How do logical error differ from syntactic and execution error ? Give example ?**

(d) **Differentiate between '=' & '==' operators in C**

(e) **Difference between DFD & flowcharts**

94. *(a)* **A C program contains the following declaration and initial assignments :**

```
int i=8,j=5,k;
float x=0.005,y=-0.01,z;
char a,b,c='c',d='d';
```

Determine the value of each of following assignment expression. Use the values given for variables.

(i) **k=c**

(ii) **i%=j**

(iii) **k=(j==5)**

(iv) **(x>y)&&(i>0)||(j<5);**

(v) **!(c==99)**

(b) **Find out the errors if any and write the corrected code :**

```
main()
{
int x=10,y=20;
if(x==y);
printf("\n %d %d",x,y);
}
```

(c) **Find output of the following code:**

```
main()
{
int I=4,j=-1,k=0,w,x,y,z;
w=i||j||k;
x=i&&j&&k;
y=i||j&&k;
z=i77j||k;
printf("w=%d,x=%d,y=%d,z=%d",w,x,y,z);
}
```

(d) **Three points are given (x1,y1), (x2,y2) & (x3,y3). Obtain a flowchart to check if they are collinear.**

95. *(a)* **A C program contains the following statements-**

```
#include<stdio.h>
int a,b;
float x,y;
```

Suppose the value 12 is to be entered into computer & assigned to a, -8 assigned to b, 0.011 to x, $-2.2*10^6$ to y.show how the input data most be conveniently be entered for each of the following scaf() statements

(*i*) scanf ("%d%d%f%f",&a, &b, &x, &y);

(ii) scanf ("%d%d%e%e",&a,&b,&x,&y);

(iii) scanf ("%2d %2d %5f %6e",&a,&b,&x,&y);

(iv) scanf ("%3d %3d %8f %8e",&a,&b,&x,&y);

(*b*) Enter five characters simultaneously and change their case from uppercase to lowercase (and vice versa) using conditional operator (without using toupper() & to lower() functions)

(*c*) A five digit number is entered through the keyboard. Write a program to obtained the reversed number and to determine wheather the original and reversed number are equal or not.

96. (*a*) How do you define "array" and where do you use these data structures and give syntax with suitable example program ?

(*b*) Write a program to generate prime number from 1 to 1000 using arrays ?

(*c*) Demonstrate call by value & call by reference with suitable example program.

(*d*) Discuss in brief all loops control statements in 'C' ?

97. (*a*) With the help of suitable example explain the difference between '=' & '==' operator in C.

(*b*) Explain the usage of '%c' , '%d', '%f'; in C

(*c*) What do ypou understand by the term algorithm. Write an algorithm for interchanging the contents of two variables ?

(*d*) Write short note on editing tools

(*e*) Differentiate between syntax & logical error.

98. (*a*) Write a program that inputs 2 numbers x & y, and then determines whether the product of 2 numbers is greater than half of 'y'

(*b*) What is a conditional operator ? Explain it with an example. Rewrite the following code using conditional statement:

```
main()
{ int high, low,p;
  scanf ("\n%d %d",&high, &p);
  if (p>high) high = p;
  else low = p;
}
```

99. Write short notes on the following :

(*i*) DFD

(*ii*) C data types

(*iii*) Nested if statements

(*iv*) Various operators used in c

(*v*) Input & output statement in c

100. (*a*) Find the output of the following code:

(*i*) main()

```
{ int x=3,y,z;
  y=x=10;
  z=x<=10; y=3>10;
  printf("\nx=%d y=%d z=%d",x,y,z)
}
```

(*ii*) main()

```
{ int bingo, schmingo,zingo;
    bingo=47; schmingo=8;
    zingo=bingo-schmingo;
    zingo=bingo*schmingo;
    printf("%d %d %d\n",bingo,schmingo,zingo);
}
```

(*b*) Point out the errors if any and write the corrected code:

```
main()
{ int c=5,d=15;
  if (c=5) && if(d=15);
printf("\n Good morning"); }
```

(*c*) Write a program to find the roots of a quadratic equation using if-else statement.

101. (*a*) Differentiate between entry controlled & exit controlled loops.

(*b*) What is the use of void ?

(*c*) What role does a return statement plays in a function ?

(*d*) What are the rules regarding the relationship between actual parameters passed to a function and its formal parameters ?

(*e*) What is the use of function prototype ?

(*f*) If array1 & array2 are declared as char arr1[10], arr2[10]; and arr1 has been initialized. What will be the effect of the statement arr2=arr1.

102. (*a*) What is the use of mentioning the return type of a function ? By default a function returns which type of a value ?

(*b*) What will be the output of the following code: (explain your answer)

(*i*) main()

```
{
    int i, j, x=0;
    for (i=0;i<3;i++)
{
    for (j=1;j<4;j++)
    {
        x = (x+i+j-1)%2;
        printf(" \n %d",x++);
    }
    printf("\n value = %d",++x);
}
}
```

(ii)
```
main()
{
    float c;
    c = check(10.2,20.5);
        printf("\n c = %f",c);
    }
    check (float i, float j)
{
    float d;
    d = i+j;
    return (d);
}
```

(iii)
```
main()
{
    int a[] = { 11, 14, 23, 31, 14};
    int *ptr[] = {a, a+1, a+2, a+3, a+4};
    int **ptr1 = ptr;

    for (i=0;i<4;i++)
    {
        if(i<3)
    {
        printf("\n %d",**ptr1);
        continue;
        printf("\n %d",*ptr);
    }
    else
        break;

    printf("\n %d",ptr);
}
    printf("\n %d %d %d",ptr, *ptr, **ptr);
}
```

(g) What purpose is served by break statement & where it is used ?

103. (a) Write a program to remove duplicate numbers from an array ?

(b) Write a program to find maximum & minimum values from a matrix ?

(c) Write a function prime that return 1, if its arguments is a prime number and return 0 otherwise.

104. (a) Write a program to apply binary search on a given set of elements.

(b) Write a program of Fibonacci series using (i) arrays & (ii) recursion

(c) What happens if number of initializing values is greater than the dimension specified for an array ?

(d) What happens if an array is initialized within its declaration and too few initialization values are specified with in the curly braces ?

105. (a) Discuss all string functions and explain with example programs ?

(b) Define "pointer in C" and demonstration with example program ?

(c) Demonstrate the relationship between arrays & pointers with suitable program.

(d) Write a program to display student details, faculty details, administrative staff details using structures ? What are the uses of structures.

□□□

APPENDIX

B

LIST OF PRACTICALS

Q. 1. Perform various operations on arrays using switch statements

(a) Traversing

(b) Insertion through position

(c) Deletion through position

(d) Searching (Linear and Binary)

(e) Bubble Sort

(f) Merging

(g) Reversing

```
#include<stdio.h>
#include<conio.h>
#include<process.h>
void insert(int*,int pos,int num);
void delet(int*,int pos);
void revers(int*);
void main()
{
int i,temp,j,n,n2,mid,lower=0,upper=5,flag=1;
int ram[6];
int hr[8];
int ch;
```

```
clrscr();
while(1)
{
printf("\nThe menu for operations are:->");
printf("\n\n1.Traversing");
printf("\n\n2.Insertion");
printf("\n\n3.Deletion");
printf("\n\n4.Reversing");
printf("\n\n5.Linear Search");
printf("\n\n6.Binary Search");
printf("\n\n7.Bubble Sorting");
printf("\n\n8.exit");
printf("\n\nenter your choice");
scanf("%d",&ch);
switch(ch)
{
case 1:
   printf("\nenter the array element");
        for(i=0;i<=5;i++)
   scanf("%d",&ram[i]);
   printf("\n\nthe array elements are :-> ");
        for(i=0;i<=5;i++)
   printf(" %d",ram[i]);
   break;

case 2:
   printf("\nenter the array element");
        for(i=0;i<=5;i++)
   scanf("%d",&ram[i]);
   printf("\n\nthe array elements are :-> ");
        for(i=0;i<=5;i++)
   printf(" %d",ram[i]);
```

```
    insert(ram,3,9);
      printf("\n\nthe array elements after insertion are:->");
       for(i=0;i<=6;i++)
      printf(" %d",ram[i]);
       break;
case 3:

   printf("\nenter the array element");
        for(i=0;i<=6;i++)
   scanf("%d",&ram[i]);
   printf("\n\nthe array elements are :-> ");
        for(i=0;i<=6;i++)
   printf(" %d",ram[i]);
   delet(ram,3);
   printf("\n\nthe array element after deletion are:->");
        for(i=0;i<=5;i++)
   printf(" %d",ram[i]);
        break;

case 4:
   printf("\nenter the array element");
        for(i=0;i<=5;i++)
   scanf("%d",&ram[i]);
   printf("\n\nthe array elements are :-> ");
        for(i=0;i<=5;i++)
   printf("%d",ram[i]);
   revers(ram);
   printf("\n\nthe array element after reversing are:->");
        for(i=0;i<=5;i++)
      printf(" %d",ram[i]);
        break;
case 5:
   printf("\nenter the array element");
        for(i=0;i<=5;i++)
```

```
    scanf("%d",&ram[i]);
    printf("\n\nthe array elements are :-> ");
        for(i=0;i<=5;i++)
    printf(" %d",ram[i]);
    printf("\n\nenter the element to be search by ls");
    scanf("%d",&n);
        for(i=0;i<=5;i++)
          {
        if(ram[i]==n)
        break;
          }
        if(i==6)
        printf("\n\nthe element %d is not in array",n);
        else
        printf("\n\nthe element %d is at %d position",n,i+1);
        break;
 case 6:
   printf("\n\nenter the element of a new array in sorted order");
       for(i=0;i<=7;i++)
        scanf("%d",&hr[i]);
       for(i=0;i<=7;i++)
        printf(" %d",hr[i]);

        printf("\n\nenter the element to be search by bs");
        scanf("%d",&n2);
       for(mid=(lower+upper)/2;lower<=upper;mid=(lower+upper)/2)
         {
          if(hr[mid]==n2)
        {
          printf("\n\nthe no. is present at %d position of
                      array",(mid+1));
          flag=0;
          break;
        }
      if(hr[mid]>n2)
         upper=mid-1;
      else
         lower=mid+1;
         }
     if(flag)
       printf("\n\nthe element is not in array");
      break;
```

```
case 7:
     printf("\n\nthe element before sorting are:->");
   for(i=0;i<=5;i++)
     printf(" %d",ram[i]);
   for(i=0;i<=4;i++)
      {
       for(j=0;j<=4-i;j++)
     {
       if(ram[j]>ram[j+1])
       {
       temp=ram[j];
       ram[j]=ram[j+1];
       ram[j+1]=temp;
       }
     }
       }
     printf("\n\nthe array after sorting is:->");
      for(i=0;i<=5;i++)
     printf(" %d",ram[i]);
      break;
case 8:
   exit(0);

   }
  getch();
    }
 }
   void insert(int*ram,int pos,int num)
      {
      int i;
      for(i=6;i>=pos;i–)
    {
     ram[i]=ram[i-1];
    }
      ram[i]=num;
      }
   void delet(int*ram,int pos)
```

```
{
int i;
for(i=pos;i<=6;i++)
{
ram[i-1]=ram[i];
}
}
void revers(int*ram)
{
int temp;
for( int i=0;i<=5/2;i++)
{
temp=ram[i];
ram[i]=ram[5-i];
ram[5-i]=temp;
}
}
```

Q. 2. Store a two D Array in memory and check whether it is stored as row measure or column measure.

```
#include<stdio.h>
#include<conio.h>

void main ()
{
int a[3][3];
int i, j, b;
clrscr();

printf("\nEnter the values of element in the array \n");

for(i=0;i<3;i++)
{
 for(j=0;j<3;j++)
 {
  printf("Enter the value in the element a[%d][%d]",i,j);
  scanf("%d",&a[i][j]);
 }
}
printf("\n the array element are:");
```

```
 for(i=0;i<=3;i++)
   {
     printf("\n");
     for(j=0;j<=3;j++)
      {
  printf("    %d",a[i][j]);
   }
   }

printf("\nCOMPARISON THROUGH ADDRESS ");
if(a == (&a[0][1] - 1))
 printf("\nROW MAJOR FORM");
else
 printf("\nCOLUMN MAJOR FORM");
getch();
}

*/ output

Enter the values of element in the array
Enter the value in the element a[0][0]1
Enter the value in the element a[0][1]5
Enter the value in the element a[0][2]5
Enter the value in the element a[1][0]45
Enter the value in the element a[1][1]5
Enter the value in the element a[1][2]7
Enter the value in the element a[2][0]8
Enter the value in the element a[2][1]9
Enter the value in the element a[2][2]7

 the array element are:
   1    5    5   45
   45   5    7    8
   8    9    7    0
   0   344    0    0
COMPARISON THROUGH ADDRESS
ROW MAJOR FORM
```

Q. 3. **Generate a try diagonal sparse matrix, store it memory and derive a formula for its address retrieval.**

```
   #include<stdio.h>
#include<conio.h>
```

```
void main()
{
 int a[10];
 int i,j;
 int *p;
 clrscr();

 printf("\nEnter the values of element in the sparse tridiagonal
matrix \n");
 for (i=0;i<10;i++)
 {
  scanf("%d",&a[i]);
 }
 printf("\nSprarse Matrix Is\n");

 for(i=0;i<4;i++)
 {
  for(j=0;j<4;j++)
  {
   if(j==i)
    printf(" %d",a[(2*i)+j]);
   else
    if(j==i-1)
     printf(" %d",a[(2*i)+j]);
    else
     if(j==i+1)
      printf("  %d",a[(2*i)+j]);
     else
      printf("    0");
  }
  printf("\n");
 }
 p=a;

 printf("\nThe address of element a[0][0] is %u ",p);
 printf("\nEnter the index of element you want to get address
[i][j]: ");
 scanf("%d %d",&i,&j);
  printf("\nThe  address  of  element  a[%d][%d]  is  %u
",i,j,p+((2*i)+j));
 getch()
```

```
/*output is
Enter the values of element in the sparse tridiagonal matrix
1
2
3
4
5
6
7
8
9
1

Sprarse Matrix Is
    1    2    0    0
    3    4    5    0
    0    6    7    8
    0    0    9    1

The address of element a[0][0] is 65502
Enter the index of element you want to get address [i][j]: 0 1

The address of element a[0][1] is 65504
*/
```

Q. 4. Generate the lower and upper sparse matrix using for loop, and store it in two one-ray. Derive a formula for its address retrieval.

```
   #include<conio.h>
#include<stdio.h>

void main()
{
 int a[4][4],b[10],i,j,ch,k=0;
 clrscr();

 printf("Do u want to enter upper triangular (press 1) or \nlower
triangular(press 2) :");
 scanf("%d",&ch);
 printf("\n\nEnter the elements of the array : \n\n");

 /* for upper matrix*/
```

```
if(ch==1)
{
 for(i=0;i<4;i++)
  for(j=0;j<4;j++)
  {
   if(i>=j)
   {
    printf("Enter the value of element a[%d][%d] : ",i,j);
    scanf("%d",&a[i][j]);
   }
   else
    a[i][j]=0;
  }

/*for column major storage in 1-D array*/

for(j=0,k=0;j<4;j++)
   for(i=0;i<4;i++)
    if(i>=j)
    {
     b[k]=a[i][j];
     k++;
    }
 }

/* for lower matrix*/

if(ch==2)
 {
  for(i=0;i<4;i++)
   for(j=0;j<4;j++)
   {
    if(j>=i)
    {
     printf("Enter the value of element a[%d][%d] : ",i,j);
     scanf("%d",&a[i][j]);
    }
    else
     a[i][j]=0;
   }

/*for column major storage in 1-D array*/
```

```
    for(j=0,k=0;j<4;j++)

     for(i=0;i<4;i++)
       if(j>=i)
      {
       b[k]=a[i][j];
       k++;
      }

    }

    printf("\n\nYour 2-D array is : \n\n");
    for(i=0;i<4;i++)
    {
     for(j=0;j<4;j++)
      printf("   %d   ",a[i][j]);
     printf("\n\n");
    }

   printf("\n\nYour 1-D array is : \n\n");
   for(k=0;k<10;k++)
    printf("  %d ",b[k]);

  printf("Enter the index of the non zero element of a[i][j]: ");
  scanf("%d %d",&i,&j);
  /*for address manupulation */
  if(ch==1)
  {
   printf("FOR UPPERCASE MATRIX ONL\n");
   printf("\nBase address is : %u\n",b);
   printf("The address of the element is : %u " ,b+(((j*(9-i))/
2)+i-j+1));
  }
  getch();

 }


 /*
  output is
Do u want to enter upper triangular (press 1) or
```

```
lower triangular(press 2) :1

Enter the elements of the array :

Enter the value of element a[0][0] : 1
Enter the value of element a[1][0] : 1
Enter the value of element a[1][1] : 1
Enter the value of element a[2][0] : 1
Enter the value of element a[2][1] : 1
Enter the value of element a[2][2] : 1
Enter the value of element a[3][0] : 1
Enter the value of element a[3][1] : 1
Enter the value of element a[3][2] : 1
Enter the value of element a[3][3] : 1

Your 2-D array is :

  1     0     0     0

  1     1     0     0

  1     1     1     0

  1     1     1     1

Your 1-D array is :

 1  1  1  1  1  1  1  1  1  1

Enter the index of the non zero element of a[i][j]: 2 3

Base address is : 65468
The address of the element is : 65486
```

Q. 5. Implement stack operation (Push and POP) using arrays.

```
#include<stdio.h>
#include<conio.h>
#define maxsize 5
struct stack
{
```

```
int ram[maxsize];
int top;
};
void inistack(struct stack*);
void push(struct stack*,int num);
int pop(struct stack*);
void main()
{
struct stack s;
int i;
clrscr();
inistack(&s);

push(&s,2);
push(&s,3);
push(&s,75);
push(&s,5);
push(&s,8);
push(&s,9);

i=pop(&s);
printf("\n\nitem is poped:%d",i);
i=pop(&s);
printf("\n\nitem is poped:%d",i);
i=pop(&s);
printf("\n\nitem is poped:%d",i);
getch();
}
void inistack(struct stack*s)
{
s->top=-1;
}
void push(struct stack*s,int num)
{
if(s->top==maxsize-1)
{
printf("\n\nstack is full");
return;
}
s->top++;
s->ram[s->top]=num;
}
```

```
int pop(struct stack*s)
{
int data;
if(s->top==-1)
{
printf("\n\nstack is empty");
return NULL;
}
data=s->ram[s->top];
s->top—;
return data;
}

/* output

stack is full

item is poped:8

item is poped:5

item is poped:75

*/
```

Q. 6. Implement queue operation (insertion and delete) using arrays.

```
#include<stdio.h>
#include<conio.h>
#define max 5
void insq(int*,int,int*,int*);
int delq(int*,int*,int*);
void main()
{
int ram[max];
int front=-1,rear=-1,i;
clrscr();
insq(ram,5,&front,&rear);
insq(ram,6,&front,&rear);
insq(ram,7,&front,&rear);
insq(ram,8,&front,&rear);
insq(ram,9,&front,&rear);
insq(ram,2,&front,&rear);
```

```
i=delq(ram,&front,&rear);
printf("\n\nitem is deleted",i);
i=delq(ram,&front,&rear);
printf("\n\nitem is deleted",i);
i=delq(ram,&front,&rear);
printf("\n\nitem is deleted",i);
getch();
}
void insq(int*ram,int num,int*pfront,int*prear)
{
if(*prear==max-1)
{
printf("\n queue is full");
return;
}
(*prear)++;
ram[*prear=num];
if(*pfront==-1)
*pfront=0;
}

int delq(int*ram,int*pfront,int*prear)
{
int data;
if (*pfront==-1)
{
printf("\n queue is empty");
return NULL;
}
data=ram[*pfront];
ram[*pfront]=0;
if(*pfront==*prear)
*pfront=*prear=-1;
else
(*pfront)++;
return data;
}
```

OUTPUT

```
item is deleted

item is deleted

item is deleted
```

Q. 7. Write a program to add two polynomials.

```
   #include<stdio.h>
#include<conio.h>
# define max 15

struct term
{
int coff;
int exp;
};

struct poly
{
struct term t[15];
int not;
};
void inipoly(struct poly*);
void polyassign(struct poly*,int c,int e);
struct poly polyadd(struct poly,struct poly);
void display(struct poly);
void main()
{
struct poly p1,p2,p3;
clrscr();

inipoly(&p1);
inipoly(&p2);
inipoly(&p3);

polyassign(&p1,1,7);
polyassign(&p1,2,6);
polyassign(&p1,3,5);
polyassign(&p1,4,4);


polyassign(&p2,1,4);
polyassign(&p2,1,3);
polyassign(&p2,1,5);
polyassign(&p2,2,2);

p3=polyadd(p1,p2);
```

```
printf("\n\nfirst polynomial is:\n\n");
display(p1);

printf("\n\nsecond polynomial is:\n\n");
display(p2);

printf("\n\nres polynomial is:\n\n");
display(p3);

getch();
}
void inipoly(struct poly*p)
{
int i;
p->not=0;
for(i=0;i<max;i++)
{
p->t[i].coff=0;
p->t[i].exp=0;
}
}
void polyassign(struct poly*p,int c,int e)
{
p->t[p->not].coff=c;
p->t[p->not].exp=e;
(p->not)++;
}
void display(struct poly p)
{
int flag=0,i;
for(i=0;i<p.not;i++)
{
if(p.t[i].exp!=0)
printf("%dx^%d+",p.t[i].coff,p.t[i].exp );
else
{
printf("%d",p.t[i].coff);
flag=1;
}
}
if(!flag)
printf("\b\b ");
```

```
}
struct poly polyadd(struct poly p1,struct poly p2)
{
int i,j,c;
struct poly p3;
inipoly(&p3);

if(p1.not>p2.not)
c=p1.not;
else
c=p2.not;
for(i=0,j=0;j<=c;p3.not++)
{
 if(p1.t[i].coff==0&&p2.t[j].coff==0)
 break;
 if(p1.t[i].exp>=p2.t[j].exp)
 {
  if(p1.t[i].exp==p2.t[j].exp)
      {
 p3.t[p3.not].coff=p1.t[i].coff+p2.t[j].coff;
 p3.t[p3.not].exp=p1.t[i].exp;
 i++;
 j++;
 }
 else
 {
 p3.t[p3.not].coff=p1.t[i].coff;
 p3.t[p3.not].exp=p1.t[i].exp;
 i++;
 }
 }
 else
 {
 p3.t[p3.not].coff=p2.t[j].coff;
 p3.t[p3.not].exp=p2.t[j].exp;
 j++;
 }
 }
 return p3;
 }

/* output
```

```
first polynomial is:

1x^7+2x^6+3x^5+4x^ +

second polynomial is:

1x^4+1x^3+1x^5+2x^ +

res polynomial is:

1x^7+2x^6+3x^5+5x^4+1x^3+1x^5+2x^ +

*/
```

Q. 8. Implement stack operation (Push and POP) using Linked List.

```
#include<stdio.h>
#include<conio.h>
#include<alloc.h>

struct node
{
   int data;
   struct node *link;
};
void push(struct node **, int);
int pop(struct node **);
void delstack(struct node **);

void main()
{
   struct node *s=NULL;
   int i;

   clrscr();

   push(&s,14);
   push(&s,-3);
   push(&s,18);
   push(&s,29);
   push(&s,31);
   push(&s,16);
```

```
    i=pop(&s);
    printf("\nItem popped:%d",i);

    i=pop(&s);
    printf("\nItem popped:%d",i);

    i=pop(&s);
    printf("\nItem popped:%d",i);

    delstack(&s);

    getch();
}

void push(struct node **top, int item)
{
    struct node *temp;
    temp=(struct node *)malloc(sizeof(struct node));

    if(temp == NULL)
        printf("\nStack is full");

        temp->data=item;
        temp->link= *top;
        *top=temp;
}
int pop(struct node **top)
{
    struct node *temp;
    int item;

    if( *top==NULL)
    {
        printf("\nStack is empty");
        return NULL;
    }

    temp= *top;
    item=temp->data;
    *top=(*top)->link;

    free(temp);
```

```
    return item;
}
void delstack(struct node **top)
{
    struct node *temp;

    if(*top == NULL)
     return;

    while(*top!=NULL)
    {
    temp= *top;
    *top=(*top)->link;
    free(temp);
    }
}

/*OUTPUT:

Item popped:16
Item popped:31
Item popped:29

*/
```

Q. 9. Implement queue operation (insertion and delete) using Linked List.

```
#include<stdio.h>
#include<conio.h>

struct node
{
    int data;
    struct node *link;
};

struct queue
{
    struct node *front;
    struct node *rear;
};

void initqueue(struct queue *);
void addq(struct queue *, int);
```

```
int delq(struct queue *);
void delqueue(struct queue *);

void main()
{
    struct queue a;
    int i;

    clrscr();

    initqueue(&a);

    addq(&a,11);
    addq(&a,-8);
    addq(&a,23);
    addq(&a,19);
    addq(&a,15);
    addq(&a,16);
    addq(&a,28);

    i=delq(&a);
    printf("\nItem extracted:%d",i);

    i=delq(&a);
    printf("\nItem extracted:%d",i);

    delqueue(&a);

    getch();
}
void initqueue(struct queue *q)
{
    q->front=q->rear=NULL;
}
void addq(struct queue *q, int item)
{
    struct node *temp;

    temp=(struct node *)malloc(sizeof(struct node));
    if (temp == NULL)
   printf("\nQueue is full");
```

```
        temp->data=item;
        temp->link=NULL;

        if(q->front == NULL)
        {
     q->rear=q->front=temp;
     return;
        }
        q->rear->link=temp;
        q->rear=q->rear->link;
}

int delq(struct queue *q)
{
     struct node *temp;
     int item;

     if(q->front==NULL)
     {
          printf("\nQueue is empty");
          return NULL;
     }
     item=q->front->data;
     temp=q->front;
     q->front=q->front->link;
     free(temp);
     return item;
}
void delqueue(struct queue *q)
{
     struct node *temp;

     if(q->front==NULL)
       return;

     while(q->front!=NULL)
     {
          temp=q->front;
          q->front=q->front->link;
          free(temp);
     }
}
```

```
/*OUTPUT
Item extracted:11
Item extracted:-8
Item extracted:23*/
```

Q. 10. Implement Circular linked list (Insertion & Deletion)

```
    #include<stdio.h>
#include<conio.h>
#include<alloc.h>
struct node
{
    int data;
    struct node *link;
};

void addcirq(struct node **, struct node **, int);
int delcirq(struct node **,struct node **);
void cirq_display(struct node *);

void main()
{
     struct node *front, *rear;

     front=rear=NULL;

     addcirq(&front,&rear,10);
     addcirq(&front,&rear,17);
     addcirq(&front,&rear,18);
     addcirq(&front,&rear,5);
     addcirq(&front,&rear,30);
     addcirq(&front,&rear,15);

     clrscr();

     printf("Before deletion:\n");
     cirq_display(front);

     delcirq(&front,&rear);
     delcirq(&front,&rear);
     delcirq(&front,&rear);

     printf("\n\nAfter deletion:\n");
```

```
    cirq_display(front);
    getch();
}

void addcirq(struct node **f, struct node **r, int item)
{
    struct node *q;
    q=malloc(sizeof(struct node));
    q->data=item;

    if(*f==NULL)
    *f=q;
    else
    (*r)->link=q;

    *r=q;
    (*r)->link=*f;
}

int delcirq(struct node **f, struct node   **r)
{
    struct node *q;
    int item;

    if(*f==NULL)
    printf("queue is empty");
    else
    {
    if(*f==*r)
    {
        item=(*f)->data;
        free(*f);
        *f=NULL;
        *r=NULL;
    }
    else
    {
        q=*f;
        item=q->data;
        *f=(*f)->link;
        (*r)->link=*f;
        free(q);
```

```
    }
    return(item);
        }
        return NULL;
}

void cirq_display(struct node *f)
{
        struct node *q=f,*p=NULL;

        while(q!=p)
        {
      printf("%d",q->data);

      q=q->link;
      p=f;
        }
}

/*OUTPUT

Before deletion:
10 17 18 5 30 15

After deletion:
5 30 15

*/
```

Q. 11. Implement Evaluation of postfix expression?

```
#include<stdio.h>
#include<conio.h>
#include<stdlib.h>
#include<math.h>
#include<ctype.h>

#define MAX 50

struct postfix
{
  int stack[MAX];
  int top, nn;
```

```
  char *s;
};

void initpostfix(struct postfix *);
void setexpr(struct postfix *, char *);
void push(struct postfix *, int);
int pop(struct postfix *);
void calculate(struct postfix *);
void show(struct postfix);

void main()
{
    struct postfix q;
    char expr[MAX];

    clrscr();

    initpostfix(&q);

    printf("\nEnter postfix expression to be evaluated:");
    gets(expr);

    setexpr(&q,expr);
    calculate(&q);
    show(q);

    getch();
}

void initpostfix(struct postfix *p)
{
    p->top=-1;
}

void setexpr(struct postfix *p, char *str)
{
    p->s=str;
}

void push(struct postfix *p, int item)
{
    if(p->top==MAX-1)
```

```
    printf("\nStack is full");
      else
      {
        p->top++;
        p->stack[p->top]=item;
      }
}

int pop(struct postfix *p)
{
      int data;
      if(p->top==-1)
      {
     printf("\nStack is empty");
         return NULL;
      }
      data=p->stack[p->top];
      p->top—;

      return data;
}

void calculate(struct postfix *p)
{
      int n1, n2,n3;
      while(*(p->s))
      {
    if(*(p->s)==' '||*(p->s)=='\t')
    {
          p->s++;
          continue;
    }
    if(isdigit(*(p->s)))
    {
         p->nn=*(p->s)-'0';
         push(p,p->nn);
    }
    else
    {
          n1=pop(p);
          n2=pop(p);
          switch(*(p->s))
```

```
    {
  case'+':
       n3=n2+n1;
       break;

  case'-':
       n3=n2-n1;
       break;

  case'/':
       n3=n2/n1;
       break;

  case'*':
       n3=n2*n1;
       break;

  case'%':
       n3=n2%n1;
       break;

  case'$':
       n3=pow(n2,n1);
       break;

  default:
        printf("Unknown operator");
        exit(1);

    }

    push(p,n3);
 }
 p->s++;
   }
}
void show(struct postfix p)
{
      p.nn=pop(&p);
      printf("Result is:%d",p.nn);
}
```

```
/*OUTPUT

Enter postfix expression to be evaluated:4 2 $ 3* 3 - 8 4 / 1 1+/+
Result is:46

*/
```

Q. 12. Implement Infix to Postfix Conversion?

```
    #include<stdio.h>
#include<conio.h>
#include<string.h>
#include<ctype.h>

#define MAX 50

struct infix
{
    char target[MAX];
    char stack[MAX];
    char *s, *t;
    int top;
};

void initfix(struct infix *);
void setexpr(struct infix *, char *);
void push(struct infix *);
char pop(struct infix *);
void convert(struct infix *);
int priority(char);
void show(struct infix);

void main()
{
    struct infix p;
    char expr[MAX];

    initfix(&p);

    clrscr();

    printf("\nEnter an expression in infix form:");
    gets(expr);
```

```
     setexpr(&p, expr);
     convert(&p);

     printf("The postfix expression is:");
     show(p);

     getch();
}

void initfix(struct infix *p)
{
       p->top=-1;
       strcpy(p->target," ");
       strcpy(p->stack," ");
       p->t=p->target;
       p->s=" ";
}

void setexpr(struct infix *p, char *str)
{
       p->s=str;
}

void push(struct infix *p, char c)
{



       if(p->top==MAX)
     printf("\nStack is full\n");
       else
       {
      p->top++;
      p->stack[p->top]=c;
       }
}

char pop(struct infix *p)
{
       if(p->top==-1)
       {
      printf("\nStack is empty\n");
      return-1;
```

```
    }
    else
    {
   char item=p->stack[p->top];
   p->top-;
   return item;
    }
}

void convert(struct infix *p)
{
    char opr;

    while(*(p->s))
    {
   if(*(p->s)==' '||*(p->s)=='\t')
   {
    p->s++;
    continue;
   }
   if(isdigit(*(p->s))||isalpha(*(p->s)))
   {
     while(isdigit(*(p->s))||isalpha(*(p->s)))
     {
          *(p->t)=*(p->s);
         p->s++;
         p->t++;
     }
   }
   if(*(p->s)=='(')
   {
       push(p, *(p->s));
       p->s++;
   }
   if(*(p->s)=='*'||(p->s)=='/'||
   *(p->s)=='%'||*(p->s)=='_'||*(p->s)=='$')
   {
       if(p->top!=-1)
       {
      opr=pop(p);
      while(priority(opr)>=priority(*(p->s)))
      {
```

```
                *(p->t)=opr;
                p->t++;
                opr=pop(p);
             }
             push(p, opr);
             push(p,*(p->s));

              }
           else
              push(p, *(p->s));
           p->s++;
         }

         if(*(p->s)==')')
         {

             opr=pop(p);
             while((opr)!='(')
              {
             *(p->t)=opr;
             p->t++;
             opr=pop(p);
              }
              p->s++;
         }
         }
         while(p->top!=-1)
         {
              char opr=pop(p);
              *(p->t)=opr;
              p->t++;

         }

         *(p->t)='\0';

    }

    int priority(char c)
    {
         if(c=='$')
         return3;
```

```
        if(c=='*'||c=='/'||c=='%')
        return2;

        else
        {
       if(c=='+'||c=='-')
           return 1;
       else
           return 0;

        }
    }
    void show(struct infix p)
    {
         printf("%s",p.target);
    }
    /*OUTPUT

    Enter an expression in infix form: 4 $ 2 * 3 - 3 + 8 / 4 / ( 1 + 1

    Stack is empty.
    The postfix expression is: 42$3*3-84/11+/+*/
```

Q. 13. **Implement Preorder, Inorder, Postorder traversing of a trees using Recursion.**

```
#include<stdio.h>
#include<conio.h>
#include<alloc.h>
struct btreenode
{
   struct btreenode *leftchild;
   int data;
   struct btreenode *rightchild;
};
void insert( struct btreenode **,int);
void inorder( struct btreenode *);
void preorder( struct btreenode *);
void postorder( struct btreenode *);

void main()
{
  struct btreenode *bt;
  int req, i=1, num;

  bt=NULL;

  clrscr();
```

```
   printf("Specify the number of items to be inserted:");
   scanf("%d", &req);
   while(i++ <= req)
   {
     printf("Enter the data:");
     scanf("%d", &num);
     insert(&bt, num);
   }
   printf("\nIn-order Traversal:");
   inorder(bt);
   printf("\nPre-order Traversal:");
   preorder(bt);
   printf("\nPost-order Traversal:");
   postorder(bt);
   getch();
}
void insert( struct btreenode **sr, int num)
{
    if( *sr==NULL)
    {
       *sr=malloc(sizeof (struct btreenode));
       ( *sr)->leftchild=NULL;
       ( *sr)->data=num;
       ( *sr)->rightchild=NULL;
       return;
    }
    else
    {
       if(num<( *sr)->data)
      insert(&(( *sr)->leftchild), num);
       else
      insert(&(( *sr)->rightchild), num);
    }
    return;
}
void inorder (struct btreenode *sr)
{
   if(sr!=NULL)
   {
      inorder(sr->leftchild);
      printf("\t%d", sr->data);

      inorder(sr->rightchild);
   }
   else
      return;
}

void preorder (struct btreenode *sr)
```

```
{
    if(sr!=NULL)
    {
        printf("\t%d", sr->data);
        preorder(sr->leftchild);
        preorder(sr->rightchild);
    }
    else
        return;
}

void postorder( struct btreenode *sr)
{
     if(sr!=NULL)
     {
   postorder(sr->leftchild);
   postorder(sr->rightchild);
   printf("\t%d", sr->data);
     }
     else
        return;
}

/*OUTPUT
Specify the number of items to be inserted:5
Enter the data:1
Enter the data:2
Enter the data:3
Enter the data:4
Enter the data:5

In-order Traversal:      1       2       3       4       5
Pre-order Traversal:     1       2       3       4       5
Post-order Traversal:    5       4       3       2       1
*/
```

❑❑❑

APPENDIX

C

GREAT FOUNDERS IN THE FIELD OF COMPUTER ENGINEERING

1. FOUNDER OF INTEL

Bob Noyce and Gorden Moore wanted to name their new company "Moore Noyce", but that was already trademarked by a hotel chain. So they had to settle for an acronym of INTegrated ELectronics as INTEL.

Intel Corporation is the world's largest semiconductor company and the inventor of the x86 series of microprocessors, the processors found in most personal computers. Founded on July 18, 1968 as Integrated Electronics Corporation and based in Santa Clara, California, USA. Intel also makes motherboard chipsets, network cards and ICs, flash memory, graphic chips, embedded processors, and other devices related to communications and computing. Founded by semiconductor pioneers Robert Noyce and Gordon Moore, and widely associated with the executive leadership and vision of Andrew Grove, Intel combines advanced chip design capability with a leading-edge manufacturing capability. Originally known primarily to engineers and technologists, Intel's successful "Intel Inside" advertising campaign of the 1990s made it and its Pentium processor household names.

Intel was an early developer of SRAM and DRAM memory chips, and this represented the majority of its business until the early 1980s. While Intel created the first commercial microprocessor chip in 1971. The 2007 rankings of the world's 100 most powerful brands published by Millward Brown Optimor showed the company's brand value falling 10 places – from number 15 to number 25.

GORDEN MOORE

BOB NOYCE

Fig. 1. Founders of Intel.

2. FOUNDERS OF MICROSOFT

Fig. 2. Picture taken when Microsoft was started, Bill Gates is shown in a square in the above picture.

Microsoft Corporation is a multinational computer technology corporation that develops, manufactures, licenses, and supports a wide range of software products for computing devices. Headquartered in Redmond, Washington, USA, its best selling products are the Microsoft Windows operating system and the Microsoft Office suite of productivity software.

Originally founded to develop and sell BASIC interpreters for the Altair 8800. One commentator notes that Microsoft's original mission was "a computer on every desk and in every home, running Microsoft software"—it is a goal near fulfillment. The company also markets both computer hardware products such as the Microsoft mouse as well as home entertainment products such as the Xbox, Xbox 360, Zune and MSN TV.

2006–present: Vista and other transitions

On June 27, 2008, Bill Gates retired from day-to day activities in the company, following a two year transition period from his role as Chief Software Architect, which was taken by Ray Ozzie, but remained the company's chairman, head of the Board of Directors and would act as an adviser on key projects. Windows Vista, released in January 2007, is Microsoft's latest operating system and has sold 140 million copies to date. Microsoft Office 2007, released at the same time, features a "Ribbon" user interface which is a significant departure from its predecessors.

3. FOUNDERS OF APPLE COMPUTER

They were three months late in filling a name for the business because they didn't get any better name of their new company. So one day they told to the staff: "If I'll not get better name by 5 o'clock today, our company's name will be anything he likes….." so at 5 o'clock nobody comeup with better name, and they were eating apple that time..... so they keep the name of the company "Apple Computers".

Apple Inc., formerly Apple Computer Inc., is an American multinational corporation which designs and manufactures consumer electronics and software products. The company's best-known hardware products include Macintosh computers, iPod portable media players, and the iPhone. Apple software includes the Mac OS X operating system, the iTunes media browser, the iLife suite of multimedia and creativity software, and Final Cut Studio, a suite of professional audio- and film-industry software products. Established in Cupertino, California on April 1, 1976 and incorporated January 3, 1977, the company was called "Apple Computer, Inc."

Fig. 3. Steve Woznaik(sitting) and Steve Jobs of APPLE Computer.

Fig. 4. The Apple I, Apple's first product. Sold as an assembled circuit board, it lacked basic features such as a keyboard, monitor, and case. The owner of this unit added a keyboard and a wooden case.

4. FOUNDER OF LINUX OPERATING SYSTEM

Linus Torvalds originally used Minix OS on his system which he replaced by his OS. Hence the working name was Linux (Linus' Minix). He thought the name to be too egotistical and planned to name it Freax (free + freak + x). His friend Ari Lemmk encouraged Linus to upload it to a network so it could be easily downloaded. Ari gave Linus a directory called linux on his FTP server, as he did not like the name Freax. Linus like that directory name and he kept the name of his new OS to LINUX....

LINUS TORVALDS

Fig. 5. Founder of Linux Operating

Linux is a Unix-like computer operating system family based on the Linux kernel. Linux is one of the most prominent examples of free and open source software development; typically all the underlying source code can be used, freely modified, and redistributed by anyone.

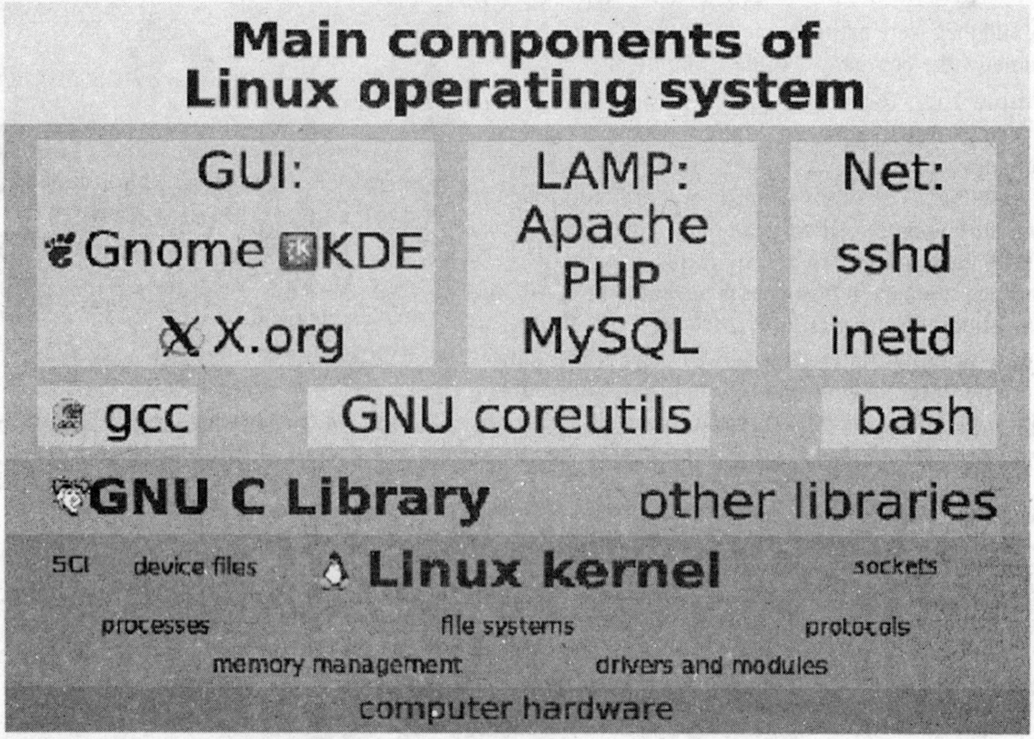

Fig. 6.

Linux distributions are predominantly known for their use in servers, although they are installed on a wide variety of computer hardware, ranging from embedded devices and mobile phones to supercomputers, and their popularity as a desktop/laptop operating system has been growing lately due to the rise of netbooks and the Ubuntu distribution of the operating system.

5. FOUNDERS OF SUN MICROSYSTEM

Fig. 7. Andreas Bechtolsheim, Bill Joy, Scott Mc Nealy and Vinod Khosla(left to right) of SUN(Stanford University Network) MicroSystem.

SUN MicroSystem is founded by four Stanford University Students, Andreas Bechtolsheim build a microcomputer; Vinod Khosla recruited him; Bill Joy to develop a UNIX-based OS for the computer; and Scott Mc Nealy to manufacture computer based on it.... **SUN** is the acronym for **Standard University Network.**

Sun Microsystems, Inc. is a multinational vendor of computers, computer components, computer software, and information technology services, founded on 24 February 1982. The company is headquartered in Santa Clara, California (part of Silicon Valley), on the former west campus of the Agnews Developmental Center.

Products include computer servers and workstations based on its own SPARC processors as well as AMD's Opteron and Intel's Xeon processors; storage systems; and, a suite of software products including the Solaris Operating System, developer tools, Web infrastructure software, and identity management applications. Other technologies of note include the Java platform and NFS.

Sun is a proponent of open systems in general and Unix in particular, and a major contributor to open source software.

6. FOUNDERS OF HP (HEWLETT-PACKARD)

Behind them in the picture is the famous HP Garage. Bill Hewlett and Dave Packard tossed a coin to decide whether the company they founded would be called Hewlett-Packard or Packard-Hewlett and the winner was NOT Bill but the winner is Dave...

Fig. 8. Bill Hewlett(L) and Dave Packard(R) of HP.

The Hewlett-Packard Company, commonly referred to as HP, is a technology corporation headquartered in Palo Alto, California, United States. HP is the largest technology company in the world and operates in nearly every country. HP specializes in developing and manufacturing computing, storage, and networking hardware, software and services. Major product lines include personal computing devices, enterprise servers, related storage devices, as well as a diverse range of printers and other imaging products. Other product lines, including electronic test equipment and systems, medical electronic equipment, solid state components and instrumentation for chemical analysis were spun off as Agilent Technologies in 1999.

HP markets its products to households, small to medium size businesses and enterprises both directly, via online distribution, consumer-electronics and office-supply retailers, software partners and major technology vendors.

7. FOUNDERS OF UNIX and C

Dennis Ritchie improved on the 'B' programming language and called it 'New B'. 'B' was created by Ken Thompson as a revision of the Bon programming language (named after his wife Bonnie). He latter called it C.

Unix (officially trademarked as UNIX, sometimes also written as Unix with small caps) is a computer operating system originally developed in 1969 by a group of AT&T employees at Bell Labs, including Ken Thompson, Dennis Ritchie, Douglas McIlroy, and Joe Ossanna. Today's Unix systems are split into various branches, developed over time by AT&T as well as various commercial vendors and non-profit organizations.

In computing, C is a general-purpose computer programming language originally developed in 1972 by Dennis Ritchie at the Bell Telephone Laboratories for use with the Unix operating system.

C is block structured, and designed for procedural and imperative programming. Although C was designed for implementing system software, it is also widely used for developing application software.

It is widely used on a great many different software platforms and computer architectures, and several popular compilers exist. C has greatly influenced many other popular programming languages, most notably C++, which originally began as an extension to C.

Fig. 9. Ken Thompson(L) and Dennis Ritchie(R) creators of UNIX.

8. FOUNDER OF WWW (WORLD WIDE WEB)

The World Wide Web (commonly shortened to the Web) is a system of interlinked hypertext documents accessed via the Internet. With a Web browser, one can view Web pages that may contain text, images, videos, and other multimedia and navigate between them using hyperlinks. Using concepts from earlier hypertext systems, the World Wide Web was begun in 1989 by English scientist Tim Berners-Lee, working at the European Organization for Nuclear Research (CERN) in Geneva, Switzerland. In 1990, he proposed building a "web of nodes" storing "hypertext pages" viewed by "browsers" on a network, and released that web in 1992. Connected by the existing Internet, other websites were created, around the world, adding international standards for domain names & the HTML language. Since then, Berners-Lee has played an active role in guiding the development of Web standards (such as the markup languages in which Web pages are composed), and in recent years has advocated his vision of a Semantic Web.

TIM BERNERS LEE

Fig. 10. Founder of World Wide Web (WWW).

The World Wide Web enabled the spread of information over the Internet through an easy-to-use and flexible format. It thus played an important role in popularising use of the Internet, to the extent that the World Wide Web has become a synonym for Internet, with the two being conflated in popular use.

WWW prefix in Web addresses

The letters "www" are commonly found at the beginning of Web addresses because of the long-standing practice of naming Internet hosts (servers) according to the services they provide. So for example, the host name for a Web server is often "www"; for an FTP server, "ftp"; and for a USENET news server, "news" or "nntp" (after the news protocol NNTP). These host names appear as DNS subdomain names, as in "www.example.com".

This use of such prefixes is not required by any technical standard; indeed, the first Web server was at "nxoc01.cern.ch", and even today many Web sites exist without a "www" prefix. The "www" prefix has no meaning in the way the main Web site is shown. The "www" prefix is simply one choice for a Web site's host name.

However, some website addresses require the www. prefix, and if typed without one, won't work; there are also some which must be typed without the prefix. Sites that do not have Host Headers properly setup are the cause of this. Some hosting companies do not setup a www or @ A record in the web server configuration and/or at the DNS server level.

Some Web browsers will automatically try adding "www." to the beginning, and possibly ".com" to the end, of typed URLs if no host is found without them. All major web browsers will also prefix "http://www." and append ".com" to the address bar contents if the Control and Enter keys are pressed simultaneously. For example, entering "example" in the address bar and then pressing either Enter or Control+Enter will usually resolve to "http://www.example.com", depending on the exact browser version and its settings.

9. FOUNDERS OF GOOGLE

Fig. 11. Larry Page(L) and Sergey Brin(R), founders of Google.

Google was originally named "Googol", after founders (Stanford Graduates) Sergey Brin and Larry Page presented their project to an angel investor.... They received a cheque made out to "Google" !.. So they kept the name as GOOGLE.

Google Inc. is an American public corporation, earning revenue from advertising related to its Internet search, e-mail, online mapping, office productivity, social networking, and video sharing services as well as selling advertising-free versions of the same technologies. The Google headquarters, the Googleplex, is located in Mountain View, California. As of 30 September 2008 the company has 20,123 full-time employees.

Google was co-founded by Larry Page and Sergey Brin while they were students at Stanford University and the company was first incorporated as a privately held company on 4 September 1998. The unofficial company slogan is "Don't be evil", although criticism of Google includes concerns regarding the privacy of personal information, copyright, censorship and discontinuation of services. According to Millward Brown it is the most powerful brand in the world.

10. FOUNDATION OF INFOSYS (INFY)

Fig. 12. Picture taken when INFY was started.

The above picture was found in the album of the clerk who took this picture. The picture was with that clerk only because it was his birthday and he just told everyone to stand together at one place to take a picture. He borrow a camera from his friend and as he can not tell any of his boss to take picture, so he took the picture by himself.

Infosys Technologies Limited is a multinational information technology services company headquartered in Bangalore, India. It is one of India's largest IT companies with over 100,306 professionals as of Sep 30, 2008.It has nine development centers in India and over 30 offices worldwide.

Infosys was founded on July 2, 1981 in Pune by N. R. Narayana Murthy and six others: Nandan Nilekani, N. S. Raghavan, Kris Gopalakrishnan, S. D. Shibulal, K. Dinesh and Ashok Arora, with Raghavan officially being the first employee of the company. N.Murthy started the company by borrowing INR 10,000 from his wife Sudha Murthy. The company was incorporated as "Infosys Consultants Pvt Ltd.", with Raghavan's house in Matunga, north-central Mumbai as the registered office.

Infosys went public in 1993. In 2001 it was rated Best Employer in India by Business Today. Infosys won the Global MAKE (Most Admired Knowledge Enterprises) award, for the years 2003, 2004 and 2005, being the only Indian company to win this award and is inducted into the Global Hall of Fame for the same.

❑❑❑

INDEX

INDEX

A

B

C

D

K

L

M

N

O

T

U

V

W

❑❑❑

Win Prizes

ATTENTION READERS

Win Prizes

We request for your frank assessment regarding various aspects of the book as given below :

A COURSE IN
DATA STRUCTURES USING C

By : DEEPAK GUPTA, R.K. TIWARI

Please Fill up the Sheet in Neat and Clean Writing

(i) What topics of your syllabus, important for your examination point of view, are not covered in this book ?

....

....

....

....

(ii) What are the chapters wherein the treatment of the subject matter is not systematic or organised or updated ?

....

....

....

....

(iii) Have you come across misprints/mistakes/factual inaccuracies in the book ? Please specify the chapters and the page numbers.

....

....

....

....

(iv) What other books on the same subject have you found/heard better than the present book ? Please specify in terms of price and quality.

....

....

....

....

(v) Further suggestions and comments for the improvement of this book.

....

....

....

....

(PS : If need be please attach a separate sheet to write your views)

OTHER DETAILS

(i) Who recommended you this book ? (Please tick (✓) in the box near the option relevant to you.

☐ Teacher ☐ Friends ☐ Bookseller

....

....

....

....

(ii) Name of the recommending teacher, his designation and address:

....

....

....

....

(iii) Name and address of the bookseller you purchased the book from :

....

....

....

....

(iv) Name and address of your institution (Please mention the University or Board as the case may be)

....

....

....

....

(v) Your name and complete address :

....

....

....

....

(vi) Course you are enrolled in :

....

....

....

....

The best assessment will be awarded each month : The award will be in the form of our publications as decided by the Editorial Board.

Please mail the filled up coupon at the vearliest to :

S.K. KATARIA & SONS®

4760-61/23, Ansari Road,

Daryaganj, New Delhi-110002

☎ 23243489, 23269324, 22733315,

Fax : 91-11-23243489

email : katariabooks@yahoo.com